PRECISION SURVEYING

PRECISION SURVEYING

The Principles and Geomatics Practice

JOHN OLUSEGUN OGUNDARE, PH.D.
Instructor of Geomatics Engineering
Department of Geomatics Engineering Technology
School of Construction and the Environment
British Columbia Institute of Technology (BCIT)-Burnaby

WILEY

Published by John Wiley & Sons, Inc., Hoboken, New Jersey
Published simultaneously in Canada

Library of Congress Cataloging-in-Publication Data:

Ogundare, John Olusegun.
Precision surveying : the principles and geomatics practice / John Olusegun Ogundare, Ph.D., Instructor of Geomatics Engineering, Department of Geomatics Engineering Technology, School of Construction and the Environment, British Columbia Institute of Technology (BCIT)-Burmaby.
pages cm
Includes bibliographical references and index.
ISBN 978-1-119-10251-9 (hardback)
1. Surveying. I. Title.
TA545.O364 2015
526.9–dc23

2015021577

Typeset in 10pt/12pt TimesLTStd by SPi Global, Chennai, India

Printed in the United States of America

10 9 8 7 6 5 4 3

3 2016

CONTENTS

ABOUT THE AUTHOR

John Olusegun Ogundare received his B.Sc. and M.Sc. degrees in surveying engineering from the University of Lagos, Nigeria, and M.Sc.E. and a Ph.D. in high precision and deformation analysis from the University of New Brunswick (UNB) in Canada. He has been in the field of geomatics for over 30 years, as a surveyor in various survey engineering establishments in Africa and Canada and also as a surveying instructor or teaching assistant in universities and polytechnic institutions in Africa and Canada.

For over 8 years, he has been serving as a special examiner for the Canadian Board of Examiners for Professional Surveyors (CBEPS) with the responsibility that includes setting and marking exams on the subject "Map Projections and Cartography" and then on the subject "Coordinate Systems and Map Projections." As a subject-matter expert in those subjects, he has also had the opportunity to serve as a consultant to the Canadian Council of Land Surveyors (CCLS) in 2007 and 2009 in reviewing and making recommendations to a Joint Syllabus Development Task Force on the subject description, prerequisites, texts, and references and in developing learning outcomes and study guides for the two subjects. The material that he has developed on these subjects is currently being used in accrediting university programs and in granting equivalencies to technical schools for related courses and also in assisting the professional associations in evaluating the credentials of candidates for professional membership in surveying. He is also a representative on the CBEPS Board of Directors and the CBEPS Exemptions and Accreditation Committee. The CBEPS Board establishes, assesses, and certifies the academic qualifications of individuals who apply to become land surveyors or geomatics professionals or both, in Canada, while the CBEPS Exemptions and Accreditation Committee is responsible for evaluating courses offered by post-secondary institutions in terms of their equivalence to individual CBEPS Syllabus items.

Dr. John Olusegun Ogundare has been working as an instructor of geomatics technology (in the diploma and degree programs) for about 20 years at the British Columbia Institute of Technology (BCIT), Canada, where he teaches subjects such as Advanced Topics in Precision surveys, Geodetic Positioning, Least Squares Adjustments, Mathematical Cartography. He also mentors the Bachelor of Technology students in their technical projects and reports. Some of his BCIT-funded works included providing manuals for CBEPS-accredited courses, which he developed and teaches to full-time and distance-learning students. Some of those funded courses are the Advanced Topics in Precision Surveys, Geodetic Positioning, Special Topics in Least Squares Adjustment, and Mathematical Cartography. Apart from being an instructor, Dr. John Olusegun Ogundare has also served for over 10 years as a member of the quality committee of the BCIT School of Construction and the Environment and for over 5 years as a member of the School Research committee. His current main professional interest and expertise include monitoring and analysis of deformations in engineering and geoscience projects; precision engineering surveys; geodetic control surveys, analysis and optimization; spatial data analysis and adjustments; coordinate systems and map projections; rock mechanics instrumentation, ground subsidence in mining areas and GPS applications.

FOREWORD

"Precision Surveying comes as a very needed textbook in North America. It fills the gap between existing textbooks dealing with basic principles of surveying and textbooks dealing with the theory of geodetic science. Theory of advanced surveying techniques, their proper use in engineering and geoscience projects and thorough accuracy analysis have been missing in the contemporary technical literature in geomatics. Dr. John Olusegun Ogundare, the author of the book, was one of my best graduate students at University of New Brunswick about 20 years ago. He was a hard working, young scientist, eager to learn, and very thorough in his work. This is reflected in this textbook, which brings enormous amount of information on modern surveying techniques of high precision, their proper use, and very detailed analysis and evaluation of surveying projects. Setting out and high-precision alignment of engineering structures, advanced techniques in mining and tunneling surveys, and structural and ground deformation monitoring and analysis are covered in this book with several case studies and practical examples. Readers at all levels of their knowledge in geomatics will certainly benefit from this textbook. My congratulations go to the author.

Adam Chrzanowski, Ph.D., Drh.c., P.Eng.
Director of Canadian Centre for Geodetic Engineering
University of New Brunswick

PREFACE

Precision surveying is not a specific area of discipline such as geodesy, hydrography, remote sensing, and photogrammetry. It is a geomatics engineering practice that applies any appropriate field of geomatics to projects in order to achieve a desired accuracy or precision; it deals with important aspects of real-world problems, such as designing and monitoring human-made infrastructures for millimeter-level movements, alignment of large machines in industrial environment, and so on. Some of the concepts and techniques involved have been developed over several decades, and some have just been accomplished recently. Although the basic concepts and techniques have not changed significantly and are not likely to change in the next several years, they are still not popular and are mainly understood by researchers or academic experts. This is partially due to the complex theoretical background involved, which are usually difficult for students and practicing surveyors/geomatics engineers to grasp.

My primary motivation to write this book came from my over 15 years of experience in teaching related courses to the Bachelor of Geomatics engineering technology students at the British Columbia Institute of Technology (BCIT) Canada, and my 8 years of being a special examiner for the Canadian Board of Examiners for Professional Surveyors (CBEPS) on Coordinate Systems, Map Projections, and Cartography subjects. My involvement in 2007 and 2009 as a consultant to the Canadian Council of Land Surveyors (CCLS)/CBEPS to develop learning outcomes, study guides, and reference materials for one of the subjects they use as entrance requirements toward becoming a Canadian professional surveyor also gave me an invaluable insight into a definite need for a comprehensive textbook on precision surveying. One of the most difficult tasks I have had is finding appropriate books on Precision (Advanced) Surveying to recommend to students; to the best of my knowledge, no comprehensive and dedicated books are available for this subject. I also wrote this book as a framework for learning underlying principles and procedures of precision surveying with examples that are simple enough for the geomatics students and the practicing surveyors/engineers to understand and to help them develop their interest in precision surveying and the interdisciplinary aspects.

I had two main goals in writing this text: to satisfy the need for a comprehensive textbook on precision surveying that would deal with the totality of precision surveying principles and practice, including the recent developments in geodetic surveying and the interdisciplinary collaborations with other fields; and to demystify various aspects of precision surveying so that practicing surveyors/geomatics engineers can apply them to real-world problems. My initial effort toward realizing a comprehensive precision book was in developing a manual titled "Precision Surveying: The Principles and Practice," funded by BCIT, which I have been using in delivering my precision surveying courses to students at BCIT. This manual has evolved over a number of years with many updates based on suggestions and corrections from students, academic colleagues, and those from the industry. Recently, during my 1 year professional development leave to the Canadian Centre for Geodetic Engineering (CCGE) at the University of New Brunswick (UNB) in Canada, I updated the manual to include more undergraduate and graduate courses, such as Survey Design and Analysis (or Geomatics Network Design and Analysis), Precision Surveying, Engineering Surveying, Mining and Tunneling Surveying, and Industrial Metrology.

In comparison with other geomatics books, this book is considered unique because of its in-depth treatment of many specialized topics and modern trends in geomatics that have only been discussed, up till now, in articles, journals, and conference papers. Although the book places more emphasis on concepts and principles to prevent its contents from aging too quickly, some theoretical discussions and complex derivations of formulae are avoided when they are not relevant to the understanding of the concepts being presented. Moreover, this book does not include descriptions of measuring techniques and some basic instrumentation, which can be found in elementary surveying books.

This book consists of 14 chapters and 4 appendixes. Chapter 1 explains the main properties of precision surveys with regard to basic survey procedures and different traditional measurement techniques; it distinguishes the properties of the main classes of precision surveys, examines general terms in the precision geodetic survey techniques, and presents some safety issues and their management in relation to precision survey projects.

Chapter 2 discusses survey observables, measuring instruments, and the theory of observation errors, including the application of the concepts of confidence regions, the importance of equipment testing and calibration and the statistical analysis tools for survey measurements and parameters. In Chapter 3, an in-depth discussion is given on various standards and specifications available for geomatics projects, including their representations, interpretations, relationships with quality assurance/quality control measures, and their use in geomatics projects.

Accuracy analyses and evaluations of survey measurements and their measurement systems, including error sources and their treatment are presented in detail in

Chapters 4–6. Chapter 4 deals with angle measurement and the measurement systems; Chapter 5 describes electronic distance measurements and the measurement systems; and Chapter 6 analyses elevation difference and coordinate difference measurements and the relevant equipment, such as geodetic leveling and Global Navigation Satellite System (GNSS) equipment.

Chapter 7 discusses survey design and analysis, including the main purpose, the steps involved, the elements and problems of network design, and the issues related to deformation monitoring schemes. The description of commonly used three-dimensional coordinate reference systems, their needs, and the common models for three-dimensional coordinating systems are presented in Chapter 8. Also presented in this chapter are detailed explanation on the concepts, features, and accuracy limitations of some coordinating systems, such as electronic theodolite coordinating system, GNSS, airborne laser, and terrestrial laser scanning systems.

Comprehensive discussions on deformation monitoring techniques and analysis with regard to operating principles of relevant instruments, design elements of deformation monitoring schemes, data gathering, data processing, and data analyses, including comparisons of different techniques and their main advantages and limitations are given in Chapters 9–11. Chapter 9 discusses the traditional geodetic techniques; Chapter 10 covers modern high-definition surveying (HDS) and remote sensing techniques while Chapter 11 carefully evaluates geotechnical and structural techniques. Some of the discussions in Chapter 10 include the essential properties and features of HDS techniques, such as laser scanning, ground-based interferometric synthetic aperture radar (GBInSAR) and Light Detection And Ranging (LiDAR) systems; and the satellite-based InSAR. Chapter 11 identifies the differences between geotechnical and geodetic deformation monitoring schemes, analyses geotechnical deformation measurements, and explains the accuracy specifications for various geotechnical instrumentations with regard to deformation monitoring and how the geotechnical monitoring techniques complement geodetic monitoring techniques. This chapter is presented from the geomatics point of view so as to inform and acquaint the geomatics specialists with the relevance of geotechnical monitoring techniques to their practice.

Chapters 12 and 13 describe the main elements of mining and tunneling surveys. Chapter 12 starts with the definitions of some mining terminology, discusses the problems and various techniques of orientation transfer in mining and tunneling surveys, and evaluates the sources of systematic and random errors in alignment and underground surveys, including how the errors are minimized. In Chapter 13, the basic elements and methods of tunneling surveys are described. This includes a discussion on approximate effects of lateral atmospheric refraction on alignment surveys, horizontal and vertical design and simulation of tunneling surveys, error analysis of underground traverse surveys, and the determination of grid azimuth from gyro azimuth measurement for underground traverse surveys.

Chapter 14 gives a comprehensive description of the main techniques of precision alignment, such as direct laser alignment, conventional surveying techniques, optical tooling, laser interferometric techniques, and polar measurement techniques; the chapter also explains the main sources of error and the advantages and limitations of the different techniques. The book ends with four appendices: Appendices I–III containing sample tables for use in statistical analyses of data, and Appendix IV presents some commonly used units.

Since this book is based on the manual that has already been used for several courses taught by the author at the undergraduate level, it can be considered to have been tested through teaching on the bachelor degree level. Certain features of the book are designed to aid in the learning and teaching activities: the chapter objectives, which provide an overview of the material contained in that chapter; and a number of example problems with suggested solutions, which are to assist readers in understanding the principles discussed. The use of this book, however, is recommended for third and fourth year technological and university undergraduate courses as well as for graduate courses. Some aspects of the book, however, can be adapted for use in second year courses if the topics of the courses are well organized with the method of least squares adjustment course taken concurrently. In general, a good understanding of elementary surveying, geodesy, and the method of least squares adjustment are recommended prerequisites to understanding some of the concepts discussed in this book.

Apart from being appropriate for use as textbook in college and university classes, this book is also a valuable tool for readers from a variety of surveying backgrounds, including practicing surveyors/engineers who are interested in precision surveys, geomatics researchers, software developers for geomatics, and so on.

John Olusegun Ogundare
Burnaby, B.C., Canada
9 July 2015

ACKNOWLEDGMENTS

The author would like to acknowledge and thank all of those bodies and individuals who have contributed in any way to the formation and updating of this book. The author is particularly indebted to British Columbia Institute of Technology (BCIT), Canada, for providing the funding for the development of the manual on which this book is based; without this funding, this book would not have been possible.

Special thanks are due to Dr. Adam Chrzanowski (Professor Emeritus, University of New Brunswick in Fredericton), the author's graduate study mentor and teacher, who provided the author with a vast material resource on various aspects of the studies resulting in this book, and for his constructive and valuable criticism. Dr. Chrzanowski is particularly acknowledged for his help in facilitating the author's professional development leave to the Canadian Centre for Geodetic Engineering (CCGE) at the University of New Brunswick in Canada, where the author wrote a substantial part of this book. In addition, a thank you to Dr. Anna Chrzanowski and Maciej Bazanowski of the CCGE for their invaluable support, friendship, and encouragement. The author also gratefully acknowledges the help received from the many papers, books, seminars, lecture notes, and reports, which were written by many other specialists in the area of precision surveys.

The author wishes to recognize the assistance of Mr. John Fletcher of the New Brunswick Power (NB Power) Generation, Mactaquac Generating Station, N.B., Canada, who devoted several hours of his time to taking the author round the Mactaquac dam monitoring systems and for providing source material on geotechnical instrumentations at the dam. The author is grateful to him and to the other NB Power staff for willingly responding to his various requests for information and for tirelessly answering the author's endless questions.

Other individuals/corporations who contributed to this book in one way or another are Dr. James Secord of the University of New Brunswick, who provided the author with valuable comments, suggestions, and reference materials; Mr. Brian Roulston and other staff members of the Potash Corporation, Sussex mine, N.B., who helped the author in understanding the workings of the underground mine; Dr. Tomas Beran and other staff members of Measurand Inc., Fredericton, N.B., Canada, who helped in clarifying the workings of a particular MEMS system and for reviewing the related section of this book; the staff of RST Instruments Ltd, Coquitlam, B.C., Canada, for providing the author with useful information on their geotechnical instrumentations; the Canadian Board of Examiners for Professional Surveyors (CBEPS) for giving the author the permission to reproduce some of their past Exam questions on Advanced Surveying subject in this book; and Alistair Boakes of BCIT Learning and Teaching Centre, who helped in the design of the cover page for this book. The author is grateful to all of them and also to the reviewers, who pointed out problems and identified some areas of improvement to this book.

Finally, the author is grateful to his wife, Eunice, and his children, Joy and Isaac, for their patience, understanding, and encouragement.

1

PRECISION SURVEY PROPERTIES AND TECHNIQUES

OBJECTIVES

After studying this chapter, you should be able to

1. Explain the main properties of precision survey procedure with respect to basic survey procedure
2. Discuss the properties of the main classes of precision surveys
3. Explain different traditional measurement techniques used in precision surveys
4. Discuss the uses of different coordinate systems for precision surveys
5. Discuss the geodetic challenges of some precision survey projects
6. Evaluate some safety issues relating to precision survey projects

1.1 INTRODUCTION

Precision surveying is not a specific area of discipline like geodesy, photogrammetry and remote sensing. It is about applying appropriate field(s) of surveying to projects in order to achieve a desired accuracy (or precision). Ordinary measurements to a few millimetres are sufficiently precise in some projects such as construction of buildings

Precision Surveying: The Principles and Geomatics Practice, First Edition. John Olusegun Ogundare.

and bridges; but greater precision may be required for alignment of prefabricated steel structure or members, and for deformation monitoring. For example, an alignment of magnets of accelerator facilities may be required to a tolerance of up to 0.1 mm or better; in monitoring and deformation surveys, strict requirements on observations and data handling methods are imposed in order to achieve desired accuracy; and in long tunnel surveys, the critical factor is usually to minimize lateral breakthrough error which requires special methods of network design that are different from those applied to ordinary geodetic networks. Precision surveys are done by educated specialists who are able to determine the appropriate instrumentation, evaluate sources of error and prescribe suitable error-mitigating procedures, for a given project.

The most significant properties distinguishing precision surveys from ordinary surveys can be summarized as follows:

1. Precision surveys require the use of precise and expensive instrumentations.
2. Precision surveys require stricter observations and data handling methods, which require directly proportionate increase in time and effort of the surveyor and also increase in cost of the surveys.
3. Precision surveys involve collecting a larger number of observations. In order to obtain accuracies in the millimetre range, a high degree of redundancy is required in the survey network which, in practice, translates into a large number of observations. Redundant observations are needed in order to be able to assess the accuracy and reliability of the results.
4. Precision surveys require more rigorous mathematical treatment for error evaluation. Errors in data handling, from observation stage to final processing can often contribute significant errors in final results. Reducing the magnitudes of these errors in data handling as well as in processing the data can significantly improve the accuracy of the survey.

It is the duty of the surveyor to maintain a degree of precision as high as can be justified by the purpose of the survey, but not higher. For the surveyor to achieve an appropriate degree of precision for a survey, the surveyor must have possessed a thorough understanding of the following:

(a) The intended use of the survey measurements.
(b) Sources of errors and types of errors in survey measurements.
(c) Design of appropriate survey scheme to aid in choosing appropriate survey instruments.
(d) Field survey procedures (including the amount, type, and survey data acquisition techniques) for keeping the magnitude of errors within allowable limits. The procedures should also include performing instrument setup or calibration or both.

(e) Methods of adjustment and analysis of the acquired measurements which will include providing an indication of the quality and reliability of the results.

1.2 BASIC CLASSIFICATION OF PRECISION SURVEYS

It should be mentioned that the classification being attempted in this section is subjective and may not be generally accepted; it is made to facilitate the understanding of various aspects of precision surveys. For the purpose of this book, the high precision survey will be classified to include the following:

1. Geodetic control network surveys
2. Monitoring and deformation surveys
3. Geodetic engineering surveys
4. Industrial metrology
5. Surveys for research and education

1.2.1 Geodetic Control Network Surveys

Geodetic control network survey is a survey process which takes into account the true shape and size of the earth; it employs the principles of geodesy and is generally conducted over large areas with precise instruments and precise surveying methods. The survey is conducted in order to establish horizontal and vertical positions of points as well as three-dimensional positions of points. A geodetic control network is a series of widely-spaced, permanent and interconnected monuments whose positions (or coordinates) and elevations are accurately known. The agencies of governments, such as the Geodetic Survey Division (GSD) of Canada, are primarily responsible for conducting geodetic surveys. Relatively few engineers and surveyors are involved in geodetic control surveys but the resulting data are usually of great importance since they provide precise points of reference to which a multitude of surveys of lower precision may be tied.

Geodetic control survey is typically carried out in order to provide:

1. Basic framework (e.g., the Canadian reference framework and the Canadian Spatial Reference Systems (CSRS), the American National Spatial Reference System (NSRS), the European Spatial Reference System (ESRS)) for detailed site plan topographic mapping, boundary demarcation (international, and inter-state or inter-provincial), mapping natural resources, and so on. Generally, it provides control for large geopolitical areas where there is a need to accurately connect with adjacent political areas, and also for the purpose of controlling inter-state transportation corridors, such as highways, pipelines, railroads, and so on.

2. Primary reference for subsequent engineering and construction projects (e.g., building of bridges, dams, tunnels, highways, pipelines, etc.).
3. Reference for positioning marine construction vessels (continuous positioning of dredges and survey boats).
4. Reference for effectively and efficiently monitoring and evaluating deformations of large extent, which may include tectonic plate, land slide, dams, and so on.

1.2.2 Monitoring and Deformation Surveys

Monitoring and deformation surveys are essentially for the purpose of modeling and analysing natural phenomena (earthquakes, landslides, crustal movement) and man-made structures (bridges, buildings, tunnels, dams, and mines). The accuracy requirements of the surveys can differ significantly from those of control or legal surveys. In monitoring and deformation surveys, stricter requirements on observation and data handling methods are usually imposed in determining the relative positions of the monitored or observed stations.

Geodetic control surveys are different from geodetic deformation surveys. In *geodetic control surveys*, the determination of absolute positions (coordinates) of points is of interest while in the *geodetic deformation surveys*, one is interested only in the determination of changes of positions (displacements). Some specific monitoring and deformation surveys projects are as follows:

- Deformation measurements of Flaming Gorge concrete dam on the Green River in Utah (Roehm, L.H., 1968).
- Monitoring Earth filled dams in Southern California (Duffy et al., 2001).
- Monitoring exposed pit walls at the Highland Valley Copper mine in British Columbia, Canada (Wilkins et al., 2003).

Other projects requiring deformation monitoring surveys are as follows:

- Measurement of deformation on buildings exposed to some particular mechanical or thermal strain. Accuracy requirements may be in the order of millimetres for object dimensions of more than 100 m (e.g. cooling towers, chimneys, dams, sluices, cranes, historical buildings, etc.).
- Deformation of concrete tanks used for galvanizing and electroplating may need to be measured under working conditions. The tanks are constructed from special concrete and in operation, are slowly filled with liquid of several tons. The tank walls are subject to critical deformations which may need to be observed at regular intervals.
- Deformation analysis of rotary cement kiln. A rotary kiln is a cylindrical vessel made of steel plate and lined with firebrick. The vessel slowly rotates about its axis between 0.5 and 5 revolutions per minute and continues to run 24 hours a day and only stop a few days once or twice a year for essential maintenance.

The kiln must be monitored for safety reason. By measuring the surface of the vessel, critical areas of the kiln can be detected and deformation monitored.

- Tunnel profile measurement requires measuring tunnel interiors for shape and deformation analysis.

1.2.3 Geodetic Engineering Surveys

Geodetic (or precision) engineering surveys apply rigorous geodetic methods to control and support construction and building projects which include construction and maintenance of tunnels, bridges, hydroelectric power stations, railways, and so on. Unlike in geodetic positioning, geodetic engineering surveys are based on local coordinate systems and relative positioning of objects are of more importance than absolute positioning. Many of today's engineering surveys require relative positional accuracies in the order of 1:100,000 or better. Most first order national geodetic networks, however, may not be suitable for controlling engineering projects where high precision is required because of possible distortions in the national geodetic networks. What is usually appropriate is to adopt appropriate geodetic model and local coordinate system.

Engineering Surveys deals with special survey techniques and precision measurement techniques developed for three purposes:

1. Positioning the construction elements of large engineering works such as dams, tunnels, pipelines, deep mine shafts, high-rise office buildings, and bridges;
2. Deformation monitoring of these works and their surrounding (ground subsidence and slope stability) and their analysis;
3. Positioning and alignment of machinery and scientific apparatus.

"*Mining surveying* is an important branch of engineering surveying dealing with rock stability control and protection of underground and surface structures that may be influenced by ground subsidence" (Chrzanowski, 1999). Actual mining surveying consists of undermining and controlling caving of the ore; it is also necessary that the position of the workings at one level be known precisely at the next level above. Mine surveying are done in cramped areas, with irregular routes, no reference objects such as sun or star to provide azimuth.

Land surveying is a highly specialized branch of geodetic engineering surveying that focuses on establishing boundary lines of real property ownerships, which include establishing new boundaries as may be required in re-establishing the original boundaries or in land partitioning; it also deals with the determination of areas of land tracts. With regard to construction projects, the land surveying problem usually arises when costly land acquisition is involved, such as in pipeline surveys.

For convenience and simplicity, engineering and land surveys are usually made as if the surveys are done on a plane earth surface. In this case, plane local coordinate system (requiring map projection process) is commonly used. Since a local coordinate system is an isolated system with respect to other types of coordinate system such as geodetic coordinate (latitude, longitude, ellipsoid height) systems, it is impossible to directly correlate one engineering survey with others when large areas are involved. Moreover, local coordinate systems cannot be extended too much from their origins since the extension may introduce some unacceptable distortions to the surveys.

Some of the geodetic engineering challenges that may be encountered in geodetic engineering surveys include the following:

1. With regard to pipeline projects, for transportation of oil and natural gas, over a long distance, for example, *Trans Mountain Pipeline (TMPL) project* from the oil fields in Alberta to British Columbia (Hamilton, 1951; Chrzanowski, 1999), the following geodetic engineering challenges are encountered:
   - Choosing the best possible route for the pipeline with consideration for the environmental impact of the project as well as the possible presence and impact of subsidence and geological fault lines on the functioning of the pipelines. This will require consulting other geoscientists and using appropriate tools, such as topographic maps, Geographic Information system (GIS), Google Earth tools and LIDAR system, to identify the best route.
   - Acquiring the right-of-way, which may involve relocating and settling the owners of the acquired landed properties; this will require carrying out legal surveys for the route. Today, traditional surveys with theodolite and chains are giving way to the use of modern technology, such as total station equipment and Global Positioning System (GPS).
   - Providing the desired grades of pipelines, since pipelines are sensitive to grades which are very important in the calculation of pumping facilities and attaining appropriate pressures in the pipelines. Today, in establishing grades for pipelines, the use of traditional differential leveling procedure is still common.
   - Ensuring that all necessary safety regulations at all government levels are complied with and that the environmental impact of the pipeline project is minimized.
2. With regard to construction of large dams, such as hydroelectric dams, the following geodetic aspects are usually involved:
   - Preliminary reconnaissance surveys using large-scale (1:50,000 or larger) topographic maps in order to identify and tentatively select the extent of the dam, the reservoir and tail-water areas.
   - Establishing permanent precision survey control stations around the dam site.
   - Mapping the topography beneath the dam with high precision for the purpose of designing the dam and estimating the quantities of materials involved.
   - Mapping the corridors for the layout of power lines; and carrying out other surveys needed for the drawing of general layout plans and the setting out of concrete forms.

- Carrying out precise monitoring surveys to detect and measure any deformation during the dam construction and during the loading and unloading of the dam.
- Carrying out surveys for the positioning of the generating equipment, and the related penstocks and outflow conduits.

Further information on geodetic surveys for large dam construction project can be found in Williams (1958), Moreau and Boyer (1972), and Chrzanowski (1999). Examples of transportation tunneling surveys is the survey for the 14.5 km long railway tunnel at the Rogers Pass in British Columbia, Canada (Lachapelle et al., 1984, 1985, and 1988) and the survey of 50.5 km Channel Tunnel transportation system connecting Britain and France in Europe; and an example of tunneling surveys for scientific research is the tunneling surveys for the Superconducting Super Collider (SSC) project in Texas involving 4.2 m diameter, 87-km-long tunnel (Chrzanowski et al., 1993; Chrzanowski, 1999; Robinson et al., 1995; and Dekrom, 1995).

1.2.4 Industrial Metrology

Metrology, in general, is the science of performing accurate measurement. Industrial metrology is the use of precision measuring techniques for positioning and aligning industrial machinery and scientific apparatus. It deals with aligning components of large antennas (parabolic, flat, etc.), checking aircraft dimensional quality of the various subassemblies which form the structure of the aircraft (aerospace alignment), making geometrical checks on finished components in ship and car buildings, alignment and positioning of magnets of colliders, alignment of accelerator facilities, setting up and aligning machines in the industries, *in-situ* calibration of industrial robots, and so on. These types of project usually require that tight tolerances be satisfied and the work is done in the environment where there are a lot of vibrations and unpleasant conditions. The commonly employed techniques (which are different from those used in conventional geodetic surveys) are based mainly on special mechanical and optical tools such as jig transits, optical squares, aligning telescopes, optical micrometers, laser interferometry.

Nowadays, geodetic measuring techniques are increasingly used in the industry (because of the advent of electronic theodolites which are easily interfaced with computers), where three-dimensional micro-triangulation surveys can be carried out in real-time positioning of industrial components with accuracies satisfying the requirements of industry. For example, in the Chalk River Nuclear Laboratory of the Atomic Energy of Canada in 1987, the University of New Brunswick (UNB) Canada team used 3D coordinating system to align over 40 magnets in a cramped laboratory space over a distance of about 40 m with accuracies better than 0.1 mm in the transverse and vertical directions and better than 0.2 mm in the longitudinal direction (Chrzanowski, 1999).

Industrial metrology or industrial surveying has another specialized component known as *optical tooling* (or *optical alignment*). It is a method of making extremely accurate measurements for manufacturing processes where small tolerances are required. Measurements are usually made by a person interpreting a scale or optical micrometer by looking through an alignment telescope, or the lines and planes are created by a laser with digital measurements.

1.2.5 Surveys for Research and Education

Surveys for research and education deal with scientific experimentation of ideals. They provide theoretical and practical testing procedures for different measurement systems. Some of the examples of such research projects are as follows:

- Photogrammetric and terrestrial deformation surveys for Turtle Mountain (Fraser and Gruendig, 1985; Chapman, 1985).
- Integrated analysis of ground subsidence in a coal mining area: a case study (Chrzanowski and Szostak-Chrzanowski, 1986).
- Implementation of the UNB generalized method for the integrated analysis of deformations at the Mactaquac generating station in Canada (Ogundare, 1990).
- Use of GPS in integrated deformation surveys (Chrzanowski, et al, 1990).

1.3 PRECISION GEODETIC SURVEY TECHNIQUES

Generally, specifications for precision geodetic survey techniques include the least angular count of instruments to be used, number of observations, rejection criteria of observations, spacing of major stations, and the expected angular and positional tolerances. To obtain precise measurements, the surveyor must use precision equipment and precision techniques. Many of the techniques used in precise surveys are adapted from the conventional geodetic positioning methods and instrumentation, but with some differences in the field survey procedures and with the stretching of instrument performance to the limit of accuracy. Conventional (non-Global Navigation Satellite System, non-GNSS) horizontal and vertical survey techniques using traditional ground survey instruments (theodolites, electromagnetic distance measurement (EDM), total stations, levels) and the GPS survey techniques are used.

1.3.1 Positioning using Global Navigation Satellite System

Global Navigation Satellite System (GNSS) currently refers to the United States' *GPS*, the Russian Federation's *GLobal Orbiting NAvigation Satellite Systsem (GLONASS)*, the European Union's *Galileo system* and China's *Compass system.* GPS, however, is currently the predominant satellite surveying system in use; GLONASS is operational, but the full constellation of the satellites is yet to be launched; Galileo and Compass are still under development. All these satellite positioning systems are known collectively as GNSS. The GNSS positioning techniques are now generally used for most horizontal control surveys performed for mapping frameworks. The current trend is to use GNSS in precision surveys, but conventional terrestrial techniques are still required in local and isolated monitoring schemes, especially for economy and relative accuracy. The surface control for large tunnels, such as the 87 km long main Collider tunnel for the SSC in Texas was established by means of GPS surveys using dual frequency equipment (Chrzanowski, et al., 1993). Control stations established using GPS techniques will inherently have the potential for higher orders of accuracy in control surveys.

Selection of the right GNSS receiver for a particular project is critical to the success of the project. Receiver selection must be based on a number of criteria, which include the applications for which the receiver is to be used, accuracy requirements and signal processing requirements. GNSS receivers range from high-end, high-cost, high-accuracy *geodetic quality* through moderate cost, meter-level accuracy *mapping grade*, to low-end, low-cost, low-accuracy *resource grade* or *recreational models*. Geodetic quality type is used mainly in high precision surveys.

There are two general types of GNSS receivers: *code phase* and *carrier phase*. Geodetic quality receivers process both code and carrier phases. The receivers and their auxiliary equipment can cost several thousands of dollars. A code phase receiver requires access to the satellite navigation message of the P- or C/A-code signal to function, while carrier phase receiver utilizes the actual GNSS signal to calculate position. There are two general types of carrier phase receivers: *single frequency* and *dual frequency*. The single-frequency receivers track the L1 frequency signal and are not very accurate in resolving long baselines where ionospheric effects are very high. Dual frequency receivers track both the L1 and L2 frequency signals and will effectively resolve baselines longer than 20 km where ionospheric effects have a larger impact on calculations. All geodetic quality receivers are multi-channel, in which a separate channel is tracking each satellite in view. Some of the qualities to look for in GNSS geodetic receivers are as follows:

1. In the case of dual frequency receivers, the receivers must provide at least the following time-tagged (based on time of receipt of signal referenced to the receiver clock) observables:
   - Full L1 C/A code, and L1 P-code
   - Continuous full wavelength L1 carrier phase
   - L2 P-code and continuous full wavelength L2 carrier phase
2. In the case of single frequency receivers, the receivers must provide at least the following time-tagged (based on time of receipt of signal referenced to the receiver clock) observables:
   - Full L1 C/A code
   - Continuous full wavelength L1 carrier phase
3. When the GNSS reference receiver is used with a remote one, the reference shall be capable of 10 mm + 2 ppm accuracy or better on baselines of 1–100 km in length when used in the static differential mode. The receivers shall have an accuracy of 5 mm or better on baselines less than 1 km
4. The receiver shall have L1 and L2 full wavelength carrier phase measurement accuracies of 0.75 cm (RMS) or better, exclusive of the receiver clock offset.

5. The receiver shall have an L1 C/A code phase measurement accuracy of 30 cm (RMS) or better, exclusive of receiver clock time and frequency offsets.
6. The processing software must allow baseline computations with the options of using the broadcast and precise ephemerides.

Typical equipment selection for precision GNSS surveys will include the following:

1. A minimum of two receivers (four receivers for economy and efficiency).
2. Ideally, an antenna type with the smallest sensitivity to multipath and the smallest phase center variation should be selected. Same type of antenna for all receivers on the project is recommended to minimize phase centre biases.
3. Dual frequency receivers are recommended where the ionosphere is unpredictable and irregular and also for second order accuracy or better and where the baseline lengths consistently exceed 15 km.

1.3.2 Conventional Horizontal Positioning Techniques

Typical conventional horizontal positioning techniques include triangulation, trilateration, combined triangulation and trilateration, traversing, intersection, and resection. A *triangulation* survey network consists of a series of interconnected triangles in which an occasional line is measured and the remaining sides are calculated from angles measured at the vertices of the triangles. This method of survey was originally favored for extending the first-order control since the measurement of angles (and only a few sides) could be taken more quickly and precisely than the measurement of all the distances as in trilateration. It is now possible to measure precisely the length of a triangle side in about the same length of time as was required for angle measurement. A triangulation net usually offer the most economical and accurate (first-order accuracy) means of developing a horizontal control system when extremely rough terrain is involved.

Trilateration survey network consists of interconnected triangles in which all lengths and only enough angles or directions for azimuth determination are measured. The trilateration techniques have become competitive with the triangulation techniques for establishing horizontal control since the advent of precision EDM. Usually, the triangles of a triangulation or a trilateration network should contain angles that are more than 15–25°. The EDM equipment used should yield the required standard deviations in distances and the distances must be corrected for all systematic instrumental errors and for the effects of atmospheric conditions. Trilateration techniques may be used for extending first-order horizontal control through an entire continent.

Combined triangulation and trilateration network consists of interconnected triangles in which all the angles and all the distances are measured. The combined triangulation and trilateration survey techniques produce the strongest network of horizontal control that can be established by conventional terrestrial methods. Modern terrestrial control survey practice favors the survey techniques since they ensure many redundant measurements. The combined triangulation and trilateration techniques may be used to provide first-order or primary horizontal control for the national control network and the network can be used for earth crustal movement studies, engineering projects of high precision, and so on. The combined techniques have also been used in providing surface geodetic network for tunnel construction, network for preconstruction work for dams.

A *traverse* consists of a series of straight lines connecting successive established points along the route of a survey. Distances along the lines are measured using tape or electromagnetic distance measurement (EDM) equipment and the angle at each traverse point is measured using a theodolite or a total station. Traversing is a convenient, fast method for establishing horizontal control in densely built up areas and in heavily forested regions where lengths of sights are too short to allow triangulation or trilateration. The advent of reliable and precise EDM instruments has made traverse method very important in strengthening a triangulation net and in providing control points. In surveying work for tunnels in mountainous areas, a combination of triangulation and traversing is most suitable. The underground survey is based on an open traverse measured with precision theodolite and EDM equipment with precision surveying gyroscope providing orientation. A typical fully automatic precision surveying gyroscope is GYROMAT 2000 with precision of one measurement of astronomic azimuth being ±3-in. This is a gyroscopic traversing for the purpose of guiding the boring machine during tunnel construction. Precision traversing can also be carried out in dam monitoring surveys. In this case, traverses are measured in corridors which have pillars with forced centring tribrachs. Traversing, however, have limited uses in precision surveys since it is incapable of providing sufficient redundancy required in most projects.

Intersection method provides the coordinates of unknown points based on the measurements made from at least two other points. This technique is commonly used in 3D coordinating systems, terrestrial laser scanning systems, automatic monitoring systems, and so on.

Resection method is used in determining the position and height of an instrument setup station by making measurements to at least two points whose coordinates had been previously determined. In this method, the accuracy of resected point increases with strong angular relationship (approaching 90° at the resected point) of the resected point and the observed points, the number of points observed to (creating redundant measurements) and the accuracy of the observed points. Resection has an important advantage of allowing the instrument to be located in any favorable location of choice by the instrument person so that one is not forced to set up on a known point that is in an unsatisfactory location. This procedure allows the effects of instrument centering errors on angular measurements to be minimized since one is not required to center on a particular station.

1.3.3 Geodetic Vertical Positioning Techniques

The geodetic vertical positioning surveys consist of establishing the elevations of points with reference to the geoid. The surveys are used to establish a basic network of vertical control points. From these, the elevations of other positions in surveys are determined by lower-accuracy methods. *Differential leveling* is a precise leveling technique for providing vertical control with high precision (within the limits of first- or special-order accuracy). In dam monitoring, precise leveling is performed along the crest as well as in corridors in the dam. Precision spirit levels with micrometer or digital levels, and invar rods are used in order to obtain a standard deviation of less than *1 mm/km* or better in leveling.

1.4 REVIEW OF SOME SAFETY ISSUES

A safety program should be designed as part of every survey project. In this program, the survey crews are trained or instructed to conform to some designed safety rules that will enable them to perform their duties in a safe manner. Dedicated personnel should be assigned a sole responsibility of managing and promoting the safety of work crews, which includes the following:

- Taking appropriate action in matters relating to safety of the crews
- Creating safety awareness in the crews
- Organizing regular safety meetings as may be needed, usually before starting any hazardous project.

The subjects that are usually considered as part of safety programs may include training of survey crews on the following:

1. How to recognize and avoid or respond to potential hazards, such as poisonous plants, poisonous snakes, insect bites and stings, and so on.
2. How to detect and take precautions with regard to threatening weather conditions, such as tornado, lightening, extreme temperatures, and so on.
3. How to properly use and operate equipment and tools, such as motor vehicle; transportation of tools and equipment, such as cutting tools; proper use of protective equipment and clothing suitable for a work area, which may include use of safety boots, eye protection and gloves; and in the case of working in boats, to use Coast Guard-approved life jackets; and so on.
4. First aid procedures and how to equip themselves with proper first-aid kits with appropriate medication and manuals.
5. Awareness of safety precautions, existing laws and policies with regard to ice crossing, working near traffic, and working underground and under overhead utility lines. For example, when working near traffic, personnel are to be constantly alert, wearing reflective colored vests and hats at all times; when surveying around the Federal highways, the laws concerning security must be strictly obeyed; when working on railway rights-of-way, permission should be secured

from the railway management; and so on. Typically, when working near traffic (within 15 m from the edge of the highway), there should be an appropriate sign boards (about work ahead) 250 m before the survey activity area of 1 km with 100 m buffer ahead displaying another sign board of the ongoing survey activity. There must be a display of sign board also at 100 m before the activity area, showing that "Survey work" is going on ahead. There must also be a first-aid kit in a standby vehicle in case of emergency.

2

OBSERVABLES, MEASURING INSTRUMENTS, AND THEORY OF OBSERVATION ERRORS

OBJECTIVES

After studying this chapter, you should be able to

1. Identify basic survey techniques and their typical observables
2. Explain basic modern survey instruments and their limitations
3. Discuss the error properties of measurements and how they are propagated
4. Discuss the needs for accuracy analysis and the steps for estimating accuracy of typical survey observables
5. Discuss the application of the concepts of confidence regions in uncertainty determination of measurements
6. Explain statistical tools for analysis of measurements and parameters
7. Discuss the importance of calibrating and testing survey equipment

2.1 OBSERVABLES, MEASUREMENTS AND MEASURING INSTRUMENTS

An *observable* is a physical or geometrical quantity to which a numerical value can be assigned (through measurement process) with a degree of certainty. Some of the typical geomatics measurement techniques and the corresponding observables are given in Table 2.1.

Precision Surveying: The Principles and Geomatics Practice, First Edition. John Olusegun Ogundare.

TABLE 2.1 Geomatics Measurement Techniques and the Typical Survey Observables.

Survey Techniques	Typical Observables
Differential leveling	Elevation (leveled height) differences between sections
Trigonometric leveling	Zenith (or vertical) angles, slope (or horizontal) distances, heights of instruments, heights of targets or staff readings, and horizontal directions (or angles)
Traverse	Horizontal directions (or angles), horizontal (or slope) distances, zenith (or vertical) angles, and bearings
Triangulation	Horizontal directions (or angles), zenith (or vertical) angles, baseline distances, and bearings
Trilateration	Horizontal (or slope) distances, zenith angles, and bearings
Gyro station/gyrotheodolite measurements	Astronomic azimuths (or bearings)
GPS surveys	Baseline vectors (coordinate differences of baselines) and ellipsoidal heights
Gravimetric leveling	Relative gravity values
Conventional photogrammetry	Photo coordinates of points (x, y); coordinates of fiducial center (x_0, y_0) of photo; focal length of camera (f); orientation of photo in space (if measured using gyro or inertia navigation system), such as Ω_0, Φ_0, K_0; and translations (X_0, Y_0, Z_0) if measured using GPS
Close-range photogrammetry and remote sensing	Distances in laser altimeters; phase shifts and intensity values of returned radar energy in interferometric synthetic aperture radar (InSAR); and the x, y, z coordinates (or the vertical angles, horizontal angles, and slope distances) in light detection and ranging (LiDAR) scanning systems

A *measurement or an observation* is a numerical value that is assigned to an observable. The term *measurement or observation* is also used, in practice, to refer to the actual process of assigning a numerical value to an observable. For example, the process of determining that an observable (the distance between two points) has a value of 100 m is a measurement, and the value so determined (e.g., 100 m) is also referred to as measurement. Since the exact values of observables cannot be determined, but estimated, it has become a fundamental principle of measurement in surveying that no measurement is exact and the true value of an observable can never be known. This, however, does not mean that the exact or true values of the observables do not exist, but that they cannot be determined exactly. Surveying is concerned with estimating the values of observables through measurement process.

2.2 ANGLE AND DIRECTION MEASURING INSTRUMENTS

Angle and direction measuring instruments are essentially theodolites with different forms of modifications. It should be pointed out that direction observables measured by theodolite (or total station) equipment are arbitrary; they are directions with respect to the reference zero scale point of the instrument. These directions should not be confused with azimuths (or bearings), which are directions with respect to the direction of the north (serving as reference zero point in space, not of the instrument). It should also be pointed out here that the term "angle and direction" discussed in this section can also be considered for vertical (or zenith) angles. Different types of angle and direction measuring instruments can be summarized as follows:

- Optical theodolites
- Electronic digital theodolites
- Gyrotheodolite/gyro station equipment
- Global navigation satellite system (GNSS) surveying equipment.

2.2.1 Optical Theodolites

Optical theodolites are nonelectronic theodolites. Typical examples of such theodolites are the Kern DKM3 precision theodolites with horizontal and vertical angular accuracy of 0.5″ and telescope magnification of 45× and Kern DKM2 precision theodolites with horizontal and vertical angular accuracy of 1″ and telescope magnification of 32×. The optical theodolites are of two types: *repeating theodolites* and *directional theodolites.*

The repeating theodolites are designed to allow horizontal angles to be repeated any number of times and added directly on the instrument circle. They have lower lock and tangent screw, which allows angles to be set and repeated. Excessive effort, however, is required to obtain results of sufficient accuracy. Their main advantage is that better accuracy is obtained through averaging of errors and mistakes by comparing values of single and multiple readings. Using repetition method (with m the total number of turnings of the same angle in both face left (FL) and face right (FR) positions), the average angle measurement can be expressed as follows:

$$\text{Average angle} = \frac{R_\text{f} - R_0 + 360^\circ \times q}{m} \tag{2.1}$$

where R_0 is the first direction reading (zeroing the circle), R_f is the direction reading after final (mth) repetition, and q is the number of times a complete 360° is turned on the graduation scale of the instrument. In repetition method, the cumulative angle is divided by the number of repetitions (m) with the resulting average angle having a precision that exceeds the nominal least count of the instrument used. Repetition can be up to between 6 and 12 repetitions; beyond 12 repetitions, the precision is not appreciably increased because of other error sources such as graduation errors. Figure 2.1 and the field notes in Table 2.2 show an example of the observations made of an angle AOB (when the instrument is set up at point O) by repetition twice in FL position of the telescope and twice in FR position of the telescope.

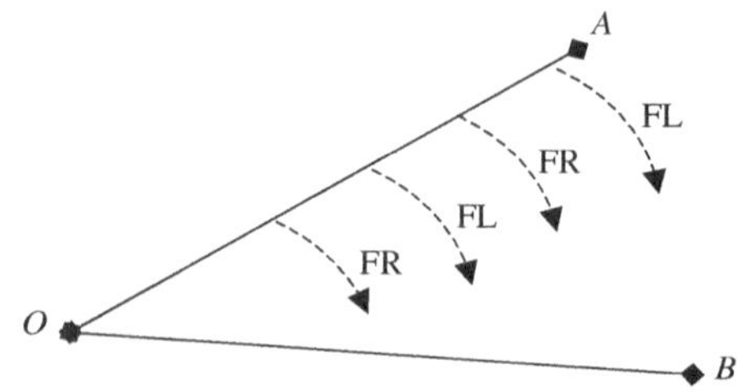

Figure 2.1 Angle measurement scheme in face left (FL) and face right (FR) positions of the telescope.

TABLE 2.2 Field Notes for Angle Measurement by Repetition Method.

Station Sighted	Repetition	Face	Circle Reading
A	0	FL	0°10′10″
B	1	FL	146°54′20″
B	4	FR	227°07′10″

The average angle is determined as follows:

The approximate angle AOB = 146°54′20″ – 0°10′10″ (or 146°44′10″).

For four repetitions, the approximate total angles turned = 4 (146°44′10″) or (586°56′40″).

Add this value to the initial direction reading (0°10′10″), giving 587°06′50″, which would have been the final direction reading if the scale graduations are limitless (more than 360°). This indicates that the circle index mark had gone past the zero (or 360°) graduation mark of the instrument once ($q = 1$ in Equation (2.1)) and 360° must be added to the final circle reading to calculate the final angle; the average angle ($\overline{\theta}$) can then be given as follows:

$$\overline{\theta} = \frac{227°07'10'' + 360° - 0°10'10''}{4} = 146°44'15''$$

Directional theodolites are traditionally nonelectronic and nonrepeating instrument that have no lower motion; they are capable of reading directions rather than angles. An example of directional theodolites is Wild T2. Angles are obtained by subtracting the first direction reading from the second direction reading. For example, using the directional method with an angle observed in n sets (i.e., one face left and one face right measurements per set), the average angle measurement ($\overline{\theta}$) can be expressed as follows:

$$\overline{\theta} = \frac{(P_{\mathrm{FS_I}} - P_{\mathrm{BS_I}})_1 + (P_{\mathrm{FS_{II}}} - P_{\mathrm{BS_{II}}})_1 + \cdots + (P_{\mathrm{FS_I}} - P_{\mathrm{BS_I}})_n + (P_{\mathrm{FS_{II}}} - P_{\mathrm{BS_{II}}})_n}{2n} \tag{2.2}$$

TABLE 2.3 Field Notes for Angle Measurement by Directional Method.

Set(1)	Station Sighted(2)	Reading Face I(3)	Reading Face II (4)	Mean Direction (5)	Mean Angle (6)
1	A	0°00′00″	0°00′00″	0°00′00″	
	B	37°30′27″	37°30′21″	37°30′24″	
	Angle	**37°30′27″**	**37°30′21″**	37°30′24″	**37°30′24″**
2	A	0°00′00″	0°00′00″	0°00′00″	
	B	37°30′26″	37°30′26″	37°30′26″	
	Angle	**37°30′26″**	**37°30′26″**	37°30′26″	**37°30′26″**
				Final average	37°30′25″

where the subscripts FS and BS represent foresight and backsight, respectively; $P_{\mathrm{FS_I}}$ is the foresight pointing in face left (or face I) position of telescope, $P_{\mathrm{BS_{II}}}$ is the backsight pointing in face right (face II) position, and the subscripts "1" to "n" denote the set number of the measurements (with each set consisting of two separate angle measurements). For example, the field notes in Table 2.3 are the observations made of an angle AOB in two sets (when the instrument is setup at point O) by directional method. Angles are computed by subtracting direction readings to A from the corresponding direction readings to B. For each set, two angles are determined, and the mean angle is given in column 6. The final average angle is the average of the mean angles (from sets 1 and 2) given in column 6.

2.2.2 Electronic Digital Theodolites

For several years, the technological progress in angle measurements has been mainly in the automation of the readout systems of the horizontal and vertical circles of the theodolites. This has resulted in the invention of electronic digital theodolites. In terms of accuracy, electronic theodolites have not brought any drastic improvements in comparison with precision optical theodolites. *Electronic digital theodolites* will automatically read and record horizontal and vertical angles. Thus, they eliminate personal reading errors due to manual reading of scales on graduated circles and provide enhanced accuracy and facility in data collection.

Electronic theodolite instruments of highest accuracy are usually designed for first-order surveys. Standard deviations of such instruments can be reduced to 0.1″ for a single reading and the instruments are usually equipped with biaxial leveling compensator. For example, electronic theodolites such as Kern E2 and Wild T3000 are equipped with microprocessor-controlled biaxial sensors (biaxial leveling compensator) or electronic tiltmeters, which can sense inclination (misleveling) of the theodolite to an accuracy of about 0.5″ and automatically correct horizontal and vertical direction readouts for the effects of the misleveling. Some of the characteristics of the electronic digital theodolites are as follows:

- Circles can be instantaneously zeroed or initialized to any value.

- Angles can be measured with increasing values either left or right.
- Angles measured by repetition can be added to provide a cumulative value that is larger than 360°.
- Mistakes in reading angles are greatly reduced.
- They are easy to operate and the speed of operation is high.

2.2.3 Gyrotheodolite/Gyro Station Equipment

Azimuths (or bearings) are not measured directly with theodolites or total station; they are derived by measuring angles to celestial bodies, such as stars and the Sun. The derived azimuths (or bearings) are known as astronomic azimuths. Currently, azimuth determination through direct measurements to celestial objects is becoming outdated. The modern technologies, such as gyrotheodolites (or gyro station) and GPS methods are capable of direct determination of astronomic azimuths without making measurements to the celestial bodies. For example, the automatic gyro station such as Sokkia GP3X and the manual gyrotheodolite such as GAK1 are capable of astronomic azimuth determination to an accuracy of 20″. Other typical precision gyrotheodolites are Gi B-23 (MOM, Hungary), MW 77 (WBK, Germany), GYROMAT 2000, and GYROMAT 3000 with compatible accuracies of about 3″; and MOM GiB-11 gyrotheodolite with an accuracy of ±5″.

2.2.4 Global Navigation Satellite System (GNSS) Survey Equipment

GNSS survey methods are gradually being used for determining geodetic azimuths since the methods can be more cost–effective, faster, accurate, and reliable than conventional (terrestrial) survey methods. GNSS methods do not require intervisibility between adjacent stations unlike conventional methods; however, they produce geodetic azimuths that are different from astronomic azimuths.

The difference between geodetic azimuths derived from GNSS surveys and those derived from astronomic observations to celestial bodies (stars and Sun) can be less than a few tens of seconds of arc. The GNSS method uses GNSS satellites to determine the coordinates of antennas located on the earth surface, which are then used to derive the needed azimuths. For example, two GNSS antennas located on the stations where the geodetic azimuth is needed are observed simultaneously and the geodetic azimuth is derived from the GNSS-determined geodetic latitude and longitude of the two stations by using, for example, the Gauss mid-latitude method. The alternative approach for deriving the geodetic azimuth is by obtaining the grid coordinates of the two stations on the basis of a reference horizontal datum; the grid azimuth obtained is then corrected for arc-to-chord and the meridian convergence to obtain the geodetic azimuth.

2.3 ELEVATION DIFFERENCE MEASURING INSTRUMENT

The precise measurement of height differences has been traditionally done by geometric or differential leveling. Differential leveling with vertical position to very high accuracy of ±1 mm over short distances (10–100 m) using precision levels is usually required. A precision level with micrometer and capable of reading elevations to 0.001 m and at least 30× magnification is commonly used in precision leveling. Three major classes of precision levels commonly used *are automatic levels, tilting levels,* and *digital levels.* Note that refraction influences can deteriorate the accuracy of leveling, thus causing systematic deviations in measurements. A dangerous accumulation of refraction error up to 15 mm for each 100 m difference in elevation may take place along moderately inclined long routes if forward and backward horizontal lines are of unequal heights above the terrain.

Digital levels are currently replacing optical levels and are being used for precision works. Digital levels make it easy to level without having to read leveling rod through the telescope and also allow electronic data recording. Sources of errors in leveling with digital levels are similar to those from leveling with automatic levels. Some of the commonly used precision spirit levels (with micrometer) and digital levels are given in Table 2.4.

In surveying, precise heights are determined from measured elevation differences obtained through geodetic leveling. Differential or trigonometric leveling techniques can be used to obtain the elevation differences with the differential leveling technique still considered the better. The elevation differences so determined, however, must first be converted into height differences before they are used in height system, knowing that height differences are usually different from measured elevation differences. Height differences are unique quantities (since they represent differences in unique height values of given points), while elevation differences (from leveling)

TABLE 2.4 Examples of Precision Leveling Instruments.

Make	Description	Accuracy (Per 1 km Double Run)
Wild N3 Precision Level	M = 42×; bubble sensitivity/div: 10″; accuracy of leveling line of sight: 0.25″	±0.2 mm
Leica NA2/NAK2	Automatic optical levels Magnification: 32×	0.7 mm (0.3 mm with parallel-plate micrometer); compensator setting accuracy of 0.3″
Leica DNA03	Digital level Magnification: 24×	1.0 mm (0.3 mm with invar)
Sokkia PL1	Tilting level Magnification: 42×	0.2 mm (0.1 mm with micrometer)
Sokkia SDL30	Digital level Magnification: 32×	1 mm (0.6 mm with invar)
Sokkia B20	Automatic level Magnification: 32×	1.0 mm (0.8 mm with micrometer)
Topcon DL-101C	Digital level Magnification: 32×	0.4 mm with invar; compensator setting accuracy of 0.3″

depend on the leveling route taken. The measured elevation differences, even starting from the same benchmark, will generally result in different heights for the end benchmark of a level circuit, depending on the leveling route. The number of possible height systems is limitless; some of them are geopotential numbers, dynamic heights, normal heights, and orthometric heights.

Height systems based solely on measured elevation differences from differential leveling (with no gravity corrections applied) and orthometric height systems based on elevation differences and gravity measurements have geometric significance because their height measurements can be likened to measurements made with a graduated scale rule in a given linear unit such as meters. In this case, points with the same height value are of the same geometric length above a reference surface. This, however, is not the case with geopotential numbers and dynamic height systems, which have no geometric significance. In geopotential numbers, geopotential units are used instead of linear unit of meters, and in dynamic height systems, the scale of measurement is incompatible with the well-known linear scales such as meters and feet.

The *orthometric height system* is the most commonly used of the height systems in precision surveys. It indirectly converts measured elevation differences obtained from geodetic leveling into uniquely defined height differences (the true geometric lengths between the geoid and the given ground surface points measured along plumb lines) by applying gravity-dependent correction known as orthometric correction. The calculated orthometric corrections are applied to known heights of starting points in order to determine orthometric heights of the unknown endpoints connected by geodetic leveling. Usually, for orthometric correction determination, gravity observations are required at every 1–2 km in the mountainous areas and at every 5–10 km in flat terrains. Orthometric correction, however, will not be necessary for short level runs in relatively flat terrains.

The concept of orthometric corrections is based on the concept of level surfaces or equipotential surfaces. The equipotential surfaces are known to correspond with the lines of sight through the telescope of a leveled surveyor's instrument so that the surfaces are perpendicular to the direction of gravity at every setup point of the instrument. The gravity field, however, increases with latitude due to earth's centrifugal force and decreases with altitude above the earth's surface. This gravity field variation causes level surfaces to converge toward the pole, instead of being parallel to each other. The orthometric correction is to account for the convergence of level surfaces for long level runs in north–south directions or runs at high elevations. In determining the orthometric heights of benchmarks, the measured elevation differences are first converted into geopotential differences by using the measured surface gravity values. For example, the geopotential differences for a geodetic leveling between points A and B can be expressed mathematically as

$$\Delta C_{AB} = \sum_{k=A}^{B} g_k \Delta h_k \tag{2.3}$$

where g_k is the average measured gravity values between turning points $k = i$ and $k = i + 1$ on the surface and Δh_k is the elevation difference between the turning points

$k=i$ and $k=i+1$. If the gravity measurements (g_k) along the leveling route are available, the geopotential difference can be evaluated by using Equation (2.3). The geopotential difference can then be converted directly into any height difference of interest. For example, the geopotential differences can be divided by the average value of gravity (G) for a given area in order to produce the dynamic height differences, which are in units of length. Using the dynamic height of a fixed point and the dynamic height difference between the point and another point, the dynamic height of the other point can be determined. For a geodetic leveling between a fixed point A and an unknown point B, the orthometric correction (OC_{AB}) to be applied to the observed (measured) elevation difference between the two points can be given (Heiskanen and Moritz, 1967) as follows:

$$OC_{AB} = DC_{AB} + \frac{\overline{g}'_A - G}{G} h_A - \frac{\overline{g}'_B - G}{G} h_B \tag{2.4}$$

where DC_{AB} is the dynamic correction expressed as

$$DC_{AB} = \sum_k \left(\frac{g_k - G}{G} \right) \Delta h_k \tag{2.5}$$

h_A and h_B are the heights of points A and B (which only need to be known approximately), respectively; G is the mean value of gravity for the region of interest; $\overline{g}'_A$ is the mean gravity value along the plumb line through point A to the geoid and $\overline{g}'_B$ is the mean gravity value along the plumb line through point B to the geoid. The main problem in using Equations (2.4) and (2.5) is that $\overline{g}'_A$ and $\overline{g}'_B$ must be predicted by using some models since they cannot be measured directly. The commonly used prediction models are discussed by Heiskanen and Moritz (1967). The choice of model determines whether Helmert orthometric heights or normal orthometric heights are determined.

The two commonly used height systems (differential leveling and orthometric) are attractive to surveyors because of their geometric significance since points having the same numerical height value will have the same geometric length from a reference datum (or level surface). For example, a point with an orthometric height value of 4.000 m will have the length from the geoid to the point as 4.000 m if measured with a scale rule along the plumb line passing through the point. The two height systems, however, have no physical significance since points on a given level surface, apart from the reference datum, will likely have different height values due to nonparallelism of level surfaces.

Geopotential numbers and dynamic height systems have physical significance but no geometric significance. In these systems, points on the same level surface will have the same height value, meaning that the surface of a lake will be represented as a flat surface in the systems. This explains the concept that a height system cannot satisfy both the geometric and the level surface properties simultaneously since the two properties together are actually incompatible with the nonparallelism of equipotential surfaces of the earth's gravity field.

The orthometric, geopotential number, and dynamic height systems are single-valued systems compared to the differential leveling height system. This means that in differential leveling systems, there will be misclosure when the height of the same point is determined following different routes due to nonparallelism of equipotential surfaces resulting from the earth's gravity variations; this is based on the assumption that no observational, environmental, instrumental, and personal errors are introduced into leveling measurements. The other systems will provide single height value for the same point irrespective of the route taken, unlike in the differential leveling system, whose leveling results are route dependent.

2.4 DISTANCE MEASURING INSTRUMENT

Different types of distance measuring instruments can be summarized as electromagnetic distance measurement (EDM) equipment and total station instruments. The two main commonly used types of EDM are electromagnetic (microwave) EDM and electromagnetic (light wave or electro-optical) EDM. The main properties of these two types of EDM are summarized in Table 2.5 with the details of the properties discussed in Chapter 5.

The *total station* instruments are electronic digital theodolites integrated with EDM instruments and electronic data collectors to replace manual field data

TABLE 2.5 Main Properties of the Two Main Types of EDM.

Property	Electro-Optical Type	Microwave Type
Accuracy of instrument	Short-range types (0.1 m to 5 km): 5 mm + 5 ppm to 5 mm + 0.1 ppm Long-range types (up to 70 km): 5 mm + 0.1 ppm to 0.1 mm + 0.1 ppm	For distances up to 150 km and depending on atmospheric refractive index: 22 mm + 5 ppm to 1 mm + 1 ppm
Operation principle	(a) Set up EDM at one end of the line being measured and a reflector at the other end of the line (b) EDM sends a modulated beam of light to the reflector (c) Reflector acts like a mirror and returns the light pulse back to EDM (d) EDM registers readings that are converted into linear distance between the EDM and the reflector (e) Requires one operator	(a) Master unit transmits a series of modulated radio waves to remote antenna in the remote instrument (b) Remote instrument interprets these signals and sends them back to the antenna of the master unit (c) Master unit measures the time required for the radio waves to make the round trip (d) Distance is computed based on the velocity of the radio waves (e) Requires one operator at each end of line

TABLE 2.6 Examples of Distance Measuring Instruments.

Make	Description	Angular/Direction Accuracy	Distance Accuracy
Kern DM502	Precision EDMs	N/A	Range: 2 km for DM502 and 5 km for DM503 3 mm ± 2.0 ppm
Kern ME3000	Precision EDM	N/A	Range: 2.5 km 0.2 mm ± 1.0 ppm
Kern ME5000	Precision EDM	N/A	Range: 8 km 0.2 mm ± 0.2 ppm
ComRad Geomensor 204DME	Precision EDM	N/A	Range: 10 km 0.1 mm ± 0.1 ppm
Leica TC2003/ TCA2003 and TC2002	Without/with ATR total station Magnification: 30×	0.5″ Resolution: 0.1″	Range: 2.5/3.5 km 1 mm ± 1.0 ppm (with one prism and average weather)
Leica TDM/ TDA 5005	Industrial total station Magnification: 30×	0.5″ Resolution: 0.1″	Range: 2–600 m 1 mm ± 2.0 ppm

recording; they are capable of providing electronic angle readings as well as distance measurements and are currently replacing theodolites, EDM, and levels. The type of EDM incorporated to the modern total station instruments is commonly of electro-optical type, using infrared and laser light as carrier signal. Examples of precision EDMs and total station equipment are shown in Table 2.6.

2.5 ACCURACY LIMITATIONS OF MODERN SURVEY INSTRUMENTS

The main instruments used by surveyors of today are the digital levels, the total stations, and the GNSS receivers. Details of the technological progress on the geodetic surveying equipment can be found in Rüeger (2003). In the past, accuracy of measurements is solely dependent on the skills of the observers and the precision of the equipment used. Today, limitations to the accuracy of measurements are mainly due to atmospheric and target conditions, equipment design and precision, and the instrument operator factor. These limitations are discussed in the following sections.

2.5.1 Atmospheric and Target Conditions

Atmospheric conditions limit the accuracy of modern survey equipment as follows:

(i) Atmospheric refraction causes the total station horizontal and vertical angle measurements to be refracted away from their ideal paths in space. The total station and EDM distance measurements are also refracted away from their ideal paths, and the distance measurements are shortened or increased in length depending on the atmospheric temperature, pressure, and humidity. When the atmospheric conditions change, the velocity of the measuring signal in the atmosphere and the resulting distance measurements are consequently changed.

(ii) Atmospheric refraction generally deteriorates the accuracy of leveling operations, causing systematic deviations in elevation difference measurements. For example, the total station distance measurement depends on the signal strength of radiation in the atmosphere.

(iii) Atmospheric temperature changes usually have higher effects on modern electronic instruments, reducing their accuracy. Because of this, most of the electronic equipment (e.g., electronic levels) need to acclimatize before use in the field.

(iv) EDM instrument may fail in tunneling surveys or may not work smoothly due to usually poor conditions in tunnels.

(v) Intervisibility between targets and instrument (e.g., GPS and receiver) are required for good results to be possible.

(vi) Reflectorless EDM or reflectorless total station equipment depends on the type of surface that is measured to and the orientation of the surface of the measuring beam in order to produce good results; they are currently not suitable for high-accuracy measurements.

One of the attempts at minimizing the effects of atmospheric refractions on EDM measurements includes developing two-color precise EDM instrument, which produces a precision of 0.5–1.0 mm (or ±0.1 ppm) for ranges between 1 and 12 km (USGS, 2010). This instrument uses two colors (red and blue) to measure the travel time of light through the atmosphere, unlike the commercially available ones that use one color (red or infrared laser) as a carrier. The difference between the travel times of red and blue wavelengths in the atmosphere is a direct function of temperature and pressure of the atmosphere between the instrument and the reflector. This difference is used to determine the average refractive index between the instrument and the reflector, which is then used to calculate the precise distance. This system is more precise than GPS at ranges less than 10 km but its range is limited, and it requires intervisibility between stations unlike in GPS. Moreover, the two-color EDM was only available commercially for a few years in the early 1980s; only a few of them were made and they cost as high as $250K. The two-color EDM was used until 2005 in Parkfield on the San Andreas Fault in California (USGS, 2010).

2.5.2 Equipment Design and Precision

Electronic theodolites have not brought any drastic improvement in accuracy compared with precision optical theodolites (old types); the precision of electronic equipment and their accessories is similar to that of the old types except that reading errors are reduced and mistakes in transferring data are reduced by the use of electronic data recorders. There are, however, some peculiarities with some of the modern survey equipment, such as the following:

(i) Laser scanners provide measurement precisions, which are dependent on the precision of direction, zenith angle, and distance measurements; they are currently not suitable for precision works. Use of laser scanners, however, provides some advantages since they are able to provide x, y, z coordinates of a large number of points. Further details on this are given in Chapters 8 and 10.

(ii) GPS provides unacceptable relative precisions for typically short baselines (<500 m) involved in structural deformation monitoring; the relative precision of measurements using GPS survey techniques is in the order of 2–5 mm, which is unacceptable.

(iii) Vibrations and internal workings of electronic components will further reduce the accuracy of measurements.

(iv) Poor calibration of electronic equipment will further compromise the accuracy of measurements; electronic equipment usually require more frequent calibration than the older types.

2.5.3 Instrument Operator Factor

Most of the modern survey instruments operate like black boxes; they are based on hardware and software components that are controlled by the manufacturers of the instruments with only little input from the operators of the instruments. Compared with traditional survey equipment, the skill requirements for an operator of modern equipment are different, such as the following:

1. Most modern instruments have digital readout and data recording units, so that the skill of being able to read plate circle scales and record measurements in a particular format is no longer important nowadays.
2. Skill of being able to perfectly level a theodolite is no more required since one can approximately level an electronic instrument and let the dual-axis compensators integrated with the instrument complete the remaining fine leveling operations.
3. With automatic target recognition (ATR) system of modern equipment, the modern equipment requires less skill in accurately bisecting survey targets; and with the motorized systems integrated with some modern instruments, the instruments are capable of automatically changing their telescope positions while making repetitive measurements.
4. Motorized total station instruments with telemetric links can recognize and track moving reflectors. The operators of such instruments, through remote controllers, are able to send instructions to the instruments to record data as they move from one point to another with reflectors. This type of instrument, which allows one-person surveys, increases greater efficiency and cost savings of survey works.
5. Robotic surveying system has made it possible to automate repetitive survey works; the instrument operator points to the reflector and then leaves it unmanned and the unmanned instrument will automatically locate and follow

the reflector. The system can be programmed for sequential self-pointing to a set of prism targets at predetermined time intervals. In this case, the system is first trained by manually pointing it to a set of targets in the desired sequence and the information is then used later by the system to find the targets again during routine measurements. This requires that the system be well calibrated in order to ensure that the results obtained are accurate.

2.6 ERROR PROPERTIES OF MEASUREMENTS

A major concern in every survey is closeness of measurements to their true values (i.e., their accuracy). The accuracy of a survey is limited because of imperfections of the measurement system (surveyor, instrument, and environment). The difference between a measurement and its true value can be due to three types of error: *blunders* (or *gross errors*), *random errors*, and *systematic errors*. No measurement is exact; a measurement is its best estimate plus the measurement uncertainty. The measurement uncertainty provides a measure of quality of the measurement by accounting for both systematic and random errors. This is a measure of how well one believes one knows the true value of the observable. Uncertainty of measurement is the doubt that one has about the validity of the outcome of a measurement. A measure of uncertainty, however, is not intended to account for mistakes and blunders.

2.6.1 Blunders (or Gross Error)

Blunder (or gross error) is a *mistake* caused by carelessness of the surveyor or by failure of the measuring equipment. The carelessness of the surveyor may include recording wrong numbers in the field notes, misreading the numbers on the measuring instrument, adding numbers incorrectly, and so on. Blunders can be eliminated by a careful checking or consistently following a self-checking procedure during measurement. All survey measurements are suspects of mistakes until the measurements have been verified. As a rule, every measurement should be immediately checked or repeated. Immediate repetition of every measurement will enable the surveyor to eliminate most mistakes and also to improve the accuracy of the measurement.

2.6.2 Random and Systematic Errors

Compared with blunders, random and systematic errors are usually very small in magnitude. The errors cannot be completely eliminated but can be minimized by following careful survey procedures and by applying appropriate corrections to measurements. There are three main sources of these errors: *people* (personal errors due to imperfect sight and touch), *instruments* (manufacturing defects, aging of instruments), and *nature* (temperature, wind, moisture, magnetic variations, etc.).

A *random* (also known as *accidental* or *compensating*) error is a type of error whose magnitude and direction are just by accident and are beyond the control of the surveyor. For instance, when a person reads a tape, they are usually not able to read it perfectly; one time they may read a value that is too large and the next time they may read a value that is too small. Since these errors are just as likely to have one sign as the other, they tend to a certain degree cancel each other or compensate for each other. Because of the imperfections in measurement systems (people, instruments, and nature), random errors are unavoidable. They cannot be mathematically modeled, but are known to follow statistical laws of probability, and they can be controlled, minimized, investigated, and estimated, but never eliminated.

A *systematic* (or *cumulative*) error is the type of error that, for constant conditions, remains the same as to sign and magnitude. For instance, if a steel tape is 0.10 m too short, each time the tape is used, the same error is made. If the full tape length is used 10 times, the error accumulates and totals 10 times the error for one measurement. Systematic errors obey mathematical or physical laws and are predictable, correctable, or avoidable. The systematic errors must be removed by following some specific observation procedures or using some mathematical models to calculate appropriate corrections to measurements.

2.7 PRECISION AND ACCURACY INDICATORS

The overall goal of a surveyor is to make measurements that are both precise and accurate. It is generally known that physical measurements acquired in the process of surveying are correct only within certain limits because of random and systematic errors. Precision and accuracy of measurements are related directly to random and systematic errors. The terms precision and accuracy are commonly used in surveying to mean the same thing, but they are not exactly the same.

Precision (or *apparent accuracy*) is the degree of closeness of one measurement to another or the repeatability of the readings. It increases when random errors decrease and decreases when the random errors increase; precision is then considered a measure of the amount of random errors present in the measurement. Precision is related to random errors due to the centering of equipment used, leveling of the equipment, pointing of telescope, atmospheric refractions, design of targets, number of repetitions of measurements, skill of observer, survey techniques, least count of instruments, and so on. Everything affecting random errors, in fact, will affect precision, since reducing random errors improves precision. Precision has to do with the method of measurement as well as the expressed value of measurement.

Accuracy refers to the degree of closeness of a measurement to its true value; it is a measure of the amount of systematic and random errors present in the measurement. Theoretically, true value of an observable exists, but it cannot be determined exactly from values based on measurements because of errors and variations in the standards and systems used to measure it. A standard value or a set of standard values must be available for comparison, for example, comparing a meter with international meter; comparing sums of angles in plane triangles with 180°; or comparing a value with a value determined by refined methods deemed sufficiently near the true value to be held as constant (like adjusted elevation of a permanent benchmark). If stable standards and systems of control are more accurate than what the surveyor can measure, accuracy will be reduced only to the effects of errors and blunders in measurements. Accuracy of measurement is determined by calibration of instruments, avoiding or removing blunders (or mistakes) and by detecting and removing systematic errors caused by the environment or instrument adjustments. Any procedure that cannot detect systematic error will not fully check accuracy.

The main steps for estimating accuracy of typical survey observables, such as spatial distances, horizontal directions and angles, height differences and zenith (or vertical) angles, are as follows:

1. Understand the procedure to be taken in the data acquisition.
2. Identify all the possible random and systematic error sources.
3. Remove the major parts of the effects of the systematic errors from the raw data by applying appropriate corrections to them. The residual systematic effects caused by uncertainty in the determination of the systematic errors are then considered random errors.
4. Compute the total effect of all the random errors and residual systematic errors on the observable by using the law of random error propagation. If one assumes that the effects of all the different types of errors on the observable are statistically independent, the variance of the observable would be equal to the sum of the squares of each individual effect.

From the foregoing, it can be seen that the surveyor can attain accuracy and precision by exercising care and patience, by using good instruments and procedures, and by applying appropriate corrections. If systematic errors have been effectively accounted for in survey measurements, one can safely take precision as being the same as accuracy.

2.8 SYSTEMATIC ERROR AND RANDOM ERROR PROPAGATION LAWS

2.8.1 Systematic Error Propagation Laws

Consider a quantity z as a function of two quantities (x and y) as expressed by the following equation:

$$z = f(x, y) \tag{2.6}$$

where z is a subject whose systematic error is to be determined, given the systematic errors of x and y as dx and dy, respectively. The differential change (dz) of z in terms of the differential changes (dx, dy) of x and y can be derived as follows:

$$dz = \frac{\partial f(x,y)}{\partial x}dx + \frac{\partial f(x,y)}{\partial y}dy \tag{2.7}$$

Equation (2.7) can be considered the rule for the propagation of systematic errors, where dx and dy are considered component systematic errors and dz is the propagated systematic error. Equation (2.7) can be expressed in matrix form:

$$dz = J \begin{bmatrix} dx \\ dy \end{bmatrix} \tag{2.8}$$

where J is the Jacobian matrix which can be given as

$$J = \begin{bmatrix} \frac{\partial f(x, y)}{\partial x} & \frac{\partial f(x, y)}{\partial y} \end{bmatrix} \tag{2.9}$$

The systematic error propagation Equation (2.7) for z expressed as a function of two variables x and y can be extended for a case of where more than two variables are involved, by appropriately increasing the terms in Equation (2.7).

2.8.2 Random Error Propagation Laws

Standard deviation s usually serves as a measure of precision of measurements or their functions affected by random errors; it is calculated as the square root of its variance, s^2. The term *covariance* with symbol s_{xy} is used as a numerical measure of the *correlation* between two quantities x and y or between two functions of the quantities. Two measurements may be correlated if the same instrument is used and there are common sources of errors that could influence both measuring procedures in a similar way. Standard deviation of a quantity (say x) is usually so small that its variance (s_{xx}^2) may be approximated by its squared differential change (dx), such as

$$s_{xx}^2 = (dx)^2 \tag{2.10}$$

Similarly, the covariance (s_{xy} or s_{yx}) of two quantities x and y may be approximated by products of their differential changes (dx and dy) as:

$$s_{xy} = (dx)(dy) = (dy)(dx) \tag{2.11}$$

Consider a quantity z as a function of two quantities (x and y) as expressed by Equation (2.6). Let z now be a subject whose variance s_{zz}^2 is to be determined, given the variance of x as s_{xx}^2, variance of y as s_{yy}^2 and the covariance of x and y as s_{xy} (or s_{yx}). According to the laws of variance–covariance propagation, the variance s_{zz}^2 can be expressed as follows:

$$s_{zz}^2 = \left[\frac{\partial f(x, y)}{\partial x}\right]^2 s_{xx}^2 + \left[\frac{\partial f(x, y)}{\partial y}\right]^2 s_{yy}^2 + 2\left[\frac{\partial f(x, y)}{\partial x}\right]\left[\frac{\partial f(x, y)}{\partial y}\right] s_{xy} \tag{2.12}$$

where it is assumed that $s_{xy} = s_{yx}$. Equation (2.12) is considered as expressing the variance–covariance (random error) propagation laws. This equation can also be expressed in matrix form as follows:

$$s_{zz}^2 = J C_{xy} J^T \tag{2.13}$$

where J is the same Jacobian given in Equation (2.8) since we are dealing with the same Equation (2.6); and C_{xy} is the variance–covariance matrix of x and y variables, given as

$$C_{xy} = \begin{bmatrix} s_{xx}^2 & s_{xy} \\ s_{yx} & s_{yy}^2 \end{bmatrix} \tag{2.14}$$

with the covariance between x and y being the same in the case of symmetric matrix, that is, $s_{xy} = s_{yx}$.

2.8.3 Confidence Regions for One-Dimensional Parameters

The one-dimensional parameters discussed here are the population means (μ) of survey data. A population mean here can also be considered as the adjusted one-dimensional parameters, such as adjusted elevations of benchmarks, and the adjusted values of survey observables, such as distances, angles, directions. In creating a confidence region for the population mean, the sample standard deviation (s) will be considered as a sufficiently good approximation to the population standard deviation (σ), provided the sample is large enough. The often quoted criterion for the required size of the sample, by the authorities in statistical inferences, is that a size larger than 30 constitutes a large sample. The population mean (μ) can be estimated in two ways: as *point estimate* and as *interval estimate*.

The *point estimate* of population mean provides a specific value ($\bar{x}$) known as sample mean as a single estimate of the population mean (μ) and stipulates the precision of this estimate at a certain probability. The uncertainty (or precision) of the estimation of $\bar{x}$ at a given probability $1-\alpha$ (where α is the statistical significance level) depends on the sample size (n). If the sample size is greater than 30 or the population variance (σ^2) is given, the *z-score* can be used to determine the *precision* of estimate (or what is sometimes known as *uncertainty, error bar* or *margin of error*) at probability $1-\alpha$ (for two-tailed case) as follows:

$$|\bar{x} - \mu|_{1-\alpha} = (\text{SE})z_{\alpha/2} \tag{2.15}$$

where SE is the standard error or the standard deviation of the error ($\bar{x} - \mu$) determined from its error propagation, which is equivalent to the standard deviation of the mean ($s_{\bar{x}}$) or the standard deviation of the adjusted quantity ($s_{\hat{x}}$) if the least squares adjustment procedure is used; in the case where the mean ($\bar{x}$) is a simple average of n measurements, the standard error, $\text{SE} = \sigma/\sqrt{n}$ for n greater than 30 with a known population standard deviation (σ), or $\text{SE} = s/\sqrt{n}$ if the standard deviation of the population is unknown but the sample standard deviation (s) is determined; $z_{\alpha/2}$ is the critical value of z at probability $1-\alpha$.

The *t-statistic* is used instead of *z-score* for a case where the sample size is less than or equal to 30 and the population standard deviation is unknown. In this case, the precision of estimate at probability $1-\alpha$ can be expressed as

$$|\bar{x} - \mu|_{1-\alpha} = (\text{SE})t_{\alpha/2,\text{df}} \tag{2.16}$$

where SE is determined based on whether σ or s is known as discussed above, $t_{\alpha/2,\mathrm{df}}$ is the critical value of t (in two-tailed case) at probability of $1-\alpha$, and df is the number of degrees of freedom. The precision of estimate in Equations (2.15) and (2.16) above can be given as $e = |\bar{x}-\mu|_{1-\alpha}$. From Equation (2.15), the probability of $|\bar{x}-\mu|_{1-\alpha}$ being less than $(\mathrm{SE})z_{\alpha/2}$ is $1-\alpha$; similarly, in Equation (2.16), the probability of $|\bar{x}-\mu|_{1-\alpha}$ being less than $(\mathrm{SE})t_{\alpha/2,\mathrm{df}}$ is $1-\alpha$. For example, from Equation (2.15), the uncertainty of an estimate (having a standard error of 0.005 m) at 99% probability will give $z_{0.01/2} = 2.58$ (Table II.1 in Appendix II), so that the uncertainty will become 2.58(0.005 m) or 0.0129 m.

The *interval estimate* places the population mean (μ) within an interval and stipulates a degree of confidence as a measure of precision of this interval estimate. The interval estimate immediately reveals the uncertainty associated with the estimation of the population mean (μ). *Confidence interval* is used to describe the amount of uncertainty associated with a sample estimate of a population parameter. Confidence intervals are random regions that contain a statistic with some confidence level $(1-\alpha)$ or $(1-\alpha)\times 100\%$ associated with it so that the true value of the parameter can be claimed to fall within these intervals. For example, if $\alpha = 0.05$, the 95% confidence interval is the range of values in which one is 95% confident that the true value of the mean or difference between the means will fall. Remember that the population parameters (μ, σ) are quantities with constant values and they cannot be treated as variables or statistics, since their values cannot change. To express a confidence interval, one needs three pieces of information:

- Confidence level $(1-\alpha)$
- Sample statistic $(\bar{x})$
- Precision of estimate (or margin of error) of the statistic given in Equations (2.15) and (2.16).

The range of confidence interval can then be defined as follows:

$$\text{Range of confidence interval} = \text{Sample statistic} \pm \text{Precision of estimate} \tag{2.17}$$

The precision of estimate (or margin of error) is considered error in confidence interval. The confidence intervals will be constructed differently depending on whether the sample size (n) is greater than 30 and if σ is known. In the case where $n > 30$, the following intervals are obtained:

$$\mu = \bar{x} \pm (\mathrm{SE})z_{\alpha/2} \tag{2.18}$$

or

$$\bar{x} - (\mathrm{SE})z_{\alpha/2} < \mu < \bar{x} + (\mathrm{SE})z_{\alpha/2} \tag{2.19}$$

where SE is the propagated standard deviation of the mean $(\bar{x})$ and $z_{\alpha/2}$ value is obtained from the standard normal distribution curve. In the case where the number of observations or sample size $n \leq 30$, the student's t-distribution value will be used as follows:

$$\mu = \overline{x} \pm (\mathrm{SE})t_{\alpha/2,\mathrm{df}} \tag{2.20}$$

or

$$\overline{x} - (\mathrm{SE})t_{\alpha/2} < \mu < \overline{x} + (\mathrm{SE})t_{\alpha/2,\mathrm{df}} \tag{2.21}$$

where $t_{\alpha/2,\mathrm{df}}$ value is obtained from the t-distribution table with the degrees of freedom (or redundancy) *df*. The *t-distribution* is similar to normal distribution, except that the degrees of freedom are now involved.

2.8.4 Confidence Regions for Two-Dimensional Parameters

The two-dimensional parameters referred to in this section are the positions of points in two dimensions, such as (x, y) coordinates of points. Confidence regions will be constructed for adjusted (x, y) coordinates of points or the adjusted coordinate differences $(\Delta x, \Delta y)$ of pairs of points. Confidence region is the area within which one has a certain degree of confidence that the true value of the quantity being determined will lie. The immediate local measure of accuracy for the adjusted coordinates of a point is the covariance matrix of the adjusted coordinates for that point. The covariance matrix of the adjusted coordinates $\widehat{p}$ of a point (using the 2×2-block covariance matrix corresponding to the point) can be given as follows:

$$C_{\widehat{p}} = \begin{bmatrix} s_{xx}^2 & s_{xy} \\ s_{yx} & s_{yy}^2 \end{bmatrix} \tag{2.22}$$

where the standard deviations (or precisions) of the adjusted coordinates $\widehat{x}$ and $\widehat{y}$ are s_{xx} and s_{yy}, respectively; and s_{xy} and s_{yx} are the covariances between $\widehat{x}$ and $\widehat{y}$. If the a priori variance factor of unit weight (σ_0^2) is unknown but set equal to 1, a new value (s_0^2) called a posteriori variance factor of unit weight must be calculated and used to scale the cofactor of the adjusted coordinates in order to obtain a more realistic covariance matrix of the adjusted coordinates. Usually, a confidence region indicating the accuracy of horizontal control survey coordinates is bounded by an ellipse. *Standard error ellipses* are generalizations of standard deviations. *Confidence error ellipses* are the 2D equivalent of the confidence intervals (for 1D cases). Three quantities (parameters) are required to define an error ellipse: the semi-major axis a, semi-minor axis b, and the bearing of the semi-major axis β. A typical error ellipse is shown in Figure 2.2.

The standard ellipse bounds a confidence region of from 30% to 39%, depending on the number of degrees of freedom (redundant observations) in the adjustment. There are two types of error ellipses depending on where the error ellipses are situated: *Absolute error ellipses* are usually situated at the station point, thus referring to that given point; *relative error ellipses* are situated in between two station points that are connected by observations and the ellipses refer to the position difference

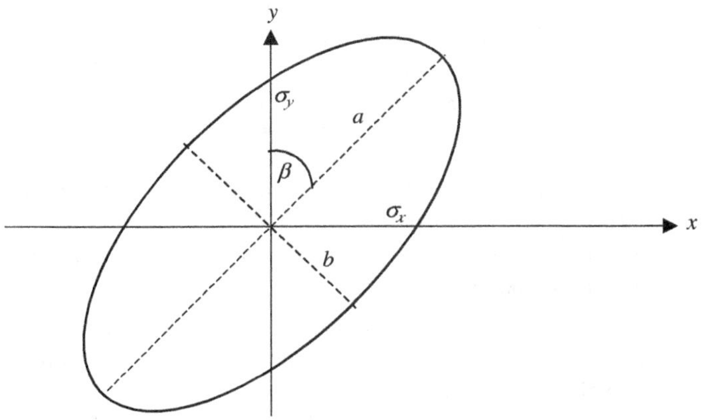

Figure 2.2 A typical error ellipse.

of the two points. An error ellipse can be constructed for a given point by using the covariance matrix ($C_{\hat{p}}$) of the adjusted coordinates of the point. If the cofactor matrix of the point is estimated (and a priori variance factor of unit weight is unknown or the standard deviations of observations not perfectly known), it must be multiplied by the estimated variance factor of unit weight (s_0^2) computed in the least squares adjustment. The parameters of an absolute error ellipse can be computed from the covariance matrix in Equation (2.22) depending on whether σ_0^2 is known or not. The steps for the computations are as follows.

- Compute the eigenvalues λ_1 (maximum value) and λ_2 (minimum value) from the covariance matrix $C_{\hat{p}}$ of the adjusted coordinates of the point (from Equation (2.22)) as follows:

$$\lambda_1 = \frac{1}{2}(s_{xx}^2 + s_{yy}^2 + z) \tag{2.23}$$

$$\lambda_2 = \frac{1}{2}(s_{xx}^2 + s_{yy}^2 - z) \tag{2.24}$$

  where

$$z = \left[\left(s_{xx}^2 - s_{yy}^2\right)^2 + 4s_{xy}^2\right]^{1/2} \tag{2.25}$$

- In the case where σ_0^2 is known, the parameters of the confidence error ellipse (at $1 - \alpha$ confidence level) can be given using $\chi^2_{\alpha,\mathrm{df}=2}$(upper-tail area) distribution with the degrees of freedom df = 2 and the 2 representing two coordinates (x, y) associated with the point as follows:

$$a_{(1-\alpha)100\%} = \sqrt{\lambda_1 \chi^2_{\alpha,\mathrm{df}=2}(\text{upper-tail area})} \tag{2.26}$$

$$b_{(1-\alpha)100\%} = \sqrt{\lambda_2 \chi^2_{\alpha,\mathrm{df}=2}(\text{upper-tail area})} \tag{2.27}$$

$$\beta = \arctan\left(\frac{s_{xy}}{\lambda_1 - s^2_{xx}}\right) \tag{2.28}$$

where $a_{(1-\alpha)100\%}$, $b_{(1-\alpha)100\%}$, and β are the semi-major axis, semi-minor axis, and the bearing of the semi-major axis of the $(1-\alpha)100\%$ confidence error ellipse, respectively. For example, $\chi^2_{\alpha,\mathrm{df}=2}$(upper-tail area) = 5.99 and 9.21 for $\alpha = 0.05$ and 0.01 respectively.

- In the case where σ_0^2 is unknown and s_0^2 is used in scaling the cofactor of the adjusted coordinates, the parameters of the confidence error ellipse (at $1-\alpha$ confidence level) can be given as follows:

  The $F_{\alpha,\mathrm{df}_1=2,\mathrm{df}_2=n-u}$(upper-tail area) distribution is used with the degrees of freedom $\mathrm{df}_1 = 2$, $\mathrm{df}_2 = n - u$, where 2 represents the two coordinates (x, y) associated with the point, n is the number of observations, and u is the number of unknown parameters (coordinates) determined in the original adjustment.

$$a_{(1-\alpha)100\%} = \sqrt{2\lambda_1 F_{\alpha,\mathrm{df}_1=2,\mathrm{df}_2=n-u}(\text{upper-tail area})} \tag{2.29}$$

$$b_{(1-\alpha)100\%} = \sqrt{2\lambda_2 F_{\alpha,\mathrm{df}_1=2,\mathrm{df}_2=n-u}(\text{upper-tail area})} \tag{2.30}$$

$$\beta = \arctan\left(\frac{s_{xy}}{\lambda_1 - s^2_{xx}}\right) \tag{2.31}$$

where $a_{(1-\alpha)100\%}$, $b_{(1-\alpha)100\%}$, and β are the semi-major axis, semi-minor axis, and the bearing of the semi-major axis of the $(1-\alpha)100\%$ confidence error ellipse, respectively.

Relative error ellipses are constructed for coordinate differences (Δx, Δy) between pairs of stations and are usually drawn at the midpoint of the two stations involved. In this case, the variance–covariance matrix of the coordinate differences between the two points will be used to construct the relative error ellipses. For example, the relative error ellipse between two stations 1 and 2 in Figure 2.3 can be constructed as follows:

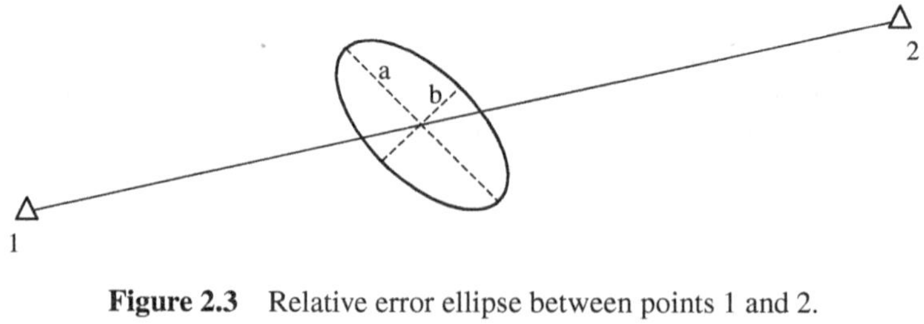

Figure 2.3 Relative error ellipse between points 1 and 2.

Let the variance–covariance matrix (from the least squares adjustment) for the two stations 1 and 2 be given as follows:

$$C_{\hat{x}} = \begin{bmatrix} s_{x_1}^2 & s_{x_1y_1} & s_{x_1x_2} & s_{x_1y_2} \\ s_{y_1x_1} & s_{y_1}^2 & s_{y_1x_2} & s_{y_1y_2} \\ s_{x_2x_1} & s_{x_2y_1} & s_{x_2}^2 & s_{x_2y_2} \\ s_{y_2x_1} & s_{y_2y_1} & s_{y_2x_2} & s_{y_2}^2 \end{bmatrix} \tag{2.32}$$

For a symmetric matrix, upper diagonal elements are the same as the corresponding lower diagonal elements, for example, $s_{x_1y_1} = s_{y_1x_1}s_{y_1x_1}$, and so on. The coordinate differences between the two points 1 (x_1, y_1) and 2 (x_2, y_2) can be given as follows:

$$\Delta x = x_2 - x_1 \tag{2.33}$$

$$\Delta y = y_2 - y_1 \tag{2.34}$$

By variance–covariance propagation law on Equations (2.33) and (2.34), the relative covariance matrix ($C_{\Delta 12}$) for the two points can be given as follows:

$$C_{\Delta 12} = BC_{\hat{x}}B^T \tag{2.35}$$

where B is the Jacobian of Equations (2.33) and (2.34) with respect to the coordinates of points 1 (x_1, y_1) and 2 (x_2, y_2):

$$B = \begin{bmatrix} \dfrac{\partial \Delta x}{\partial x_1} & \dfrac{\partial \Delta x}{\partial y_1} & \dfrac{\partial \Delta x}{\partial x_2} & \dfrac{\partial \Delta x}{\partial y_2} \\ \dfrac{\partial \Delta y}{\partial x_1} & \dfrac{\partial \Delta y}{\partial y_1} & \dfrac{\partial \Delta y}{\partial x_2} & \dfrac{\partial \Delta y}{\partial y_2} \end{bmatrix} = \begin{bmatrix} -1 & 0 & 1 & 0 \\ 0 & -1 & 0 & 1 \end{bmatrix} \tag{2.36}$$

Using Equations (2.32) and (2.36) in Equation (2.35) gives

$$C_{\Delta 12} = \begin{bmatrix} s_{\Delta x}^2 & s_{\Delta x\Delta y} \\ s_{\Delta y\Delta x} & s_{\Delta y}^2 \end{bmatrix} \tag{2.37}$$

where

$$s_{\Delta x}^2 = s_{x_1}^2 + s_{x_2}^2 - 2s_{x_1x_2} \tag{2.38}$$

$$s_{\Delta y}^2 = s_{y_1}^2 + s_{y_2}^2 - 2s_{y_1y_2} \tag{2.39}$$

$$s_{\Delta x\Delta y} = s_{x_1y_1} + s_{x_2y_2} - s_{x_1y_2} - s_{y_1x_2} \tag{2.40}$$

To compute the parameters of the relative error ellipse between points 1 and 2, use the relative covariance matrix ($C_{\Delta 12}$) in Equation (2.37) as follows:

$$\lambda_1 = \frac{1}{2}(s_{\Delta x}^2 + s_{\Delta y}^2 + R) \tag{2.41}$$

$$\lambda_2 = \frac{1}{2}(s_{\Delta x}^2 + s_{\Delta y}^2 - R) \tag{2.42}$$

$$\theta = \arctan\left(\frac{s_{\Delta x\Delta y}}{\lambda_1 - s_{\Delta x}^2}\right) \tag{2.43}$$

$$R = \left[\left(s_{\Delta x}^2 - s_{\Delta y}^2\right)^2 + 4s_{\Delta x\Delta y}^2\right]^{1/2} \tag{2.44}$$

where λ_1 and λ_2 are the maximum and minimum eigenvalues of the relative covariance matrix $C_{\Delta 12}$ (here also note that λ_1 is always greater than λ_2); θ is the bearing of the major semi-axis of the relative error ellipse. The confidence relative error ellipses can be obtained similarly as in the case of the absolute error ellipses, by substituting the eigenvalues in the appropriate equations into Equations (2.26)–(2.31):

$$a_{(1-\alpha)100\%} = \sqrt{\lambda_1 \chi^2_{\alpha,\mathrm{df}=2}(\text{upper-tail area})} \tag{2.26}$$

$$b_{(1-\alpha)100\%} = \sqrt{\lambda_2 \chi^2_{\alpha,\mathrm{df}=2}(\text{upper-tail area})} \tag{2.27}$$

$$\beta = \arctan\left(\frac{s_{\Delta x\Delta y}}{\lambda_1 - s_{\Delta x}^2}\right) \tag{2.28}$$

$$a_{(1-\alpha)100\%} = \sqrt{2\lambda_1 F_{\alpha,\mathrm{df}_1=2,\mathrm{df}_2=n-u}(\text{upper-tail area})} \tag{2.29}$$

$$b_{(1-\alpha)100\%} = \sqrt{2\lambda_2 F_{\alpha,\mathrm{df}_1=2,\mathrm{df}_2=n-u}(\text{upper-tail area})} \tag{2.30}$$

$$\beta = \arctan\left(\frac{s_{\Delta x\Delta y}}{\lambda_1 - s_{\Delta x}^2}\right) \tag{2.31}$$

2.9 STATISTICAL TEST OF HYPOTHESES: THE TOOLS FOR DATA ANALYSIS

The type of hypothesis testing discussed in this section is based on the null hypothesis (H_0) probability distribution in which it is assumed that H_0 is true (with an error of judgment of α, known as significance level). This hypothesis testing does not include a distribution based on the alternative hypothesis (H_A) being true (so that the probability $1-\beta$, the power of test, is not considered).

TABLE 2.7 Formulated Hypotheses.

	Null Hypothesis	Alternative Hypothesis
One-tailed test	$H_0 : \mu = \delta$	$H_A : \mu > \delta$ (or $H_A : \mu < \delta$)
Two-tailed test	$H_0 : \mu = \delta$	$H_A = \mu \neq \delta$

2.9.1 Observations of One Observable: Test on the Mean

The statistical test of the mean of the observations of one observable is a case in which one has to decide if a population mean (μ) is equal to a known standard value (δ). In this test, it is required to find if the sample mean ($\bar{x}$) is consistent with the population mean that is assigned a standard value ($\mu = \delta$). The hypotheses in Table 2.7 can be formulated for one-tailed and two-tailed tests:

At a selected significance level α and a given sample size n, the decisions in Table 2.8 are possible.

In Table 2.8, SE is the propagated standard deviation of the mean ($\bar{x}$); $t = (\bar{x} - \mu)/\text{SE}$ and $z = (\bar{x} - \mu)/\text{SE}$. The critical values $t_{\alpha,\text{df}}$ or z_α for one-tailed tests and $t_{\alpha/2,\text{df}}$ or $z_{\alpha/2}$ for two-tailed tests are extracted from the appropriate statistical distribution curves. Remember that if H_0 is accepted, it is being accepted against the alternative hypothesis H_A. In the case of two-tailed tests, if $\bar{x}$ is significantly less than μ, we must accept H_A.

2.9.2 Observations of Two Observables: Test on the Difference of Their Means

The statistical test of the difference of the means of observations from two observables is a case in which one is trying to decide if two population means (μ_1 and μ_2) for two observables are equal. For example, if two survey crews independently determined the elevation of a benchmark (as $\bar{x}_1$ and $\bar{x}_2$) based on their leveling run from different

TABLE 2.8 Decisions on a Single Population Mean.

	Decision
One-tailed test	Accept H_0 if the following are satisfied: For $n \leq 30 : t < t_{\alpha,\text{df}}$ (or $t > t_{\alpha,\text{df}}$) or $\bar{x} - \mu < (\text{SE})t_{\alpha,\text{df}}$ For $n > 30 : z < z_\alpha$ (or $z > z_\alpha$) or $\bar{x} - \mu < (\text{SE})z_\alpha$
Two-tailed test	Accept H_0 if the following are satisfied: For $n \leq 30 : \bar{x} - t_{\alpha/2}\text{SE} < \mu < \bar{x} + t_{\alpha/2}\text{SE}$ or $\lvert\bar{x} - \mu\rvert < (\text{SE})t_{\alpha/2,\text{df}}$ For $n > 30 : \bar{x} - z_{\alpha/2}\text{SE} < \mu < \bar{x} + z_{\alpha/2}\text{SE}$ or $\lvert\bar{x} - \mu\rvert < (\text{SE})z_{\alpha/2}$

starting points and along different routes, one may want to decide if $\bar{x}_1$ and $\bar{x}_2$ are statistically equal or they are from the same population, that is, μ_1 and μ_2 are equal. The hypotheses can be formulated as follows:

$$\text{One-tailed test}: \quad H_0 : \mu_1 = \mu_2 \quad \text{versus} \quad H_A : \mu_1 \neq \mu_2 \tag{2.45}$$

$$\text{Two-tailed test}: \quad \mathrm{H}_0 \ : \mu_1 - \mu_2 = 0 \quad \text{versus} \quad \mathrm{H}_A = \mu_1 - \mu_2 \neq 0 \tag{2.46}$$

For this test, the t-statistic is used if the sample sizes n_1 or $\leq n_2 \leq 30$ and z-score used when the sample sizes $n_1, n_2 > 30$. The decisions in Table 2.9 can be made according to the above hypotheses.

In Table 2.9, the standard error (SE) is propagated from the difference $(\bar{x}_1 - \bar{x}_2)$ using the corresponding variances and covariances of the two means $(\bar{x}_1, \bar{x}_2)$ and following the variance–covariance propagation laws (refer to Section 2.8.2); the t-statistic and the z-score are determined from the following equations:

$$t = \frac{(\bar{x}_1 - \bar{x}_2)}{\mathrm{SE}} \tag{2.47}$$

$$z = \frac{(\bar{x}_1 - \bar{x}_2)}{\mathrm{SE}} \tag{2.48}$$

Considering the two-tailed test further, it can be shown from Equations (2.47) and (2.48) that the expected critical value of the difference between the two sample means at $(1 - \alpha)$ confidence level will be

$$|\bar{x}_1 - \bar{x}_2| = (\mathrm{SE})t_{\alpha/2,\mathrm{df}} \tag{2.49}$$

or

$$|\bar{x}_1 - \bar{x}_2| = (\mathrm{SE})z_{\alpha/2} \tag{2.50}$$

TABLE 2.9 Decisions on the Difference Between Two Population Means.

	Decision
One-tailed test	Accept H_0 if the following are satisfied: For n_1 or $n_2 \leq 30 : t < t_\alpha$ (or $t > t_\alpha$) or $\bar{x}_1 - \bar{x}_2 < (\mathrm{SE})t_{\alpha,\mathrm{df}}$ For $n_1, n_2 > 30 : z < z_\alpha$ (or $z > z_\alpha$) or $\bar{x}_1 - \bar{x}_2 < (\mathrm{SE})z_\alpha$
Two-tailed test	Accept H_0 if the following are satisfied: For n_1 or $n_2 \leq 30 : (\bar{x}_1 - \bar{x}_2) - t_{\alpha/2}\mathrm{SE} < 0 < (\bar{x}_1 - \bar{x}_2) + t_{\alpha/2}\mathrm{SE}$ or $|\bar{x}_1 - \bar{x}_2| < (\mathrm{SE})t_{\alpha/2,\mathrm{df}}$ For $n_1, n_2 > 30 : (\bar{x}_1 - \bar{x}_2) - z_{\alpha/2}\mathrm{SE} < 0 < (\bar{x}_1 - \bar{x}_2) + z_{\alpha/2}\mathrm{SE}$ or $|\bar{x}_1 - \bar{x}_2| < (\mathrm{SE})z_{\alpha/2}$

where the standard error (SE) is propagated from the difference $(\bar{x}_1 - \bar{x}_2)$ using the corresponding variances and covariances of the two means $(\bar{x}_1, \bar{x}_2)$ following the variance–covariance propagation laws (refer to Section 2.8.2).

There is a common relationship between the $z_{1-\alpha/2}$ critical values from the normal distribution curve and the Chi-square $(\chi^2_{\alpha,\mathrm{df}=1}(\text{upper-tail area}))$ critical values for one-dimensional cases (upper-tail areas) from the Chi-square distribution curve, which can be expressed as follows:

$$z_{1-\frac{\alpha}{2}} = \sqrt{\chi^2_{\mathrm{df}=1,\alpha}} \tag{2.51}$$

where df is the number of degrees of freedom, df = 1 for one-dimensional cases and α is the level of significance (for upper-tail areas). If Equation (2.51) is substituted into Equation (2.50), the following expression can be used to test if the difference between two parameters $(\bar{x}_1 - \bar{x}_2)$ is significantly different from zero value:

$$|\bar{x}_1 - \bar{x}_2| = (\mathrm{SE})\sqrt{\chi^2_{\alpha,\mathrm{df}=1}(\text{upper-tail area})} \tag{2.52}$$

Example 2.1 The line between two survey markers P and Q was measured repeatedly by survey crew A and the adjusted distance obtained was 1500.030 m; survey crew B obtained the adjusted distance for the same line as 1500.042. If the standard error of the adjusted distance by each crew is 4 mm (considered well known), determine if the expected critical value of the difference in the two distances has exceeded at 80% confidence level. Based on your result, are the two distances significantly different at 80% confidence level?

Solution:

Difference in measurements, $|\bar{x}_A - \bar{x}_B||\bar{x}_A - \bar{x}_B| = 1500.042 - 1500.030$ (or 12 mm)

By the error propagation of the difference, SE = $4\,\mathrm{mm}\sqrt{2}$ (or 5.66 mm)

Significance level: $\alpha = 0.20$

Using Equation (2.50), if $|\bar{x}_A - \bar{x}_B| \leq (\mathrm{SE})z_{\alpha/2}$ is satisfied, then the distances are not significantly different at 80% confidence level.

Since the standard deviations are considered well known, *z-score* will be used:

$$z_{\frac{0.20}{2}} = 1.28 \quad \text{(two-tailed test)}$$

From Equation (2.50), is 12 mm ≤ 5.66(1.28)? or is 12 mm ≤ 7.2 mm? Since this condition is not satisfied, the two distances are significantly different at 80% confidence level.

2.9.3 Observations of One Observable: Test on the Variance

The statistical test on the variance of the observations of one observable is a case in which one is to decide if the sample standard deviation (s) compares with the

TABLE 2.10 Decisions on a Population Variance.

	Decision
One-tailed test	Accept H_0 if the following are satisfied: $\chi^2 < \chi^2_{\alpha,\mathrm{df}}$ (or $\chi^2 > \chi^2_{1-\alpha,\mathrm{df}}$) or $s \leq \sqrt{\dfrac{\chi^2_{\alpha,\mathrm{df}}\sigma^2}{\mathrm{df}}}$
Two-tailed test	Accept H_0 if the following is satisfied: $\chi^2_{1-\frac{\alpha}{2},\mathrm{df(upper\ area)}} < \chi^2 < \chi^2_{\frac{\alpha}{2},\mathrm{df(upper\ area)}}$

published precision (or population standard deviation) σ. The hypotheses can be formulated as follows:

$$\text{One-tailed test:} \quad H_0: s^2 \leq \sigma^2 \quad \text{versus} \quad H_A: s^2 > \sigma^2 \tag{2.53}$$

$$\text{Two-tailed test:} \quad H_0: s^2 = \sigma^2 \quad \text{versus} \quad H_A: s^2 \neq \sigma^2 \tag{2.54}$$

The test statistic for this type of test is the χ^2 *statistic (or Chi-square statistic)* given as

$$\chi^2 = \frac{(\mathrm{df})s^2}{\sigma^2} \tag{2.55}$$

If $\chi^2_{\alpha,\mathrm{df}}$, $\chi^2_{1-\alpha,\mathrm{df}}$, $\chi^2_{\alpha/2,\mathrm{df}}$, and $\chi^2_{1-\alpha/2,\mathrm{df}}$ (with df as the degrees of freedom) are the critical values from the Chi-square distribution curve (upper area type), the decisions given in Table 2.10 can be made with regard to the above hypotheses.

Considering the one-tailed test (in Table 2.10) further; it can be shown from Equation (2.55) that the expected critical value of the sample standard deviation at $(1-\alpha)$ confidence level will be

$$s_\alpha \leq \sqrt{\frac{\chi^2_{\alpha,\mathrm{df}}\sigma^2}{\mathrm{df}}} \tag{2.56}$$

where s_α is the critical standard deviation. Usually, the sample standard deviation must be less than or equal to this critical standard deviation in order to accept that the sample standard deviation (s) compares with the published value (σ) according to the one-tailed hypothesis test formulated earlier. In this type of problem, the one-tailed test seems to be more reasonable than the two-tailed test since having a smaller standard deviation (s) than the published one (σ) is usually not critical.

Example 2.2 The standard deviation of measuring a 1000.000-m-long baseline with the Leica TPS 1203 equipment is 1.8 mm (according to the manufacturer's specification). After calibrating the equipment on the baseline, the calculated standard deviation is 2.5 mm based on 15 measurements of the baseline. Determine, statistically at

95% confidence level, if the equipment is performing according to the manufacturer's specification.

Solution:

$$\sigma = 1.8\,\text{mm}; s = 2.5\,\text{mm}; \alpha = 0.05; \text{df} = 15 - 1$$

$$H_0 : s^2 \leq 1.8^2\,\text{mm}^2 \quad \text{versus} \quad \text{H}_\text{A} : s^2 > 1.8^2\,\text{mm}^2$$

$$\chi^2_{\text{df}=14,\alpha=0.05} = 23.685$$

$$s \leq \sigma\sqrt{\frac{\chi^2_{\text{df}=14,\alpha=0.05}}{\text{df}}} \rightarrow s \leq 1.8\sqrt{\frac{23.685}{14}}$$

$$2.5\,\text{mm} \leq 2.34\,\text{mm}$$

Since 2.5 mm is not less than or equal to 2.3 mm, we are 95% certain that the equipment is not performing according to the manufacturer's specification.

2.9.4 Observations of Two Observables: Comparison of Their Standard Deviations

Comparison of standard deviations of observations of two observables deals with testing if two experimental standard deviations, s_1 and s_2, for the two observables as determined from their different samples of measurements belong to the same population (σ) at the confidence level $1 - \alpha$. The two samples will be considered different if (1) the samples are collected using the same instrument but different observers, (2) the samples are collected using different instruments with the same observer, or (3) the samples are collected at different times using the same instrument with the same observer. The statistical tests can be expressed as follows:

$$H_0 : \sigma_1^2 = \sigma_2^2 \quad H_\text{A} : \sigma_1^2 \neq \sigma_2^2 \tag{2.57}$$

The corresponding H_0: $\sigma_1^2 = \sigma_2^2$ is not rejected if the following condition is satisfied:

$$F_{1-\alpha/2,\text{df}_1,\text{df}_2}(\text{upper-tail area}) \leq \frac{s_1^2}{s_2^2} \leq F_{\alpha/2,\text{df}_1,\text{df}_2}(\text{upper-tail area}) \tag{2.58}$$

where the smaller of the two variances is used as the numerator in Equation (2.58); df_1 and df_2 are the degrees of freedom for determining s_1 and s_2, respectively; and $F_{1-\alpha/2,\text{df}_1,\text{df}_2}(\text{upper-tail area})$ and $F_{\alpha/2,\text{df}_1,\text{df}_2}(\text{upper-tail area})$ are the Fisher distribution values that can be extracted from the F-distribution curve for α being the upper-tail area of the F-distribution curve. Note that it is assumed in Equation (2.58) that s_1 is smaller than s_2, otherwise, they should be switched around and also their corresponding degrees of freedom. Generally, $F_{1-\alpha/2,\text{df}_1,\text{df}_2}(\text{upper-tail area}) = 1/F_{\alpha/2,\text{df}_2,\text{df}_1}(\text{upper-tail area})$, taking note of the flipping around of the degrees of freedom in the denominator as well as the change in the significance level.

2.10 NEED FOR EQUIPMENT CALIBRATION AND TESTING

Calibration is the process of establishing the accuracy performance of an instrument within some stated and limited criteria. It is the act of checking or adjusting by comparison with a standard or reference, the accuracy of a measuring instrument. It involves comparing the output of an instrument being tested with a known standard in order to determine some conversion factor or a constant (both systematic and random effects) that can be applied to the instrument output to make the output more accurate. The manufacturer's claimed accuracies of instruments, however, usually represent in general the average situations and may be significantly different from the actual situations under which observations are being made, hence the need to independently estimate accuracies of measurements. The calibration procedures to be adopted must conform to an acceptable standard and be within statistically stated rules in order for the results to be valid. A *standard* or a *reference* in this case can be taken as an instrument or a method that will measure more accurately and precisely the desired quantity than the measuring instrument itself. For an example, a laser interferometer can measure more accurate distances (relative displacements) than an EDM does, so it is considered a standard or a reference instrument for calibrating the EDM.

Testing is a simpler process used to find out if the instrument is performing according to the manufacturer's specification. This process will not require comparison with a set of standards; it simply determines the random component of the accuracy measure (i.e., the precision that can be expected under similar conditions of testing). Testing procedures usually exclude the influences of external factors such as atmosphere, targeting devices, or observers. If the specification claimed by the manufacturer is not satisfied, then it may be possible to calibrate the instrument so that it does. Usually, instruments are calibrated less often than they are field tested; calibration is done by the manufacturer or by the accredited calibration laboratory, while testing is done by the instrument users.

Calibration and testing of precision instruments are important in investigating if the precision in use of the measuring equipment is appropriate for the intended survey project. A priori knowledge of accuracies of proposed observations in the project is needed at the design stage in order to understand how the project and the final results are to be affected by both the instruments and the environment. Note that precision is used as a measure of accuracy and standard deviation is the expected precision of one measurement based on the use of the given procedure. To arrive at a reliable standard deviation for a measurement, a test must be done, using several repetitions (about 15 or 20) of a measurement, simulating the field conditions to be used later.

Before calibrating and testing the measuring equipment, the equipment must be in known and acceptable states of permanent adjustment as specified by the manufacturer, and the equipment must be used with recommended supporting equipment. It should also be noted that results of tests are influenced by meteorological conditions, especially by the gradient of temperature. An overcast sky and low wind speed will guarantee the most favorable weather conditions. Notes should also be taken of the actual weather conditions at the time of measurement and the type of surface above which the measurements are made. Laboratory tests will be most preferred since such tests are almost unaffected by atmospheric influences but are too costly and are not practicable for most users. Laboratory tests also yield precisions that are much higher than those that can be obtained under field conditions.

In Chapters 4–6, the field procedures for determining and evaluating the accuracy (precision) of survey equipment when used in surveying measurements will be specified. Rigorous procedures for testing distance measuring equipment (EDM or total station instruments), direction and angle measuring equipment (precision theodolites), elevation difference measuring equipment (precision levels), and the GPS survey equipment are considered in those chapters. The procedures adopted in these chapters are to create awareness of the existence of the internationally accepted standards, such as the International Organization for Standardization (ISO) standards and the German Deutsches Institut für Normung (DIN) standards.

The ISO and DIN standards specify field procedures to be followed each time the achievable precision (or accuracy) for a given surveying instrument used together with its supporting equipment (tripod, staffs, etc.) has to be determined. These procedures, which are not to discredit the equipment manufacturers' quoted precisions for their equipment, are to help the surveyor investigate if the precision given by the measuring equipment is appropriate for the intended project. Moreover, the procedures are to provide a means of associating precision (accuracy) to different survey equipment and in the process help in classifying equipment, such as 5″- and 3″-instruments. With this, the surveyor is able to know those instruments that are in the same category and those that are not.

A number of recommendations concerning the needs for calibration and testing of survey equipment have been made (Becker, 2002) to partners (surveyors, survey institutions, etc.) that are responsible for maintenance and quality specifications of survey instruments. The recommendations to the surveyors include the following (Becker, 2002):

- Always require a calibration document from the manufacturer at the delivery of equipment.
- Be familiar with the equipment and read the technical documents in order to understand the possibilities and limitations of the equipment.
- Follow the manufacturer instructions for proper handling of the equipment.
- Check the instrument performance regularly for its repeatability and suitability.
- Monitor continuously the instrument health in a logbook from the time of its delivery.
- Check before each project the functionality and suitability of the equipment.
- Use appropriate equipment for each specific work type.
- Report all changes, weaknesses, errors, and so on to the manufacturer, owner and other users, of the equipment.

It can be generally understood from the above list that it is important that surveyors be familiar with the ISO standards and their procedures in order to determine the precision of their measuring system and to monitor the health of their equipment.

Some of the recommendations (Becker, 2002) to the surveyor training institutions are as follows:

- Test all new and current equipment.
- Report about the possibilities, limitations, and weaknesses of equipment to all partners.
- Report also about how to operate and how to minimize the error budget when using different equipment (i.e., the best use practice).
- Ensure that the students are trained to carry out routine checks and calibrations in accordance with existing standards and regulations.
- Make the students aware about the error sources and their minimization.
- Spread the importance of guidelines, standards, and so on.
- Collaborate with users, manufacturers, and ISO to upgrade guidelines and standards.

The recommendations go further to encourage students to be more involved in the standardization work in order to better understand the needs for standards, maintenance, and calibration of instruments. The overall interest in error testing, however, seems to be relatively low among surveying professionals. A surveyor needs to have very good understanding of how errors are investigated through calibration and testing. Assumptions, or manufacturers' statements as to precision, can be considerably far from reality, and the geometry and atmospheric conditions of the survey affect the errors much more than many realize. Depending on how the instrument is used, the measurement accuracy may be higher or lower than the specified value. The ISO or DIN accuracy (or precision) values indicated for instruments should, therefore, be used with caution.

3

STANDARDS AND SPECIFICATIONS FOR PRECISION SURVEYS

OBJECTIVES

After studying this chapter, you should be able to

1. Explain various standards available for geomatics projects
2. Discuss accuracy standards and specifications for precision surveys
3. Explain how the concepts of confidence regions are applied in accuracy standards and survey specifications
4. Interpret the common standards used in conventional horizontal control surveys
5. Interpret the common standards used in conventional vertical control surveys
6. Apply various standards to geomatics projects
7. Determine network and local accuracy values and use them to classify geomatics projects
8. Discuss and apply the various specifications for precision leveling and GPS surveys
9. Discuss the differences between quality assurance (QA) and quality control (QC) as applied in geomatics
10. Develop QA/QC checklists for some geomatics projects

Precision Surveying: The Principles and Geomatics Practice, First Edition. John Olusegun Ogundare.

3.1 INTRODUCTION

Standards are limits, requirements, or rules approved as minimum acceptable benchmarks or a list of technical specifications describing the important characteristics (the quality) of a service or a deliverable. If a service or a deliverable satisfies the given standards, the service or the deliverable will be said to have quality according to the standards. In this case, the quality of any work is defined by some standards that are ideally dependent on the generally accepted characteristics of the work. According to the American Congress on Surveying and Mapping (ACSM), four types of standards can be identified as (ACSM, 2002) *precision standards*, *accuracy standards*, *content standards*, and *performance standards*. They are discussed in the following sections.

3.1.1 Precision Standards

In order to understand what precision standards are, the concept of precision must be understood first. Precision is the level of closeness of agreement of a set of measurement results of the same observable among themselves. It can also be referred to as the *repeatability* or *reproducibility* of the measurement results of the observable. *Repeatability* of results of measurements is defined as precision of measurement results in which repeated measurements of the same observable are made over very short time intervals under the same conditions such as same measurement procedure, observer, measuring instrument, location, and environment.

Reproducibility of results of measurements is the same as repeatability of results of measurements except that measurements of the same observable are repeated over long time intervals at different conditions, such as different measurement principle, method, observer, location, or environment. With regard to the meanings of precision and standards, *precision standards* can be defined as approved limits with which precisions of measurement results can be compared for conformance. Quality of instrument operation or the degree of perfection in instrument and the method used in making measurements are determined by using the precision standards. The allowable discrepancy between independent forward and backward leveling runs between benchmarks for the vertical control surveys in Canada and in the United States (discussed in Section 3.3) can be taken as an example of precision standards.

3.1.2 Accuracy Standards

Understanding what accuracy standards are starts with the understanding of what accuracy is. *Accuracy* of measurement refers to closeness of mean of measurement results to the true value and the degree of agreement within individual measurement results. This is a measure of combined effect of systematic and random errors in a measurement. A measurement that is affected only by random errors is considered accurate to within the precision of the measurement. If systematic errors are present in the measurement, the accuracy of the measurement cannot be based on the precision alone, but on the combined effects of systematic errors and the precision. In determining the accuracy of measurements, however, the focus is usually on identifying and eliminating systematic errors since precision is random in nature and cannot be eliminated but can only be minimized.

With regard to the meanings of accuracy and standards, *accuracy standards* can be defined as accepted values (considered to be close to their true values) with which measurement results can be compared for conformance or the maximum acceptable uncertainties in a result. They are a measure of *quality of end results*. Accuracy standard describes the standard for classifying results; in this case, accuracy can be seen as closeness of an estimated or measured value to an accuracy standard. Accuracy of a survey, for example, cannot be determined solely from measurements; a standard value or set of standard values must be available as a reference for comparison somewhere during the accuracy determination. A reference for comparison, for example, could be 180° for the sum of angles in a triangle, the internationally accepted standard unit values for the conventional unit of measurements, a value determined by refined methods and deemed sufficiently near the ideal or true value to be held constant as reference for other similar determination, and so on.

The main component of accuracy standard is the *positional accuracy,* which deals with how closely the coordinate descriptions of features compare with their actual location. Typical standards based on positional accuracy are standards for geodetic control networks for determining the quality of geodetically surveyed points (discussed in Sections 3.3.2 and 3.4.3) and those designed to allow users of maps and geospatial data to determine if their maps or data are suitable for use, such as National Map Accuracy Standards (NMAS), the American Society for Photogrammetry and Remote Sensing (ASPRS) standard and the National Standard for Spatial Data Accuracy (NSSDA) (discussed in Section 3.6). Other example of accuracy standards is the accuracy standards for vertical control in the United States in which the order of accuracy is determined by using the standard deviations from least squares processes of elevation differences between directly connected points (discussed in Section 3.3). Generally, the standards for geodetic control networks are to provide common methodology for determining and reporting the positional accuracy for all geodetic control points represented by permanent monuments (FGDC, 1998a). With the standards, the accuracy of coordinate values of some points determined from GPS surveys, for example, can be compared with the accuracy of coordinate values of corresponding points based on conventional terrestrial survey methods.

3.1.3 Content Standards

Content standards specify the amount of features to be measured and represented on a deliverable and describe issues with attribute accuracy, extent to which geometric problems and drafting inconsistencies are taken care of, sources of data and data processing steps, and completeness of data representation.

3.1.4 Performance Standards

Performance standards specify steps to follow in a survey operation, which may go beyond purely technical operations of the survey. They define the levels of performance to be made available to clients and cover issues, which include accuracy standards and precision standards.

3.1.5 General Comparison of Standards

Precision and *accuracy standards* deal with quality in technical ways, which are more meaningful to practitioners. *Content* and *performance standards* deal with steps to be taken in order to complete a project by establishing the scope of work for both the practitioner and the client. These standards are conceptual in nature and are of more interest to clients who see them as being less complex than the technical standards. In general, all the standards present the specific requirements and basic characteristics of an acceptable quality system. In order to help meet the requirements of the standards, some accepted technical *specifications and guidelines* are usually designed to provide survey options, methods, procedures, tolerance limits, equipment, technologies, and so on to be used in order to be able to achieve the given standards.

3.1.6 Standards and Specifications

Specifications or *survey specifications* describe the field operations and procedures required in order to attain a particular accuracy standard. They prescribe precision and allowable tolerances for data collection, appropriate network geometry, field procedures, instrumentation, calibration procedures, office procedures, monumentation, and description of survey points. Specifications are not substitutes for instrument manuals that give recommended field operations and procedures for achieving the specified accuracy of the instrument. Before an instrument is chosen for any survey, one must be sure that the instrument will meet the precision requirements of the specifications. *Accuracy specifications* will be considered a means of quantifying and documenting accuracy. Some of the advantages of specifications can be summarized as follows:

- They help the surveyor in understanding the techniques to be used for a particular project.
- They provide an outline of the practices and standards of how work is to be carried out and how it is to be presented.
- They help the surveyor in managing the client expectations; the surveyor is then able to focus on what a client actually needs.
- They help the surveyor to be accountable with regard to the survey process.

There are international standards, which can be considered as specifications or guidelines for field procedures, such as International Organization for Standardization (ISO) standards (e.g., *ISO 17123* standards) and the German Deutsches Institut für Normung (DIN) standards (e.g., DIN 18723). The *ISO 17123* standards specify field procedures to be adopted when determining and evaluating the precision of geodetic instruments and their ancillary equipment when used in surveying measurements. This type of standard provides standard deviation that is repeatable for particular equipment for the specified measuring procedure. The procedures constitute the first step in the process of evaluating the accuracy of a surveying instrument. The ISO standards, for example, make it possible to compare the achievable precision of different instruments or the precision of one instrument at different times.

3.2 STANDARDS AND THE CONCEPT OF CONFIDENCE REGIONS

The use of standards requires a fundamental understanding of statistics and adjustments, while specifications are based on considerable practical experience. The precision and accuracy standards are based on the concepts of standard deviation and confidence region estimation. Confidence region is a region where one has a specified level of confidence (e.g., 95% confidence) that a true value of quantity being estimated will lie. For example, a 95% confidence region about an adjusted point is a region within which the probability is 0.95 that the true coordinate position (vertical or horizontal) of the point lies relative to the selected point (or group of points) used as datum in the survey network. The concept of confidence regions is used for control survey specifications in Canada.

Generally, for classifying survey projects, 95% confidence region (or $\alpha = 0.05$) is used for specifications. For example, in a loop traverse survey, 95% confidence region may be computed for the unclosed traverse to check the actual misclosure against what is expected for that level of confidence. If the 95% confidence region does not enclose the starting position, then there is a probability that either blunder or bias or both may exist in the measurements. If such a blunder or bias is indicated to exist and an investigation cannot disclose and correct the error, there will be a need to do the survey all over again.

The application of the concept of confidence regions in surveying can be summarized as follows:

1. Vertical control surveys specifications in Canada require that uncertainty of the discrepancy between independent forward and backward leveling runs between survey benchmarks at 95% confidence level be used to assess the leveling runs. The formula for estimating this uncertainty value ($e_{0.95}$) is given in Equation (2.15), where SE is the propagated standard error of the discrepancy and z-value at 95% probability (two-tailed) is obtained from the standard normal distribution curve. Equation (2.16) can also be used depending on the situations surrounding the field measurements.
2. Horizontal control surveys specifications in Canada require that 95% confidence region be used as the basic criterion for assessing the accuracy of horizontal control. In this case, the related observations are statistically assumed to be normally distributed and the confidence region indicating the accuracy of horizontal control survey coordinates is bounded by a 95% relative confidence error ellipse discussed in Equations (2.41)–(2.44).
3. The following should be considered with regard to constructing any type of confidence regions for horizontal control surveys:
   a. In the case of *minimal constraint adjustment* (where only one station is fixed), if good estimates of standard deviations of observations are available, the

a priori variance factor of unit weight ($\sigma_0^2 = 1$) should be used in determining the variance–covariance matrix of the adjusted coordinates and the Chi-square statistics should be used in Equations (2.26) and (2.27) in order to determine the confidence ellipses.

b. In the case of *overconstrained adjustment* (where more than one station is fixed), if good estimates of standard deviations of observations are available, the a posteriori (or computed) variance factor of unit weight (s_0^2) should be used in determining the variance–covariance matrix of the adjusted coordinates and the Chi-square statistics should be used in Equations (2.26) and (2.27) in order to determine the confidence ellipses.

c. In the case of *minimal constraint or overconstrained adjustment*, if good estimates of standard deviations of observations are *not available*, the a posteriori (or computed) variance factor of unit weight (s_0^2) should be used in determining the variance–covariance matrix of the adjusted coordinates and the F-statistics should be used in Equations (2.29) and (2.30) in order to determine the confidence ellipses.

3.3 STANDARDS FOR TRADITIONAL VERTICAL CONTROL SURVEYS

3.3.1 Accuracy Measure of Vertical Control Surveys

The inherent precision of differential (or spirit) leveling has made it the most commonly used geodetic measurement system in vertical control surveys. The measurement system can reliably be designed to enhance the precision of height determination of survey points, considering the sources of systematic and random errors and minimizing or eliminating their effects. The majority of the field specifications and instrumental requirements in differential leveling are to eliminate or minimize possible systematic errors; and the statistically independent random errors associated with the leveling procedures are generally controlled through redundant measurements and randomization procedures. The concepts of accuracy measure of vertical control surveys can be summarized as follows:

1. The accuracy of leveling a line of length L (km) is influenced by random and systematic errors of measurements. Generally, the influence of systematic errors is much smaller than that of random errors if leveling lines do not exceed a few kilometers and if leveling specifications are followed in field measurements. The effect of systematic errors (σ) in L (km) of leveling accumulates as follows (Bomford, 1980):

$$\sigma = \sigma_{\text{syt}} L \tag{3.1}$$

where σ_{syt} is the systematic error accumulating in proportion to length L (km). The effect of random errors (σ) in L (km) of leveling accumulates as follows (Bomford, 1980):

$$\sigma = \sigma_{\text{ran}} \sqrt{L} \tag{3.2}$$

where σ_{ran} is the random error accumulating over 1 km of leveling (the standard deviation of elevation difference over 1 km). The accumulation of the total errors may also be proportional to the number (n) of setups or time spent on the work. According to Bomford (1980), these quantities are both roughly proportional to length L. It is the recognition of the potentially large error contribution from systematic effects that has dictated many of the procedural requirements specified for geodetic leveling, as listed in Section 3.3.2. The total systematic and random errors in leveling a line of L (km) is given (Bomford, 1980) as

$$\sigma_{\text{Total}} = (\sigma_{\text{syt}}^2 L^2 + \sigma_{\text{ran}}^2 L)^{1/2} \tag{3.3}$$

2. According to Bomford (1980), the effect of random errors (Equation (3.2)) predominates for a short distance (about 1–5 km) of leveling, while the effect of systematic errors (Equation (3.1)) predominates over a long distance (greater than 5 km) of leveling. Over a short distance of leveling, Equation (3.2) can be considered the *standard deviation of the difference in elevation between the benchmarks in a single-run section* (assuming the systematic error effects are minimized). The main sources of random errors are due to centering the spirit level, reading the leveling rods, and variations in refractions.
3. If the leveling is run round a loop of length L (km), Equation (3.2) can be considered as representing the standard deviation of the loop closure. From Equations (3.2) and (2.15), the *loop closure,* which is the precision of estimate at 95% confidence level for one-way leveling in a loop of length L (km), can be given as follows:

$$\Delta_{\text{loop}} = \sigma_{\text{ran}} \sqrt{L} \times 1.96 \tag{3.4}$$

where the standard error (SE) of the discrepancy is expressed by Equation (3.2) and the standard normal distribution value at 95% confidence level ($z_{\alpha=0.05/2}$) is 1.96.
4. If one levels between two benchmarks separated by L (km) (once forward (F) and once backward (B)), the standard deviation of the discrepancy ($\Delta = F - B$) between forward (F) and backward (B) leveling runs can be determined. Assuming the error in forward and backward leveling is each expressed by Equation (3.2) and using variance–covariance propagation laws on the discrepancy, the propagated standard deviation for the discrepancy over the line of length L (km) can be given as follows:

$$\sigma_{\Delta} = \sqrt{2} \times \sigma_{\text{ran}} \sqrt{L} \tag{3.5}$$

From Equations (3.5) and (2.50), the maximum discrepancy or the *section closure* (Δ_{section}) between two leveling runs over the line of length L (km) can be given as

$$\Delta_{\text{section}} = \sqrt{2} \times \sigma_{\text{ran}} \sqrt{L} \times 1.96 \tag{3.6}$$

where the standard error (SE) of the discrepancy is expressed by Equation (3.5) and $z_{\alpha=0.05/2} = 1.96$.

5. The accuracy specifications for vertical control in Canada and in the United States are given (NRC, 1978; Blachut et al., 1979) in Table 3.1, where L (km) is the approximate distance between benchmark positions measured along the leveling route. (Note: L is one-way distance in a section or the distance round the loop in the case of a loop.) The table provides the maximum discrepancies of leveling for different orders. In each order, the choice of value for σ_{ran} and the spacing L (km) will vary in order to maintain fairly consistent expected maximum discrepancy in all orders. In this case, if the value for σ_{ran} is reduced, then the spacing between the benchmarks in that order must be increased. On this basis, higher order benchmarks have greater separation than lower order ones; in the same way, the higher order leveling requires higher precision than its lower order counterpart.
6. The precision of the vertical distances between points depends on the spacing between the points. According to Blachut et al. (1979), the common separations between benchmarks are as follows:
   - First-order control points are spaced 2–4 km (with an average of 3 km).
   - Second-order control points are spaced 0.5–1 km (with an average of 0.75 km).
   - Third-order control points are spaced 0.1–0.3 km (with an average of 0.2 km). For example, the third-order benchmarks are spaced at 200-m intervals in the core city and at 500-m intervals in suburban areas.

   For example, given (Table 3.1) the specification for first-order vertical control as $\pm 4\,\text{mm}\sqrt{L}$, where $L = 3$ km, the maximum discrepancy expected will be 6.9 mm; and for second-order vertical control, the specification is $\pm 8\,\text{mm}\sqrt{L}$, where $L = 0.75$ km, the maximum discrepancy expected will be 6.9 mm. It can be seen that the precisions of leveling runs in the first-order and second-order surveys are different and the separations between the corresponding benchmarks are correspondingly varied in order to maintain consistent maximum discrepancy in leveling.

TABLE 3.1 Accuracy Specifications for Vertical Control in Canada and the United States.

Order of Accuracy (Canada)	Order of Accuracy (USA)	Allowable Discrepancy between Independent Forward and Backward Leveling Runs between Benchmarks
Special order	First-order, Class I	$\pm 3\,\text{mm}\sqrt{L}$
First order	First-order, Class II	$\pm 4\,\text{mm}\sqrt{L}$
Second order	Second-order, Class II	$\pm 8\,\text{mm}\sqrt{L}\left(\text{USA Class I: } \pm 6\,\text{mm}\sqrt{L}\right)$
Third order		$\pm 24\,\text{mm}\sqrt{L}\left(\text{USA third order } \pm 12\,\text{mm}\sqrt{L}\right)$
Fourth order		$\pm 120\,\text{mm}\sqrt{L}$

7. With regard to step 6 , $\Delta_{section} = 6.9$ mm between two first-order benchmarks having an average separation of $L = 3$ km; if we use these values in Equation (3.6) and solve for σ_{ran}, we will have $\sigma_{ran} = 1.44$ mm/km. Similarly, $\Delta_{section} = 6.9$ mm between two second-order benchmarks having an average separation of $L = 0.75$ km; if we use these values in Equation (3.6) and solve for σ_{ran}, we will have $\sigma_{ran} = 2.87$ mm/km. These results are consistent with the general conclusion (Blachut et al., 1979) that the accuracies of higher order networks are usually at least twice as high as that of lower order networks. From this, it can be seen that by leveling over a 1 km section, the standard deviation of second-order leveling will be two times as high as that of first-order leveling. Similarly, it is likely that the third order will be about four times as high as the first order, and so on.
8. Generally, it can be stated that there are more accumulated errors in lower order benchmark elevations than in higher order benchmark elevations; this means that the lower order benchmark elevations are less accurate than the higher order benchmark elevations, in absolute term. The precision of the vertical distances between the third-order benchmarks will be due to three sources: the leveling errors of the third-order network itself, errors due to the second-order network, and errors due to the first-order network.

3.3.2 Specifications and Guidelines for Vertical Control Surveys

Specifications for leveling are based on the different orders of vertical control, which are defined in terms of the allowable discrepancy between independent forward and backward leveling runs between benchmarks (refer to Table 3.1). Special-order leveling surveys are the most precise type and are usually conducted for monitoring earth movement. Fourth-order surveys are the lowest order type, which are conducted to support construction works. If recommended procedures and equipment are used in each survey type, it is expected that the above-specified allowable discrepancies will not exceed in approximately 95% of the sections over the course of a level line. Those sections exceeding the allowable discrepancy must be releveled. If loop misclosures are to be used, the allowable discrepancy is not to be exceeded by taking L (km) as the length along the level route around the loop. In this case, long, narrow loops should be avoided in order to maintain the specified accuracy.

Note that the discrepancy between the forward and backward leveling runs will not detect systematic errors that remain the same in the forward and backward leveling runs; the classifications in Table 3.1 cannot be referred to as accuracy standards, but as part of field specifications. They are specifications since achieving these values alone does not actually guarantee the accuracy of the job except all of the other field specifications stated in the following list are satisfied. For example, it is possible to achieve the numerical value specified for a special-order job by using an inappropriate field procedure (e.g., using wooden staff, engineer levels, and observing readings below 0.5 m on the rod); however, it is obvious that the value so obtained is not a confirmation that the job has been precisely done. There is obviously no attempt in this type of procedure to remove possible systematic errors and to minimize random errors, making the job unacceptable for the special order even though the value for the order is achieved.

Some of the typical specifications for the differential leveling field procedures, which must be followed together with the specifications in Table 3.1, are discussed as follows. The emphasis is being placed on the special-order and the first-order geodetic leveling runs since they require the highest possible level of care. To achieve the standards of accuracy set out for the special-order and the first-order leveling runs in Table 3.1, the following procedures are recommended (NRC, 1978):

1. Level each section once forward and once backward independently using different instrument men, and if possible, different instruments under different weather conditions and at different times of the day. This is referred to as *double-run* leveling procedure. Redundancy is introduced through double running and through the use of double-scale rods, making measurements more precise and blunder free. Since the procedure forward is about the same as that of backward, the random error is reasonably assumed to accumulate about the same way in forward and backward. If the forward and backward runs are done on alternate days, there are possibilities that random effects of refraction, movement of tripod during setups, and gradual movement of turning pins/plates between setups might be minimized.
2. After a section is double run, check that the elevation differences from the two runs agree within the allowable discrepancies specified in Table 3.1. This process of checking for the agreement is usually referred to as “closing the section”. Each leveling section will be complete if the agreement is achieved. Otherwise, the section must be releveled. It should be mentioned that the misclosures between the forward and backward runs in double-run leveling provides a measure of systematic errors, but does not provide any direct insight into the source of the errors. Significant misclosures may be due to blunders or the occurrence of crustal deformation during the completion of a run, so that misclosures alone cannot be taken as the overall indicator of systematic errors in leveling.
3. The following rejection steps should be carried out if the allowable discrepancy is not satisfied in step 2 . Note that this rejection test is not a substitute for the overall test to check the compliance with the allowable discrepancy specified in Table 3.1. This test is only an intermediate test for deciding which of the forward and the backward leveling runs to be used for the final compliance test with regard to Table 3.1:
   (i) After three or more runs of a section, check agreement again.
   (ii) Compute the mean ($\bar{x}$) of all the runs (disregarding signs) including those that have been rejected previously.
   (iii) Compute the differences between the mean and each running, $(x - \bar{x})$.
   (iv) Perform the following leveling rejection test:

$$|x - \bar{x}| > 3.7\,\text{mm}\sqrt{L} \quad \text{(For special order)} \tag{3.7}$$

$$|x - \bar{x}| > 4.7\,\text{mm}\sqrt{L} \quad \text{(For first order)} \tag{3.8}$$

(v) Remove the one that fails the rejection test and compute new mean, excluding the failed one and performing the test again.

(vi) After all have been tested, if there are at least two forward runs and two backward runs passing the rejection test (even though there is no check between the forward running and backward running), the releveled section is said to be complete.

(vii) If only two forward runs and no backward run passed, rerun the leveling for the section; include the new section run with all of the previous runs (including those previously rejected) and start the test from step (ii).

(viii) Some runs rejected previously may now pass after the number of runs has increased; this is acceptable since the mean of sample has also improved.

4. The mean elevation difference for forward (Δh_{f}) and backward (Δh_{b}) runs between two benchmarks are given as follows (while retaining their negative or positive signs):

$$\text{Mean} = \frac{\Delta h_{\mathrm{f}} - \Delta h_{\mathrm{b}}}{2} \tag{3.9}$$

5. All sections must have an even number of setups. This is to cancel out the effect of the zero-point offsets of two leveling staffs used.
6. Difference between backsight (BS) and foresight (FS) distances at each setup and their total for each section must not exceed 5 m for special order or 10 m for first order. This is to minimize the effects of collimation error of leveling instrument, collimation change due to refocusing of telescope, and the refraction effects.
7. Alternate readings of backsight and foresight at successive setups must be adopted, for example, backsight–foresight, foresight–backsight, backsight–foresight, and so on. This will minimize the effects due to the sinking of instrument/tripods between measurements.
8. Maximum length of sight is 50 m for special order or 60 m for first order, with weather conditions and terrain permitting. This has been found to have improved precision of leveling.
9. Line of sight must not be less than 0.5 m above the ground. This is to minimize the effect of refraction, which might be higher when the line of sight is closer to the ground.
10. Rod reading must consist of mean of center-wire reading on each scale after applying constant; if three-wire method is used in the case of first-order leveling, mean of the readings for the three wires must be used. The mean of redundant measurements is more precise than that of individual measurements.
11. Benchmark stability must be checked by carrying out two-way leveling between the starting and an adjacent benchmark and comparing the new difference of elevation with the original difference. The two benchmarks must be far enough apart so that any disturbing influence is not the same on both benchmarks. If the check is within the allowable discrepancy for the order of leveling, both benchmarks are assumed to be stable. Otherwise, other

benchmarks must be used for the check until an agreement is obtained with respect to the allowable discrepancy. This will help check blunders due to the occurrence of crustal deformations that may be misconstrued as random misclosure.

In order to achieve the standards of accuracy set for precise leveling, the following equipment is recommended by Natural Resources Canada (1978) for special-order and first-order leveling works:

1. Self-leveling instrument equipped with parallel-plate micrometer, telescope magnification of at least 40× for special order (and 32× for first order), and a high-speed compensator with sensitivity equal to or better than a 10″/2-mm-level vial; or spirit-level instrument equipped with parallel-plate micrometer, telescope magnification of at least 40× for special order (and 32× for first order), and a 10″/2 mm or better level vial. The compensator is to take care of under- or overcompensation, collimation error due to collimation fluctuations with temperature or collimation change due to refocusing of telescope.
2. Invar, double-scale rods with line graduations of width 1–1.6 mm (invar rods of checkerboard design with smallest graduations not less than 1 cm and with check graduations on the reverse side is also acceptable for first-order jobs).
3. Rod supports for special order (not required for first order).
4. Circular levels permanently attached to the rods.
5. Foot plates or steel pins for turning points.
6. Sun shade and instrument cover.
7. Calibration of rods to check rod scale error; and in abnormal temperature, thermal expansion corrections to leveling rods must be made.

Parallel glass plate micrometer is usually fitted in front of the objective of a precise or geodetic level. The plate is to enable the interval between the crosshair and the nearest staff division to be read directly to 0.1 mm. The plate is tilted till a full reading of the staff coincides with the crosshair; this will result in a certain displacement, which gives the fractional reading that can be obtained directly from the micrometer drum. It is required that when employing the parallel-plate method of leveling for special-order or first-order leveling, double-scale line-graduated rods be used. The spacing of the smallest graduations must be equivalent to the displacement of the parallel-plate micrometer. Using the three-wire method for first- or second-order leveling requires that rods with checkerboard design be used.

3.3.3 Typical Field Procedure for Precise Differential Leveling

Three-wire leveling is a differential leveling method applied in geodetic or precision work. In ordinary (nongeodetic) leveling procedure, the leveling staff is read against only the middle horizontal crosshair, whereas in three-wire leveling procedure, leveling staff is read against all the three horizontal crosshairs [upper (u), middle (m), and

TABLE 3.2 Sample Field Notes for Three-Wire Leveling Method (Forward Run).

Station (1)	Backsight (BS+) (2)	Stadia Interval (Stadia Unit) (3)	Stadia Distance (m) (4)	Foresight (FS−) (5)	Stadia Interval (Stadia Unit) (6)	Stadia Distance (m) (7)
BMA						
(*u*)	0819			1034		
(*m*)	~~0753~~	86	8.6	0951	83	8.3
	0733					
(*l*)	0648	85	8.5	0869	82	8.2
	2200/3	**171**	**17.1**	2854/3	**165**	**16.5**
Mean	**+0733.3**			**0951.3**		
TP1						
(*u*)	1052			1140		
(*m*)	0982	70	7.0	1069	71	7.1
(*l*)	0913	69	6.9	0997	72	7.2
	2947/3	**139**	**13.9**	3206/3	**143**	**14.3**
Mean	**+0982.3**			**1068.7**		
TP2						
(*u*)	2009			1365		
(*m*)	1941	68	6.8	1293	72	7.2
(*l*)	1873	68	6.8	1222	71	7.1
	5823/3	**136**	**13.6**	3880/3	**143**	**14.3**
Mean	**+1941.0**			**1293.3**		
BMB						
SUM	**3656.6**	**446**	**44.6**	**3313.3**	**451**	**45.1**

lower (*l*) cross hairs] and recorded as shown in the sample field notes in Table 3.2. In the table, for example, *u*, *m*, and *l* crosshair readings are recorded for the backsight in column 2 and *u*, *m*, and *l* crosshair readings for the foresight in column 5. The crosshair readings are considered the stadia readings. These stadia readings can be used to determine the approximate distance (known as the *stadia distance*) between the instrument and the staff sighted to if the stadia factor of the instrument is known (usually the stadia factor is 100).

For example, referring to Table 3.2, the stadia readings are made in stadia unit (in this case, millimeters); the stadia intervals $(u - m)$ and $(m - l)$ are given in columns 3 and 6; assuming the stadia factor is 100, half of the stadia distance between the instrument and the rod is the corresponding stadia interval (in columns 3 and 6) multiplied by 100; the sum of two halves for a given setup gives the approximate distance between the instrument and the staff sighted to. Half the stadia distances are recorded in columns 4 and 7 for the backsight and foresight staffs, respectively (assuming the stadia factor is 100). For example, in Table 3.2, half stadia interval for the BS reading on BMA is $(u - m) = (0819 - 0733)$ or 86 mm; half stadia distance to BMA is 100(86 mm) or 8.6 m. Similarly, the other half stadia distance to BMA is 8.5 m; the total stadia distance between the instrument and the backsight staff at BMA is 17.1 m (shown in column 4).

The surveyor must guide against blunders in field notes. Before the stadia readings on a given staff can be accepted, the readings must be checked using a number of procedures such as

1. The interval values $(u-m)$ and $(m-l)$ must agree within one or two of the smallest units being recorded (e.g., ± 2 mm) or repeat observations.
2. The average $(u+m+l)/3$ must be close to m reading within the last digit (± 1 mm).
3. If steps 1 and 2 are not satisfied, you must do the measurement again.

Assuming, for some reasons, the blunders were not detected and removed immediately in the field, you can still do some minor alterations on the field measurements; in this case, steps 1 and 2 will still be performed for each set of readings in a setup to be followed by the following additional steps:

4. If steps 1 and 2 are not satisfied, adjust just one of the digits in only one of the stadia readings (u, m, or l). For example, if you are adjusting u, do not adjust m and l; if you are adjusting m, then u and l should be left as they are, and so on. In Table 3.2, the original BS readings (0819, 0753, 0648) to BMA do not satisfy step 1 (stadia intervals 66 and 105 are obtained); if 0753 is changed to 0735 (note that 3 and 5 are transposed here as a possible mistake), step 1 will still not be satisfied (stadia intervals 84 and 87 are obtained) even though the sum of the stadia distances will be close to that of FS readings $(8.3+8.2)$; changing 0753 to 0733 (assuming that 5 in 0753 is a typo) will satisfy step 1 as shown in Table 3.2.
5. Continue with step 4 until steps 1 and 2 are satisfied (making sure also that the BS and FS stadia distances are the most identical, assuming the surveyor made a good attempt at balancing the BS and FS distances in the field). In Table 3.2, the new stadia distance to BMA is 17.1 (still identical to that from the FS readings and also identical to the other trials in step 4).
6. If there are too many blunders in the field notes, it would be safer for the surveyor to go back to the field and redo the measurements. The above procedure should only be used in fixing the data if the blunders are obvious and few. The fixed data can then be used in addition to the other mistake-free data in the field data reduction process.

If during calibration of the leveling equipment, it is found that there is a collimation error, the elevation difference in a leveling section must be corrected for the effect of this collimation error. This will be necessary if the BS distances are not the same as the corresponding FS distances. The amount of correction to be added to the observed elevation difference in a leveling section can be given as

$$\text{Correction} = C \times \left(\sum_{1}^{n} d_{\text{BS}_i} - \sum_{1}^{n} d_{\text{FS}_i} \right) \tag{3.10}$$

where C is the collimation factor (or C-factor) in mm/m or mm/stadia unit (be sure to confirm the units of the C-factor for your equipment), n is the number of instrument setups in the leveled section, and d_{BS_i} and d_{FS_i} are the BS and FS distances, respectively, at a given setup number i. The corrected elevation difference over a leveled section can be given as

$$\Delta h(\text{corrected}) = \Delta h(\text{observed}) + \text{Correction} \tag{3.11}$$

where Δh(observed) is the observed elevation difference.

3.3.3.1 ***Electronic Leveling*** Due to the advancement of technology, precision differential leveling is now possible electronically using digital level instrument with bar-code rods. In this type of instrument, the electronic eye does the reading instead of optical reading. A typical example of a digital level is Leica DNA03, which is capable of electronic measurement with a standard deviation per kilometer double-run (ISO 17123-2) of 0.3 mm (when used with bar-code invar rods). Leica DNA03 is considered suitable for first-order and high-precision jobs. The instrument has a distance range of 1.8–110 m for electronic measurements. In electronic leveling, it has been suggested in FGCS (2004) that a minimum of three readings with a standard deviation less than or equal to 1.0 mm be taken to obtain a complete observation to a bar-code rod.

3.3.4 Accuracy of Height Differences

Height differences should be distinguished from elevation differences: height differences are derived from the least squares adjusted heights of the leveling network points, while elevation differences are those derived from direct differential leveling measurements. The USA accuracy standards for vertical control are given (FGCC, 1993) in Table 3.3. In this table, L (km) is the approximate distance between benchmark positions traced along existing level routes (L is one-way distance in a section or the distance round the loop in the case of a loop), and the standard deviation is for the elevation difference between survey control points obtained by error propagation in a correctly weighted least squares adjustment procedure. The least squares

TABLE 3.3 Accuracy Standards for Vertical Control in the United States (Accuracy of Height Difference).

Order of Accuracy	Relative Accuracy between Directly Connected Points or Benchmarks (Standard Deviation of Elevation Difference)
First order, Class I	$\pm 0.5\,\text{mm}\sqrt{L}$
First order, Class II	$\pm 0.7\,\text{mm}\sqrt{L}$
Second order, Class I	$\pm 1.0\,\text{mm}\sqrt{L}$
Second order, Class II	$\pm 1.3\,\text{mm}\sqrt{L}$
Third order	$\pm 2.0\,\text{mm}\sqrt{L}$

adjustment procedure allowed for the modeling of some typical systematic errors and checking for blunders and gross errors in the leveling measurements. Remember that the least squares adjustment is only done after the job has satisfied some leveling field specifications (which include satisfying some section and loop misclosure specifications similar to the Canadian version in Table 3.1). The values given in Table 3.3 are accuracy standards since the compliance test of the measured leveling network will fail if systematic errors in the measurements are not thoroughly accounted for; the process of statistical blunder detection in least squares adjustment of the leveling network is to help identify and eliminate the blunders that were not detected by following the specified leveling procedures consistent with the order of the leveling.

The elevation difference accuracy pertains to all pairs of points; the standard deviations were not chosen based on any special theoretical concepts, but by the experience of the National Geodetic Survey agencies. For example, if the distance between two leveling points is 5 km, first-order, Class I accuracy of the vertical relationship between the two points will be $\pm 0.5\,\text{mm}\sqrt{5}$ or 1.1 mm.

The classification standards of the horizontal and vertical control networks in the United States are based on accuracy (or the ability of that survey to duplicate already established control values), not the observation closures within a survey. The standards take into account all the known systematic effects that may influence the survey measurements.

3.3.5 Vertical Control Surveys Examples

Example 3.1 Consider Figure 3.1, where line AD was not leveled. The accuracy of the vertical relationship between points A and D can be derived based on the leveling route A-B-C-D (10 km) as $\pm 0.5\,\text{mm}\sqrt{10}$ or 1.6 mm.

Example 3.2 Consider a differential leveling with the Leica NA2 automatic level with the telescope magnification of 32× and a compensator setting accuracy of $\sigma_v = 0.3''$ and the standard deviation of mean elevation difference of 0.7 mm/km (double run). Determine the standard deviation of elevation differences over 1 km (for single) and the section closure and the loop closure over $L = 3$ km.

Solution: Given for double leveling run, the standard deviation of mean elevation difference as 0.7 mm/km, the following can be determined.

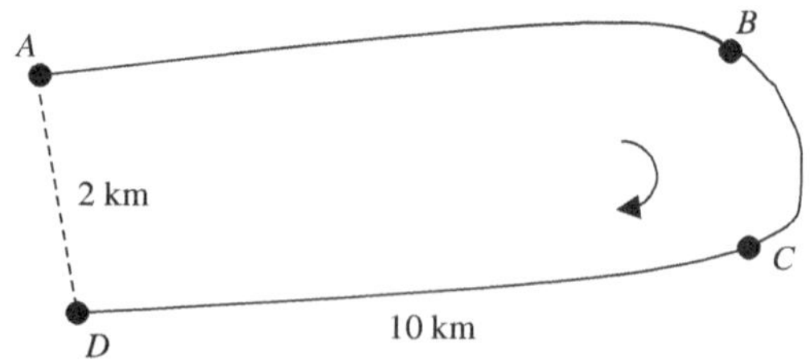

Figure 3.1 Sample leveling network.

For single run: The leveling accuracy (double run) is propagated for the mean elevation difference from Equation (3.9) as follows:

$$\bar{x} = \frac{\Delta h_{\mathrm{f}} - \Delta h_{\mathrm{b}}}{2}$$

Error propagation on this equation gives:

$$\sigma_{\bar{x}}^2 = \frac{1}{4}(\sigma_{\Delta h_{\mathrm{f}}}^2 + \sigma_{\Delta h_{\mathrm{b}}}^2)$$

Assuming $\sigma_{\Delta h_{\mathrm{f}}}^2 = \sigma_{\Delta h_{\mathrm{b}}}^2 = \sigma_{\Delta h}^2$ and simplifying $\sigma_{\bar{x}} = \frac{\sigma_{\Delta h}}{\sqrt{2}}$, where $\sigma_{\Delta h}$ is the standard deviation of single leveling (for elevation difference in one way) over 1 km; $\sigma_{\bar{x}} =$ 0.7 mm; $\sigma_{\Delta h} = 0.7\,\mathrm{mm} \times \sqrt{2}$ (or 1.0 mm/km).
Hence, from Equation (3.6), standard deviation of elevation differences over 1 km (for single run), $\sigma_{\mathrm{ran}} = 1.0\,\mathrm{mm/km}$.
Section closure (Equation (3.6)):

$$\Delta_{\mathrm{section}} = \sqrt{2} \times \sigma_{\mathrm{ran}} \sqrt{L} \times 1.96 \quad \sigma_{\mathrm{ran}} = 1.0\,\mathrm{mm/km}; \quad L = 3\,\mathrm{km}$$

$$\Delta = \sqrt{2} \times 1.0\sqrt{3} \times 1.96\,\mathrm{mm} \quad (\text{or } 4.8\,\mathrm{mm})$$

Loop closure (Equation (3.4)):

$$\Delta = \sigma_{\mathrm{ran}} \sqrt{L} \times 1.96 \quad \sigma_{\mathrm{ran}} = 1.0\,\mathrm{mm/km}; \quad L = 3\,\mathrm{km};$$

$$\Delta = 1.0\sqrt{3} \times 1.96\,\mathrm{mm} \quad (\text{or } 3.4\,\mathrm{mm})$$

Example 3.3 The error of 5 mm in difference in elevation between the third-order benchmarks (with an average separation of 200 m) is usually accepted as the maximum allowable error at 95% confidence level (refer to Blachut et al., 1979). Assuming the standard deviation of higher order leveling is twice as high as the lower order leveling, determine the standard deviation of leveling a 1-km section based on first-order procedure.

Solution: Based on the concept of confidence intervals (Section 2.8.3), the precision of estimate at 95% confidence can be given from Equation (2.18) or (2.15) as

$$\text{Maximum error(at 95\%)} = (\mathrm{SE}) \times z_{\alpha/2=0.025} \quad (z_{\alpha/2=0.025} = 1.96)$$

$$5\,\mathrm{mm} = (\mathrm{SE}) \times 1.96 \rightarrow \mathrm{SE} = 2.55\,\mathrm{mm}$$

Since the average separation between the third-order benchmarks is 200 m (or 0.2 km), the error in one section (2.55 mm) will be propagated over five independent sections making up 1 km to obtain the propagated error over 1 km as $2.55\,\mathrm{mm}\sqrt{5}$. The standard deviation of third-order leveling is due to three sources:

- Standard deviation (σ_3) of the third-order leveling
- Standard deviation (σ_2) of the second-order leveling
- Standard deviation (σ_1) of the first-order leveling.

Total standard deviation $\sigma^2_{\text{Total}} = \sigma_1^2 + \sigma_2^2 + \sigma_3^2$
Since the standard deviation of the higher order leveling is twice as high as the lower order leveling, the following relationships can be established:

$$\sigma_2 = 2\sigma_1;\ \ \sigma_3 = 2\sigma_2 \text{ or } (\sigma_3 = 4\sigma_1)$$

Substituting into total standard deviation gives $\sigma^2_{\text{Total}} = 21\sigma_1^2 \rightarrow \sigma_1 = \frac{\sigma_{\text{Total}}}{\sqrt{21}}$.

Substitute $\sigma_{\text{Total}} = 2.55\,\text{mm}\sqrt{5} \rightarrow \sigma_1 = \frac{2.55\sqrt{5}}{\sqrt{21}} = 1.24\,\text{mm/km}$.
The standard deviation of leveling a 1-km section based on first-order procedure is 1.24 mm/km.

Example 3.4 Referring to Table 3.2, determine the elevation difference between BMA and BMB and apply the corrections due to collimation errors on the elevation difference, assuming the *C*-factor is +0.5 mm/stadia unit and the stadia factor is 100. Express the difference in elevation in meters.

Solution (forward run):

$$\Delta h_{\text{f}} = \text{Sum(BS)–Sum(FS)}$$
$$= 36,566 - 33,133 \rightarrow 03,433\,\text{mm (or 0.3433)}$$

Difference in Sum of Stadia Intervals

= Sum(BS Stadia Interval) – Sum(FS Stadia Interval)

(Remember that the stadia distance divided by the stadia factor gives the stadia interval.)

Difference in Sum of Stadia Intervals = 446 – 451 → –5 stadia unit

C-correction = +0.5 mm/stadia unit × –5 stadia unit

C-correction = –0.3 mm(or – 0.0003 m)

Adjusted Elevation difference Δh(adjusted) = 0.3433 – 0.0003 → **0.3430 m**

Total Forward Distance = 44.6 + 45.1 → 89.7 m

Example 3.5 Continuing from Example 3.4, if the backward leveling run gives the corrected elevation difference as –0.3420 m, determine the mean elevation difference between BMA and BMB and check if the leveling satisfies the first-order specification.Referring to Equation (3.9):

$$\text{Mean} = \frac{\Delta h_{\text{f}} - \Delta h_{\text{b}}}{2} \tag{3.9}$$

where $\Delta h_{\text{f}} = 0.3430\,\text{m}$ and $\Delta h_{\text{b}} = -0.3420\,\text{m}$, giving the mean elevation difference as 0.3425 m. The misclosure is 0.001 m (or 0.3430 − 0.3420 m) and the allowable discrepancy over the distance of 89.7 m is 1.2 mm or $4\,\text{mm}\sqrt{0.0897}$; it can be seen that the first-order specification is satisfied since the misclosure of 1 mm is less than the allowable discrepancy of 1.2 mm.

Example 3.6 Canadian Special Order Leveling procedures require that "… difference between backsight and foresight distances at each setup and their total for each section not to exceed 5 m … " with maximum lengths of sight of 50 m. Normally, invar double-scale rods and a level [$M \cdot 40\times$, sensitivity · 10"/div] with parallel-plate micrometer are used. How well would the lengths of sight have to be determined (i.e., σ_s)? How would they be measured? Interpret "not to exceed" as being at 99%.

(Reproduced by with permission of CBEPS.)

Suggested Solution: Maximum discrepancy between backsight and foresight (Δs) is based on the concept of confidence regions in Equation (2.52), where the maximum discrepancy between backsight and foresight distances ($\Delta s = 5$ m) can be given as being equivalent to the 99% confidence interval, which is given as

$$\Delta s = \sigma_{\Delta s}\sqrt{\chi^2_{\text{df}=1,\alpha=0.01}} \rightarrow \Delta s = \sigma_{\Delta s}\sqrt{6.63}$$

$$5 = \sigma_{\Delta s} \times 2.575 \rightarrow \sigma_{\Delta s} = 1.9417$$

where $\sigma_{\Delta s}$ is the standard deviation for the discrepancy Δs. Applying the error propagation law on the discrepancy expressed as $\Delta s = s_{\text{b}} - s_{\text{f}}$ (with equal contribution from backsight and foresight distances, s_{b} and s_{f}, respectively):

$$\sigma_{\Delta s} = \sigma_s\sqrt{2}$$

Substituting this into $\sigma_{\Delta s}$ above gives

$$\sigma_s = \pm\frac{1.9417}{\sqrt{2}} \rightarrow \mathbf{\pm 1.373\,m}$$

This value was obtained by error propagation backward from discrepancy. The standard deviation of sight measurement should therefore be less than 1.4 m. How they would be measured can be given as follows:

- Careful pacing will give an accuracy of ±1/100 or about ±0.4 m if consistent over uniform terrain.
- Stadia method will give an accuracy of ±1/300 or better.
- Taping will give an accuracy of ±1/1000 or better.

These can be applied to the length of sight to see whether the method is appropriate, keeping in mind the conditions, especially the nature of the terrain.

3.4 STANDARDS FOR HORIZONTAL CONTROL SURVEYS

3.4.1 Accuracy Standards for Traditional Horizontal Control Surveys

The various measurements for horizontal geodetic control demand different levels of positional accuracy. In Canada, horizontal control surveys are classified as first, second, third, or fourth order according to standard of accuracy (NRC, 1978). A survey station of a network is classified according to whether the semi-major axis (a) of the 95% confidence region, with respect to other stations of the network, is less than or equal to

$$a = C(d + 0.2)\,\text{cm} \tag{3.12}$$

where d is the distance (in kilometers) to any station and C is a factor assigned according to the order of survey. Example of accuracy standards for horizontal control surveys for Canada is given (NRC, 1978) in Table 3.4. For example, from the table, if two stations are 1 km apart, the semi-major axis of the 95% confidence region of one station relative to the other must be less than or equal to $a = 2.4$ cm in order to classify the stations as first order. Care should be taken to ensure that neighboring survey stations, particularly those not directly connected by measurements, meet this criterion.

The testing of a network computation can be done by computing the semi-major axis values (a_{95}) of the relative error ellipses between pairs of points based on Equation (2.26) and the relative eigenvalues (λ_1) computed by using Equation (2.41). The computed a_{95} values are then compared with the "a" values (Equation (3.12)) based on the accuracy standards given in Table 3.4; if the computed a_{95} values are less than the given standards, then the associated network computations are said to satisfy the given accuracy standards.

TABLE 3.4 Accuracy Standards for Horizontal Control Surveys in Canada.

Order	For Distance d = 1.0 km		
	C	a (cm)	Ratio
First	2	2.4	1/41,700
Second	5	6.0	1/16,700
Third	12	14.4	1/6,900
Fourth	30	36.0	1/2,800

TABLE 3.5 Horizontal Accuracy Standards in the United States.

Order of Accuracy	Maximum Closure (1:*r*)
First order	1:100,000
Second order, Class I	1:50,000
Second order, Class II	1:20,000
Third order, Class I	1:10,000
Third order, Class II	1:5,000

Example of accuracy standards for horizontal control surveys in the United States is given (FGCC, 1993) in Table 3.5. The standards are expressed as distance accuracy ratio 1:r, which is computed from a minimally constrained, correctly weighted, least squares adjustment by

$$r = \frac{d}{S} \tag{3.13}$$

where S is the propagated standard deviation of the distance between survey points obtained from the least squares adjustment and d is the distance between the survey points. Using distance accuracy to represent the accuracy of horizontal coordinates is like saying that the coordinates of that control point bear a distance relation of the specified accuracy to the coordinates of all other points in the horizontal control network. For example, if a distance of 10,000 m is measured between two points, the first-order horizontal accuracy of the distance (from Table 3.5) is 10,000 m/100,000 (or 0.1 m).

Consider the distance measurement in Figure 3.2, where there is no direct connection between network points B and C:

The first-order accuracy of distance AB is 15,000 m/100,000 or 0.15 m.

The first-order accuracy of distance AC is 10,000 m/100,000 or 0.10 m.

The first-order accuracy of distance BC can be derived as follows:

$$\sqrt{(0.15)^2 + (0.10)^2} \quad \text{or} \quad (0.18\,\text{m})$$

Generally, *first-order* (or *Primary*) control is used to establish geodetic points and to determine the size, shape, and movements of the earth; *second order*, *Class I* (or *Secondary*) or *second-order* control is used for network densification in urban areas

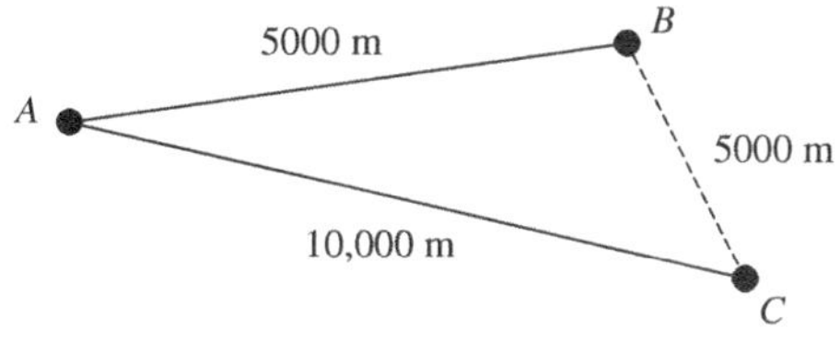

Figure 3.2 Indirect distance measurement.

and for precise engineering projects; and lower order controls are used for network densification in nonurban areas and for surveying and mapping projects.

3.4.2 Accuracy Standards and Specifications for Traverse Surveys

Engineering and construction traverse surveys are normally specified and classified based on the horizontal (linear) point closure ratio standard. The minimum closure accuracy standards (traverse misclosure or precision) for N number of traverse angle stations are summarized in Table 3.6 (FGDC, 2002). The closure standard for low-precision engineering construction is typically of fourth order.

Two types of traverse discussed in this section are closed and open traverses. Closed traverses can be divided into loop and connecting traverses as discussed as follows:

1. *Loop traverse.* In this traverse type, position misclosure usually reveals measurement blunders and internal loop errors, but will not disclose systematic errors or external inaccuracies in the control point coordinates. The closure of the traverse can be given as $(n-2)\times 180°$ for n number of internal angles and $(n+2)\times 180°$ for n number of external angles.
2. *Connecting traverse.* This traverse type usually starts on a station of known position and terminates on a different station of known position. The traverse is capable of detecting and eliminating systematic errors and position inaccuracies as well as blunders and accidental errors of measurements.

Open traverses are very seldom used in topographic surveys. They start on known stations and terminate on stations of unknown positions, and they usually provide no checks to determine blunders, accidental errors, or systematic errors that may occur in measurements.

In a traverse survey, the ratio of the resultant error of closure for the traverse to the total length of the traverse provides an indication of the accuracy of the survey on a local scale and is often referred to as the *ratio of misclosure* (ROM) or the *relative accuracy ratio.* For example, if the resultant closure of a traverse is 0.20 m for a traverse having a total length of 2000 m, the ROM for this traverse is 0.20 m/2000 or 1 part in 10,000 (or 1:10,000). This provides the relative accuracy of the traverse but not the absolute accuracy in position for each station in the traverse. The techniques of error propagation are employed to determine the covariance matrix for each point in the traverse in order to estimate the accuracies possible at specific traverse stations.

TABLE 3.6 Minimum Closure Accuracy Standards for Traverse Surveys.

Closure Standard	First Order	Second Order		Third Order		Fourth Order
		Class I	Class II	Class I	Class II	
Distance ratio	1:100,000	1:50,000	1:20,000	1:10,000	1:5,000	1:2,500
Angle closure	$2\sqrt{N}$ sec s	$3\sqrt{N}$ sec s	$5\sqrt{N}$ sec s	$10\sqrt{N}$ sec s	$20\sqrt{N}$ sec s	$60\sqrt{N}$ sec s

To achieve a desired relative accuracy for a given traverse, specifications are provided to govern the traverse field operations and the types of equipment allowed.

For a connecting (nonloop) traverse, the resultant closure is caused by random errors in observations as well as uncorrected systematic errors in distance and direction measurements. When blunders or uncorrected systematic errors in distance or directions are present, the closure and consequent relative accuracy ratio will be very large. For a loop traverse, the resultant closure depends on random errors in observations and uncorrected systematic errors in angles or directions. Any systematic errors in distance-measuring equipment will cancel out and will not be revealed by the mathematical closure of the traverse. Moreover, it is possible for the entire polygon to be rotated about the starting point (due to a constant systematic error in a direction or angle measurement) without any noticeable effect on the traverse computations. This effect, however, will only be revealed if there is a second tie to a line of known bearing or azimuth. Generally, it can be said that in loop traverse computation, error of closure cannot detect systematic errors in distances, which then means that error of closure does not check the accuracy of the work but the precision. In fact, the error of closure in this case will be the same whether systematic errors in distances are corrected or not.

In a traverse survey, the horizontal control standard is a number corresponding to the radius of a relative error circle (or semi-major axis of the confidence ellipse) with a probability of 0.95. The accuracy of a traverse survey can be categorized into two depending on whether one is evaluating networks or local surveys. For network accuracy, the error circle (or the semi-major axis of the confidence ellipse) is determined by error propagation in a least squares adjustment between the traverse points and the geodetic datum (such as the Canadian Active Control System (CACS)). For local accuracy, the error circle (or the semi-major axis of the confidence ellipse) is determined by error propagation in a least squares adjustment between known control points connected by the local survey.

Remember that taking one survey station of the project as the origin of the coordinate system and one line to another survey station to provide orientation forms a local grid coordinate system. This local system should always be connected by additional surveys to points of the national or regional geodetic control network, which are usually of higher order, even though the national or regional network may sometimes be less accurate than the local network from the point of view of relative positioning. Some of the reasons for doing so are given as follows:

1. To calculate some geodetic corrections to local observations, for example, convergence of meridians in gyro azimuth measurements.
2. To calculate transformation parameters between local and national systems.
3. To integrate local surveys into the regional mapping and geographic information system for future applications.

In township surveying, relocation work can be achieved in many ways, including using large-scale maps supplemented by original field survey sketches and using coordinates of all points. In some cases, coordinates of points are used as the only evidence for the positioning and relocation of land details, including property boundaries. This, however, requires higher density and accuracy requirements of horizontal control for the area. The geodetic horizontal control points are usually spaced in such a way that surveyors are able to tie detailed surveys with one or two instrument setups; the orders of control are dependent on the spacing between the control points with first order having the longest and the lowest order having the shortest. The concept of orders of horizontal control is discussed by Blachut et al. (1979) as being based on the need for surveyors to be able to locate corners of properties in urban areas to within 25 mm (taken as a positional error at 95% confidence level). The 25 mm is accepted (Blachut et al., 1979) as the maximum positional error in a relocation survey. In this case, the accuracy of a surveying network is fully defined if errors of relative positions between any two points in the network are known at a certain confidence level (usually 95% confidence level) as required in accuracy standards (refer to Table 3.4). In relocating a point by using independent coordinate surveys, the maximum positional error (at 95% confidence level) consists of three partial errors (Blachut et al., 1979):

1. Errors of relative positioning of the control network points (given as the covariance matrix of the points) if the original and relocation surveys are tied to different points of the network. If the same points of the network are used in both the original and the relocation surveys, the errors in this step will be zero.
2. Errors of the original connecting survey.
3. Errors of the connecting surveys in the relocation procedure (based on the order of survey).

If each factor of the aforementioned list has the same influence, the accuracy of the control surveys would be in the order of $25/\sqrt{3}$ mm (or 14 mm) in terms of the semi-major axis value of the relative error ellipse at the 95% confidence level. According to Blachut et al. (1979), if 200 m is accepted as an average spacing between control points in urban areas, the required relative accuracy becomes 14/200,000 (or 1:14,000) for the lowest order control. The higher order control points are more accurate so that when held fixed for the adjustment of lower order surveys, the lower order control will not be significantly distorted as a result. The order-based classifications with listed accuracies of control networks are recommended for use with purely numerical system of the integrated survey system, based on coordinates of boundaries as the primary evidence in property surveys. If the third-order job (refer to Table 3.4 with $C = 12$) is satisfied in the connecting surveys in the relocation procedure and assuming 150 m is the average spacing between the survey points, the accuracy (at 95% confidence level) of the survey can be calculated by using Equation (3.12) as 4.2 cm. If each of the aforementioned factors will have approximately the same influence, the total maximum positional error (at 95% confidence level) in the relocation survey can be determined through error propagation as $4.2\sqrt{3}$ mm or 7.3 cm. The relative positioning error in terms of the semi-major axis of the standard error ellipse can then be determined from Equation (3.16), giving 7.3/2.45 cm or 3 cm. The limiting accuracy for relocating a point by using independent coordinate surveys is then 3 cm.

Assuming the relocation surveys were tied to the second-order control network with an average spacing of 3 km, the expected relative positional error between a pair of control points will be calculated from Equation (3.12) as 16 cm. This error will propagate to any point in the traverses even if the connecting traverses are errorless.

3.4.3 Accuracy Standards and Specifications for GNSS Surveys

Accuracy standards for Global Navigation Satellite System (GNSS) surveys are not based on the technical training or ability of the surveyor but are based on the capabilities of GNSS measurement systems. The original GPS geodetic control networks classifications are based on distance-dependent accuracy standards, such as

$$a = \sqrt{e^2 + (0.1 \times k \times L)^2} \tag{3.14}$$

where a is the maximum allowable error (geometric relative position accuracy standard) in centimeters between a pair of control points at 95% confidence level, e is the base error (at 95% confidence level) from 0.3 cm (highest order) to 5 cm (lowest order), and k is minimum geometric relative position accuracy standard (at 95% confidence level) from 0.01 ppm (highest order) to 100 ppm (lowest order), and L is the distance in kilometers between any two stations. Equation (3.14) applies to both one-dimensional traditional terrestrial techniques and three-dimensional GPS relative positioning techniques. The survey point in a given network is classified based on whether the propagated error of the station at 95% confidence level is less than or equal to the maximum allowable error (a) chosen for the project; the standard deviation that is independently determined from the survey is multiplied by a factor of 1.96 (in one-dimensional case) or 2.79 (in three-dimensional case) in order to convert it into error at 95% confidence level. The base error (e) is usually associated with the sources of errors, such as antenna setup (plumbing, centering, and measurement of height of antenna phase center above the station mark); antenna phase center stability; and signal multipath. The parts per million (ppm) values of constant k in Equation (3.14) can be given as

$$k = \left(\frac{\text{Horizontal separation of a pair of control points}}{\text{Relative positional error of a pair of control points}} \right) \tag{3.15}$$

Specifications for GPS field procedures will be common for all precision surveys, no differences in the field procedures for higher and lower order surveys. Note that if one or more of the stations in a project network are continuously reoccupied during each session, these stations are generally called "master" or "fiducial" stations. In this observing scheme, the observations for the "master" stations are common to most or all the other observing sessions for the project. Some of the specifications for GPS field survey procedures were extracted from FGCC (1989) and given as shown in Table 3.7.

TABLE 3.7 Specifications for GPS Field Survey Procedures.

Procedures	Items
1. Two frequency (daylight) observations required	Yes
2. Recommended number of receivers observing simultaneously, not less than	5 to 4
3. Period of observing session (observing span) (with 4 or more simultaneous satellite observations not less than 25% of the observing period):	Not less than 240 to 120 min Not less than 180 to 60 min
• Processing carrier phase data using single, double, nondifferencing, or other comparable precise relative positioning techniques	
• Continuous observations (data collected that have no breaks involving all satellites or those with occasional breaks for individual satellites caused by obstructions)	
4. Data sampling rate, maximum time interval between observations	15–30 s
5. Maximum angle above horizon for obstructions such as buildings, trees, fences, human beings, vehicles	10–20°
6. Antenna setup with independent heavy weight plumb bob check (if optical plummet used in centering) is required	B-M-E to B-E
• Number of antenna phase center height measurements per session, not less than (measured in meters and feet at the beginning (B), midpoint (M), and end (E) of each station occupation)	
7. Meteorological observations (at beginning (B), midpoint (M), and end (E))	B-M-E to B-E 30–60 min
• Per observing session, not less than	
• Sampling rate (measurement interval), not less than:	

3.5 UNIFIED STANDARDS FOR POSITIONAL ACCURACY

As discussed in Section 3.3.2, the usual accuracy standards for traditional triangulation networks or traverse surveys used to be based on proportional distance-dependent standards; and the accuracy of GPS surveys are based on a different standard using positional covariance matrices. In order to allow comparison of coordinate values from different survey techniques, the National Geodetic Survey of the United States (FGCS, 1998) and Geodetic Survey of Canada (1996) came up with a unified methodology for reporting the accuracy of horizontal and vertical coordinate values. The unified accuracy standards are based on two types of accuracy that can be estimated for geodetic coordinates of latitude, longitude (horizontal coordinates), and ellipsoidal height: *network accuracy* and *local accuracy*.

3.5.1 Network Accuracy

Network accuracy is the absolute accuracy (or station error ellipse) of the coordinates of a point at the 95% confidence level, with respect to the geodetic datum. It is an indication of how accurately a point is positioned relative to the geodetic datum, such as the Canadian Spatial Reference System (CSRS), the National Spatial Reference System (NSRS) for the USA or the European Terrestrial Reference System 89 (ETRS89). The network accuracy provides the positional tolerance associated with a set of computed coordinates of a point. For example, the network accuracy of a newly positioned point in CSRS will depend on the network accuracy at the known point and the relative accuracy within the new work. Since points in the CACS and the Canadian Base Network (CBN) may be considered to approach an error-free realization of the CSRS, the accuracy with respect to these monumented points in the national CSRS network may be interpreted as an expression of network accuracy. Network accuracy, therefore, can be considered a measure of how well the given coordinates approach an idea, error-free datum. The accuracies of the horizontal coordinates and ellipsoidal heights of points in the CSRS are computed from the elements of a covariance matrix of the adjusted parameters. The covariance matrix is from the least squares adjustment where the known CSRS control coordinate values have been weighted using their one-sigma network accuracies. The semi-axes (major axis, a_{95}; and the minor axis, b_{95}) of the 95% confidence ellipse representing the network accuracy at a given point are generally computed as follows:

$$a_{95} = 2.45a \tag{3.16}$$

$$b_{95} = 2.45b \tag{3.17}$$

where a and b are the semi-major and semi-minor axes of the standard absolute error ellipse for the given two-dimensional network point.

3.5.2 Local Accuracy

Local accuracy of a control point is a number (e.g., mean, median, etc.) that represents the uncertainty, at the 95% confidence level, in the coordinates of a control point with respect to the coordinates of other directly connected, existing primary control points. The coordinates of the primary control points are weighted by using their one-sigma network accuracies in the least squares adjustment of the network measurements. Local accuracy is an indication of how accurately a point is positioned with respect to other adjacent points in the network. It provides practical information for users conducting local surveys between control monuments of known positions. For horizontal coordinate accuracy of a point, the local accuracy of the point is the average of the major semi-axes of the 95% relative confidence ellipses between the point and other adjacent points. For ellipsoidal height accuracy, the local accuracy is the average of the 95% relative confidence intervals between the point and other adjacent points. Note that high or low individual local accuracies are not considered in computing the average local accuracy of a control point.

Local accuracy depends on the positioning method used to establish a point. If very precise instruments and techniques are used, local accuracies related to the point will be very good. Local accuracy is best adapted to check relations between nearby control points. For example, a surveyor checking closure between two CSRS points is mostly interested in a local accuracy measure. The local accuracy is especially important for surveys that are designed to meet high-accuracy requirements such as surveys for establishment of a precision primary network, deformation measurement investigations (crustal motion, subsidence monitoring, motion of structures, etc.), and other special precision surveys.

3.5.3 Accuracy Classification

The network and local accuracies may be classified by comparing the 95% confidence ellipse for horizontal coordinate accuracy and the 95% confidence interval for ellipsoidal height accuracy, against a set of standards. To classify control points in a survey, the survey must be properly connected to existing datum points with established network accuracy values, and the control points must be verified as being consistent with all other points in the network, not merely those within that particular survey. The procedure leading to classification involves four steps (FGCS, 1998):

1. Survey measurement systems (measurements, field records, sketches, and other documentations) are ensured to be in accordance with specifications. If specifications are not followed, the expected accuracy may be modified at this stage.
2. Minimally constrained, least squares adjustment of survey measurements is performed to ensure correct weighting of observations and correct removal of possible blunders.
3. Local and network measures computed by random error propagation are used in determining the provisional accuracy. These accuracy measures are to be computed by weighting datum values in accordance with the network accuracies of the existing network control.
4. The survey accuracy is checked by comparing minimally constrained adjustment results with established control. This comparison takes into account the network accuracy of the existing control, as well as systematic effects such as crustal motion or datum distortion. If the comparison fails at a 95% confidence level, then both the survey and the network measurements must be scrutinized to determine the source of the problem.

The classification standard for geodetic networks is based on accuracy. The accuracies are categorized separately according to Table 3.8 for geodetic elements, such as horizontal, ellipsoid height, and orthometric height (Geodetic Survey of Canada, 1996). The standards apply to both conventional and GPS geodetic network surveys. In the case of GPS surveys, the surveys must be performed by relative positioning techniques in which two or more receivers are simultaneously collecting carrier phase measurements. It should also be mentioned that long observation times

TABLE 3.8 Accuracy Classification Standards (Horizontal, Ellipsoid Height, and Orthometric Height).

Accuracy Classification	Upper Class Boundary (Less Than or Equal to) 95% Confidence
1 cm	0.010 m (or 0.005–0.010 m)
2 cm	0.020 m (or 0.010–0.020 m)
5 cm	0.050 m (or 0.020–0.050 m)
1 dm	0.100 m (or 0.050–0.100 m)
2 dm	0.200 m (0.100–0.200 m)
5 dm	0.500 m (0.200–0.500 m)
1 m	1.000 m (0.500–1.000 m)
2 m	2.000 m (1.000–2.000 m)
5 m	5.000 m (2.000–5.000 m)
10 m	10.000 m (5.000–10.000 m)

are necessary to establish geodetic control. Techniques such as rapid static, fast static, kinematic, and real-time kinematic are not acceptable to establish control that meets the geodetic-level standards, such as millimeter accuracies.

The National Geodetic Survey of the United States of America uses similar accuracy standards (FGCS, 1998) as Canada. Their standards include the following classes: 1 mm (or 0.001 m), 2 mm (or 0.002 m), and 5 mm (or 0.005 m). The classification standards are recommended for use during the survey design and evaluation phases of a positioning project. The classification process provides an opportunity to assess the reliability of the results of a positioning project and to assign accuracy classes accordingly. The global and regional geodynamics measurements, deformation measurements, and some precision engineering surveys will require that 1-mm to 5-mm local accuracy standards are met. When providing geodetic point coordinates, a statement should be provided that the data meet a particular accuracy standard for both the local accuracy and the network accuracy. For example, it can be stated that these geodetic control data meet the 2-cm local accuracy standard for the horizontal coordinate values and the 5-cm local accuracy standard for the vertical coordinate values (heights) at the 95% confidence level. A similar statement should also be provided while reporting the network accuracy.

Example 3.7 Consider the network in Figure 3.3 in which two control points H and M are related to a datum (CSRS) point CSRS-1. If the network accuracy of station H is $N_{\mathrm{H}} = 3$ unit and that of station M is $N_{\mathrm{M}} = 4$ unit, determine the local accuracy between H and M represented as $L_{\text{H-M}}$.

Since points H and M are not connected (as shown in Figure 3.3), only the local accuracy between them can be determined as follows:

$$L_{\mathrm{H-M}} = \sqrt{(N_{\mathrm{H}})^2 + (N_{\mathrm{M}})^2}$$

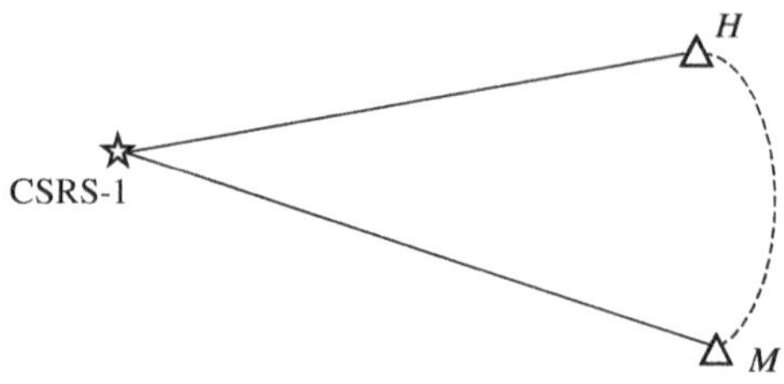

Figure 3.3 Local accuracy between control points.

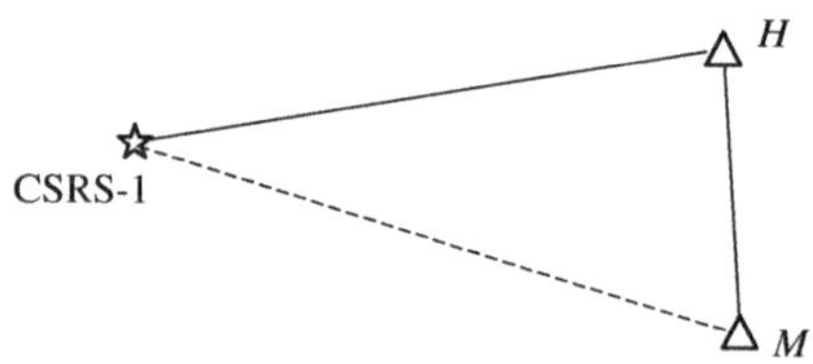

Figure 3.4 Network accuracy between a control point and a datum.

or

$$L_{\text{H-M}} = \sqrt{(3)^2 + (4)^2} \rightarrow 5\,\text{units}$$

Example 3.8 Consider the network in Figure 3.4 in which control point H is well connected to the CSRS point CSRS-1 with a network accuracy of $N_{\text{H}} = 3$ unit and the local accuracy from point H to M as $L_{\text{H-M}} = 5$ units. Calculate the network accuracy for station M.

Points CSRS-1 and M are not connected. The network accuracy can be given as

$$N_{\text{M}} = \sqrt{(N_{\text{H}})^2 + (L_{\text{H-M}})^2}$$

$$N_{\text{M}} = \sqrt{(3)^2 + (5)^2} \rightarrow 5.83\,\text{units}$$

Example 3.9 A new survey point is tied to one of the national geodetic control monuments using GPS RTK survey procedure. The new point is 5 km away from the control monument whose published network accuracy is 0.030 m; the specification for the RTK survey is such that the standard deviation of a baseline is 1 cm ± 2 ppm. Determine the local accuracy, the network accuracy, and the accuracy classification for the new survey point.

Solution:

$$\text{Standard deviation over 5 km} = \sqrt{(10)^2 + (2 \times 10^{-6} \times 5 \times 10^{6})^2} = 14.1\,\text{mm}$$

From Equation (3.17), the local accuracy = 14.1 mm × 2.45 = 34.6 mm (or 3.5 cm)

$$\text{Network accuracy} = \sqrt{(34.6)^2 + (30.0)^2} = 45.8\,\text{mm} \quad (4.6\,\text{cm})$$

From Table 3.8, it can be seen that the survey satisfies horizontal network accuracy of 5 cm and a local accuracy of 5 cm.

3.6 MAP AND GEOSPATIAL DATA ACCURACY STANDARDS

The map and geospatial data accuracy standards are designed to allow users of maps and geospatial data that comply with the standards to determine if those maps are accurate enough for them to use. These standards apply to all features on maps and spatial data but do not apply to abstract features such as cadastral boundaries, survey networks, or geodetic network points. Three map and geospatial data accuracy standards are common:

- The National Map Accuracy Standards (NMAS) by the U.S. Bureau of Budget (1947)
- The American Society for Photogrammetry and Remote Sensing (ASPRS) standard by the ASPRS specifications and standards committee (1990)
- The National Standard for Spatial Data Accuracy (NSSDA) by the FGDC (1998b)

In each of the standards, the accuracy of dataset is checked by comparing coordinate values of locations in the test dataset with coordinate values of locations that can be assumed to be the same in the independent source of higher accuracy, such as geodetic terrestrial surveys, GPS surveys, and maps of larger scale and better accuracy. It is recommended (ASPRS specifications and standards committee, 1990; FGDC, 1998b) that at least 20 well-defined and well-distributed points by independent source of higher accuracy be used as checkpoints for comparing the coordinate values. If ground survey control points are to be used as independent source of higher accuracy, according to the NMAS, those points must be established to an accuracy of three times the allowable error of plotted points. The typical features whose locations are checked are buildings, roads, contours, and spot elevations.

The three map and geospatial data accuracy standards are different in their statistical means and methodology for presenting accuracies. The usually reported accuracy value based on the standards assumes that systematic errors have been eliminated as best as possible so that the accuracy value reflects all uncertainties, including those introduced by geodetic control coordinates, map compilation, data conversion, and data manipulation (FGDC, 1998b). The NSSDA, however, provides the best language for reporting accuracy, which makes it easier for users to evaluate the quality of their dataset. This standard, however, is not really a true map standard in the same sense as in the NMAS and ASPRS standards, but it is considered a general guideline that provides a well-defined statistical estimation and testing methodology for evaluating and reporting positional accuracy of points on maps and in digital geospatial data. The other main elements of the three accuracy standards are given in Tables 3.9 and 3.10.

3.6.1 Positional Accuracy Determination Based on NSSDA

On the basis of NSSDA, positional accuracy is usually determined in two separate components: horizontal accuracy and vertical accuracy. The horizontal accuracy is determined by comparing the planimetric (x, y) coordinates of well-defined points in the dataset with the (x, y) coordinates of the same points from an independent source of higher accuracy (at 95% confidence level) and can be expressed as (FGDC, 1998b)

$$\text{Accuracy}_x = 2.4477 \times \text{RMSE}_x \tag{3.18}$$

$$\text{Accuracy}_y = 2.4477 \times \text{RMSE}_y \tag{3.19}$$

$$\text{Accuracy}_\text{h} = 2.4477 \times \text{RMSE}_\text{h}/\sqrt{2} \tag{3.20}$$

where the value 2.4477 is obtained from the Chi-square statistical distribution $(\sqrt{\chi_{\text{df}=2,\alpha=0.95}})$ for the degrees of freedom df = 2 and the lower tail area $\alpha = 0.05$; Accuracy_x and Accuracy_y are the accuracies of x and y coordinates, respectively; Accuracy_h is the horizontal positional accuracy;

$$\text{RMSE}_x = \sqrt{\frac{\sum_i^n (x_{\text{map},i} - x_{\text{ground},i})^2}{n}} \tag{3.21}$$

$$\text{RMSE}_y = \sqrt{\frac{\sum_i^n (y_{\text{map},i} - y_{\text{ground},i})^2}{n}} \tag{3.22}$$

$x_{\text{map},i}$, $y_{\text{map},i}$ are the coordinates of the ith checkpoint in the map; $x_{\text{ground},i}$, $y_{\text{ground},i}$ are the coordinates of the ith checkpoint in the independent source of higher accuracy; n is the number of checkpoints tested; and i is an integer ranging from 1 to n;

$$\text{RMSE}_\text{h} = \sqrt{\frac{\sum_i^n (x_{\text{map},i} - x_{\text{ground},i})^2 + (y_{\text{map},i} - y_{\text{ground},i})^2}{n}} \tag{3.23}$$

or

$$\text{RMSE}_\text{h} = \sqrt{(\text{RMSE}_x)^2 + (\text{RMSE}_y)^2} \tag{3.24}$$

TABLE 3.9 Main Features of NMAS, ASPRS Accuracy Standard, and NSSDA – Part I.

	NMAS	ASPRS Accuracy Standards	NSSDA
Scope	Suitable for large- and small-scale photogrammetric mapping; focused on paper or hardcopy maps with accuracy values based on the published map scales	Suitable for large-scale topographic and engineering-grade maps (maps of 1:20,000 scale or larger)	Suitable for all types of maps and geospatial data (digital or printed form) derived from aerial photographs, satellite imagery, ground surveys, or maps and can be used for map scales smaller than 1:20,000
Methodology (how accuracies are estimated)	Accuracy is based on the residual between position of a feature on a hardcopy map and its corresponding spatial position on the earth	It uses the statistical root mean square error (RMSE) to estimate positional accuracy for x, y, z coordinate values, individually; at least 20 well-defined and well-distributed points by an independent source of higher accuracy are used in computing RMSE	It uses RMSE to estimate positional accuracy for x, y, z coordinate values, individually; at least 20 well-defined and well-distributed points by an independent source of higher accuracy are used in computing RMSE
Confidence level of accuracy	Based on 90% confidence level for both horizontal and vertical	Based on RMSE (or one standard deviation), which can be scaled to 95% confidence level	Positional accuracy is reported in ground distances at 95% confidence level
Sample accuracy reporting	This map complies with NMAS of 1947 for horizontal accuracy (or for vertical accuracy or for both)	This map was compiled to meet the ASPRS standard for Class (I, II, III) map accuracy	Tested___(meters, feet) horizontal accuracy at 95% confidence level, ___ (meters, feet) vertical accuracy at 95% confidence level

TABLE 3.10 The Main Features of NMAS, ASPRS Accuracy Standard, and NSSDA – Part II.

	NMAS	ASPRS Accuracy Standards	NSSDA
Pass/fail criterion for accuracy of horizontal locations	Threshold accuracy values are defined at map units Residuals between measured checkpoints and mapped features not to be more than 0.8 mm or 1/30″ for map scales larger than 1:20,000; and not more than 0.5 mm or 1/50″ for map scales of 1:20,000 or smaller	Threshold accuracy values are defined at ground units Maximum allowable RMSE or accuracy limiting RMSE (in meters) range from 0.0125 to 5.00 for map scales 1:50 to 1:20,000, respectively, for Large-scale maps, Class I Class II has RMSE values twice as those allowed for Class I maps; Class III has three times RMSE values allowed for Class I	Does not depend on map scales and does not define threshold accuracy values. It provides statistical measure but does not specify a pass/fail RMSE Data and map producers are expected to determine what accuracy exists or is achievable for their data and report it according to NSSDA
Pass/fail criterion for accuracy of vertical locations of well-defined points	The following are applicable on all publication scales for well-defined points: *For contour maps*: within one-half of contour interval (CI) (and within one full CI at 100% confidence level) *For spot elevations*: within one-fourth of CI (and within one-half of CI at 100% confidence level)	*Contour maps*: maximum allowable errors (limiting RMSE) relative to contour interval (CI): Class I is CI/3; Class II is (2 × CI)/3; Class III is CI *Spot elevation*: maximum allowable errors (or limiting RMSE): Class I is CI/6; Class II is CI/3; and Class III is CI/2	Same as in horizontal accuracy; it does not determine pass/fail criterion, which is left to the users. It gives only statistical measure but does not specify RMSE

The vertical positional accuracy is determined by comparing the elevations in the dataset with elevations of the same points as determined from an independent source of higher accuracy (at 95% confidence level); this can be considered the margin of error, expressed as

$$\text{Accuracy}_z = 1.96 \times \text{RMSE}_z \tag{3.25}$$

where 1.96 is the normal distribution value at 95% confidence level.

$$\text{RMSE}_z = \sqrt{\frac{\sum_{i}^{n} (z_{\text{map},i} - z_{\text{ground},i})^2}{n}} \tag{3.26}$$

$z_{\text{map},i}$ is the vertical coordinate of the ith checkpoint in the dataset, $z_{\text{ground},i}$ is the vertical coordinate of the ith checkpoint in the independent source of higher accuracy, n is the number of checkpoints tested, and i is an integer ranging from 1 to n.

3.6.2 Relationship between Standards

3.6.2.1 NSSDA and NMAS Horizontal Accuracy Standards NMAS standards are commonly interpreted as the limiting size of error of which 90% of the ground positions will not exceed. The circular map accuracy standard (CMAS) corresponds to the 90% confidence level circular map error defined in the NMAS (FGDC, 1998b) as follows:

$$\text{CMAS} = 2.1460 \times \text{RMSE}_x = 2.1460 \times \text{RMSE}_y \tag{3.27}$$

or

$$\text{CMAS} = 2.1460 \times \frac{\text{RMSE}_\text{h}}{\sqrt{2}} = 1.5175 \times \text{RMSE}_\text{h} \tag{3.28}$$

where 2.1460 is the same as $\sqrt{\chi^2_{\text{df}=2,\alpha=0.90}}$ for α being the lower tail area of Chi-square distribution and df = 2 as the number of degrees of freedom. Using Equations (3.18) and (3.27), the CMAS can be converted into accuracy (Accuracy_x) reported according to NSSDA, as

$$\text{Accuracy}_x = \frac{2.4477}{2.1460} \times \text{CMAS} = 1.1406 \times \text{CMAS} \tag{3.29}$$

The NMAS horizontal accuracy reported according to the NSSDA can be expressed for map scales larger than 1:20,000 with the CMAS given as $S/(30 \times 12)$ feet or $0.00278 \times S$ feet, where S is the map scale denominator. The CMAS can then be used in Equation (3.29) to obtain Accuracy_x according to NSSDA; for map scales of 1:20,000 or smaller, the CMAS can be given as $S/(50 \times 12)$ feet or $0.00167 \times S$ feet with S as the map scale denominator.

3.6.2.2 NSSDA and NMAS Vertical Accuracy Standards NMAS specifies the maximum allowable vertical tolerance to be one-half the contour interval, at all contour intervals. Therefore, the Vertical Map Accuracy Standard (VMAS) based on NMAS (at 90% confidence level) is estimated by the following formula (FGDC, 1998b):

$$\text{VMAS} = 1.6449 \times \text{RMSE}_z \tag{3.30}$$

where 1.6449 is the same as $\sqrt{\chi^2_{\text{df}=1,\alpha=0.90}}$ (lower area of Chi-square distribution). The VMAS can be converted into Accuracy_z, the accuracy reported according to the NSSDA as follows:

$$\text{Accuracy}_z = \frac{1.9600}{1.6449} \times \text{VMAS} = 1.1916 \times \text{VMAS} \tag{3.31}$$

The NMAS vertical accuracy reported according to the NSSDA can be expressed for well-defined points for contour maps with VMAS given as CI/2 or $0.5 \times \text{CI}$. The VMAS can then be used in Equation (3.31) to obtain Accuracy_z according to NSSDA, as $0.5958 \times \text{CI}$, where CI is the contour interval.

3.6.2.3 NSSDA and ASPRS Standards NSSDA standard is directly derived from the ASPRS standard but with the ASPRS coordinate-based standard converted into a 95% radial (circular) error statistic and the vertical from one-sigma (68%) to 95% standard (linear error), giving the following:

$$\text{Radial Accuracy(NSSDA)} = 2.4477 \times \text{RMSE(ASPRS for } X \text{ or } Y) \tag{3.32}$$

$$\text{Vertical Accuracy (NSSDA)} = 1.96 \times \text{RMSE (ASPRS for } Z) \tag{3.33}$$

3.7 QUALITY AND STANDARDS

Quality is the degree to which survey products (services or deliverables or both) are satisfactory to the clients. The survey product will be considered to have an acceptable level of quality if it satisfies some precision standards or some accuracy standards or both. In this case, the standards ensure quality and are considered components of quality. *Quality assurance* (QA) is a set of activities put in place for ensuring a desired level of quality in the processes involved in providing survey products, while *quality control* (QC) is a set of activities for verifying a desired level of quality in the survey products. Some of the elements of QA/QC are given in Tables 3.11–3.13.

A sample QA checklist for a typical control survey can be given as follows:

- List all the testing standards to be used in each phase of survey.
- Train project personnel in all aspects of the survey project.
- Plot all existing geodetic stations and proposed locations considered for the project as an overlay on a topographic map for use in reconnaissance and survey planning.

TABLE 3.11 Some of the Elements of QA/QC (Part I).

	Quality Assurance (QA)	Quality Control (QC)
Main concerns	Administrative and procedural activities to help prevent or minimize errors in observables and survey products: • Assuring the clients of the ability of the industry to deliver on contractual promises	Identifying errors in finished products and recommending how to correct the errors: • An error-detection system for uncovering errors so that decision can be made as to whether to accept or reject the product
Goal at design stage of project	Defining the standards and specifications to be followed in order to achieve the set requirements (so that errors will be eliminated or minimized)	Not applied at the design stage
Goal at the process and finished product stages	Verifying compliance of processes with set standards, specifications, and requirements: • When testing procedure is applied to the process rather than the finished product, it is considered QA procedure; this procedure is done in order to control the process • It ensures the right processes are being followed in the right way. Some of the process parameters that can be controlled will be checked for rejection so as to achieve the overall QC objective of providing error-free product or service	Validating compliance of finished product with set standards to identify errors in the product or assign proper quality to the product
Overall goal	Ensure: • Known inconsistencies and uncertainties in data are minimized • Errors and omissions in data are identified and taken care of • Data are correct and complete • Reported data and conclusions are justifiable	Ensure: • Products obtained are according to expectation • Sources of quality problems are identified • Results obtained agree with the expected values

TABLE 3.12 Some of the Elements of QA/QC (Part II).

	Quality Assurance (QA)	Quality Control (QC)
How goals are achieved	• It ensures quality through good project management, good training, use of proper tools, careful planning, good documentation, continued testing of procedures, immediate provision of corrective actions, and so on • It assures that a sound process is being followed	• It physically verifies or tests final products for compliance with standards and takes corrective steps by reclassifying the quality of the product • It ensures that the products obtained are what are expected • It tests for quality by controlling it but does not assure quality
Who can provide it	Everyone on the team, managers, clients, or third-party reviewers, such as the International Organization for Standardization (ISO) 9000 (NRC, 1996)	Specific teams of experts who perform tests on the final product or perform reclassification of the products
Application of statistics	When statistical testing procedures are applied to processes (observables and intermediate parameters), they are still considered part of QA, but known as statistical process control (SPC) Examples of statistical process control are • Checking the acceptability of each measurement in repeated sets of measurements • Comparing outcome of station adjustment with what is expected • Blunder detection in least squares adjustment	When statistical testing procedures are applied to finished products (process outputs), they become part of QC, known as statistical quality control (SQC). Examples of statistical quality controls are • Post-least squares analysis of positional accuracy • Statistical testing of calibration parameters • Precision standard testing in control surveys • Accuracy standard testing in control surveys In the case of project management, quality control requires that the project manager and the project team inspect the completed work to ensure its conformance with the project scope

TABLE 3.13 Some of the Elements of QA/QC (Part III).

	Quality Assurance (QA)	Quality Control (QC)
Verifiable features	Management demonstrating that: 1. They are committed to quality through mission statement and quality policy 2. They have management skills with regard to budgets, milestone events, client service 3. They have needed resources, including qualified personnel, field and office equipment, advanced technologies, training policy for staff, and so on 4. They have good project work plans, including flowchart of activities, framework for major projects, established procedures for project implementation, reporting methodology, safety policies, familiarity with existing legislation and codes, and so on Professionals demonstrating that: 1. System calibration parameters are properly applied 2. Data collection meets project accuracy requirements and adequately covers the project area 3. Results match the checks derived by an alternative technology 4. Results meet datum, map projection, feature symbology, project format criteria, and so on 5. Adequate measurements and results are acquired to verify the internal accuracy of the applied technology and process	Team of experts must demonstrate that finished products conform to standards, such as: 1. Survey or mapping criteria, including review and checking formats 2. Standards, such as NMAS, ASPRS, and NSSDA standards 3. Office technical production procedures, such as drafting and CAD standards, which include final map format, mapping limits, feature location and attribute requirements, scale, contour interval, sheet layout, and so on 4. Accuracy reporting: when providing geodetic point coordinates data, a statement should be provided that the data meet a particular accuracy standard for both the local accuracy and network accuracy. For example, "These geodetic data meet the 2-cm local accuracy standard for the horizontal coordinate values and the 5-cm local accuracy standard for the vertical coordinate values (heights) at the 95% confidence level" 5. Checking traverse closure and compliance with standards Instrument calibration statistical testing (if the goal is to determine the quality of the instrument)

- Make available all needed well-calibrated equipment and data recorder for the project.
- Adhere to the milestones as indicated on the project Schedule and Task Order Statement of Work.
- Make daily QA reviews and conduct daily progress meetings.
- Conduct internal team meetings on a minimal weekly basis to monitor progress.
- Ensure survey work is done under the supervision of a local licensed land surveyor trained and qualified in geodesy and in the use of the equipment and software.
- Download daily all field measurements from the data collector to a field computer.
- Back up all downloaded data daily onto a secure server site.
- Archive the raw data and use a copy of that data for processing and adjustment.

Typical checklist of process control (sometimes considered as the QC aspect of QA) for a typical control survey can be given as follows:

- Check tripods for good working order and calibrate bubble levels prior to movement to the field.
- Check tripods for plumb at start, during, and end of each observing session.
- Enter any unusual occurrences in the remarks section of the observation log.
- Verify station descriptions and provide a station mark rubbing at every station occupation.
- Check field forms for accuracy and completeness.
- Check and initial all manual computations.
- Check manual data computer entries.
- Check all reports and deliverable data for accuracy and completeness.
- Check field measurements recorded on the observation forms against data retrieved from the data collector.
- Perform a series of adjustments (both horizontal and vertical and both free and constrained) of all project data to ensure that all project data are free of blunders.

The QC aspect on a typical survey product can be stated as follows:

- Perform postanalysis of the least squares adjusted results to ensure that all project data meet project accuracy standards.
- Present final data and final report details according to the office standards or map accuracy standards.

4

ACCURACY ANALYSIS AND EVALUATION OF ANGLE MEASUREMENT SYSTEM

OBJECTIVES

After studying this chapter, you should be able to

1. Discuss the sources of errors in angle measurements and how their influence can be minimized or eliminated
2. Adjust survey instruments and measurements for the effects of systematic errors
3. Analyze the accuracy of horizontal direction (angle) measurements, including sources of errors and the appropriate error budgets
4. Formulate error propagation for horizontal direction (including azimuth and bearing) and angle measurements
5. Evaluate the precision of geodetic theodolite instrument under field conditions

4.1 SOURCES OF ERRORS IN ANGLE MEASUREMENTS

The main instruments for measuring directions and angles are theodolites and total stations; the main error sources in angle measurements are associated with them. Two types of errors in measuring horizontal direction and angle observables can be given as follows:

Precision Surveying: The Principles and Geomatics Practice, First Edition. John Olusegun Ogundare.

- *Internal or instrumental errors*, which consist of theodolite axial (construction) errors, pointing, reading, and instrument leveling (due to compensator or defective level bubble) errors.
- *External errors*, which consist of errors in manually leveling and centering the instrument and targets on survey markers, and the errors due to lateral and vertical atmospheric refraction.

The other important source of error is the operator of the survey instrument. Due to personal differences, instrument operators tend to introduce some errors into measurements during the measuring process. The theodolite axial errors and atmospheric refraction are the main sources of systematic error. The random errors are unavoidable and can be due to all of the aforementioned sources of error.

4.2 SYSTEMATIC ERRORS ELIMINATED BY MEASUREMENT PROCESS

In order to understand the different types of systematic errors that can be eliminated by measurement procedure, typically known as double centering (making measurements at the face-left and face-right positions of the theodolite telescope), the relationships among the three axes of a theodolite instrument must be well understood. The axes of a typical theodolite are illustrated in Figure 4.1.

In Figure 4.1, *VV* represents the vertical axis of the theodolite, *HH* is the tilting (horizontal) axis, *XX* is the optical or line-of-sight (collimation) axis, and *LL* is the plate level axis (the straight line tangent to the longitudinal curve of the plate level tube at its center, which is supposed to be perpendicular to the vertical axis when the instrument is leveled). The expected relationships among the axes after the instrument has been constructed are such that *VV* must be perpendicular to *LL*, otherwise there will be standing axis error; *HH* must be perpendicular to *VV*, otherwise there will be tilting axis error; *XX* must be perpendicular to *HH*, otherwise there will be horizontal collimation error. All these errors are collectively referred to as *axial errors*.

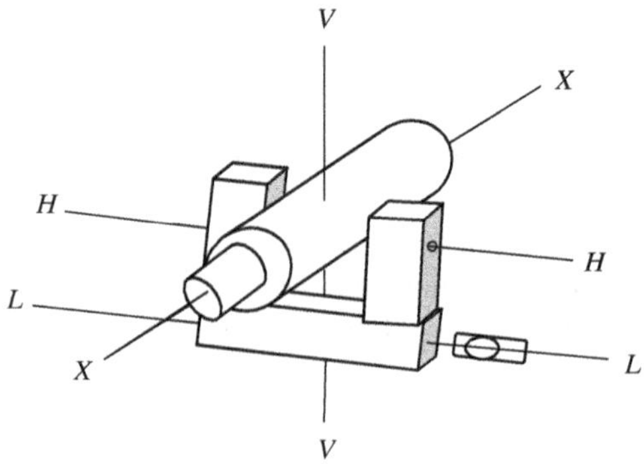

Figure 4.1 Relationship among the axes of a theodolite.

Other possible instrumental errors are vertical-index (vertical collimation) error, instrument circle graduation error, and compensator index error (if a theodolite is equipped with a compensator). When the theodolite is equipped with the compensator, the compensator will automatically compensate for the leveling error that may occur after the operator has approximately leveled the instrument. The zero index of the compensator, however, may be out of alignment with the direction of gravity, producing what is known as compensator index error. The effects of horizontal collimation error, vertical collimation error, tilting axis error, compensator index error, and circle graduation error are systematic and must be eliminated from theodolite measurements.

4.2.1 Horizontal Collimation (Line-of-Sight) Error

Horizontal collimation error is a defect due to the line of sight not being constructed perpendicular to the tilting axis of the theodolite. This defect (c) is illustrated for a theodolite in Figure 4.2. In the figure, the defective theodolite will have its circle reading aligned with line RR while the line of sight through the telescope is inclined at an angle c along line XX. This means a positive value of angle c must be added to the circle reading in order to make the reading correspond with the direction in which

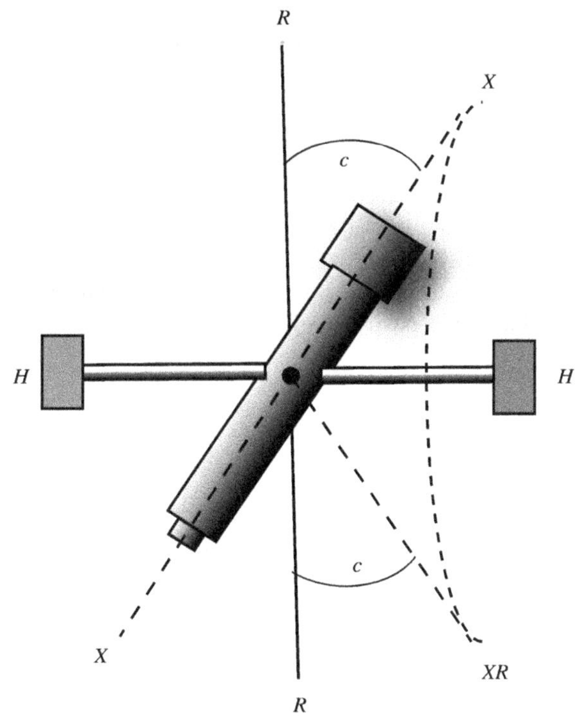

Figure 4.2 An illustration of a horizontal collimation error and its effect on angle measurement.

the telescope is currently pointing (i.e., along line *XX*). In this case, the construction defect or collimation error (c) is negative when the line of sight through the telescope is to the right of the perpendicular line *RR* (with the telescope in the face left position). If the telescope is rotated in the vertical plane (about the horizontal axis *HH*), the telescope will not move along line *RR* (as expected) but along the curve X-XR, as shown in Figure 4.2.

If a theodolite with a construction defect (or collimation error) of c is used to measure an angle, the angle will be in error (ε_{c}), which can be expressed as

$$\varepsilon_{\mathrm{c}} = \frac{c}{\sin(z)} \tag{4.1}$$

where z is the zenith angle reading. Since ε_{c} is an error contribution to a particular horizontal circle reading, it should be subtracted from the reading in order to obtain the corrected horizontal reading. It can be seen from Equation (4.1) that the influence of horizontal collimation error on horizontal circle-reading depends on the zenith angle (z), and this influence varies from the horizon ($z = 90°$) to the zenith ($z = 0°$).

Horizontal collimation error (c) can be determined in the field by observing to a well-marked target (that is close to the horizon) in face left and face right positions of the telescope. The expression for the horizontal collimation error at a particular zenith angle ($z = 90°$) can then be given as

$$c = \frac{\mathrm{Hz_I} - (\mathrm{Hz_{II}} - 180)}{2} \tag{4.2}$$

where $\mathrm{Hz_I}$ and $\mathrm{Hz_{II}}$ are the horizontal direction readings in the face left and face right positions of the telescope to the target located in the horizon. The instrument can be adjusted to remove this collimation defect by loosening the capstan screws and moving the crosshair ring left or right to eliminate the error, that is, make lines *RR* and *XX* coincide. This systematic error, however, will cancel out if all horizontal angles are measured at the same zenith angle position or if all the angles are measured in the face left and face right positions of the telescope and their averages taken as measured angles.

4.2.2 Vertical Collimation (Index) Error

Vertical collimation (or vertical index) error (v) is a defect due to the zero point of the vertical scale reading not being aligned perfectly with (or parallel to) the standing axis of the instrument, as shown in Figure 4.3. This error can be determined in the field as follows:

- With the telescope in the face left position, measure the zenith angle (z_{I}) to a well-defined point.
- With the telescope in the face right position, measure the zenith angle (z_{II}) to the same point.
- The vertical index (or collimation) error (v) can be given as

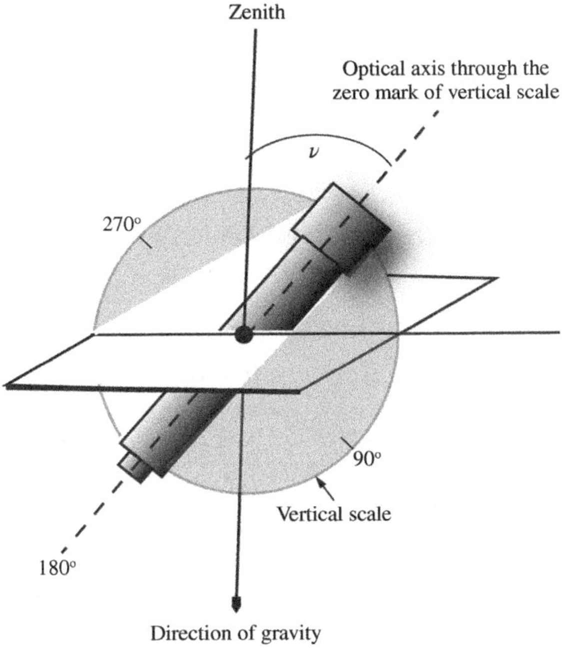

Figure 4.3 An illustration of a vertical collimation error of a theodolite.

$$v = \frac{z_{\rm I} + z_{\rm II} - 360}{2} \tag{4.3}$$

The vertical collimation error is a systematic error that will cancel out if the zenith angle is measured in face left and face right positions of the telescope with the average (z) taken as the measured zenith angle; this average can be given as

$$z = \frac{z_{\rm I} + (360 - z_{\rm II})}{2} \tag{4.4}$$

Example 4.1 An optical line of sight makes an angle 89°59′00″ (measured clockwise) with the horizontal (tilting) axis of the instrument. In turning a horizontal angle, if the line of sight on the backsight is horizontal and on the foresight, it is inclined with a zenith angle 60°00′00″, determine the error in the observed horizontal angle due to lack of adjustment.

Solution: Horizontal collimation error, $c = 89°59'00'' - 90°$ (giving the instrument defect $c = -60''$, and the line of sight is to the right of the perpendicular to the horizontal axis).
Apply Equation (4.1) with $c = -60''$ to the backsight (BS) reading as follows. Since the line of sight to the backsight is horizontal, zenith angle will be $z = 90°$; for the

foresight, $z = 60°$.

$$\text{Corrected backsight reading} = \text{BS} - \frac{-60''}{\sin(90)} \text{or BS} + 60''$$

$$\text{Corrected foresight reading} = \text{FS} - \frac{-60''}{\sin(60)} \text{or FS} + 69.3''$$

The corrected horizontal angle is equal to the difference between the corrected foresight reading and the corrected backsight reading:

$$\begin{aligned}\text{Corrected horizontal angle} &= \text{FS} + 69.3'' - (\text{BS} + 60'') \\ &= (\text{FS–BS}) + 9.3''\end{aligned}$$

The error in the observed angle is –9.3″ (following the convention that error has opposite sign to correction).

4.2.3 Tilting (or Horizontal) Axis Error

Tilting (or horizontal) axis error (t) is a defect due to the tilting axis (or the horizontal axis of theodolite) not being constructed perpendicular to the standing axis (vertical axis) of the instrument, as illustrated in Figure 4.4. In the figure, the horizontal axis HH is tilted to the right by angle t with respect to the horizontal plane of the instrument or inclined by angle $(90° + t)$ with the vertical axis through the zenith.

To determine if an instrument has a tilting axis error, the following steps can be taken:

- With the telescope in the face left position, sight a high point A with the telescope, then drop the line of sight to the ground level and mark the point on the ground as point A'.
- Reverse the telescope and sight the same high point A again, and drop the line of sight to the ground level and mark this point as point A''.
- If points A' and A'' are not the same point, there is tilting axis error.

Correcting a theodolite for tilting axis error requires moving the adjustable end of the horizontal axis up or down in order to eliminate the tilt angle t shown in Figure 4.4. The error can cancel out by measuring all angles at the same vertical angle position, if possible, or by measuring in the face left and face right positions of the telescope and taking the averages as the measurements. The error contribution (ε_t) of the tilting axis error (t) on horizontal angle readings can be determined mathematically as

$$\varepsilon_t = \frac{t}{\tan(z)} \tag{4.5}$$

where t is the tilting axis error of the instrument or the angle by which the tilting axis deviates from the horizontal plane when the vertical axis is aligned with the direction of gravity. From Equation (4.5), it can be seen that the influence of tilting axis error

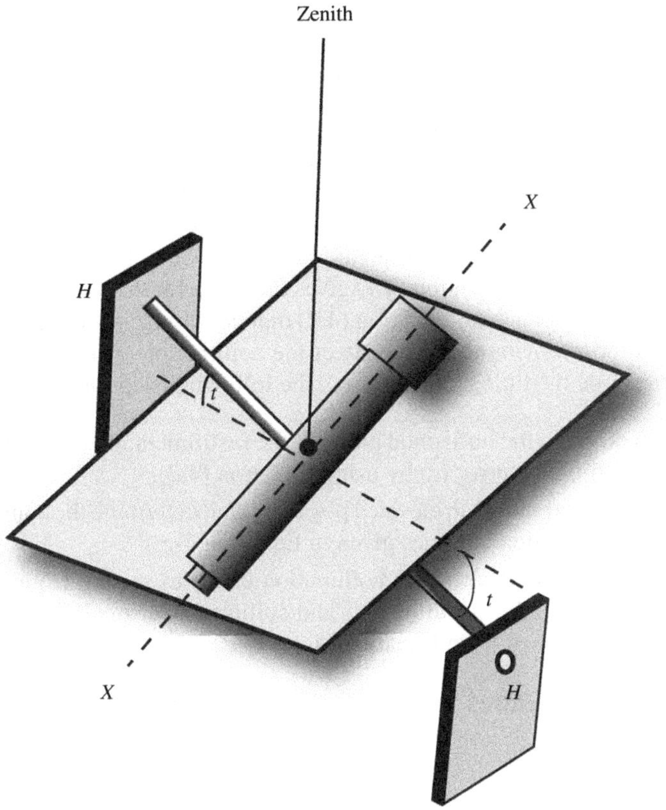

Figure 4.4 An illustration of tilting axis error of a theodolite.

on the horizontal circle reading is dependent on the zenith angle (z), and this influence varies from zero on the horizon and increases toward the zenith. The error ε_t can be applied to the horizontal circle readings as in the case of horizontal collimation error ε_c. The tilting axis error (t), if the telescope is in the face left position, is positive when the telescope is tilted to the right of the telescope (the angle of inclination of the tilting axis with the vertical axis will be greater than 90° in this direction). The tilting axis error t of an instrument can be determined through a calibration procedure in the field. At this time, the combined effect of collimation and tilting axis errors at a particular zenith angle (z) can be determined based on the procedure adopted for determining the horizontal collimation error. By following the procedure for determining the horizontal collimation error, if the horizontal angles are not observed in the horizontal plane, the result obtained will be due to combined tilting axis and collimation errors, which can be expressed as follows:

$$\text{Combined tilting axis and collimation errors} = \frac{\text{Hz}_\text{I} - (\text{Hz}_\text{II} - 180)}{2} \tag{4.6}$$

or

$$\frac{t}{\tan(z)} + \frac{c}{\sin(z)} = \frac{\mathrm{Hz_I} - (\mathrm{Hz_{II}} - 180)}{2} \tag{4.7}$$

where c is the collimation error of the instrument, t is the tilting axis error of the instrument, z is the zenith angle, $\mathrm{Hz_I}$ is the horizontal direction reading in face I, $\mathrm{Hz_{II}}$ is the horizontal direction reading in face II, and the first and second terms on the left of Equation (4.7) are the tilting axis error and collimation error components, respectively. It can be seen from Equation (4.7) that when the target observed to is on the horizon ($z = 90°$), Equation (4.7) gives the value for the horizontal collimation error. It can also be seen from Equation (4.7) that the tilting axis error (t) can only be determined after removing the influence of the collimation error (c). In this case, in order to determine the tilting axis error (t), the following steps are to be taken:

1. Select a target in the horizontal plane of the instrument, then determine the horizontal collimation error (c) by using Equation (4.2).
2. Select another target at an elevated position, then determine the combined tilting axis and collimation errors as given in Equation (4.6).
3. Using the already determined collimation error (c) in step (1) and the zenith angle and the combined tilting axis and collimation errors in step (2), solve for the tilting axis error (t) in Equation (4.7).

Example 4.2 The circle readings to targets *A* and *B* in Table 4.1 were recorded with a theodolite.

a. Calculate the vertical collimation (index) error for this theodolite and the adjusted vertical circle readings to targets *A* and *B*.

Solution: For Target A: Use Equation (4.3) to determine the vertical collimation error (v) as follows:

$$v = \frac{z_\mathrm{I} + z_\mathrm{II} - 360}{2}$$

$$\text{Vertical collimation error, } v = \frac{60°00'20'' + 300°00'20'' - 360}{2} = +20''$$

$$\text{Adjusted vertical circle reading to target } A: \ 60°00'20'' - (20'') = 60°00'00''$$

TABLE 4.1 Circle Readings to Targets *A* and *B*.

	A	*B*
Horizontal circle reading in face I	12°23′30″	74°33′50″
Horizontal circle reading in face II	192°23′50″	254°34′10″
Vertical circle reading in face I	60°00′20″	90°00′20″
Vertical circle reading in face II	300°00′20″	–

Alternatively, the adjusted zenith angle to target *A* is given from Equation (4.4) as follows:

$$\bar{z} = \frac{60°00'20'' + (360 - 300°00'20'')}{2} = 60°00'00''$$

For Target B: The same vertical collimation error applies, giving the adjusted vertical circle reading to target *B* as $90°00'20'' - (20'') = 90°00'00''$.

b. Calculate the horizontal collimation error and the tilting axis error for this theodolite.

Solution: The adjusted zenith angle to target *B* is $90°00'00''$ so that when substituted into Equation (4.7) gives

$$c = \frac{\mathrm{Hz_I} - (\mathrm{Hz_{II}} - 180)}{2}$$

Substituting the horizontal circle readings to target *B* in the equation gives

$$c = \frac{74°33'50'' - 74°34'10''}{2} = -10''$$

Substituting the horizontal circle readings to target A and $c = -10''$ into Equation (4.7) gives

$$\frac{t}{\tan 60°00'00''} + \frac{-10''}{\sin 60°00'00''} = \frac{12°23'30'' - 12°23'50''}{2}$$

$$\frac{t}{\tan 60°00'00''} - 11.55'' = -10''$$

or

$$t = 1.55 \tan(60°00'00'') \text{ or } t = 2.7''$$

4.2.4 Compensator Index Error and Circle Graduation Error

The compensator index error is due to the zero point of the compensator not being in alignment with plumb line. The error comes in if the instrument has a compensator for correcting the vertical axis (standing axis) error. The compensator will calculate the influence of the vertical axis error (i) on the horizontal reading (i_H) and on the vertical reading or along the telescope axis (i_V) and apply them accordingly. The residual errors after these corrections have been applied are the compensator-index errors. With a dual-axis compensator, the index error of the compensator is divided into two components: *alongside error* with the telescope and *crosswise error* to the telescope. The alongside error component is similar to the vertical index error (affecting the vertical angle only); the crosswise error is similar to the horizontal index error (affecting the horizontal angle only).

Compensators usually have specified setting accuracies of 0.3″–6″ with working ranges of 2′–6′. This means that compensators are capable of correcting circle readings for standing axis effect with a precision of 0.3″–6″ if the standing axis of the theodolite is within 2′–6′ of being vertical.

Note that the *circle graduation error* is negligible with today's total station equipment. Because of this, it will not be discussed any further.

4.2.5 Eliminating Systematic Errors by Double-Centering: Example

Double-centering or double-sighting procedure consists of making a measurement with a theodolite once with the telescope in the face left position and once with the telescope in the face right position; the two measurements will have equal and opposite axial errors. A typical example of using double-centering method to eliminate some axial errors is in the extension of a straight line. For example, consider a case in which a straight line *AB* is to be extended to *C* as shown in Figure 4.5. The double-centering steps for extending line *AB* to *C* can be described as follows:

1. Set up the instrument at station B, sight to station A in face I (face left) position of the instrument, then plunge the instrument telescope to face II (face right) position, and sight to the direction of *C*; mark the image of the reticule as C_1.
2. While still in face II, rotate the telescope to sight to station A and plunge the telescope again to face I to sight in the direction of *C*; mark the image of the reticule as C_2.
3. The middle of the two marks C_1 and C_2 is the location of station C, forming part of the extended line. Note that angle C_1-B-C_2 is equal to four times the collimation error ($4\varepsilon_c$) expressed by Equation (4.1).

By locating station C in the aforementioned steps, the influence of instrument axial errors is eliminated. Since no angles are set out (i.e., circle readings are not taken) in the process, the problem of possible circle graduation errors, which are negligible with today's total stations, will not arise. If the line of sight, however, is not horizontal, the standing axis error will affect the location of point *C* since it cannot be removed by double-centering procedure. For a case involving inclined sights, the appropriate procedure for aligning point *C* with *A* and *B* can be given as follows:

1. With the telescope in face I position, sight to station A, then turn off 180°, and sight in the direction of *C* and mark the point as C_1. The compensator of the instrument must be on during this process.

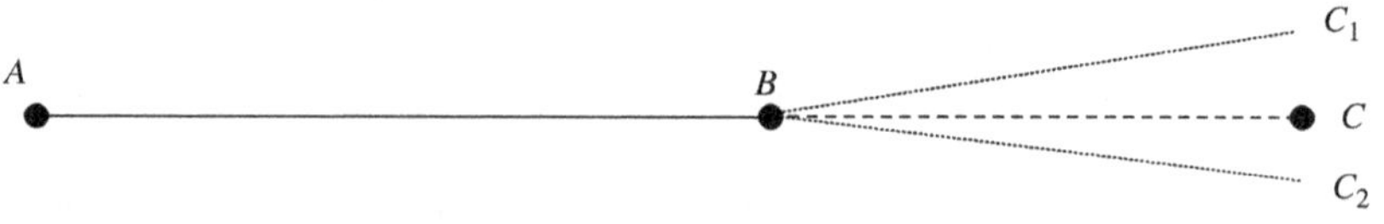

Figure 4.5 Extending a straight line by double-centering method.

2. Now sight to station A in face II position, then turn the telescope by 180° toward station C and mark the point as C_2.
3. The average of the readings in both faces results in the correct straight line extension even for steep sightings (both directions to C_1 and C_2 should practically coincide if the instrument has been properly calibrated).

Note: In order for the compensator in an instrument to correct the circle readings, the instrument must be rotated physically; the corrections are only applied as the angular readings are being changed.

Example 4.3 The line of sight of a theodolite is out of adjustment by a collimation error of 12″. In prolonging a line by plunging the telescope between backsight and foresight, but not double-centering, what angular error is introduced and what off-line linear error results on a foresight of 500 m (assuming flat terrain)?

Solution: Collimation error of instrument (or instrument defect) is 12″:

Referring to Figure 4.5, the angle C_1-B-C_2 represents $\alpha = 4\varepsilon_c$ (the accumulated collimation error of direction measurements in double-centering procedure).

By the double-centering method, the angle $\alpha/4$ represents the collimation error on a single-direction measurement; by assuming flat terrain condition, $z = 90°$ in Equation (4.1) so that $c = \varepsilon_c$ and $\alpha = 12'' \times 4$ (or 48″). The angular error introduced in aligning A-B-C_1 is $\alpha/2$ (or 24″):

Linear error (or distance C-C_1) = Angular error (in radians) × length B-C_1

$$= \frac{24''}{206,265''} \times 500\,\text{m} \rightarrow 0.058\,\text{m}$$

Example 4.4 Referring to Figure 4.5, if the distance from the instrument setup point B to points C_1 and C_2 is 600 m each and the distance between C_1 and C_2 is 10 cm, calculate the possible collimation error (to one decimal arc second) of the instrument, assuming the instrument tilting axis error is zero and the measured vertical angles to points A, C_1, and C_2 are +15°.

Solution: Using the linear error approach in Example 4.3:

$$0.10 = \frac{\alpha}{206,265''} 600\,\text{m}$$

$$\alpha = \frac{0.10 \times 206,265}{600} \rightarrow 34.4''$$

Collimation error effect on a single-direction measurement is $\alpha/4$ or 8.6″. The collimation error of instrument can be determined from Equation (4.1) using the error in a single-direction measurement (since tilting axis error is negligible) so that:

$$\frac{c}{\sin(z)} = 8.6'' \text{ and } c = 8.6 \sin(75) \text{ or } c = 8.3''$$

The collimation error of instrument is 8.3″.

4.3 SYSTEMATIC ERRORS ELIMINATED BY ADJUSTMENT PROCESS

Typical systematic errors that can only be removed by adjusting the instrument or by mathematically correcting the angle measurements are due to the following error sources: *plummet error, standing axis error, plate bubble error, atmospheric refraction, deflection of the vertical* (by comparing the geoid with the reference ellipsoid).

4.3.1 Plummet Error

Plummet error or centering error is an instrument defect due to the optical axis of the plummet not being aligned in the direction of the vertical axis of the instrument; this may be due to the wearing out of tribrach or plummet is out of adjustment. With this defect, accurate horizontal angles cannot be determined. The effect of this error is similar to that of standing axis error (discussed in what follows).

Testing an instrument for plummet error depends on whether the plummet is mounted on the upper part of the instrument alidade and can be rotated about the vertical axis or the plummet is located on the tribrach. In the case where the plummet is located on the instrument alidade, the plummet error can be checked as follows:

- Secure a piece of paper on the ground below the instrument (after it has been leveled on its tripod) and mark where the plummet intersects it.
- Rotate the theodolite 180° and mark second point where the plummet intersects the paper; if the second point coincides with the first point, the plummet is in adjustment.

In the case where the plummet is located on the tribrach, check the plummet error as follows:

- Carefully lay the theodolite on its side on a table.
- Look through the optical plummet to a piece of paper on a wall about 1.5–2 m away.
- Mark the point on the paper, where the line of sight through the plummet hits the paper.
- Rotate the tribrach 180°, and mark again the point where the line of sight through the plummet hits the paper.
- If the optical plummet is out of adjustment, its line of sight will form a circle on the paper when the tribrach is rotated round.
- Measure the diameter and the radius of the circle formed.
- Calculate the angle subtended at the instrument by the radius of the circle.

In the case of laser plummets, set the theodolite on its tripod, level the instrument, and switch on the laser plummet; mark the center of the laser spot on the ground, slowly rotate the instrument through 360° while observing the positions of the laser spot; if the center of the laser spot makes a circular movement of more than 1–2 mm instead of remaining stationary, the plummet needs adjustment.

To correct the instrument for the plummet error, use an adjustment tool to raise or lower the three corners of the "bulls-eye" bubble, until there is no circle scribed out on the paper sighted to through the optical plummet. The effect of this error on angular measurement can be cancelled out by measuring an angle with one position of the tribrach, turning the tribrach 180° on the tripod, measuring the angle again, and taking the average of the two readings as the actual reading.

4.3.2 Standing Axis Error

The standing axis error (i) is a setup error, not an instrument error. This error is due to the observer not perfectly centering the bubble so that the standing axis of the instrument is not aligned with the plumb line (gravity) direction. In this case, the vertical axis (standing axis) is inclined. If the instrument is turned around its standing axis (assuming the plumb line does not correspond with the standing axis), the instrument is actually not turned around the vertical axis or the plumb line direction as it should be. This creates an error (similar to tilting axis error) whose value is zero on the horizon and varying with zenith angle. Apart from changing with zenith angle, standing axis error also changes with change in horizontal direction. By changing the horizontal directions, the horizontal measurements are affected by the effects of the misleveling.

The standing axis error affects the theodolite by longitudinal (i_{V}) and traverse (i_{T}) tilts expressed as follows:

$$i_{\mathrm{V}} = i \cos \alpha \tag{4.8}$$

$$i_{\mathrm{T}} = \frac{(i \sin \alpha)}{\tan z} \tag{4.9}$$

where i_{V} is the tilt along the direction of the telescope, i_{T} is the tilt in the direction perpendicular to the telescope, i is the standing axis error (or the amount of displacement on the plate level bubble in the vial or the angle between the standing axis and the direction of gravity), z is the zenith angle measurement, and α is the angle turned between the plane containing the inclined axis and the direction of the telescope. Theodolites with dual-axis compensators compute and display i_{V} and i_{T} and also internally correct the horizontal and zenith angle measurements with these values as follows:

$$\mathrm{Hz}' = \mathrm{Hz} - i_{\mathrm{T}} \tag{4.10}$$

$$z' = z - i_{\mathrm{V}} \tag{4.11}$$

where i_{T} is the transverse tilt defined by Equation (4.9), Hz$'$ is the corrected horizontal direction, Hz is the measured horizontal direction, z' is the corrected zenith angle, z is the measured zenith angle, and i_{V} is the longitudinal tilt defined by Equation (4.8).

If a telescope is in face left position and the standing axis is inclined to the left, the standing axis error will be taken as positive. It can be seen from Equation (4.11) that by changing the zenith angle from the horizon, the zenith angle measurements are affected mainly by the longitudinal tilt (i_{V}) component of the misleveling (using the compensator). The effects of standing axis error on horizontal direction and zenith angle measurements are automatically corrected for if compensator is activated in the total station equipment used. The dual-axis compensator will correct standing axis tilt in the direction of the telescope and measure how much the tilting axis is out of level and correct horizontal circle readings automatically for this error. The other way to correct this error is to level the instrument using proper procedure as follows:

- Align the level vial along two of the leveling screws, turn the telescope 180°, and observe the amount of movement of the bubble.
- In this position, if the bubble is off the center, bring it half way back and turn the telescope again 180°; repeat this step until the bubble remains centered when turned 180°.
- Now align the bubble vial in the direction of the third foot screw and center the bubble.
- Turn the telescope 180° and observe the movement of the bubble; if off the center, bring it back half way; repeat this step until the bubble remains centered when turned 180°.
- At this location, the telescope can be pointed in any direction and the bubble will still remain centered.

If the bubble vial of the instrument is not sensitive enough to provide the desired accuracy, an additional level, such as striding level, of higher sensitivity must be used on the horizontal axis of the telescope. If the amount of inclination and the direction of inclination of the vertical axis are known, mathematical formulas (Equations (4.10) and (4.11)) can be applied to correct each reading to its proper value. An alternative way of determining the corrected horizontal direction measurement is to count the number of graduations (NR) the leveling bubble is off the center to the right and the number of graduations off the center to the left (NL). The corrected horizontal direction can then be given as

$$\mathrm{Hz}' = \mathrm{Hz} - \frac{(\mathrm{NR} - \mathrm{NL}) \times v''}{2 \tan z} \tag{4.12}$$

where Hz is the measured horizontal direction, Hz′ is the corrected horizontal direction, and v'' is the sensitivity of the bubble. Considering Figure 4.6 for example, if the bubble to the left NL = 2, that to the right, NR = 4, the sensitivity of the bubble

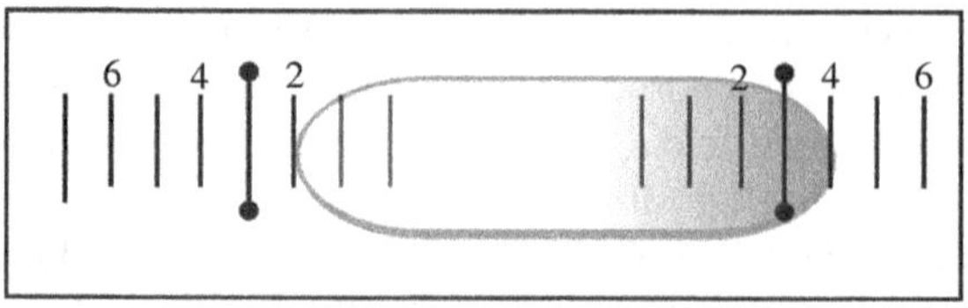

Figure 4.6 Typical plate bubble vial.

is 20″/div, and the zenith angle of the line of sight is 75°, the error to be subtracted from the measured horizontal direction will be 5.4″. The position of the bubble in this example is illustrated in Figure 4.6.

(Remember that standing axis error is a temporary error that changes in magnitude each time that the instrument is leveled.)

4.3.3 Plate Bubble Error

Plate bubble error is an instrument error due to the plate bubble axis not being perpendicular to the vertical axis of the instrument or the vertical axis of the instrument is not being aligned with the plumb line (gravity) direction after centering the bubble. When the bubble is centered, it runs out when rotated 180° in azimuth. Normally, if the plate bubble is in adjustment, its axis (axis *LL* in Figure 4.1) will be at right angles to the vertical axis (line *VV* in Figure 4.1); otherwise, the instrument will be set to measure horizontal angles with an inclined vertical axis, that is, angles will be measured on an inclined plane. In this case, if the instrument is turned around its standing axis (assuming the plumb line does not correspond with the standing axis), the instrument is actually not turned around the vertical axis or the plumb line direction as it should be. The effect of this error is similar to that of standing axis error and will not cancel out by finding averages of measurements taken in face left and face right positions of the telescope. An instrument can be tested in the field for plate bubble error by centering the plate bubble of the instrument in two positions (at right angles); rotate the instrument through 180°, if the bubble moves off centre in this third position, the plate level is out of adjustment.

The correction of the theodolite for plate bubble error will require raising or lowering one end of the level vial until the bubble ends up in the center when testing the instrument. If the amount of inclination and the direction of inclination of the vertical axis are known, mathematical formulas (Equations (4.10) and (4.11) or Equation (4.12)) can be applied to correct each reading to its proper value. Most total stations today have the "automatic compensation" software built into the processor, which automatically applies calculated corrections to each reading. Dual-axis compensators will correct both the vertical and horizontal angles for the plate bubble error if the inclination angle is within the working range of the compensators; single-axis compensators, however, will only correct the vertical angles.

Example 4.5 A line was prolonged by reversing the telescope from the backsight (BS) position to foresight (FS) position with the zenith angles for both backsight and foresight being 60° and the vertical axis being inclined 30″ in the direction perpendicular to the direction to BS. What is the measured angle between the backsight and the foresight (assuming no collimation error and no tilting axis error)?

Solution: This is a standing-axis problem (Equation (4.9)); take the bearing (clockwise) of BS direction as $\alpha = 270°$; the bearing of FS direction will be $\alpha = 90°$; $i = 30''$; $z = 60°$.

Substitute the values into Equation (4.9) and apply the correction to each direction as follows:

$$\text{Corrected backsight reading} = \text{BS} - \frac{(30'' \sin 270^\circ)}{\tan 60^\circ} = \text{BS} + 0^\circ 00' 17.3''$$

$$\text{Corrected foresight reading} = \text{FS} - \frac{(30'' \sin 90^\circ)}{\tan 60^\circ} = \text{FS} - 0^\circ 00' 17.3''$$

$$\text{The measured angle} = (\text{FS–BS}) - 0^\circ 00' 17.3'' - 0^\circ 00' 17.3''$$

$$= (\text{FS} - \text{BS}) - 0^\circ 00' 34.6''$$

The angle at the station at which the instrument was set would be in error of 34.6″; angle set out will be 179°59′25.4″ instead of being the true value of 180°. If the prolonged line from the set-up station to the forward station is 100 m, the established line would depart from the true direction by

$$\text{Departure} = 100 \times \frac{34.6}{206,265} = 0.017\,\text{m}$$

4.3.4 Atmospheric Refraction

Whenever one is sighting through an instrument telescope to a target, one is observing the natural wave or radiation (e.g., white light) emitted by the target. Varying densities of the atmospheric air along the path of any wave propagation will cause the speed and direction of the wave to change. The change, in either speed or direction, is referred to as refractivity. This means that the optical path of radiation in the atmosphere is curved due to atmospheric refraction. The curvature of the optical path may differ from a given point A to another point B along the line of sight AB as shown in Figure 4.7.

When light ray passes from colder air to warmer air, the ray will bend in a concave toward the direction of the gradient. This is to say that generally, light rays will move toward warmer air, where the propagation speed is greater. For the horizontal direction measurement from point A to B in Figure 4.7, an angular error $d\delta$ is introduced. Even though point B' is sighted to, point B is actually located; the linear error in sighting to the wrong point is e.

Atmospheric refraction is dangerous to any optical measurements. The refraction effects are most pronounced in leveling and zenith angle measurements, especially when the line of sight is near (about 2 m or less) the ground surface with the temperature of the layers of air above the surface being significantly different. The horizontal effects of refraction may also be dangerous if the line of sight of the observed horizontal direction runs parallel and very close (like say 1 m) to prolonged objects of a different temperature, such as walls of structures or soil exposed to the Sun's radiation, walls of tunnels, galleries of long dams, turbines, transformers, and so on. Generally, if the temperature gradient $\partial t/\partial y$ across the line of sight (Figure 4.7) is constant at all points of the line, then the line will be refracted along a circular curve

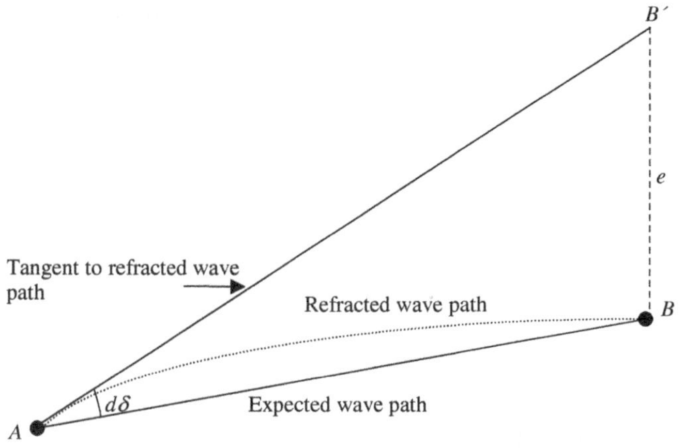

Figure 4.7 Refracted and expected wave propagation paths.

producing a linear error (e) of pointing to a survey target. This error can be given (US Army Corps of Engineers, 2002) as

$$e = \frac{3.9 \times PS^2}{T^2} \frac{\partial t}{\partial y} \times 10^{-5} \tag{4.13}$$

where

S = distance between the stations;

P = barometric pressure (mbar);

T = atmospheric temperature in Kelvin (273.15 + t °C)

t = atmospheric temperature in °C;

$\partial t/\partial y$ = temperature gradient in the direction perpendicular to the direction of wave propagation.

Usually, the temperature gradient differs from one point to another, producing an irregular shape of the refracted line of sight. The temperature gradient as shown in Figure 4.7 must be determined, either horizontally for lateral refraction (for horizontal direction measurement) or vertically for vertical refraction (for vertical angle measurement), in a direction perpendicular to the optical path or line of sight. This suggests that horizontal or vertical gradients of temperature and barometric pressure should be measured at a number of points along the line of sight of the survey. Note that this effect of refraction calculated using Equation (4.13) is usually not applied directly to measurements, but is mainly used to quantify the expected effect of refraction so that the surveyor can avoid any unacceptable atmospheric condition for the measurements. This is usually the case since the distribution of horizontal temperature gradients is difficult to measure precisely.

Equation (4.13) gives positional error in the units of distance S; the directional error in arc seconds can be derived from Equation (4.13). If the directional error is

$d\delta$ (radians) and the distance is S, then $e = S \times (d\delta)$ so that from Equation (4.13)

$$S(d\delta) = \frac{3.9 \times PS^2}{T^2} \frac{\partial t}{\partial y} \times 10^{-5} \tag{4.14}$$

or

$$d\delta = \frac{3.9 \times PS}{T^2} \frac{\partial t}{\partial y} \times 10^{-5} \times 206,265''(\text{s}) \tag{4.15}$$

or

$$d\delta = \frac{8'' \times PS}{T^2} \frac{\partial t}{\partial y} \tag{4.16}$$

Equation (4.16) assumes that a uniform temperature gradient persists over the whole length S of the line of sight. For example, if a gradient of 0.2 °C/m persists over a distance of 250 m at $P = 1000$ mbar and $t = 25$ °C, from Equation (4.13), the positional error (e) will be calculated as 5.5 mm. Using Equation (4.16), the directional error will be 4.5 arcsec.

Effects of atmospheric refraction are systematic, and they must be applied as corrections (known as meteorological corrections) to the raw data before use in network adjustment. Over a line of sight of length S, the effects of atmospheric refraction on horizontal direction measurement d can be given by Equation (4.16) or expressed in terms of the horizontal coefficient of refraction (k_h) as (Torge, 2001):

$$\delta d = \frac{k_\text{h} S}{2R} \tag{4.17}$$

Refer to Section 5.3.3 for further discussion on atmospheric refractions. For a measured horizontal angle (θ), which is the difference between two measured directions d_{li} and d_{lj} (Figure 4.8), the effects of the atmospheric refraction can be given from Equation (4.17) as

$$\delta\theta = \frac{1}{2R}(k_\text{h}^{lj} S_{lj} - k_\text{h}^{li} S_{li}) \tag{4.18}$$

where k_h^{lj} and k_h^{li} are the lateral coefficients of refraction of the lines of sight lj and li, respectively; S_{lj} and S_{li} are the distances along the lines of sight lj and li, respectively.

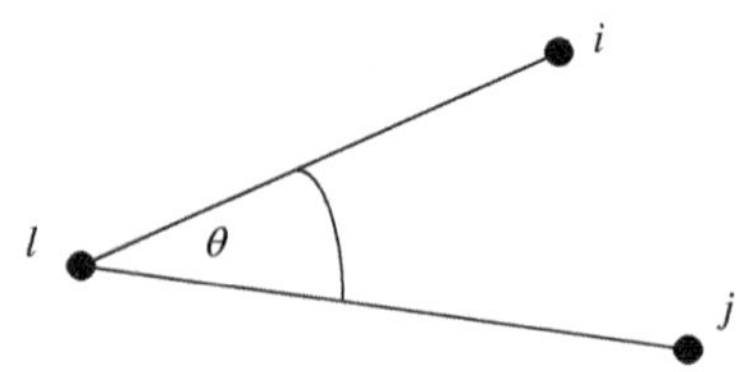

Figure 4.8 Representation of a horizontal angle (θ) between survey points.

The systematic effects of atmospheric refraction on a zenith angle Z can be deduced from Equation (4.17) as

$$\delta Z = \frac{k_v S}{2R} \tag{4.19}$$

where k_v is the vertical coefficient of refraction, S is the sight length, and R is the mean radius of the earth.

Example 4.6 During a city survey, it was necessary to run a traverse line at ~0.4 m away from the south face of several buildings over a distance of ~270 m. The temperature at the building face was about 40 °C and, at ~1 m away, was 30 °C. This line is one of the four traverse lines supposedly closing around the block. Explain what might affect the angular misclosure and suggest by how much and whether the misclosure would appear to be larger or smaller than it would be without the influence(s). What would you do about it or them?

(Reproduced by permission of CBEPS.)

Solution: The particular dangers to measurements:

a. Atmospheric refraction – gradient of air temperature $\partial t/\partial y$ in the direction perpendicular to the line of sight being the main parameter
b. Human errors of pointing telescope
c. Centering errors (more pronounced for short distances).

The gradient of air temperature: $\dfrac{\partial t}{\partial y} = \dfrac{40-30}{1\,\text{m}} = 10°\text{C/m}$

Interpolated temperature along the traverse line: $40\text{–}0.4\,\text{m}' \times 10°\text{C/m} = 36°\text{C}$

Refraction Error from Equation (4.16):

$$d\delta = \frac{8'' \times PS}{T^2}\left(\frac{dt}{dy}\right); \quad T = (273.15 + t°\text{C}) \quad \text{or} \quad T = 309.15°\text{C}$$

Substituting $P = 1000\,\text{mbar}$, $S = 270\,\text{m}$, and $T = 309.15\,°\text{C}$, gives: $d\delta = 226''$ (or $3'\,46''$).
This gives positional error of the target as

$$((270\,\text{m} \times d\delta\,(\text{rad})) = 0.296\,\text{m} \quad (\text{or } 270\,\text{m} \times 1.09568\text{E} - 3\,\text{rad})$$

What to do: Avoid the refraction condition – perform measurements when radiation is stable.
Perform short-distance measurements or perform reciprocal angle measurements.

Example 4.7 The effects of lateral refraction can be quantified more to recognize when conditions should be avoided rather than to apply as a correction. Along the south face of a block of buildings, temperature readings were taken at the wall surface and at 1 m away. The average values were 35 and 30 °C, respectively. A traverse around the block had to be run with the lines offset by 0.5 m from the building faces. The block is 300 m^2. By how much is the effect of refraction along this one side likely to contaminate the misclosure of the block, assuming standard pressure?

(Reproduced by permission of CBEPS.)

Solution:

$$\text{Refraction error: } d\delta = \frac{8'' \times PS}{T^2}\left(\frac{dt}{dy}\right) \quad \frac{\partial t}{\partial y} = \frac{35-30}{1\text{ m}} \quad \text{or } 5\,°\text{C/m}$$

Given $t = 32.5$ °C, $T = (273.15 + t$ °C) or $T = 305.65$ °C, $P = 1013.25$ mbar, $S = 300$ m

$$d\delta = 8''\frac{1013.25 \times 300 \times 5}{305.65^2} = 130.15''$$

$d\delta = 130.15''$ (or $2'10''$) smaller at each edge; giving a total misclosure of $260''$ (or $4'20''$) ($130''$ smaller at each of the SE and SW corners).

4.4 SUMMARY OF SYSTEMATIC ERROR ELIMINATION

It should be mentioned that direction measurements by double-centering (face left and face right positions of precision electronic theodolites) procedure must always be obeyed in order to eliminate systematic errors caused by mechanical misalignment of the theodolite's axial system. This should be done even if the manufacturer claims that the errors are automatically taken care of. Those errors that cannot be eliminated by double-centering procedure must also be accounted for as discussed earlier. It should be noted that Equations (4.8) and (4.9) are both applicable to standing axis, plummet, and plate bubble errors. A summary of the sources of systematic errors that are eliminated by double-centering procedure and those that are not is given in Table 4.2.

4.5 RANDOM ERROR ESTIMATION

Sources of random errors are pointing, reading, leveling, and centering of measuring instrument as well as centering of the target and the effects of residual atmospheric refraction.

4.5.1 Pointing Error

Pointing error (σ_P) is the error in aligning or aiming the crosshair of the instrument's telescope with the target and is due to a number of factors such as the following:

- Optical qualities of telescope, such as telescope magnification and focusing error of instrument
- Limited human vision when using the instrument, including the visibility and brightness conditions

TABLE 4.2 Summary of Systematic Error Elimination.

Error Type	Affected Angles		Eliminated by Two-Face Measurement
	Horizontal	Vertical	
Horizontal collimation error (c)	Yes		Yes
Tilting axis error	Yes		Yes
Vertical index (vertical collimation) error		Yes	Yes
Compensator index error (l and t)	Yes	Yes	Yes
Standing axis error	Yes	Yes	No
Plummet error	Yes	Yes	No
Plate bubble error	Yes	Yes	No
Atmospheric refraction	Yes	Yes	No

- Variations of the atmospheric conditions (heat waves or thermal turbulence, fog, etc.)
- Target conditions and design, such as size, shape, distance to the target, background, and illumination of the target point
- Width of crosshairs.

According to Chrzanowski (1977), "*... with a properly designed target and in average visibility and thermal turbulence conditions the standard deviation of one pointing over short distances is equal to*" the following:

$$\sigma_{\mathrm{p}} = \frac{C}{M} \tag{4.20}$$

where C is a constant, which can vary from $C = 30''$ to $C = 60''$, and M is the telescope magnification of the instrument. Unless otherwise specified, the average value ($C = 45''$) is commonly used in computations. For example, an objective lens with 30× magnification would have a pointing error (σ_{p}) of approximately 1.5″ if $C = 45''$. The actual pointing error is likely to be larger than the estimated value if the visibility is poor or the thermal turbulence is large. The standard deviation due to pointing error (σ_{d}) of a single direction measured in "n" sets (i.e., one face left and one face right measurements per set) can be determined as follows:

$$\sigma_{\mathrm{d}} = \frac{\sigma_{\mathrm{p}}}{\sqrt{2n}} \tag{4.21}$$

If the pointing error of one-direction measurement is the same as the pointing error of the other, the pointing error for *n sets* of angle (θ) measurement will contain $2n$ pointings in the backsight direction and another $2n$ pointings in the foresight direction so that the pointing error in an angle (θ) can be propagated as follows:

$$\sigma_{\theta\mathrm{p}} = \frac{\sigma_{\mathrm{p}}}{\sqrt{n}} \tag{4.22}$$

or

$$\sigma_{\theta\mathrm{p}} = \sigma_{\mathrm{d}}\sqrt{2} \tag{4.23}$$

where σ_p is the pointing error of a direction measured only once. As an example, if σ_p is 1.8″, and an angle is measured four times (or two sets) using the directional method, the expected error in the angle measurement due to the pointing error can be given as

$$\sigma_{\theta p} = \frac{1.8''}{\sqrt{2}} = 1.3'' \tag{4.24}$$

Pointing error for a given instrument can be determined empirically as follows (cf. Nickerson, 1978):

(a) Set and level the instrument and the target according to standard techniques.
(b) Point the theodolite crosshairs on the target and record the direction reading.
(c) Move the crosshairs off the target, then point and make another direction reading on the target.
(d) Repeat the pointing procedure at least 20 times to gather a sufficient number of direction measurements for calculating a mean error value from the data.
(e) Compute the standard deviation of the resulting data (in arcsec). This gives the combined pointing and reading error ($\sigma_p + \sigma_r$) with σ_r as the reading error.
(f) Calculate the pointing error for the instrument as follows:

$$\sigma_p = \sqrt{(\sigma_p + \sigma_r)^2 - \sigma_r^2} \tag{4.25}$$

The reading error (σ_r) is determined independently from either the standard deviation of a series of 20 direction readings (read at separate times) of the same pointing of the theodolite with the instrument's motion locked or extracted from the instrument specifications. In the case of total station equipment, the power should be switched on and off (or clicking the measure key) each time to take each reading.

The pointing error can be minimized by observing survey targets under high magnification. Some instruments, such as Leica T3000 Electronic precision theodolite, are equipped with interchangeable eyepieces that provide up to $59 \times M$ (where M is the telescope magnification). In the absence of greater lens magnification, the technique of averaging repeated sets of angle measurements can be used to reduce the instrument pointing error.

4.5.2 Reading Error

The reading error (σ_r) is due to an inability of the observer to repeat the same reading. The reading error for one sighting can be given (Chrzanowski, 1977) as

$$\sigma_r = \psi d'' \tag{4.26}$$

where d'' is the least count (s) of the theodolite; for theodolites with coincidence micrometers with least count of $d = 1''$ or $d = 0.5''$, $\psi = 2.5$; and for theodolites with a microscope and a least count of 10″ to 1′, $\psi = 0.3$. Reading error in an angle (θ)

can be propagated depending on the measuring method with a theodolite, such as *repetition* and *directional* methods.

4.5.2.1 ***Repetition Method*** Using repetition method (with m total number of turnings of the same angle in both face left and face right positions), the average angle measurement can be expressed as

$$\text{Average angle} = \frac{R_f - R_0 + 360° \times q}{m} \tag{4.27}$$

where q is the number of times the zero index mark of the instrument is passed on the horizontal circle reading scale, R_0 is the first direction reading (zeroing the circle), and R_f is the direction reading after the final (mth) pointing. If the reading error of the first direction reading is the same as the reading error of the final direction reading, the reading error for the angle (θ) can be derived from Equation (4.27) as

$$\sigma_{\theta r} = \frac{\sigma_r \sqrt{2}}{m} \tag{4.28}$$

or

$$\sigma_{\theta r} = \frac{\sigma_r}{n\sqrt{2}} \tag{4.29}$$

where σ_r is the reading error of a direction measured only once, $m = 2n$ with n as the number of sets of readings (one set consisting of measurements made in face left and face right positions). As an example, if $\sigma_r = 1.5''$, and an angle is measured four times using the repetition method, the expected error in the angle due to the reading error can be given as

$$\sigma_{\theta r} = \frac{1.5'' \sqrt{2}}{4} = 0.6'' \tag{4.30}$$

4.5.2.2 ***Directional Method*** In directional method of angle measurement, if n is the number of sets of measurements, $2n$ readings are made in the backsight direction and another $2n$ readings in the foresight direction, similar to the case of pointing error. The propagated error due to reading will be similar to that due to pointing error and can be given as follows. If the reading error of one-direction reading is the same as the reading error of the other, the reading error for the angle (θ) can be given as

$$\sigma_{\theta r} = \frac{\sigma_r}{\sqrt{n}} \tag{4.31}$$

where the reading error for each direction for n set is given as

$$\sigma_{dr} = \frac{\sigma_r}{\sqrt{2n}} \tag{4.32}$$

and σ_r is the error of one single reading in one direction. As an example, if $\sigma_r = 1.5''$, and an angle is measured four times (or two sets) using the directional method, the expected error in the angle due to the reading error can be given as

$$\sigma_{\theta r} = \frac{1.5''}{\sqrt{2}} = 1.1'' \tag{4.33}$$

4.5.3 Instrument Leveling Error

Instrument leveling error depends on the sensitivity of the tubular spirit level used in leveling the instrument. According to Chrzanowski (1977), if an instrument is well protected from any possible heat sources, "*a careful observer and a well-adjusted spirit level may give the standard deviation of levelling,*" which can be given as

$$\sigma_v = 0.2v'' \tag{4.34}$$

where σ_v is the estimated standard deviation of leveling the instrument and v'' is the level bubble sensitivity per division. For a split bubble leveling instrument, which is centered by a coincidence reading system, the accuracy for leveling the instrument can be given (Kuang, 1996) as

$$\sigma_v = 0.02v'' \tag{4.35}$$

The tubular bubbles on instruments provide higher accuracy than bull's eye bubbles (having the usual 8′ sensitivity). Instruments of high accuracy occasionally have *striding level* placed directly onto the horizontal axis of the instruments to level the axis or to determine its tilt during measurements involving inclined sights (especially for directions with slope angles larger than 5°). The standard deviation (σ_{dL}) of any measured horizontal direction due to the effect of the leveling error can be given as

$$\sigma_{dL} = \sigma_v \cot z \tag{4.36}$$

where z is the measured zenith angle to the target and σ_v can be considered as the inclination of the standing axis of the theodolite with the direction of gravity. In the case of zenith (vertical) angle measurement, Equation (4.36) is not used, but the leveling error for zenith (vertical) angle measurement can be given simply

$$\sigma_{dL} = \sigma_v \tag{4.37}$$

Typical electronic theodolite with biaxial leveling compensator can sense the inclination (misleveling) of the theodolite to an accuracy of about 0.5″ and automatically correct vertical and horizontal direction readouts. In optical theodolites, inclination is controlled only by a spirit level so that leveling error of several seconds of arc in horizontal direction measurements can be produced when measuring along steeply inclined lines of sight. Leveling errors affect the accuracy of horizontal angle measurements mainly when observing over steep vertical angles. For precision surveys, electronic theodolites with biaxial leveling compensators should be used. This situation is common in monitoring embankment dams where targets set on the crest of the dam are observed from the top of the structure or vice versa. The instrument misleveling error on an angle (θ) can be given from Equation (4.36) as

$$\sigma_{\theta L} = \sqrt{\left[\sigma_v \cot\left(z_b\right)\right]^2 + \left[\sigma_v \cot\left(z_f\right)\right]^2} \tag{4.38}$$

where

σ_v = fractional division of the bubble × sensitivity of the bubble.

z_b = zenith angle to the back station.

z_f = zenith angle to the forward station.

The leveling error is considered random if the leveling bubble is carefully monitored and attempts are made to keep it centered while turning angles. For n repetition of an angle or direction (leveling the instrument each time), the corresponding leveling error will be reduced by a factor of $\sqrt{n}$, so that Equation (4.36) will become

$$\sigma_{dL} = \frac{\sigma_v \cot z}{\sqrt{n}} \tag{4.39}$$

and Equation (4.38) will become

$$\sigma_{\theta L} = \frac{\sqrt{[\sigma_v \cot(z_b)]^2 + [\sigma_v \cot(z_f)]^2}}{\sqrt{n}} \tag{4.40}$$

Example 4.8 What error in misleveling can be expected from a sun shot if the following are given?

Bubble sensitivity = 30″/div

Leveled to within 0.5 fraction of a division of the plate bubble

Backsight zenith angle = 91° 30′ 45″

Foresight zenith angle = 55° 15′ 30″

Solution:

$$\sigma_{\theta L} = \sqrt{\left[\sigma_v \cot\left(Z_b\right)\right]^2 + \left[\sigma_v \cot\left(Z_f\right)\right]^2}$$

σ_v = fractional division of the bubble × sensitivity of the bubble

$\sigma_v = 0.5 \times 30''$ (or 15″)

$$\sigma_{\theta L} = \sqrt{\left[15'' \cot\left(91°30'45''\right)\right]^2 + \left[15'' \cot\left(55°15'30''\right)\right]^2}$$

$$\sigma_{\theta L} = \sqrt{[-0.4]^2 + [10.4]^2} = 10.4''$$

The error in misleveling expected from the sun shot is 10.4″.

4.5.4 Instrument and Target Centering Errors

Centering is a process of setting an instrument or a target over a survey marker so that the vertical axis of the instrument or of the target passes through the marker. Centering error is due to an inability of the observer to perfectly make the vertical axis of a well-leveled and well-adjusted instrument or of a well-leveled and well-adjusted target pass through the center of the survey marker. The magnitude of the centering error depends on the method and equipment used. For example, using a well-adjusted optical plummet, a well-adjusted laser plummet or plumbing rod, with well-defined station marker, will give centering error:

$$\sigma_{c3} = 0.5\,\text{mm/m} \times (\text{Height of Instrument in meters}) \tag{4.41}$$

The string plumb bobs in windless weather condition will give a centering error of 1 mm per height of instrument in meters; it will be quite worse in windy weather. Given a well-defined station marker, the forced centering (or self-centering) method, which requires leaving the tribrachs attached to the tripods and exchanging only the instrument and the targets, will give centering error:

$$\sigma_{c3} = 0.1\,\text{mm/m} \times (\text{Height of Instrument in meters}) \tag{4.42}$$

If the forced-centering is done on a dedicated survey pillar (as in the calibration baselines), the centering error can be taken as 0.1 mm.

Centering error can be divided into two parts: error due to *target miscentering* and error due to *instrument miscentering*. The *effect of target miscentering* error is shown in Figure 4.9, where the erroneous position of target B is likely to be B' or B'' due to the target centering error of $\pm\sigma_{c1}$.

In Figure 4.9, instead of observing direction AB, the observed direction is likely AB' or AB$''$ with a centering error of $\pm\sigma_{c1}$ (meters). If the length of AB is S_1 (meters) and there is no instrument centering error, the target miscentering error on the direction measurement can be given as

$$\sigma_t(\text{direction}) = \frac{\sigma_{c1}}{S_1}(206,265)(\text{arcsec}) \tag{4.43}$$

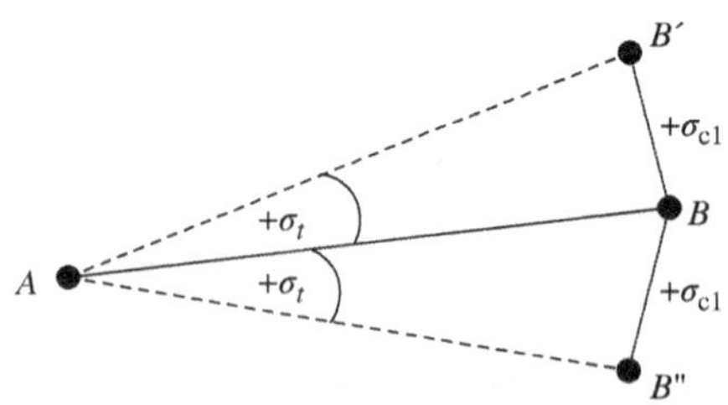

Figure 4.9 Error in direction measurement due to target miscentering.

Considering two targets that are of distances S_1 and S_2 meters away from the instrument station A, the error of target miscentering on angle (θ) at point A can be deduced from Equation (4.43) as

$$\sigma_{\theta t} = (206,265)\sqrt{\left(\frac{\sigma_{c1}}{S_1}\right)^2 + \left(\frac{\sigma_{c2}}{S_2}\right)^2} \text{ (arcsec)} \tag{4.44}$$

where σ_{c1} and σ_{c2} are the miscentering errors of first and second targets, respectively. It should also be understood that the miscentering error is systematic for an individual setup and it occurs on every pointing. Target miscentering error cannot be reduced in size by multiple pointings, but the error will appear random over multiple setups on a point. In this case, the effect of centering will be randomized and reduced by the square root of the number of independent recentering done in the multiple setups while measuring the observable. For example, if an instrument is recentered four times over a point while measuring an angle, the centering error in the angle will be reduced by 2.

Example 4.9 If hand-held targets are centered over a station to within 0.005 m, what is the error in angle due to target miscentering? Assume $S_1 = 75.00$ m and $S_2 = 50.00$ m.

Solution: Using Equation (4.44):

$$\sigma_{\theta t} = (206,265)\sqrt{\left(\frac{\sigma_{c1}}{S_1}\right)^2 + \left(\frac{\sigma_{c2}}{S_2}\right)^2}$$

$$\sigma_{\theta t} = (206,265)\sqrt{\left(\frac{0.005}{75}\right)^2 + \left(\frac{0.005}{50}\right)^2} = 25''$$

The effect of instrument miscentering error on a direction measurement is similar to that due to target miscentering. Similar to Equation (4.43), the instrument miscentering error [σ_i(direction)] on a direction measurement can be given as

$$\sigma_i(\text{direction}) = \frac{\sigma_{c3}}{S_1}(206,265)\text{(arcsec)} \tag{4.45}$$

where σ_{c3} is the amount of error (miscentering error of instrument) by which the vertical axis of the instrument is out of alignment with the survey mark on which the instrument is set up and S_1 is the distance from the instrument setup point to the target being observed to.

The effect of instrument miscentering error on angle (θ) is illustrated in Figure 4.10, where there is an error in every direction due to instrument miscentering error $\pm\sigma_{c3}$. In the figure, the measured angle is either θ' or θ'' instead of θ due

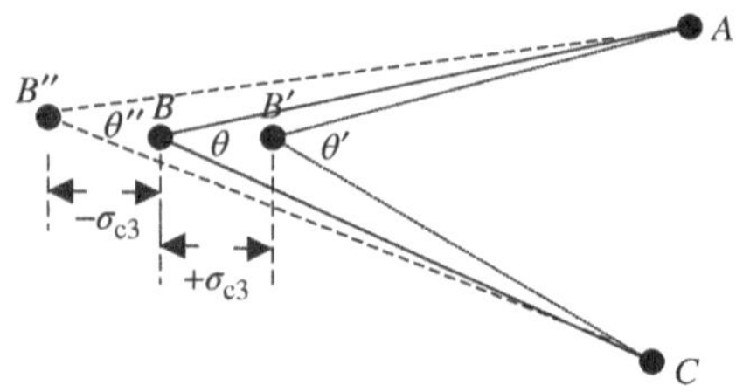

Figure 4.10 Effect of instrument miscentering on angle measurement.

to instrument miscentering error of $\pm\sigma_{c3}$. The error in angle due to instrument miscentering can be given as

$$\sigma_{\theta i} = (206,265)\sqrt{\frac{S_1^2 + S_2^2 - 2S_1S_2\cos\theta}{S_1^2S_2^2}}\sigma_{c3}(\text{arcsec}) \tag{4.46}$$

where S_1 and S_2 are the distances to the first and second targets, respectively.

The error of instrument miscentering is systematic for a particular setup, but appears random with multiple setups and multiple stations.

Example 4.10 What is the error in a 50° angle due to instrument miscentering, if the setup is within 0.002 m and $S_1 = 50.00$ m and $S_2 = 75.00$ m?

Answer:

$$\sigma_{\theta i} = (206,265)\sqrt{\frac{S_1^2 + S_2^2 - 2S_1S_2\cos\theta}{S_1^2S_2^2}}\sigma_{c3}(\text{arcsec})$$

$$\sigma_{\theta i} = (206,265)\sqrt{\frac{50^2 + 75^2 - 2(50)(75)\cos 50}{(50)^2(75)^2}}(0.002) = 6.3''$$

Generally, if the centering error of instrument is σ_{c3}, and the centering errors of the targets are σ_{c1} and σ_{c2}, the influence of the errors on a measured horizontal angle (θ) is derived by combining Equations (4.44) and (4.46) as follows:

$$\sigma_\theta = (206,265'')\sqrt{\frac{\sigma_{c1}^2}{S_1^2} + \frac{\sigma_{c2}^2}{S_2^2} + \frac{\sigma_{c3}^2}{S_1^2S_2^2}(S_1^2 + S_2^2 - 2S_1S_2\cos\theta)} \tag{4.47}$$

where S_1 and S_2 are the slope distances from the instrument station to the two targets involved. The combined influence of the target and instrument centering errors (σ_{c3}, σ_{c1}) on a horizontal direction measurement can be obtained by combining Equations (4.43) and (4.45) as follows:

$$\sigma_d(\text{direction}) = \frac{(206,265'')}{S_1}\sqrt{\sigma_{c3}^2 + \sigma_{c1}^2} \tag{4.48}$$

Notes: The centering error (for instrument or target) is a constant for a setup; the mean direction or mean angle has the same centering error as a single direction or single angle, so that this error is not reduced by taking several repetitions. Also, the effects of centering errors on zenith (or vertical) angle measurements are negligible and can be taken as zero.

4.5.5 Random Atmospheric Refraction Error

By applying the error propagation laws on Equations (4.17) and (4.18), the standard deviations (or random errors) of measurements due to atmospheric refraction corrections (inaccuracy of determining the temperature gradient) can be given as follows. For direction d, the random error (σ_d) due to inaccuracy of determining the horizontal temperature gradient is

$$\sigma_d = \frac{S}{2R}\sigma_{k_\mathrm{h}} \tag{4.49}$$

where σ_{k_h} is the standard error of determining the coefficient of lateral refraction, which is related to the standard error of the lateral temperature gradient ($\sigma_{\partial t/\partial y}$). For zenith angle z, the random error (σ_ref) due to inaccuracy of determining the vertical temperature gradient is

$$\sigma_\mathrm{ref} = \frac{S}{2R}\sigma_{k_\mathrm{V}} \tag{4.50}$$

where σ_{k_V} is the standard error of determining the coefficient of vertical refraction, which is related to the standard error of the vertical temperature gradient ($\sigma_{\partial t/\partial y}$). It should be mentioned that temperature gradients are difficult to measure; usually, regions where significant air temperature variations may occur should be avoided. In order to minimize the effects of the atmosphere, observations should be made under more favorable conditions, such as in the early morning hours, on overcast days, and during cooler seasons.

4.5.6 Random Error Propagation for Angle Measurements

It should be mentioned that the random error components of the atmospheric refractions are omitted in the following error propagation formulas since the values of these error components are usually ignored in reality as they are generally unknown. To minimize the random errors due to refractions, however, measurements must be repeated at significantly different atmospheric conditions. In the error propagation formulas for angle measurements, it is assumed that the errors in the backsight direction measurements are the same as those of the foresight direction measurements.

4.5.6.1 Horizontal Direction Measurements Variance of a horizontal direction measurement d can be given as follows (with refraction effect ignored):

$$\sigma_d^2 = \sigma_\mathrm{p}^2 + \sigma_\mathrm{r}^2 + \sigma_\mathrm{L}^2 + \sigma_{dc_\mathrm{i}}^2 + \sigma_{dc_\mathrm{t}}^2 \tag{4.51}$$

where σ_{p}, σ_{r}, σ_{L}, $\sigma_{d\mathrm{c_i}}$, and $\sigma_{d\mathrm{c_t}}$ are as defined, respectively, for Equations (4.20), (4.26), (4.36), (4.45), and (4.43), respectively; $\sigma_{d\mathrm{c_i}}$ and $\sigma_{d\mathrm{c_t}}$ are the instrument and target miscentering errors respectively, for the direction measurement d. Traditionally, horizontal directions are measured in multiple sets with two pointings and two readings (in face I and face II positions) of the same horizontal (Hz) direction within each set. This measurement procedure is mainly to minimize the instrumental effects on measurements as discussed in Sections 4.2–4.4. If n sets of the same direction are measured, the mean direction $\overline{d}$ can be calculated as follows (using directional theodolite method):

$$\overline{d} = \frac{1}{n}\sum_{i=1}^{n}\left(\frac{\mathrm{Hz}_{\mathrm{I}_i} + \mathrm{Hz}_{\mathrm{II}_i} - 180^\circ}{2}\right) \tag{4.52}$$

where $\mathrm{Hz}_{\mathrm{I}_i}$ and $\mathrm{Hz}_{\mathrm{II}_i}$ are the horizontal direction measurements in face I and face II positions of the instrument for each set i, respectively. If no releveling and no recentering of the instrument and targets are done between sets, the variance of the mean direction measurement is calculated based on the error propagation laws as follows:

$$\sigma_{\overline{d}}^2 = \frac{\sigma_{\mathrm{p}}^2 + \sigma_{\mathrm{r}}^2}{2n} + \sigma_{\mathrm{L}}^2 + \sigma_{d\mathrm{c_i}}^2 + \sigma_{d\mathrm{c_t}}^2 \tag{4.53}$$

with the notations as defined for Equation (4.51).

4.5.6.2 Horizontal Angle Measurements Horizontal angles are derived as differences of pairs of directions. On this basis, the variance of a horizontal angle (θ) derived from two horizontal directions (Figure 4.8) can be given as follows:

$$\sigma_\theta^2 = 2(\sigma_{\mathrm{p}}^2 + \sigma_{\mathrm{r}}^2 + \sigma_{\mathrm{L}}^2) + \sigma_{\theta\mathrm{c_i}}^2 + \sigma_{\theta\mathrm{c_t}}^2 \tag{4.54}$$

where $\sigma_{\theta\mathrm{c_i}}$ and $\sigma_{\theta\mathrm{c_t}}$ are the instrument and target centering errors of the angle θ, respectively, defined in Equations (4.46) and (4.44); and all other notations are as defined in Equation (4.51). If the mean angle $\overline{\theta}$ is derived from the means of horizontal directions measured in n sets (referring to the directional method of angle measurements of Equations (4.22) and (4.31)), the variance of the mean angle can be given as follows:

$$\sigma_{\overline{\theta}}^2 = \frac{\sigma_{\mathrm{p}}^2 + \sigma_{\mathrm{r}}^2}{n} + 2\sigma_{\mathrm{L}}^2 + \sigma_{\theta\mathrm{c_i}}^2 + \sigma_{\theta\mathrm{c_t}}^2 \tag{4.55}$$

Equation (4.55) is based on the assumption that there are no releveling and no recentering of instrument and targets between n sets. If the instrument and targets are releveled and recentered between the n sets, the leveling and the centering variance components will be reduced by n as follows:

$$\sigma_{\overline{\theta}}^2 = \frac{\sigma_{\mathrm{p}}^2 + \sigma_{\mathrm{r}}^2 + 2\sigma_{\mathrm{L}}^2 + \sigma_{\theta\mathrm{c_i}}^2 + \sigma_{\theta\mathrm{c_t}}^2}{n} \tag{4.56}$$

4.5.6.3 Zenith (or Vertical) Angle Measurements Zenith angle can be considered as the difference in the vertical axis direction of the instrument and the line of sight through the instrument telescope to the target. The measured quantity is the direction of the optical axis. The types of errors in measuring *zenith (or vertical) angle observable* are similar to those in measuring horizontal direction observable. Those errors include errors due to pointing ($\sigma_{\rm p}$), reading ($\sigma_{\rm r}$), leveling ($\sigma_{\rm v}$), and vertical atmospheric refraction ($\sigma_{\rm ref}$). The effects of centering errors on zenith (or vertical) angle measurements are negligible. The variance of any zenith angle z can be given as follows:

$$\sigma_z^2 = \sigma_{\rm p}^2 + \sigma_{\rm r}^2 + \sigma_{\rm v}^2 \tag{4.57}$$

where $\sigma_{\rm p}$, $\sigma_{\rm r}$, and $\sigma_{\rm v}$ are expressed in Equations (4.20), (4.26), and (4.34) or (4.35), respectively, with $\sigma_{\rm v}$ accounting for the leveling error of the vertical circle index. If a zenith angle is measured n sets, the variance for the mean zenith angle $\bar{z}$ is given as follows:

$$\sigma_{\bar{z}}^2 = \frac{\sigma_{\rm p}^2 + \sigma_{\rm r}^2 + \sigma_{\rm v}^2}{2n} \tag{4.58}$$

where the vertical circle index is assumed to be releveled for each observation.

Example 4.11 A survey instrument having the smallest display resolution of 1″ (for angle measurement) and 1 mm (for distance measurement) is specified by the manufacturer as having the accuracies of 5″ (according to ISO17123-3) for angle and 2 mm ± 2 ppm (according to ISO17123-4) for distance measurements. Interpret the manufacturer's specifications in terms of what standard deviation to associate with an angle measurement.

Solution: It should be mentioned that a manufacturer's specifications for accuracy of angle measurement (based on ISO17123-3 standard or DIN18723 standard) are based on two-face measurements of a direction. These specifications can be interpreted to mean that when one horizontal direction is measured in two face (direct and reverse) positions of the telescope, the standard deviation for the mean of that direction is 5″; for one face measurement of a direction, the standard deviation of the measurement (using error propagation laws) will be $5''\sqrt{2}$ (or 7.1″). The actual standard deviation of the average angle at a point (within two lines), if four face positions (two face positions in each line) were taken, will be $5''\sqrt{2}$ (or 7.1″) and not 5″.

Example 4.12 A horizontal angle was measured six times by an observer with a total station having an ISO17123-3 value of ±5″. What is the estimated error in the angle, assuming the measurements were made in one face position?

Solution: According to the ISO17123-3 standards, the standard deviation for single direction measured with both faces of instrument is $\sigma_{\rm ISO2} = \pm 5''$. The standard deviation for single-direction measurement in one face position (σ_{ISO1}) can be given as

$$\sigma_{\text{ISO1}} = \sigma_{\text{ISO2}}\sqrt{2} \tag{4.59}$$

From Equation (4.59), the propagated error for an angle measurement will be $\sigma_{\text{ISO1}}\sqrt{2}$ or $\sigma_{\text{ISO2}}\sqrt{2}$. For n measurements of an angle (in one face), the standard deviation of the angle will be reduced by $\sqrt{n}$, which can be given as

$$\sigma_\theta = \frac{2\sigma_{\text{ISO2}}}{\sqrt{n}} \tag{4.60}$$

Substituting the given information into Equation (4.60) gives the estimated error in the angle as

$$\sigma_\theta = \frac{2 \times 5''}{\sqrt{6}} = 4.1''$$

Example 4.13 A direction was observed in one set using a "single-second" theodolite, for example, a Wild T2. Determine the realistic standard deviation of the direction measurement if the line of sight is 500 m with the inclinations of ±30°. For Wild T2 instrument, the magnification (M) is 30×, bubble sensitivity of the plate level is 20″/2 mm run, and micrometer reading is 1″/div.

(Reproduced by permission of CBEPS.)

Solution: T2 specifications: Magnification (M) = 30×
Bubble sensitivity of the plate level = 20″/2 mm run Direction reading = 1″.

Centering error (1 set): $\sigma_c = \pm 0.5$ mm HI for centering error with optical plummet
For height of instrument, HI = 1.5 m; $\sigma_c = \pm 0.75$ mm.
Assuming the same instrument and target miscentering errors, Equation (4.48) can be used as follows:

$$\sigma_{\delta_c}^2 = \frac{(206,265'')}{S_1}\sqrt{\sigma_{c3}^2 + \sigma_{c1}^2}$$

$$\sigma_{\delta_c} = \frac{206,265\sqrt{0.75^2 + 0.75^2}}{500,000} \rightarrow \sigma_{\delta_c} = 2.12132 \times 206,265''(\text{or } 0.44'')$$

Leveling error (1 set):
From Equation (4.31), $\sigma_v = \pm 0.2v''$ leveling error with v = sensitivity/division

$$\sigma_v = \pm 0.2 \times 20'' \rightarrow \sigma_v = \pm 4''$$

From Equation (4.36):

$$\sigma_{\delta_1}^2 = \sigma_v^2 \cot^2 z$$

$$\sigma_{\delta_1}^2 = 4^2 \cot^2 60 \rightarrow \sigma_{\delta_1} = 2.31''$$

Pointing error (1 set):

From Equation (4.20), the pointing error in one-direction measurement is $\sigma_{\delta_p} = \left[\pm\frac{45''}{M}\right]$; in one set (consisting of two measurements), the error will be reduced by $\sqrt{2}$, giving the following:

$$\sigma_{\delta_p} = \frac{1}{\sqrt{2}}\left[\pm\frac{45''}{M}\right]$$

$$\sigma_{\delta_p} = \frac{1}{\sqrt{2}}\left[\pm\frac{45''}{30}\right] \rightarrow \sigma_{\delta_p} = 1.06''$$

Reading error (1 set):

From Equation (4.26), the reading error in one-direction measurement is $\sigma_{\delta_r} = 2.5d''$; in one set (consisting of two measurements), the error will be reduced by $\sqrt{2}$ from Equation (4.32), giving the following (for $d = 1''$):

$$\sigma_{\delta_r} = \frac{1}{\sqrt{2}}(2.5d'')$$

$$\sigma_{\delta_r} = \frac{1}{\sqrt{2}}[\pm 2.5''] \rightarrow \sigma_{\delta_r} = 1.77''$$

Total error in direction measurement (1 set):

$$\sigma_T = \sqrt{\sigma_{\delta_c}^2 + \sigma_{\delta_l}^2 + \sigma_{\delta_p}^2 + \sigma_{\delta_r}^2}$$

$$\sigma_T = \sqrt{0.44^2 + 2.31^2 + 1.06^2 + 1.77^2} \rightarrow \sigma_T = \sqrt{9.7862}$$

$$\sigma_T = 3.13''$$

4.5.7 Error Analysis of Azimuth Determination

Errors in determining the azimuth of a line (based on solar or stellar observations) can be divided into two parts:

(a) Errors in measuring the horizontal angle from the reference line to the celestial object (Sun or star)
(b) Errors in determining the azimuth of the celestial object.

Errors in measuring the horizontal angle are similar to those in any other field angle measurements, except for the pointing errors to the Sun. *Leveling error* is very critical in solar observations; the higher the vertical angle to the sun or the reference object, the more the error in angle measurement. The *pointing error* is also important; the width of a theodolite crosshair in relation to the sun may range from 2 to 3 arcsec and trailing Sun's edge can be pointed within this width, thereby introducing more error than pointing to the backsight reference mark. Errors in determining the sun's azimuth is a function of errors in obtaining UT1 time and the errors in scaling latitude and longitude from large-scale map. The magnitude that these errors will contribute to the total error is in turn dependent on the observer's latitude, declination of the sun, and the time from local noon. The interpolation of the Greenwich hour angle (GHA) and the declination of the sun data have to be done to the time of solar observation; time is therefore the most critical element in an hour angle solar observation.

Latitude is critical in both the hour angle and altitude solar observations because we are solving a large PZS astronomic triangle where the co-latitude is one of the sides. Longitude is also critical in the hour angle solar computation. Errors in scaling latitude and longitude will be constant for all data sets of an observation; each computed azimuth of the sun will contain a constant error; errors in time affect the azimuth in a similar manner. Increasing the number of data sets will not appreciably reduce the sun's azimuth error. But an increase in data sets will improve the horizontal angle accuracy only on the azimuth.

For direction accuracy requirements of about 10″ or less, a star observation will be required. For the middle latitudes of the Northern hemisphere, Polaris is the preferred star to observe for azimuth. Polaris is much less critical for time as long as the observer is at a convenient observing latitude (between 25° and 55°). Polaris requires only interpolation of GHA and the declination for each day since the data change only very slowly in each day for stars, unlike for the sun. At near-pole and near-equator latitudes, a star other than Polaris should be selected (Polaris cannot be seen in the southern latitudes); near the pole, time becomes very critical in determining the azimuth. Polaris, however, may not be visible near the equator and *horizontal refraction may be a problem.* Hour angle method is the most general and convenient method of azimuth determination.

Leveling error is critical in Polaris observation since it is always at a significant vertical angle; this is by far the most significant contributor to error in the Polaris azimuth. In order to minimize leveling error, the instrument must be carefully leveled. In general, the accuracy of an azimuth determination (based on Polaris observations) will depend on the following:

- Instrument and personal errors (such as pointing, reading, leveling, centering). These will affect the measurement of horizontal angle from the line to the Polaris.
- Time of pointing on the Polaris (which also will affect the local hour angle).
- Scaling of the observer's latitude. It is possible to scale the latitude from a good large-scale map to an accuracy of about 10″ (or 300 m).
- Scaling of the observer's longitude; this can also be scaled to an accuracy of 10″ (or 270 m at Latitude 30° or 150 m at Latitude 60°).

In North America, disregarding the effect of errors in horizontal angle from mark to the star, accuracy of 0.5″ is possible. Since errors in scaling latitude and longitude will be constant for all data sets of a stellar observation, each computed azimuth of the star will contain a constant error. Errors in time affect the azimuth in a similar manner. As a result of this, increasing the number of data sets will only improve the horizontal angle accuracy (and the overall azimuth accuracy of the reference line), but will not appreciably reduce the star's azimuth error. Other corrections that may contribute about 0″–0.5″ are as follows (in the order of importance): standing axis error, diurnal aberration, eccentric station, curvature, refraction, Polar variation, and skew normal.

Note that linear interpolation of GHA and declination is highly accurate for stars than for Sun (in solar observations). Solar observations for azimuth determination are not accurate enough for precision projects. Also remember that gyrotheodolites provide astronomic azimuths without any need of observation to any celestial object (star or sun); for example, Sokkia GP1–2A Gyro Station combined with a SET3110 electronic total station, making the system fully automatic, will give astronomic azimuth to 20″ accuracy.

4.5.8 Check of Angular Closure of a Traverse

Example 4.14 Assume that each of the angles shown in Figure 4.11 and given in Table 4.3 was observed in three sets (a set being one face left and one face right measurement) and the estimated standard deviation of measuring an angle at each station was 5″. Does this traverse meet acceptable angular closure at a 95% level of confidence?

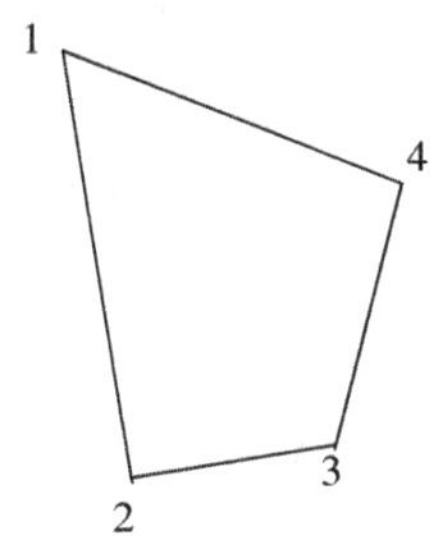

Figure 4.11 Example of a looped traverse survey.

TABLE 4.3 Field Measurements.

Angle	Observed Value
1	60°40′50″
2	91°59′45″
3	107°09′55″
4	100°09′10″

Solution: Formula for misclosure:

$$\hat{w} = \theta_1 + \theta_2 + \theta_3 + \theta_4 - 360^\circ \tag{4.61}$$

Actual angular misclosure, $\hat{w} = -20''$
Two approaches will be used to check if this closure is acceptable at 95% level.

Approach No. 1: If we assume that the allowable discrepancy between each measurement and the mean measurement must be satisfied at each station, we can determine the 95% probable error at each station and then propagate the errors as sum of squares, summing to 95% error values. Each station has 5 redundant measurements (three sets times two measurements per set minus one unknown angle) since one measurement is actually needed.
The allowable discrepancy between the mean measurement and a given measurement at a station can be expressed from Equation (2.49) in Chapter 2, as

$$|\Delta_i| \leq \sigma_{\Delta_i} \times t_{\alpha/2,\mathrm{df}} \tag{4.62}$$

where

$$\sigma_{\Delta_i} = \sigma_{\theta_i}\sqrt{\frac{n+1}{n}} \tag{4.63}$$

$\sigma_{\theta_i} = 5''$ is the standard deviation of each angle measurement, $n = 6$ is the number of measurements used in determining the mean, df = 5 (for three sets of angle measurements or six angle measurements) at each station, the significance level $\alpha/2 = 0.025$ and $t_{0.025,\mathrm{df}=5} = 2.57$. Substituting the appropriate values into Equation (4.62) gives the allowable discrepancy between an angle measurement and the mean of all the measurements at each station as $\Delta_i \leq 13.88''$. For the problem, $\sigma_{\theta_1} = \sigma_{\theta_2} = \sigma_{\theta_3} = \sigma_{\theta_4} = 5''$ and all the angles in the network will have the same standard deviation. The propagated allowable discrepancy for the whole network can be given as follows:

$$\Delta = \sqrt{\Delta_1^2 + \Delta_2^2 + \Delta_3^2 + \Delta_4^2} \tag{4.64}$$

or

$$\Delta = 13.88\sqrt{4} \quad \text{or} \quad \Delta < 27.8''$$

Since $20''$ is less than $27.8''$, then the misclosure is not significantly different from zero. We can then conclude that the traverse angles are well within the range of allowable error. We cannot reject the null hypothesis that the error in the angles is not statistically equal to zero. Note that the target and instrument centering errors affect angle observations only if the instrument and targets are reset after each observation. Since this is never done, only pointing and reading errors are considered mainly for the angle measurements.

Approach No. 2: Consider mean angle at each station as single observation; this assumes that systematic errors are not present at each station (since the average of face left and face right measurements will eliminate the systematic errors due to the instrument). In this case, for each station, there will be three mean values; since only one mean value is needed, the number of degrees of freedom is two (df = 2). For this problem, the modified versions of Equations (4.62) and (4.63) can be used, where $\sigma_{\theta_i} = 5''/\sqrt{2}$ (or $\sigma_{\theta_i} = 3.53''$) is the standard deviation of the mean of each set of angle measurement, $n = 3$ is the number of sets of measurements used in determining the overall mean, df = 2 (for three sets of angle measurements) at each station, the significance level $\alpha/2 = 0.025$ and $t_{0.025,\mathrm{df}=2} = 4.30$. Substituting the appropriate values into Equation (4.62) gives the allowable discrepancy between a set and the mean of three sets at each station as $\Delta_i \leq 17.55''$. For the problem, $\sigma_{\theta_1} = \sigma_{\theta_2} = \sigma_{\theta_3} = \sigma_{\theta_4} = 3.53''$ and all the angles in the network will have the same standard deviation. The propagated allowable discrepancy for the whole network can be given as follows:

$$\Delta = \sqrt{\Delta_1^2 + \Delta_2^2 + \Delta_3^2 + \Delta_4^2} \tag{4.65}$$

or

$$\Delta = 17.55\sqrt{4} \quad \text{or} \quad \Delta < 35.1''$$

Since 20″ is less than 35.1″, then the misclosure is not significantly different from zero. The same conclusion is arrived at as in Approach No. 1.

4.6 TESTING PROCEDURE FOR PRECISION THEODOLITES

This section deals with testing procedure for precision theodolites as measurement system for angles. Usually, theodolites are calibrated by special laboratory instrumentation (using a set of collimators and precise invar scales), determining the following types of errors: horizontal collimation error, vertical (index) collimation error, tilting axis error, errors of horizontal and vertical angle measurements, setting accuracy of compensators and centering device, and circle graduation errors. The testing procedure for a precision theodolite is for determining the best achievable measure of precision (repeatability) of a particular precision theodolite and its supporting equipment under field conditions. The measure of precision of theodolites is expressed in terms of the experimental standard deviation of a horizontal direction observed once in both face positions of the telescope or of a zenith (or vertical) angle observed once in both face positions of the telescope. The testing procedures for horizontal direction and zenith angle measurements will be given separately as follows.

4.6.1 Precision of Theodolite Based on Horizontal Direction Measurements

It was understood from the previous discussions that there are three main sources of error that can contribute to the total random error in the horizontal direction measurements: *instrumental*, *personal*, and *atmospheric conditions*. These error sources were further broken down to include errors due to reading, pointing, instrument leveling, instrument centering, target centering, measuring method (repetition vs. directional method), number of repetitions, sighting conditions (sun, haze, heat waves, ground stability, etc.). Some surveyors may incorrectly think that the only source of error in direction measurement is the reading precision (or least count) of the instrument; and most often, the surveyors use the reading precision to denote the precision of the direction measurements. It should be noted, however, that precision of a theodolite, for example, cannot be inferred from the least count of the theodolite. In fact, with the advent of electronic instruments, reliance on the least count for precision is highly inadvisable.

Some of the sources of error listed in the preceding paragraph can be isolated and tested for in horizontal direction measurements, but the scope of this section is to investigate how to estimate the overall precision of the horizontal direction measuring equipment; the procedures illustrated are not necessarily the standard ones. The readers are referred to the internationally accepted standards (ISO17123-3, 2001) for details of the acceptable procedures. The measurement configuration for the illustration of how to determine the precision of horizontal direction measurement of the measuring equipment (e.g., a theodolite) is shown in Figure 4.12. The test field consists of four well-marked and well-defined targets, labeled 1–4. The targets are situated at almost regular intervals around the same horizontal plane as the instrument's telescope at point *P*. The instrument should be set up between 100 and 250 m away from those targets in order to minimize the effects of instrument miscentering on the direction measurements.

The measuring scheme given in this section is just for the purpose of illustrating how to determine the precision of the horizontal direction measurements. It consists of centering and leveling the theodolite instrument at point *P* and making a series of measurements to targets 1–4 and closing the horizon back at target 1. Assume that *S* series of measurements were made with each series consisting of *T* sets of readings to each direction *P*-1, *P*-2, *P*-3, *P*-4, *P*-1, and each set of readings consisting of two measurements (one in face left position and the second in face right position). In order to randomize the centering and leveling errors, the instrument must be recentered and releveled on point *P* at the beginning of measurement of each series. Targets are to be observed in clockwise sequence with the graduated horizontal circle to be changed by 60° after each set in the case where physical rotation of the graduated circle is possible; for digital theodolites, the theodolite itself may be turned on the tribrach by approximately 120°. Remember that a computed standard deviation for a direction measurement will only be valid when several repetitions of measurements are used in its computation.

4.6.1.1 Precision Determination of Horizontal Direction Measurement Let the index t represent the measurement set number and the index k the target number. Let one-direction measurement in face left position (L) be given by ℓ_{kt}^{L} and for the direction in face right position (R) be ℓ_{kt}^{R}. The steps for determining the standard deviation of horizontal direction measurement will be illustrated using two sets

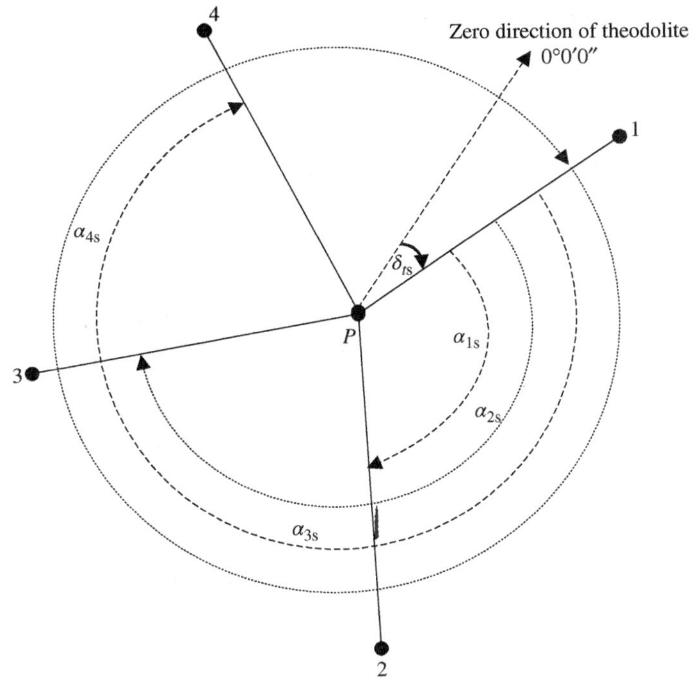

Figure 4.12 Test field for horizontal angle measurements showing the position P of theodolite and the arrangement of targets 1–4 (with subscript t representing set number and subscript s representing series number).

of measurements in only one series (with no closing of horizon for simplicity) as follows:

1. Calculate the mean horizontal direction ($\bar{\ell}_{kt}$) to each direction k in each set t as follows:

   [Recenter and relevel at the beginning of a series.]

   Set $t = 1$: graduated circle is turned only by 60° or 120°

$$\begin{aligned}
\text{To target k} = 1 : \bar{\ell}_{11} &= \frac{\ell_{11}^{\mathrm{L}} + \ell_{11}^{\mathrm{R}} \pm 180^\circ}{2} \\
\text{To target k} = 2 : \bar{\ell}_{21} &= \frac{\ell_{21}^{\mathrm{L}} + \ell_{21}^{\mathrm{R}} \pm 180^\circ}{2} \\
\text{To target } k = 3 : \bar{\ell}_{31} &= \frac{\ell_{31}^{\mathrm{L}} + \ell_{31}^{\mathrm{R}} \pm 180^\circ}{2} \\
\text{To target } k = 4 : \bar{\ell}_{41} &= \frac{\ell_{41}^{\mathrm{L}} + \ell_{41}^{\mathrm{R}} \pm 180^\circ}{2}
\end{aligned} \tag{4.66}$$

Set t = 2: graduated circle is turned only by 60° or 120°

$$\begin{aligned}
&\text{To target } k=1: \bar{\ell}_{12} = \frac{\ell^{L}_{12} + \ell^{R}_{12} \pm 180°}{2}\\
&\text{To target } k=2: \bar{\ell}_{22} = \frac{\ell^{L}_{22} + \ell^{R}_{22} \pm 180°}{2}\\
&\text{To target } k=3: \bar{\ell}_{32} = \frac{\ell^{L}_{32} + \ell^{R}_{32} \pm 180°}{2}\\
&\text{To target } k=4: \bar{\ell}_{42} = \frac{\ell^{L}_{42} + \ell^{R}_{42} \pm 180°}{2}
\end{aligned} \tag{4.67}$$

2. Use the eight mean horizontal directions calculated in step 1 to formulate the least squares parametric equations as follows:
   For set t = 1:

$$\begin{aligned}
&\bar{\ell}_{11} = \delta_{11} \quad \text{(For the first measurement)}\\
&\bar{\ell}_{21} = \delta_{11} + \alpha_{11}\\
&\bar{\ell}_{31} = \delta_{11} + \alpha_{21}\\
&\bar{\ell}_{41} = \delta_{11} + \alpha_{31}
\end{aligned} \tag{4.68}$$

For set t = 2:

$$\begin{aligned}
&\bar{\ell}_{12} = \delta_{21} \quad \text{(For the first measurement)}\\
&\bar{\ell}_{22} = \delta_{21} + \alpha_{11}\\
&\bar{\ell}_{32} = \delta_{21} + \alpha_{21}\\
&\bar{\ell}_{42} = \delta_{21} + \alpha_{31}
\end{aligned} \tag{4.69}$$

where the vector of unknown parameters is given as $x^T = \begin{bmatrix}\delta_{11} & \delta_{21} & \alpha_{11} & \alpha_{21} & \alpha_{31}\end{bmatrix}$, δ_{11} and δ_{21} are the unknown orientation of the zero index reading of telescope with respect to the first direction to target 1 (there are two since the graduated horizontal circle is changed by 60° at the beginning of each of the two sets); α_{11}, α_{21}, and α_{31} are the unknown included angles between pairs of lines (these angles are to remain the same in a series since the instrument is not releveled and not recentered). If the above procedure is repeated exactly in another series (after releveling and recentering of instrument), there will be exactly another 5 unknown parameters (giving a total of 10 unknown parameters in two series).

3. From the parametric equations in step 2, calculate the adjusted values of the unknown parameters by least squares method as follows:

$$\hat{x} = \begin{bmatrix} \delta_{11} \\ \delta_{21} \\ \alpha_{11} \\ \alpha_{21} \\ \alpha_{31} \end{bmatrix}^0 + \delta \tag{4.70}$$

where

$\begin{bmatrix} \delta_{11} \\ \delta_{21} \\ \alpha_{11} \\ \alpha_{21} \\ \alpha_{31} \end{bmatrix}^0$ is a vector of the approximate values of the unknown parameters;

δ is the vector of corrections to be applied to the approximate values of the parameters, which can be given as

$$\delta = -(A^{\mathrm{T}}A)^{-1}A^{\mathrm{T}}w \tag{4.71}$$

A is the Jacobian matrix of the 8 parametric equations with respect to the 5 unknown parameters and w is the misclosure vector.

4. Calculate the sample standard deviation ($\hat{\sigma}$) of the mean horizontal direction measurements taken at the face left and face right positions of the telescope as follows:

$$\hat{\sigma} = \sqrt{\frac{r^{\mathrm{T}}r}{\mathrm{df}}} \tag{4.72}$$

where df $= n - u$ is the number of degrees of freedom and r is the residual vector given as follows:

$$r = A\hat{\delta} + w \tag{4.73}$$

5. The sample standard deviation ($\hat{\sigma}$) computed in Equation (4.72) should be statistically tested using the Chi-square testing procedure in Chapter 2, Equation (2.56), to check if the computed sample standard deviation ($\hat{\sigma}$) is compatible with the standard deviation (σ) provided by the manufacturer of the equipment. This testing procedure is illustrated in the following Example 4.14.

Example 4.15 If the standard deviation of a horizontal direction measurement with a theodolite is provided by the manufacturer as $\sigma = 2''$, check if the experimental standard deviation ($\hat{\sigma} = 2.3''$) of the mean of the measurements made in both face left and face right positions of the telescope is smaller than or equal to the manufacturer's value at 95% confidence level. Assume that the number of degrees of freedom for the determination of $\hat{\sigma}$ is 32.

Solution: This example deals with hypothesis testing to determine if the calculated experimental standard deviation, $\hat{\sigma}$, is smaller than or equal to the manufacturer's (or some other) predetermined value σ at the confidence level $1-\alpha$. This is a typical test of hypothesis for a population variance discussed in Section 2.9. The two hypotheses involved can be given as

$$H_0 : \hat{\sigma} \leq \sigma \quad H_A : \hat{\sigma} > \sigma$$

Given $\sigma = 2''$ and $\hat{\sigma} = 2.3''$, $\alpha = 0.05$ and df = 32.
The test statistic to be used is given in Equation (2.56) in Section 2.9.3 as

$$s_\alpha \leq \sqrt{\frac{\chi^2_{\alpha,\mathrm{df}}(\text{upper-tail area})\sigma^2}{\mathrm{df}}}$$

$$\hat{\sigma} \leq \sqrt{\frac{\chi^2_{\alpha=0.05,\mathrm{df}=32} \times 2^2}{32}} = \sqrt{\frac{46.19 \times 4}{32}}$$

$$2.3'' \leq 2.4''$$

Since the above condition is fulfilled, the null hypothesis stating that the empirically determined standard deviation, $\hat{\sigma} = 2.3''$, is smaller than or equal to the manufacturer's value, $\sigma = 2''$, is not rejected at the confidence level of 95%.

4.6.2 Precision of Theodolite Based on Zenith Angle Measurements

The purpose of this section is not to give the standard testing procedures but to present the method of calculating the precision of a zenith angle, which can be used in conjunction with any standard procedures. Those who are interested in testing their equipment according to the internationally accepted standards should refer to the ISO standards (ISO17123-3, 2001) for the appropriate testing procedures. In this section, the zenith angle measurements for determining the precision of zenith angle of a theodolite will be based on the measurement configuration shown in Figure 4.13. The configuration consists of three well-marked lines (serving as targets) on a precise invar leveling rod (or on a high-rise building), labeled 1–3. The theodolite should be set up in a distance approximately 5 m from the precise invar rod (or the high-rise building) covering the range of zenith angle of approximately 60° as shown in Figure 4.13.

Assuming after setting up the instrument at point *P* and allowing the instrument to acclimatize to the ambient temperature, the following measurements are made: *S* series of measurements consisting of *T* sets of zenith angle readings to targets 1–3 with each set consisting of two measurements (one in face left position and the second in face right position). The standard deviation of one zenith angle measurement can be determined as follows.

4.6.2.1 Precision Determination of Zenith Angle Measurement Let index *t* represent the measurement set number and index *k* the target number. Let one zenith

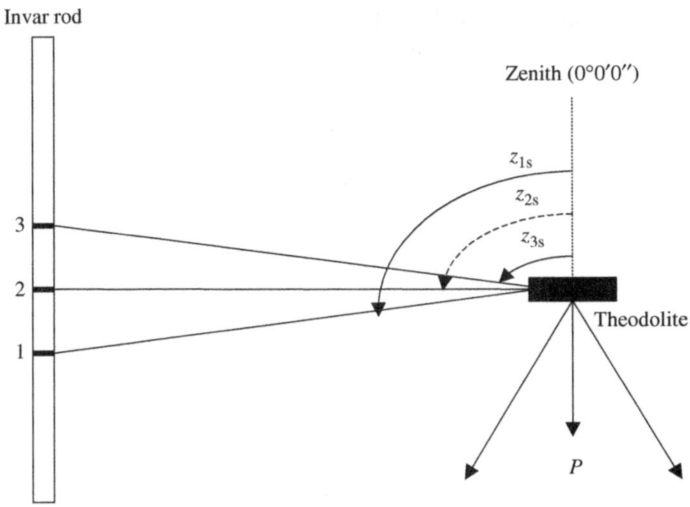

Figure 4.13 Test field for zenith angle measurements (with subscript *s* representing series number) showing the position *P* of the theodolite and the invar rod targets 1–3.

angle measurement in face left position (L) be given by ℓ_{kt}^{L} and for the zenith angle in face right position (R) be ℓ_{kt}^{R}. The steps for determining the standard deviation of zenith angle measurement will be illustrated using two sets of measurements in only one series as follows:

1. Calculate the mean zenith angle ($\overline{\ell}_{kt}$) to each target *k* in each set *t* as follows:
   Set t = 1:

$$\text{To target } k = 1 : \overline{\ell}_{11} = \frac{\ell_{11}^{\mathrm{L}} - \ell_{11}^{\mathrm{R}} + 360^\circ}{2}$$
$$\text{To target } k = 2 : \overline{\ell}_{21} = \frac{\ell_{21}^{\mathrm{L}} - \ell_{21}^{\mathrm{R}} + 360^\circ}{2} \tag{4.74}$$
$$\text{To target } k = 3 : \overline{\ell}_{31} = \frac{\ell_{31}^{\mathrm{L}} - \ell_{31}^{\mathrm{R}} + 360^\circ}{2}$$

   Set t = 2:

$$\text{To target } k = 1 : \overline{\ell}_{12} = \frac{\ell_{12}^{\mathrm{L}} - \ell_{12}^{\mathrm{R}} + 360^\circ}{2}$$
$$\text{To target } k = 2 : \overline{\ell}_{22} = \frac{\ell_{22}^{\mathrm{L}} - \ell_{22}^{\mathrm{R}} + 360^\circ}{2} \tag{4.75}$$
$$\text{To target } k = 3 : \overline{\ell}_{32} = \frac{\ell_{32}^{\mathrm{L}} - \ell_{32}^{\mathrm{R}} + 360^\circ}{2}$$

   Note that the calculated mean zenith angles are free from vertical index error.

2. Use the six mean zenith angles calculated in step 1 to formulate the least squares parametric equations as follows:

   For set t = 1:

$$\begin{aligned} \bar{\ell}_{11} &= z_{11} \\ \bar{\ell}_{21} &= z_{21} \\ \bar{\ell}_{31} &= z_{31} \end{aligned} \tag{4.76}$$

   For set t = 2:

$$\begin{aligned} \bar{\ell}_{12} &= z_{11} \\ \bar{\ell}_{22} &= z_{21} \\ \bar{\ell}_{32} &= z_{31} \end{aligned} \tag{4.77}$$

   where the vector of unknown parameters is given as $x^{\mathrm{T}} = \begin{bmatrix} z_{11} & z_{21} & z_{31} \end{bmatrix}$, and z_{11}, z_{21}, z_{31} are the respective zenith angles (free of index error). If the above procedure is exactly repeated in another series (after releveling the instrument), there will be another 3 unknown parameters (giving a total of 6 unknown parameters in two series).

3. From the parametric equations in step 2, calculate the adjusted values of the unknown parameters by least squares method as follows:

$$\hat{x} = \begin{bmatrix} z_{11} \\ z_{21} \\ z_{31} \end{bmatrix}^0 + \delta \tag{4.78}$$

   where

   $\begin{bmatrix} z_{11} \\ z_{21} \\ z_{31} \end{bmatrix}^0$ is the vector of the approximate values of the parameters;

$$\delta = -(A^{\mathrm{T}}A)^{-1}A^{\mathrm{T}}w; \tag{4.79}$$

   A is the Jacobian matrix of the six parametric equations with respect to the unknown three parameters, and w is the misclosure vector.

4. Calculate the sample standard deviation ($\hat{\sigma}$) of the mean zenith angle measurements taken at the face left and face right positions of the telescope, as follows:

$$\hat{\sigma} = \sqrt{\frac{r^{\mathrm{T}}r}{\mathrm{df}}} \tag{4.80}$$

where df = $n - u$ is the number of degrees of freedom and r is the residual vector given as follows:

$$r = A\hat{\delta} + w \tag{4.81}$$

5. The sample standard deviation ($\hat{\sigma}$) computed in Equation (4.80) should be statistically tested by using the Chi-square testing procedure in Section 2.9.3, Equation (2.56), to check if the computed sample standard deviation ($\hat{\sigma}$) is compatible with the standard deviation (σ) provided by the manufacturer of the equipment. The testing procedure is similar to the one illustrated in Example 4.14.

Example 4.16 On the shelf in the company's survey stores, you have found a total station that has not been used for at least 20 years. The manufacturer's claim, following DIN 18723 [or ISO 17123, now], is an angular "accuracy," horizontally or vertically, of $\pm 2''$. Since there is no record of any testing or calibration of this particular instrument, explain the steps that you would recommend following to determine whether this total station is capable of behaving as the manufacturer claimed. (Reproduced by permission of CBEPS.)

Suggested Solution: The testing procedures consist of two parts: procedure for determining the precision of horizontal direction measurements and that for determining the precision of zenith angle measurements. Refer to Section 4.6.1 for precision determination of horizontal direction measurements and note the following differences with regard to ISO17123-3 (2001) standards:

- 5 targets are used for full test procedure;
- 4 series of direction measurements with 3 sets per series are to be made to the 5 targets.
- Graduated circle is to be changed by 60° after each set or lower part of theodolite rotated by 120° on the tribrach after each set.
- Parametric least squares approach is not applied in adjusting the measurements.

With regard to precision determination of zenith angles, refer to Section 4.6.2 and take note of the following differences with regard to ISO17123-3 (2001) standards:

- Distances between instrument and targets are to be about 50 m;
- 4 series of zenith angle measurements with 3 sets per series are made to 4 targets;
- No releveling and recentering of instrument is required between sets or between series;
- Parametric least squares approach is not applied in adjusting the measurements.

5

ACCURACY ANALYSIS AND EVALUATION OF DISTANCE MEASUREMENT SYSTEM

OBJECTIVES

After studying this chapter, you should be able to

1. Describe the general properties of electromagnetic (EM) waves, including the spectrum
2. Discuss the application of EM waves to EDM including the basic principles of EDM measurement
3. Perform computations related to EM wave propagation
4. Apply velocity corrections to EDM measurements
5. Analyze the accuracy of distance measurements, including sources of errors and the appropriate error budgets
6. Formulate error propagation for distance measurement
7. Evaluate geodetic EDM under field conditions (instrumental and scale errors)

5.1 INTRODUCTION

The accuracy analysis and evaluation of distance measurement system are discussed in this chapter. The modern distance measurement system is the electromagnetic distance measurement (EDM) instrument, which is now an integral component of the

Precision Surveying: The Principles and Geomatics Practice, First Edition. John Olusegun Ogundare.

modern total station instruments. The EDM instruments use the properties of electromagnetic waves to measure spatial distances. In order to analyze and evaluate the accuracy of EDM distance measurements, it is necessary to first understand how EDM distance measurements are obtained, which includes understanding the general properties of waves.

5.2 GENERAL PROPERTIES OF WAVES

A wave is a moving disturbance in a medium that transports energy from one point to another without transporting the material of the medium. The following summarizes some of the properties of waves:

- A wave transports energy and momentum from a source through vibrations (with or without the help of a medium).
- A wave has no mass.
- Waves continue to travel after the source is turned off.
- Waves can pass through one another; after passing through, they continue on their separate ways.
- When two waves overlap, the total wave is just the sum of the two waves.
- Speed of a wave is solely determined by the characteristics of the medium, not by the frequency. For example, the speed of a particular wave in air is different from its speed in water; the frequency of the wave will remain the same in both media.
- When a wave passes into another different medium, its speed changes, but its frequency will not, that is, change in medium does not change the frequency of wave.

There are two types of waves: Transverse and Longitudinal:

- *Transverse waves* – The direction in which the particles of the conducting medium oscillate (or vibrate) is perpendicular to the direction in which the wave travels, for example, electromagnetic (EM) waves.
- *Longitudinal waves* – The direction in which the particles of the conducting medium oscillate (or vibrate) is parallel to the direction in which the wave travels, for example, sound waves.

Figure 5.1 illustrates the characteristics of waves based on the familiar water waves as an example. In the figure, a water drop into a lake produces longitudinal waves (traveling disturbances with the water molecules not traveling with the waves) that propagate on the surface of the water. The line joining all the *crests* of a wave is the *wave front*.

For distance measurements, electromagnetic (EM) waves are used. The general properties of EM waves can be seen for different waves shown in Figure 5.2.

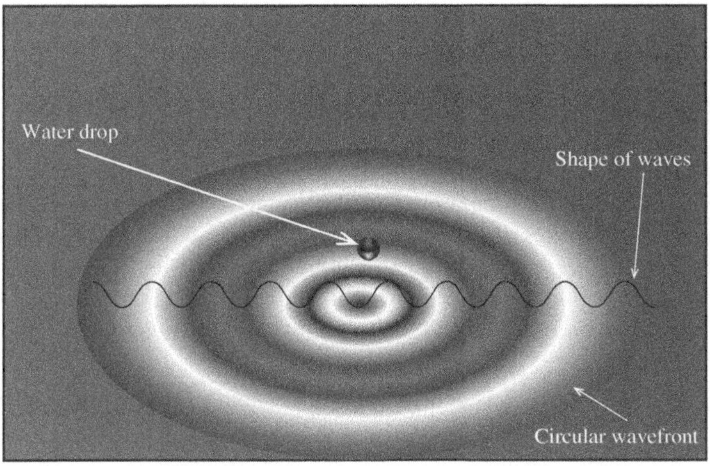

Figure 5.1 Familiar circular water waves.

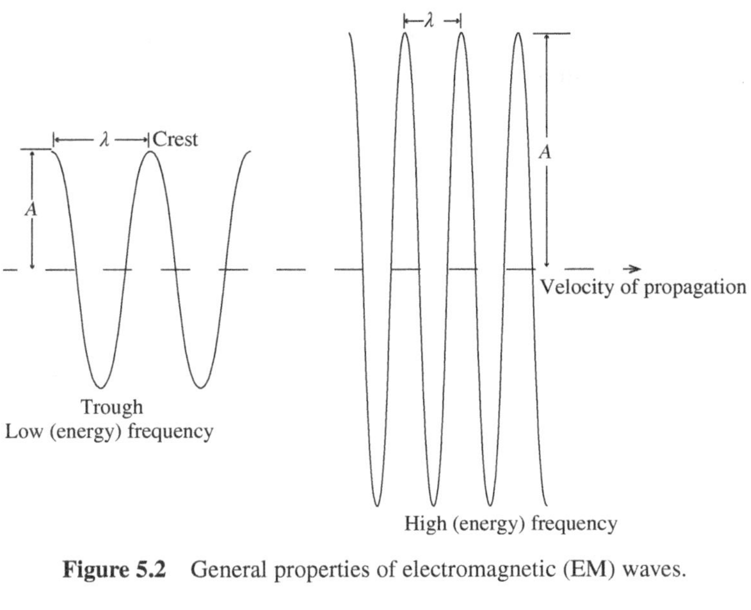

Figure 5.2 General properties of electromagnetic (EM) waves.

In Figure 5.2, λ is the wavelength and A is the amplitude. Crests are the highest points of a wave while troughs are the lowest points of a wave. Velocity of waves (c) in a vacuum is expressed as

$$c = f\lambda \tag{5.1}$$

where f is the frequency of the wave (number of complete waves produced per second measured in hertz or Hz), λ is the wavelength (the distance between any two identical points on successive waves measured in meter or m). Three very important parameters describing an EM wave are as follows:

- *Frequency*, which describes how rapidly the wave oscillates or the color or energy level of the wave; it is the number of times the particle of the medium at a given spot moves up and back to equilibrium level in 1 s.
- *Amplitude*, which is the maximum displacement from equilibrium that any point in the medium makes as the wave travels by. This indicates how much potential energy the wave transports. In another words, it indicates how bright or intense the wave or its source is. Amplitude only depends on how much energy is input and does not depend on the frequency (f), velocity of wave (v), and wavelength (λ).
- *Velocity* of wave, which is constant and equivalent to that of light in a vacuum, that is, 299,792,458 m/s $\pm$ 1.2 m/s.

The EM waves consist of both electric and magnetic fields, which travel through space together under certain circumstances. The electric and magnetic components of the wave are always perpendicular to each other and also to the direction in which the wave is traveling. As illustrated in Figure 5.3, an oscillating electric charge generates electric field (E) and magnetic field (B) perpendicular to each other with the EM waves being generated in the direction perpendicular to the two fields. These electric and magnetic fields oscillate with the same frequency as the source charges that created them. The EM waves generated, however, become independent of the source charges as they move far away from the source; at this stage, they now generate and regenerate themselves as a result of their own changing electric and magnetic fields.

The sun is able to radiate a wide range of wavelengths of EM energy. A chart showing the different classes of electromagnetic radiation according to their wavelengths with each class associated with a descriptive name is known as *EM spectrum*. A portion of the EM spectrum showing some of the classes of EM radiation that are usable in surveying is given in Figure 5.4.

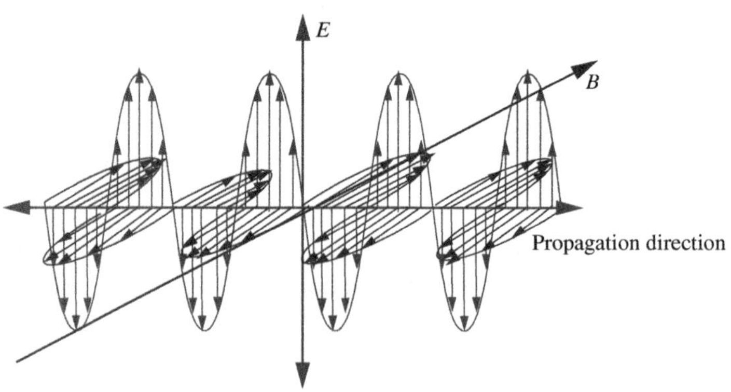

Figure 5.3 Electromagnetic (EM) wave propagation in space (E is the direction of electric field; B is the direction of magnetic field).

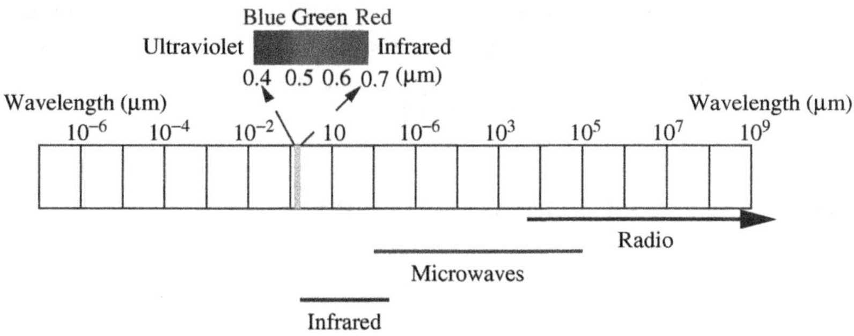

Figure 5.4 A portion of the electromagnetic spectrum.

5.2.1 Modulation of EM Waves

In order to further explain the properties of electromagnetic (EM) waves, the following terms will be described: *modulation*, *carrier wave*, *modulating signal,* and *modulated signal*. *Modulation* is a process in which an electromagnetic signal (the modulating signal) is encoded into one or more of the characteristics of another signal (the carrier wave) to produce a third signal (the modulated signal), whose properties are matched to those of the medium in which the signal is being transmitted. The main purpose of modulation therefore is to overcome any inherent incompatibilities between the electromagnetic properties of the modulating signal and those of the medium in which signal is transmitted. A carrier wave is an electromagnetic wave that is capable of carrying some data just as a long narrow steel band is capable of carrying some scale graduations for distance measurement. The data carried by the carrier wave is known as the modulating signal (just like the graduations on a steel band). Carrier waves are usually of higher frequencies than the modulating signal. The modulating signal contains the data (units of length), which the carrier wave will carry for use in distance determination, just as the steel band will carry scale graduations for use in distance measurement. When a carrier wave is modulated, the carrier wave is encoded with data, and the data can be recovered later from the modulated wave by a process called demodulation.

Modulation also means varying some parameter of the carrier wave using modulating signal. The parameters of the carrier that can be varied are its *amplitude*, *frequency,* or *phase*. Based on this, the type of modulation can be amplitude modulation, frequency modulation, or phase modulation. The instruments using infrared and visible-spectrum as carrier will employ amplitude (or intensity) modulation, those using microwave as carrier will use direct frequency modulation, and those using long radio waves as carriers use no modulation at all. The signal used for EDM is modulated by imposing on the basic carrier wave a series of modulation frequencies, which are used for measurement. The velocity of the resultant waveform is the *group velocity*. The corresponding refractive index is known as *group refractive index*.

In modern EDM instruments, time is no more directly measured, but the number of one-half of the modulation wavelength ($\lambda/2$) also known as *pattern wavelength* or *unit length* of instrument. The modulation frequency can be calibrated with an accuracy of about 0.1 ppm and can be stabilized during the use of the EDM instrument (at constant temperature), but may drift per year, making it necessary to calibrate the instrument and determine the amount of drift.

5.3 APPLICATION OF EM WAVES TO EDM

Two main types of EDM are in common use:

1. Electromagnetic (microwave) EDM, which uses the microwave and radio parts of the spectrum (with wavelengths that are greater than 10^3 μm and frequencies that are in the range of 3–30 GHz) as carrier. Since microwave has a longer wavelength, it has better penetration through haze and fog and is good for long distance measurements; distances up to 150 km can be measured with this type of EDM. An example of this type of EDM is Tellurometer with possible accuracy of ±15 mm to 5 mm/km.
2. Electro-optical (light wave) EDM, which uses the visible part of the spectrum (light with 0.4–0.7 μm wavelengths or infrared (IR) with 0.7–0.9 μm wavelengths) as carrier. Since light waves have shorter wavelengths, they produce better accuracy if visibility is good. The modern short-range types of this EDM use IR and are capable of measuring distances between 0.1 m up to 5 km depending on the number of prisms used; and the long-range types use coherent lasers and are capable of measuring distances up to 70 km with many prisms. An example of long-range type is Geodimeter (using amplitude modulation) with possible accuracy of ±10 mm to ±2 mm/km; the near IR type of the EDM is Distomat with possible accuracy of ±10 mm.

The basic principles of measurement used in the two main types of EDMs are *time* (*pulse or time-of-flight*) and *phase* (or *continuous wave*) measurement principles.

5.3.1 EDM Pulse Measurement Principle

EDM pulse measurement is based on the principle where the EDM instrument transmits a short and intensive signal and the time (ΔT) taken by the signal to travel to and from the target is recorded and used to determine the one-way distance (d) between the instrument and the target. For example, the measured distance, d, can be expressed as

$$d = \frac{v(\Delta T)}{2} \tag{5.2}$$

where v is the speed of the electromagnetic wave; ΔT is the time from the start pulse to the return pulse (measured in EDM). The speed of EM energy (v) can be expressed in terms of the well-known speed (c) of the electromagnetic wave in the vacuum:

$$v = \frac{c}{n} \tag{5.3}$$

where n is the index of refraction of the atmosphere (varying between 1.0001 and 1.0005), which is mainly a function of temperature and pressure. Using Equations (5.2) and (5.3), the EDM distance by *pulse measurement* becomes

$$d = \frac{c(\Delta T)}{2n} \tag{5.4}$$

Pulse technique is widely used in geodesy and in other applications. Some of the electro-optical applications of pulse technique include *Lunar Laser Ranging* (LLR) and Satellite Laser Ranging and Tracking (SLRT). Some of the microwave and radio waves applications of pulse technique include Radio Detection And Ranging (*RADAR*) and Satellite Radar Altimeter.

5.3.2 EDM Phase Difference Measurement Principle

Phase (or continuous wave) measurement technique of distance determination considers the phase of a wave as the level of energy from 0 to 2π (usually expressed in radians) at a point of the wave. It can be used to describe the position of a point (at a given time) in one wave relative to another wave. The EDM phase measurement principle can be explained with the continuous wave diagram given in Figure 5.5.

With the aid of Figure 5.5, the total 2-way distance traveled by a measuring wave can be generalized as follows:

$$D = \lambda M + \frac{\lambda(\Delta\varphi)}{2\pi} \tag{5.5}$$

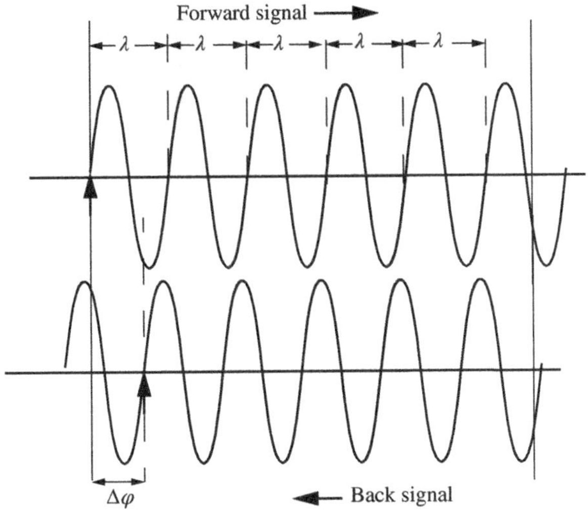

Figure 5.5 EDM phase measurement technique.

where

M is the unknown integer number of modulation wavelengths (λ) over the 2-way measuring path or the integer ambiguity;

λ is the wavelength of the EDM signal;

$\Delta\varphi$ is the phase delay in radian (measured in the EDM);

$\Delta\varphi/2\pi$ is the phase delay as a fraction of a cycle (measured in the EDM).

Using a phase resolver, a typical phase resolution is 3×10^{-4} of the modulation wavelength (λ). For a unit length ($\lambda/2$) of 10 m, the accuracy of distance measurement will be about $3 \times 10^{-4} \times 10$ m (or 0.003 m). This is to say that a phase delay in this case will only be measured to four decimal places and the equivalent distance to three decimal places. For example, it is possible for the resolver to measure a phase delay of 0.0123 of a cycle (not 0.01226 of a cycle), giving an equivalent distance (if the unit length is 10 m) as 0.123 m. Phase measurements are usually repeated several times during distance determination, and the phase error is decreased by averaging the measurements. The one-way distance (d) between the transmitter and the receiver can be deduced by dividing Equation (5.5) by 2:

$$d = \frac{M\lambda}{2} + \left(\frac{\Delta\varphi}{2\pi}\right)\frac{\lambda}{2} \tag{5.6}$$

Let the unit length of EDM be $U = \lambda/2$ and $L = (\Delta\varphi/2\pi)U$, Equation (5.6) becomes

$$d = MU + L \tag{5.7}$$

The unknowns in Equation (5.7) are the d (which is *constant for a length*) and M (which *varies depending on the unit length* of instrument used); unit length (U) is usually provided for each instrument by the manufacturer. The value of M is considered as the integer ambiguity (unknown integer number of modulation wavelengths) that must be solved for and the value of M is determined in the EDM as the EDM successively sends out (and receives back) signals at different frequencies. Some EDMs can send up to four signals of different frequencies. Four signals of different frequencies will result in four unit lengths U_i $(i = 1, 2, 3, 4)$ with some unit lengths ranging from 10 m to 10 km.

The general equation for distance measurement in EDM based on several unit lengths can be formulated from Equation (5.7) as

$$d = M_i U_i + L_i \tag{5.8}$$

where $L_i = (\Delta\varphi_i/2\pi)U_i$; $U_i = \lambda_i/2$, and λ_i is the wavelength of measuring signal modulated on the carrier wave. The phase delay $\Delta\varphi_i$ (in radians) is measured in the EDM by comparing incoming phase with an onboard (receiver) reference. Examples 5.1 and 5.2 explain how the EDM instruments indirectly fix the integer ambiguities (M_i) when measuring distances.

Example 5.1 A total station sent three different frequencies from point B to a remote prism at point A. The returned signals from the prism back to the total station at point B are shown in Figure 5.6 with the following phase delays (in fractions of a unit length):

$$p_1 = \frac{\Delta\varphi_1}{2\pi} = 0.2135 \text{ for } U_1 = 10\,\text{m}; p_2 = \frac{\Delta\varphi_2}{2\pi} = 0.0214 \text{ for } U_2 = 100\,\text{m, and}$$

$$p_3 = \frac{\Delta\varphi_3}{2\pi} = 0.1025 \text{ for } U_3 = 1000\,\text{m}$$

Determine the ambiguities M_i for the total station measurements.

Solution:

$$p_1 = \frac{\Delta\varphi_1}{2\pi} = 0.2135 \text{ for } U_1 = 10\,\text{m}; p_2 = \frac{\Delta\varphi_2}{2\pi} = 0.0214 \text{ for } U_2 = 100\,\text{m};$$

$$p_3 = \frac{\Delta\varphi_3}{2\pi} = 0.1025 \text{ for } U_3 = 1000\,\text{m}$$

Equation (5.8) can be rewritten as

$$d = M_i U_i + \frac{\Delta\varphi_i}{2\pi} U_i \tag{5.9}$$

or

$$d = (M_i + p_i) U_i \tag{5.10}$$

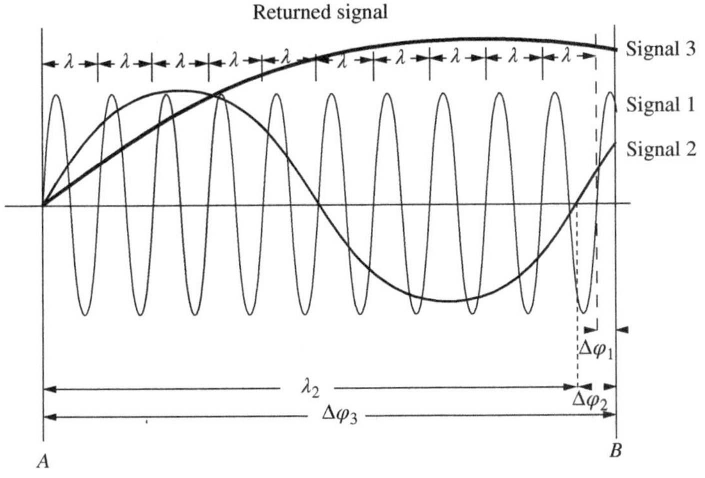

Figure 5.6 Resolving ambiguities in EDM measurements.

where $p_i = \Delta\varphi_i/2\pi$ (in fractions of a cycle). Considering Figure 5.6 and Equation (5.10), the following three equations for the three different frequencies (or wavelengths) with corresponding phases p_1, p_2, and p_3 can be formulated:

$$d = (M_3 + p_3)U_3 \tag{5.11}$$

$$d = (M_2 + p_2)U_2 \tag{5.12}$$

$$d = (M_1 + p_1)U_1 \tag{5.13}$$

Starting from Equation (5.11) and knowing from Figure 5.6 (and also understanding that an instrument cannot measure a distance that is longer than its longest unit length, U_3), the number of integer wavelength for this wave is $M_3 = 0$, for $p_3 = 0.1025$ and $U_3 = 1000$ m. We can then obtain

$$d = (0 + 0.1025) \times 1000.000 = 102.500\,\text{m} \tag{5.14}$$

(This gives the approximate distance as 102.500 m.)
Use the approximate distance (102.500 m) to obtain the approximate integer wavelength, M_2 from Equation (5.12) as

$$M_2 = \frac{102.5}{U_2} - p_2 \text{ or } M_2 = \frac{102.5}{100} - 0.0214$$

Use $M_2 = 1, p_2 = 0.0214$, and $U_2 = 100$ m back into Equation (5.12) in order to obtain a more precise distance:

$$d = (1 + 0.0214) \times 100.000 = 102.140\,\text{m} \tag{5.15}$$

The more precise distance is now 102.140 m; use this value in Equation (5.13) to obtain the approximate integer wavelength (M_1) for this wave:

$$M_1 = \frac{102.14}{U_1} - p_1 \text{ or } M_1 = \frac{102.14}{10} - 0.2135 = 10$$

Use $M_1 = 10, p_1 = 0.2135$, and $U_1 = 10$ m back into Equation (5.13) in order to obtain the most precise distance:

$$d = (10 + 0.2135) \times 10.000 = 102.135\,\text{m} \tag{5.16}$$

The most precise and final distance (recorded by the instrument) is 102.135 m. The simple way to quickly produce the measured distance from the data provided in relation to Figure 5.6 is summarized in Table 5.1. In the table, items in column (4) are obtained by multiplying corresponding items in columns (2) and (3) and aligning the digits as shown. Since p_1 has finer unit length (column 3), the whole corresponding digits given in column (4) will be considered good; as the unit length increases, the

TABLE 5.1 Simple Approach for Resolving EDM Ambiguities – Example 5.1.

Position (1)	Phase Delay, $p_i = \frac{\Delta\varphi_i}{2\pi}$ (2)	Unit Length (U_i) (3)	$L_i = p_i \times U_i$ (4)
1	0.2135	10 m	2.135
2	0.0214	100 m	02.14
3	0.1025	1000 m	102.5
		Measured distance	102.135

TABLE 5.2 Simple Approach for Resolving EDM Ambiguities – Example 5.2.

Position (1)	Phase Delay, $p_i = \frac{\Delta\varphi_i}{2\pi}$ (2)	Unit Length (U_i) (3)	$L_i = p_i \times U_i$ (4)
1	0.8253	10 m	8.253
2	0.4384	1000 m	438.4
		Measured distance	438.253

decimal parts as shown in column (4) become less reliable. The final measured distance will be 102.135 (where the 102 part is composed from 2, 02, and 102 from p_1, p_2, and p_3, respectively; the 0.135 part is composed from 0.135 from p_1 only). The underlined figures in column (4) are transferred mechanically to the distance readout of the instrument, giving 102.135 as the final distance measurement.

Example 5.2 An EDM capable of a maximum range of 1 km has two unit lengths, $U_1 = 10$ m and $U_2 = 1000$ m. Using the EDM to measure a distance *AB*, the phase delay measurements (in fractions of a unit length) are 0.8253 and 0.4384, respectively. What is the value of the distance *AB*?

Solution: Following the similar approach used in Example 5.1, Table 5.2 is obtained as follows. The measured distance will be 438.253 (the 438 part is composed from 8 and 438 from p_1 and p_2, respectively; the 0.253 part is composed from 0.253 from p_1 only). The underlined figures in column (4) are transferred mechanically to the distance readout of the instrument, giving 438.253 as the final distance measurement.

5.3.3 Effects of Atmosphere on EDM Measurements

The basic operation of an EDM is that the electromagnetic wave travels outward from the source with uniform velocity (v) in all directions. Three important parameters of the wave are its *amplitude (or intensity)*, *frequency* (f), and *phase angle* (Φ). The intensity of the signal will naturally reduce slowly along the path length as energy is dissipated when the signal is traveling through an absorbing medium, such as the earth's atmosphere. The earth's atmosphere will also vary the speed of the propagated electromagnetic energy and the shape of its path by bending (or refracting) it. Usually, the frequency (f) of the signal is a constant factor within some limits and will not change unless there is a relative movement between the source and the target. Wavelength (λ) of the signal is generally a variable, for its value depends on the velocity (v), which itself depends on the refractive index (n) of the medium.

The generally accepted value for the speed of light (c) in a vacuum is 299,792,458 m/s ± 1.2 m/s. This speed, however, is affected by temperature, pressure, and humidity in the earth's atmosphere. For example, at sea level and under standard conditions, the velocity of light (v) in the earth's atmosphere is about 299,702,532 m/s. The ratio between the velocity of light in a vacuum (c) and the actual velocity (v) is known as the refractive index n, and is computed as follows:

$$n = \frac{c}{v} \tag{5.17}$$

The atmospheric refraction introduces errors in the wavelengths of waves, which also result in systematic scale error in the measured distance. Systematic scale error is introduced into the measured distance because the actual refractive index (n_a) at the time of measurement is different from the reference refractive index (n_{REF}) set by the manufacturer for the instrument. The actual wavelength (λ_a) of a wave in space can be given as

$$\lambda_a = \frac{\lambda_0}{n_a} \text{ or } \lambda_a = \frac{c}{f}\frac{1}{n_a} \tag{5.18}$$

where c is the speed of light in a vacuum, λ_0 is the wavelength of light in a vacuum, and f is the modulation frequency. Similarly, the reference wavelength (λ_{REF}) based on the reference refractive index (n_{REF}) set by the manufacturer for a given EDM instrument can be expressed as follows:

$$\lambda_{REF} = \frac{c}{n_{REF} f} \tag{5.19}$$

The refractive index n_{REF} is usually calculated from Equation (5.19) or expressed in a simple formula by the EDM instrument manufacturer. The recommended actual refractive index n_a for *electro-optical instruments* is determined using the following equation (IUGG, 1960):

$$n_a = 1 + \frac{(n_g - 1)273.15p}{(273.15 + t)1013.25} - \frac{(11.20 \times 10^{-6})e}{(273.15 + t)} \tag{5.20}$$

where p is the measured atmospheric pressure (mbar) (valid between 533 and 1067 mbar); t is the atmospheric temperature (°C) (valid between −40 and +50 °C); e is the measured partial water vapor pressure (mbar); and n_g is the group refractive index (for all frequencies making up the wave), which can be given as

$$n_g = 1 + \left(2876.04 + \frac{48.864}{\lambda^2} + \frac{0.680}{\lambda^4}\right) \times 10^{-7} \tag{5.21}$$

with λ as the carrier wavelength (micrometer or μm unit); for example, if the given carrier wavelength is 0.45 μm, $\lambda = 0.45$ should be used in Equation (5.21), not 0.45E−6. The group refractive index n_g is the same as the refractive index determined at the standard air temperature (0 °C), standard pressure (1013.25 mbar), and dry air (humidity of zero) with 0.03% carbon dioxide. This refractive index is the most important property of wave propagation since all the group of waves (except the carrier) travel with the group velocity with the carrier waves traveling at phase velocity. Very often, the partial water vapor pressure (e) is disregarded in formulae provided by manufacturers; an approximate formula for calculating the actual refractive index for *electro-optical instrument* is simplified from Equation (5.20) as

$$n_a = 1 + \frac{n_g - 1}{1 + 0.003661t} \times \frac{p}{760} \tag{5.22}$$

where n_g = group refractive index of white light expressed in Equation (5.21)

n_a = actual refractive index of atmosphere;

t = ambient temperature (°C);

p = ambient pressure (mmHg; 1 mbar = 0.7500616 mmHg).

Note that e is disregarded in Equation (5.22) and the units used for pressure is in millimeter Mercury (mmHg) while millibar (mbar) is used for pressure in Equation (5.21); take note of these differences when using either of the equations in a calculation.

If, for example, an error in wavelength is 5 nm (or 5×10^{-9} m), the error (ppm) in the group refractive index n_g (assuming $\lambda = 0.910$ μm or 910 nm) can be determined from Equation (5.21) as follows. Equation (5.21) can be rewritten as

$$(n_g - 1) \times 10^6 = 287.604 + \frac{4.8864}{\lambda^2} + \frac{0.068}{\lambda^4}$$

and finding the partial derivatives of the equation with respect to the wavelength (λ), gives

$$dn_g \times 10^6 = [3(1.6288)(-2)\lambda^{-3} + 5(0.0136)(-4)\lambda^{-5}]d\lambda$$

or

$$dn_g \times 10^6 = [-9.7728\lambda^{-2} + -0.272\lambda^{-4}]\frac{d\lambda}{\lambda}$$

The wavelength values must be substituted into the equation in μm; $\lambda = 0.910$ μm; $d\lambda = 0.005$ μm as follows:

$$dn_g \times 10^6 = [-9.7728(0.910)^{-3} + -0.272(0.910)^{-5}] \times 0.005$$

$$dn_g \times 10^6 = [13.404527] \times 0.005 \rightarrow 0.067$$

$$dn_g = 0.067\,\text{ppm}$$

The error in the group refractive index n_g is calculated as 0.067 ppm.

It is usually more convenient to represent n in parts per million such as $N=(n-1)\times 10^6$ known as *refractive number* (or *refractivity*). For example, in the earth's atmosphere at sea level, n is of the order of 1.0003000. In this case, the N value for earth's atmosphere at sea level will be 300.0. The formula for calculating the refractivity N directly for visible light and modulated infrared light (electro-optical) can be derived from Equation (5.20) (Laurila, 1976) as

$$N = \frac{N_g p - 41.8e}{3.709T} \tag{5.23}$$

where $N_g = (n_g - 1)\times 10^6$ is the group refractive number (or refractivity), T is the temperature in Kelvin ($T = 273.15 + t$), t is the temperature in °C, p is the total pressure (mbar), and e is the partial pressure of water vapor in (mbar).

Usually, a surveyor would want to know how accurately to observe the parameters T, p, and e in order to maintain a certain required accuracy level in N. This can be addressed by differentiating Equation (5.23) with respect to T, p, and e as follows:

$$\frac{\partial N_T}{\partial T} = \frac{1}{T^2}\left(-\frac{N_g}{3.709}p + 11.3e\right) \tag{5.24}$$

$$\frac{\partial N_p}{\partial p} = \frac{N_g}{3.709T} \tag{5.25}$$

$$\frac{\partial N_e}{\partial e} = -\frac{11.3}{T} \tag{5.26}$$

Using the partial differentials in Equations (5.24)–(5.26) and assuming an average sea level atmospheric conditions of $t = 15.0\,°\mathrm{C}$, $p = 1015$ mbar, and $e = 10.0$ mbar with $N_g = 294.0$ and $N = 278.8$, the corresponding systematic changes in N due to systematic change in temperature of 1 °C, and 1 mbar systematic changes in p and e can be determined. Substituting the above values into Equations (5.24)–(5.26) gives the following:

$$\frac{\partial N_T}{\partial T} = -0.97 \quad \text{or} \quad dN_T = -0.97dT \tag{5.27}$$

$$\frac{\partial N_p}{\partial p} = 0.28 \quad \text{or} \quad dN_p = 0.28dp \tag{5.28}$$

$$\frac{\partial N_e}{\partial e} = -0.039 \quad \text{or} \quad dN_e = -0.039de \tag{5.29}$$

Since $N = (n-1)\times 10^6$, it can be deduced that $dn = dN \times 10^{-6}$ (or the value of dN in ppm). From Equations (5.27)–(5.29), it can be seen that

- a change (or error) of 1 °C in T produces a change of −0.97 in N or a dn of −0.97 ppm;
- a change (or error) in 1 mbar in p produces a change of 0.28 in N or a dn of 0.28 ppm;
- a change (or error) of 1 mbar in e produces a change of −0.039 in N or a dn of −0.039 ppm.

If an assumption is further made that a thermometer with a precision of ±1 °C (or 2 °F) is used for measuring the temperature and a barometer with a precision of ±3 mbar is used for measuring the atmospheric pressure, and taking $\sigma_N = \pm 2$ ppm, the standard deviation of measuring the partial pressure e can be determined. According to the concept of the general variance–covariance propagation, the variance of N can be determined as follows:

$$\sigma_N^2 = \left(\frac{\partial N}{\partial T}\sigma_T\right)^2 + \left(\frac{\partial N}{\partial p}\sigma_p\right)^2 + \left(\frac{\partial N}{\partial e}\sigma_e\right)^2 \tag{5.30}$$

By substituting Equations (5.27)–(5.29) into Equation (5.30), the following are obtained:

$$(0.039\sigma_e)^2 = 4 - (1 \times 0.97)^2 - (3 \times 0.28)^2$$

$$\sigma_e = \pm 39.3\,\text{mbar}$$

This example shows that to keep σ_N within ±2 ppm in close to sea level atmospheric condition, the allowable errors in T, p, and e are $\sigma_T = \pm 1\,°\text{C}$, $\sigma_p = \pm 3\,\text{mbar}$, $\sigma_e = \pm 39.3\,\text{mbar}$. This is to say that any variations in the actual values of air temperature and pressure compared with the normal values will affect the refractive index (n_{REF}) set in the EDM instrument and also the corresponding measured distance. It is recommended that humidity should be considered for more precise and over long distances when using electro-optical instruments.

The actual refractive index n_a for *microwave instruments* is determined using the following equation (IUGG, 1960):

$$(n-1) \times 10^6 = \frac{77.64}{(273.16+t)}(p-e) + \frac{64.68}{(273.16+t)}\left(1 + \frac{5748}{273.16+t}\right)e \tag{5.31}$$

where p is the measured atmospheric pressure (mbar); t is the atmospheric temperature (°C); and e is the measured partial water vapor pressure (mbar). The equation is valid for carrier wavelengths between 0.03 and 1.00 m. Following similar approach as in the case of electro-optical, in microwave instruments, the following deductions can be made:

- Error in t of 1 °C is likely to affect n and distance by 1.4 ppm.
- Error in p of 1.0 mbar is likely to affect n and distance by 0.3 ppm.
- Error in humidity (or e) of 1.0 mbar is likely to affect n and distance by 4.6 ppm.

As it can be seen in the above discussion, the critical parameter in microwave measurement is humidity. Since e cannot be precisely determined, the error due to humidity limits the accuracy of microwave instruments compared to electro-optical ones. In general, the following statements can be made:

- An accuracy better than 3 ppm in the refractive index of microwave cannot easily be achieved, even if the humidity (e) is measured very precisely at both instrument stations. This means that the microwave measurement is less accurate than the electro-optical measurement.
- During normal field measurement, the *effect of atmospheric conditions* is corrected for by entering a setting into the instrument, determined from ambient temperature and pressure measurement; this is for applying the first velocity correction.
- Some EDMs reduce all measurements automatically for the first velocity correction assuming the refractive index at the instrument is representative of the whole wave path.

5.3.3.1 Velocity Corrections to EDM Measurements There are two types of velocity corrections to be made to the EDM distance value, d', actually displayed on a distance meter: *First velocity* and *second velocity* corrections. The effects of these corrections are expressed in the following derivations. Remember that the actual refractive index (n_a) during the measurement will be different from the one (n_{REF}) input in the instrument by the manufacturer. The two different refractive index values indirectly mean that the wavelength used by the manufacturer in determining distance is different from the wavelength actually used in determining distance in the field by the instrument. From Equations (5.18) and (5.19), the change (or error) in the wavelength ($\Delta\lambda$) can be given as

$$\Delta\lambda = \lambda_a - \lambda_{REF} = \frac{c}{f}\left(\frac{1}{n_a} - \frac{1}{n_{REF}}\right) \tag{5.32}$$

or

$$\Delta\lambda = \frac{c}{f}\left(\frac{n_{REF} - n_a}{n_a n_{REF}}\right) \tag{5.33}$$

Substituting Equation (5.19) into Equation (5.33) gives

$$\Delta\lambda = \lambda_{REF}\left(\frac{n_{REF} - n_a}{n_a}\right) \tag{5.34}$$

where $\Delta\lambda$ can be considered as the correction to be applied to λ_{REF} in order to obtain the actual value of the wavelength λ_a. Equation (5.34) represents the amount of correction to be applied to each wavelength making up a distance (d') measured with the

EDM. The total correction (δ') to be made to measured distance (d') as a result of the error in the wavelength ($\Delta\lambda$) can be expressed by

$$\delta' = \left(\frac{n_{\mathrm{REF}} - n_{\mathrm{a}}}{n_{\mathrm{a}}}\right) d' \tag{5.35}$$

or

$$\delta' = \left(\frac{\Delta n}{n_{\mathrm{a}}}\right) d' \tag{5.36}$$

where $\Delta n = n_{\mathrm{REF}} - n_{\mathrm{a}}$, and δ' is known as the first velocity correction (in the unit of the distance, e.g., meters). The distance corrected for first velocity can be given as

$$s = d' + \left(\frac{n_{\mathrm{REF}} - n_{\mathrm{a}}}{n_{\mathrm{a}}}\right) d' \tag{5.37}$$

Equation (5.37) can be simplified to

$$s = \left(\frac{n_{\mathrm{REF}}}{n_{\mathrm{a}}}\right) d' \tag{5.38}$$

or since n_{a} is approximately equal to 1, the simplified equation can be given as

$$s \approx n_{\mathrm{REF}} d' \tag{5.39}$$

An alternative approach is to use refractivity to compute the first velocity correction to be applied to the measured distance as follows (assuming n_{a} is approximately equal to 1.0):

$$\delta' = (N_{\mathrm{REF}} - N_{\mathrm{a}}) d' \times 10^{-6} \tag{5.40}$$

where

$$N_{\mathrm{REF}} = (n_{\mathrm{REF}} - 1) \times 10^{6} \tag{5.41}$$

$$N_{\mathrm{a}} = (n_{\mathrm{a}} - 1) \times 10^{6} \tag{5.42}$$

Note that Equation (5.40) is exactly the same as Equation (5.35) when n_{a} is assumed to be equal to 1.0. The quantity ($N_{\mathrm{REF}} - N_{\mathrm{a}}$) in Equation (5.40), which can be seen as a correction (ppm) to be applied to the measured distance, can be replaced by a differential change $\Delta N = N_{\mathrm{REF}} - N_{\mathrm{a}}$. This correction given by Equation (5.35) or (5.36) or Equation (5.40) is also known as the *first velocity correction*.

After correcting the measured distance for first velocity correction (δ'), the corrected distance will follow the curvature of the earth (with radius R), which is different from the actual wave path, due to *second velocity* effect (δ''). The second velocity correction accounts for the nonuniformity of the curvature of the propagated wave path due to the heterogeneous refractive index along the wave path. This correction is negligible for electro-optical instruments, but can be significant for microwave instruments.

If the spherically layered atmosphere is assumed, the mean refractive index at both terminals (B and E) of the wave path would be valid along the circular curve with its radius of curvature (R) being the mean radius of curvature of the earth along the path. The mean refractive index (n_a) based on the refractive indices n_B and n_E at the terminals can be given as

$$n_a = \left(\frac{n_B + n_E}{2}\right) \tag{5.43}$$

This assumption, however, is not valid since the actual wave path will have a radius of curvature ρ that is different from the earth curvature with the radius of curvature (R) of the earth being smaller than that of the wave path. This means that the ray path falls into the lower and warmer atmosphere with a greater refractive index n than the mean value n_a determined in Equation (5.43). This requires that additional correction be applied to the first velocity corrected distance. The correction is referred to as the *second velocity correction,* which is given (Rüeger, 1980) as

$$\delta'' = -(k - k^2)\frac{(d')^3}{12R^2} \tag{5.44}$$

with k being the coefficient of lateral refraction (or curvature of the optical path relative to the earth curvature); δ'' is the small systematic scale correction in the distance introduced by the *second velocity correction*; d' is the measured distance, displayed on instrument; and R is the mean radius of curvature of the earth along the line measured. Equation (5.44) is similar to arc-to-chord correction in which the correction is added to the "arc distance" in order to obtain the "chord distance." The coefficient of refraction (k) is usually expressed as

$$k = \frac{R}{\rho} \tag{5.45}$$

The usually adopted mean values of k (under normal conditions) for EDM lines that are high above the ground is $k = 0.13$ for light waves and $k = 0.25$ for microwaves. These values, however, should be used with great caution as k can have values as low as -1.0 or less and as high as $+1$ for measurements made close to glacier surfaces or near hot ground (Chrzanowski, 1977). Moreover, the values of k may vary considerably during the night, at sunrise, or at sunset. It is recommended that in a project requiring very high precision, simultaneous reciprocal zenith angle measurements be made for the sole purpose of determining k. The second velocity correction can be ignored for most of engineering applications. The wave path distance (corrected for first and second velocity corrections) can be given as

$$d_1 = d' + \delta' + \delta'' \tag{5.46}$$

5.3.3.2 *Geometric Correction: Wave Path to Chord Correction* This is another correction that can be ignored in most engineering applications. The correction reduces the measured distance along a curved wave path to the chord distance between two terminals of the wave path. The correction is given (Rüeger, 1980) by

$$\delta''' = -k^2 \frac{(d_1)^3}{24R^2} \tag{5.47}$$

where d_1 is the measured distance (with the first and second velocity corrections already applied as shown in Equation (5.46)). Generally, to improve the accuracy of distance measurements by EDM, measurements should be made in daylight and at nighttime with the meteorological conditions at intermediate points measured for calculating appropriate corrections.

Example 5.3 The formula given in a manufacturer's instruction manual for computing the atmospheric refractive index for an electro-optical distance measurement is

$$n = 1 + \frac{0.00029433513}{0.003661t} \times \frac{P}{1013.25}$$

t = ambient atmospheric temperature in K (where K = t °C + 273);
P = ambient atmospheric pressure (mbar).

The ambient atmospheric temperature was $t = 12$ °C. If the field barometer is in error by +24 mbar, determine the change in refractive index due to this error and the corresponding correction to be made to a distance measurement of 2999.100 m.

Solution: Given:

$$n_a = 1 + \frac{0.00029433513}{0.003661t} \times \frac{p}{1013.25}$$

Based on the concept of partial derivatives, the partial derivative of n_a with respect to pressure, p, can be given as follows:

$$dn_a = \frac{\partial n_a}{\partial p} dp$$

This will give

$$dn_a = \frac{0.00029433513}{0.003661t} \times \frac{dp}{1013.25}$$

$$dn_a = \frac{0.00029433513}{0.003661 \times 285} \times \frac{dp}{1013.25}$$

$$dn_a = 2.784075 \times 10^{-7} dp$$

$$dn_a \times 10^6 = 0.2784075 dp$$

With $dp = -24$ mbar, the change in refractive index can be given as

$$dn_a = -0.2784075 \times 24 \times 10^{-6} \rightarrow -6.682 \times 10^{-6}$$

The correction (negative of error) to the measured distance using the first velocity correction Equation (5.36) and assuming that n_a is approximately equal to 1.0:

$$\delta' = \left(\frac{dn}{n_a}\right) d'$$
$$= 2999.100 \times 6.682 \times 10^{-6} \rightarrow 0.020\,\text{m}$$

Example 5.4 For visible and NIR radiation and neglecting the effects of water vapor pressure, refractive index, n, can be determined by

$$n - 1 = \frac{0.269578[n_0 - 1]}{273.15 + t} p$$

The first velocity correction is in the sense that $S = S_1 + k'S_1$ with $k' = [n_0 - n]/n$. The uncorrected distance S_1 measured by the instrument is 1600 m, the reference refractive index set in the instrument by the manufacturer is $n_0 = 1.000294497$ and the average temperature and pressure during the measurements are 30 °C and 950 mbar.

a. What is the first velocity correction to the distance?

Solution: First velocity correction to the distance

$$n - 1 = \frac{0.269578[n_0 - 1]}{273.15 + t} p, t = 30^\circ\text{C}, p = 950\,\text{mbar},$$
$$n_0 = 1.000294497, S_1 = 1600\,\text{m}$$
$$n - 1 = \frac{0.269578[1.000294497 - 1]}{273.15 + 30}(950)$$
$$n - 1 = 2.48789103258\text{E} - 4 \text{ (or } n = 1.0002487891)$$
$$k' = \frac{[n_0 - n]}{n} = \frac{0.0000457079}{1.0002487891}$$
$$k' = 4.56965312011\text{E} - 5$$

The first velocity correction to the distance ($k'S_1$) is

$$k'S_1 = 4.56965312011\text{E} - 5 \times 1600\,\text{m(or } 0.073 \text{ m)}$$

b. What is the true distance (distance corrected for first velocity)?

Solution: The distance corrected for the first velocity correction:

$$S = S_1 + k'S_1$$
$$S = 1600 + 0.073 \text{(or } 1600.073\,\text{m)}$$

Example 5.5 An electro-optical EDM instrument was used to measure the distance between two stations A and C. The reference manual for the EDM gives the following equation for use in correcting distance readings for atmospheric effects:

$$N = \frac{108P}{273 + t}$$

where P is pressure in mmHg; t is temperature in °C; N is refractivity, expressed as $(n-1) \times 10^6$; n is refractive index. Given that the distance AC is 5021.845 m measured with the EDM when the temperature is 28 °C, the pressure is 750 mmHg, and the calibration refractivity is 300, what is the corrected distance from A to C?

Solution: Corrected distance from A to C:Use Equation (5.41) to determine refractive index of calibration (n_{REF}):

$$N_{\text{REF}} = (n_{\text{REF}} - 1) \times 10^6; N_{\text{REF}} = 300$$

Use the reference manual formula to determine refractivity of measurement (N_a):

$$N_\text{a} = \frac{108 \times 750}{273 + 28} \Rightarrow 269.103$$

Use Equation (5.40) to determine the correction to the distance measurement:

$$\delta' = (N_{\text{REF}} - N_\text{a})d' \times 10^{-6}$$

$$\delta' = (300.000 - 269.103)5021.845 \times 10^{-6} = +0.155\,\text{m}$$

Corrected distance for first velocity is 5021.845 + 0.155 (or 5022.000 m).

5.4 EDM INSTRUMENTAL ERRORS

EDM instrumental or internal errors consist of zero error (or system constant), cyclic error, phase measurement error, phase drifts, long-term variations in EDM modulation frequency, vertical tilt axis error (affecting the centering of instrument). The internal errors of microwave and electro-optical instruments are basically same. The internal sources of EDM errors that are systematic in nature are the vertical tilt axis error, system constant or zero error, cyclic error, phase drift, and long-term variations in modulation frequency. The most important of the errors are the zero, cyclic, and phase measurement errors.

The *zero error* (*or system constant*) is due to the inaccurate knowledge of the difference between electronic and mechanical centers of EDM and the difference between the optical and mechanical centers of the reflector. Light-wave instruments usually have small zero errors while the errors can be significant in microwave instruments. Zero error of an instrument is a constant value that is usually provided by the manufacturer or determined through the calibration of the instrument on known baselines. A measured distance must be corrected for a zero error before use.

For example, a distance measured with an instrument having zero error (z_0) should be corrected by applying the correction ($-z_0$).

Cyclic error of an electro-optical instrument is caused primarily by electric cross-talk within the instrument. The error, which is a function of internal phase measurement of an EDM, repeats itself for every unit length contained within a measured distance. Modern instruments are designed such that this type of error is minimum or negligible.

The *phase measurement error* will depend on the accuracy of phase resolver used. The *variation in EDM modulation frequency* depends on the stability of frequency generation. For example, if the actual frequency (f_2) is significantly different from the nominal frequency (f_1) for which the instrument is designed, the measured distance (S_1) can be corrected for scale errors, giving the *frequency correction* (δs_f) to be added to the measured distance as

$$\delta s_f = S_1 \frac{(f_1 - f_2)}{f_1} \tag{5.48}$$

The accuracies of phase measurements and of the modulation frequencies are usually very high. Measured distances must be corrected for systematic errors before they are used in any analysis or computations. After removing some of the systematic errors, those that are not removed will be random in nature and will become random errors. These random errors are usually due to the inability to determine the systematic effects exactly. The following internal sources of EDM errors are random:

- Leveling errors of EDM instrument, assuming compensators, which are integrated with the EDM are used for leveling the instrument.
- Errors in the manufacturer's determination of the velocity of light, modulation frequency, and refractive index; the combined effect is usually expressed in the form of instrument's accuracy specifications, such as $\pm$(2 mm + 3 ppm).
- Error in reading vertical angles of theodolite for slope reduction, assuming the digital readout unit of the instrument is used.

The random errors that will have the largest effect on the level of uncertainty of EDM measurements are the random errors expressed in the form of the instrument's accuracy specifications.

5.5 EDM EXTERNAL ERRORS

The external sources of EDM errors are atmospheric conditions, refraction and earth curvature (for long distances), centering and leveling of EDM instrument and prism on survey markers, reading atmospheric conditions, reading vertical angles of theodolite for slope reduction, and EDM/theodolite/prism height relation. Those error sources that are systematic in nature are

- effects of atmospheric conditions, which change the speed of signal propagation in the atmosphere;

- EDM/theodolite/prism height relation, which results in optical pointing error;
- effects of refraction and earth curvature (for long distances).

The following error sources are random:

- Centering and leveling errors of EDM instrument and the prism
- Error in manually reading vertical angles of theodolite for slope reduction
- Error in reading atmospheric conditions.

The random errors that will have the largest effect on the level of uncertainty of EDM measurements are the centering errors of the EDM and prism. The uncertainties from misreading of atmospheric conditions would normally be negligible.

5.6 RANDOM ERROR PROPAGATION OF EDM DISTANCE MEASUREMENT

All common EDM instruments used in surveying are based on phase difference ($\Delta\varphi$) method as discussed in Section 5.3.2. The equation for distance based on the phase difference measurement method is given from Equation (5.6) as follows:

$$d = \frac{M\lambda_{\mathrm{REF}}}{2} + \left(\frac{\Delta\varphi}{2\pi}\right)\frac{\lambda_{\mathrm{REF}}}{2} \tag{5.49}$$

or

$$d = MU + L \tag{5.50}$$

where

d is the EDM measured (uncorrected) distance;

$\Delta\varphi$ is the measured phase difference between the transmitted and the reflected waves in radians;

λ_{REF} is the manufacturer's specified modulation wavelength based on the manufacturer's set EDM modulation frequency (f);

M is the integer number of wavelength in twice the distance, which is resolved by introducing more than one wavelength for the EDM measurement;

U is the unit length given as $\lambda_{\mathrm{REF}}/2$;

L is a fraction of unit length (U) measured in the EDM, which can be expressed as $L = (\Delta\varphi/2\pi)(\lambda_{\mathrm{REF}}/2)$.

All the EDM instruments used in surveying use modulated signals for distance measurements. The wavelength (λ_{REF}) of the modulated signal is called a *reference wavelength*, and it is used in creating a *unit for the measurement* (also referred to as *unit length*). Different instruments use different patterns of wavelengths that range from a few decimeters to a few hundred meters.

The usual order of correcting EDM measurements is as follows:

- Apply system constant correction.
- Apply scale difference correction.
- Apply first and second velocity corrections.
- Apply geometric corrections.

The aforementioned corrections will now be applied to the distance measurement given in Equation (5.50) as follows. If n_{REF} is the refractive index specified by the manufacturer for the EDM and n_a is the refractive index of the atmosphere during the measurement, the EDM measured distance can be corrected for the difference in refractive indices by applying the first velocity correction in Equation (5.35) to the measured distance; the distance corrected for the first velocity effect is given from Equation (5.38) as

$$s = \left(\frac{n_{REF}}{n_a}\right) d \tag{5.51}$$

where s is the corrected distance. The following can be shown from Equation (5.50):

$$s = \left(\frac{n_{REF}}{n_a}\right) MU + \left(\frac{n_{REF}}{n_a}\right) L \tag{5.52}$$

Substituting $U = \lambda_{REF}/2$ and $\lambda_{REF} = c/n_{REF}f$ (from Equation (5.19)) into Equation (5.52) gives

$$s = \frac{c}{f}\frac{M}{2n_a} + \left(\frac{n_{REF}}{n_a}\right) L \tag{5.53}$$

where c is the velocity of propagation of electromagnetic radiation in a vacuum, f is the EDM modulation frequency, and the other symbols are as defined previously. Adding the zero correction or system constant (Z_0) and the other corrections, including the second velocity correction and the geometric correction to reduce the distance to a datum surface (Δs), the final corrected distance (s_0) can be given from Equation (5.53) as follows:

$$s_0 = \frac{c}{f}\frac{M}{2n_a} + \left(\frac{n_{REF}}{n_a}\right) L + Z_0 + \Delta s \tag{5.54}$$

If the actual frequency varies from the manufacturer's set frequency (due to instability of the modulation frequency), the frequency correction must be added to Equation (5.54). Applying the rules of error propagation on Equation (5.54) gives

$$\sigma_{s_0}^2 = \left(\frac{\partial s_0}{\partial L}\right)^2 \sigma_L^2 + \left(\frac{\partial s_0}{\partial c}\right)^2 \sigma_c^2 + \left(\frac{\partial s_0}{\partial f}\right)^2 \sigma_f^2 + \left(\frac{\partial s_0}{\partial n_a}\right)^2 \sigma_{n_a}^2 + \left(\frac{\partial s_0}{\partial Z_0}\right)^2 \sigma_{Z_0}^2 + \left(\frac{\partial s_0}{\partial \Delta s}\right)^2 \sigma_{\Delta s}^2 \tag{5.55}$$

Assuming that $n_{\mathrm{REF}}/n_{\mathrm{a}} = 1$ in Equation (5.54) and evaluating Equation (5.55) give

$$\sigma_{s_0}^2 = \sigma_L^2 + \left(\frac{M}{2n_{\mathrm{a}}f}\right)^2 \sigma_c^2 + \left(\frac{cM}{2n_{\mathrm{a}}f^2}\right)^2 \sigma_f^2 + \left(\frac{cM}{2n_{\mathrm{a}}^2 f}\right)^2 \sigma_{n_{\mathrm{a}}}^2 + \sigma_{Z_0}^2 + \sigma_{\Delta s}^2 \tag{5.56}$$

where

σ_L is the standard deviation of the fractional distance measurement, which is mainly a function of the phase difference determination;
σ_c is the standard deviation of the velocity of the signal propagation in the vacuum;
σ_f is the standard deviation of the modulation frequency;
$\sigma_{n_{\mathrm{a}}}$ is the standard deviation of measuring the refractive index;
σ_{Z_0} is the standard deviation of the zero correction determination; and
$\sigma_{\Delta s}$ is the standard deviations of the other corrections to the distance.

Substituting an approximation $2S = M\lambda = cM/n_{\mathrm{a}}f$ into Equation (5.56) gives the following:

$$\sigma_{s_0}^2 = \sigma_L^2 + S^2\left[\left(\frac{\sigma_c}{c}\right)^2 + \left(\frac{\sigma_f}{f}\right)^2 + \left(\frac{\sigma_{n_{\mathrm{a}}}}{n_{\mathrm{a}}}\right)^2\right] + \sigma_{Z_0}^2 + \sigma_{\Delta s}^2 \tag{5.57}$$

Equation (5.57) can be related to precisions usually specified by manufacturers for their EDM instruments. In this case, an EDM precision specified for a distance S can be expressed in terms of a constant error (a) and a distance-dependent error (b) as follows:

$$\sigma_{s_0}^2 = a^2 + b^2 S^2 \tag{5.58}$$

or

$$\sigma_{s_0} = \pm a \pm bS \tag{5.59}$$

Relating Equation (5.57) to Equations (5.58) and (5.59) gives

$$a^2 = \sigma_L^2 + \sigma_{Z_0}^2 \tag{5.60}$$

and

$$b^2 = \left[\left(\frac{\sigma_c}{c}\right)^2 + \left(\frac{\sigma_f}{f}\right)^2 + \left(\frac{\sigma_{n_{\mathrm{a}}}}{n_{\mathrm{a}}}\right)^2\right] \tag{5.61}$$

Usually $\sigma_{\Delta S}$, which is primarily due to the error in determining the coefficient of refraction, the error in making geometric reduction of the distance to the reference datum, and the error in centering the EDM instrument and the targets, is determined separately and added later to σ_{s_0}. A more meaningful way of expressing the precision of a distance measurement is in the form of what is known as *relative precision* or *accuracy ratio* (at 95% confidence level). For example, if $\sigma_s = 3\,\mathrm{mm}$ and $S = 1000\,\mathrm{m}$, the accuracy ratio of the distance measurement (at 95% confidence level) can be expressed generally as

$$\frac{1.96 \times \sigma_s}{S} = \frac{1.96 \times 3}{1 \times 10^6} (\text{or } 6\,\text{ppm}) \tag{5.62}$$

It can be seen from Equations (5.60) and (5.61) that "a" accounts for the effects of *zero error* (additive constant of the instrument and of the reflector), *cyclic error*, and *phase measurement error*; and "b" accounts mainly for the effects of the uncertainties in determining the velocity of light in the vacuum, uncertainties in determining the *refractive index*, and the errors in calibrating *the modulation frequency*. The velocity of light in the vacuum can be precisely determined and the modulation frequency can be calibrated with very high accuracy and can also be stabilized during the use of the EDM instrument. However, the modulation frequency can drift, creating up to 10 ppm error in distance measurement, due to aging of the control crystals or when the EDM is not allowed to acclimatize before use. The frequencies of EDM should therefore be checked as frequently as possible. A significant contribution to scale errors may also be due to refractive index (n) determination if the appropriate reduction formulas are not used or if the devices for measuring the weather conditions at the time of use of the EDM equipment are not precise enough. The main limiting factors in the accuracy of distance measurement will therefore be due to those relating to the constant error (a), the uncertainties in frequency modulation, and the measurement of weather conditions. A repetition of measurement in different atmospheric conditions (resulting in different determination of values for refractive index n), such as making one set of measurements in the daylight and another at night, may improve the accuracy of distance measurement, assuming the effects of constant error and the modulation frequency are minimum.

Microwave instruments may yield much larger deviations compared with electro-optical instruments from the values of the parameters "a" and "b" listed by the manufacturers. This is because the microwave instruments are more vulnerable to ground reflections and are more affected by the uncertainties in the determination of the relative humidity of air. For short range (several kilometers) measurements, electro-optical EDM instruments with visible or near infrared continuous radiation are used widely in engineering surveys. In very-high-precision EDM instruments such as Kern ME5000, "a" is 2 mm to 0.2 ppm based on a high modulation frequency and high resolution of the phase measurements in the instrument. Over distances longer than a few hundred meters, the prevailing error in all EDM instruments is due to the difficulty in determining the refractive index.

5.6.1 Numerical Examples

Example 5.6 Assuming the accuracy of the velocity of propagation of an EDM is 0.1 km/s; the accuracy of determination of index of refraction in the laboratory is $2 \times 10^{-6} n$; the accuracy of determining the modulation frequency of the EDM is $1.5 \times 10^{-6} f$. Determine the combined effect of the errors on range measurements by this EDM if the speed of light is taken as 299,792.5 km/s.

Solution: The ratio of errors to distance can be given as follows:

$$\frac{\sigma_c}{c} = \frac{0.1}{299792.5} \rightarrow 0.3\,\text{ppm}$$

$$\frac{\sigma_f}{f} = 1.5 \times 10^{-6} \rightarrow 1.5\,\text{ppm}$$

$$\frac{\sigma_n}{n} = 2 \times 10^{-6} \rightarrow 2.0\,\text{ppm}$$

The combined effect of errors on range measurement (from Equation (5.61)):

$$b = \sqrt{\left[\left(\frac{\sigma_c}{c}\right)^2 + \left(\frac{\sigma_f}{f}\right)^2 + \left(\frac{\sigma_n}{n}\right)^2\right]}$$

$$b = \sqrt{[(0.3)^2 + (1.5)^2 + (2.0)^2]} \rightarrow 2.5\,\text{ppm}$$

Example 5.7 An EDM has a combined error due to the velocity of light, atmospheric condition and the modulation frequency as $b = 3$ ppm. Assume that the EDM can measure the phase difference with a standard deviation of 2.0 mm, and the zero correction can be determined with a standard deviation of 1.5 mm.

a. Calculate the factor "*a*" for the EDM and the standard deviation of the measured distance of 500 m.
b. If the centering error of the EDM instrument and the prism is 0.8 mm each, calculate the accuracy of the distance measurement.

Solution:

a. Factor "*a*" for the EDM and the standard deviation of the measured distance

$$b = 3\,\text{ppm} \quad \sigma_U = 2.0\,\text{mm} \quad \sigma_{Z_0} = 1.5\,\text{mm}$$

$$a = \sqrt{\sigma_U^2 + \sigma_{Z_0}^2} \rightarrow \sqrt{2^2 + 1.5^2} \quad a = 2.5\,\text{mm}$$

Standard deviation of distance of 500 m:

$$\sigma_S = \sqrt{a^2 + b^2 S^2} \rightarrow \sigma_S = \sqrt{(2.5\,\text{mm})^2 + (3 \times 10^{-6})^2 (0.5 \times 10^6\,\text{mm})^2} = 2.9\,\text{mm}$$

b. Accuracy of the distance measurement:
Centering error of EDM = 0.8 mm
Centering error of prism = 0.8 mm
Combined standard deviation, $\sigma_S = \sqrt{(2.9\,\text{mm})^2 + (0.8\,\text{mm})^2 + (0.8\,\text{mm})^2} =$ 3.1 mm

Example 5.8 Two distances $S_1 = PQ = 2000\,\text{m}$ and $S_2 = PR = 2500\,\text{m}$ were measured (as shown in Figure 5.7) using two different EDM instruments. The distance S_1 was measured with an instrument with a specified standard deviation of $\sigma_{S_1} = 3\ \text{mm} + 2\,\text{ppm}$ and S_2 was measured with an instrument with a specified standard deviation of $\sigma_{S_1} = 5\ \text{mm} + 2\,\text{ppm}$. The two distances were measured independently (one at daylight and the other at night) so that the correlation between the two measurements can be ignored.

Answer the following:

a. What is the accuracy (amount of random error) of the short distance $QR = 500\,\text{m}$, which is calculated as a difference $\Delta S = S_2 - S_1$?

Solution: Refer to Section 2.8.2 for the review of variance–covariance propagation laws. The two measurements (S_1 and S_2) are not correlated since they were measured independently and the instruments were different. Equation for calculating the distance can be given as

$$\Delta S = S_2 - S_1 \tag{5.63}$$

Through variance–covariance propagation (Section 2.8.2):

$$\sigma_{\Delta S}^2 = \left(\frac{\partial \Delta S}{\partial S_2}\right)^2 \sigma_{S_2}^2 + \left(\frac{\partial \Delta S}{\partial S_1}\right)^2 \sigma_{S_1}^2 \tag{5.64}$$

or

$$\sigma_{\Delta S}^2 = (1)^2 \sigma_{S_2}^2 + (-1)^2 \sigma_{S_1}^2 \tag{5.65}$$

Using the listed standard deviation for the corresponding EDM instrument:

$$\sigma_{S_1}^2 = (3\ \text{mm})^2 + (2 \times 10^{-6} S)^2 = 25\,\text{mm}^2 \quad (\text{for } S_1 = 2000\,\text{m}) \tag{5.66}$$

$$\sigma_{S_2}^2 = (5\ \text{mm})^2 + (2 \times 10^{-6} S)^2 = 50\,\text{mm}^2 \quad (\text{for } S_2 = 2500\,\text{m}) \tag{5.67}$$

Substituting Equations (5.66) and (5.67) into Equation (5.65) gives the following:

$$\sigma_{\Delta S}^2 = 50 + 25 \tag{5.68}$$

$$\sigma_{\Delta S} = \sqrt{50 + 25} = 9\ \text{mm}$$

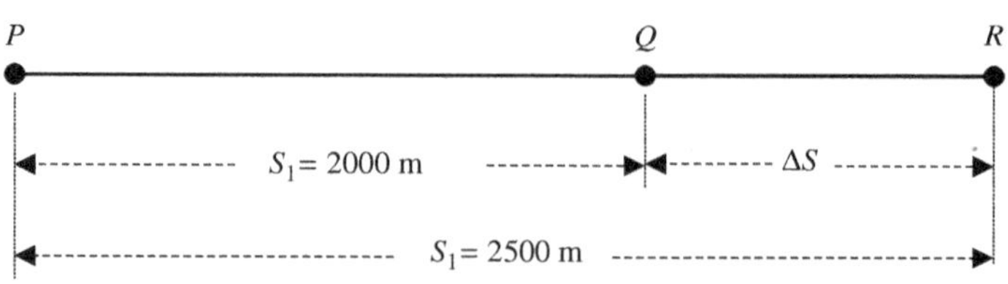

Figure 5.7 Baseline measurements with two different EDM instruments.

b. If the two EDMs have systematic errors of 1.5 mm and 2.0 mm, respectively, what is the systematic error on the short distance *QR*?

Solution: Total systematic error on the short distance *QR*: $\delta_1 = 1.5\,\text{mm}$; $\delta_2 = 2.0\,\text{mm}$. Finding the partial derivatives of Equation (5.63) and referring to systematic error propagation laws (Section 2.8.1):

$$\delta_{\Delta S} = \frac{\partial \Delta S}{\partial S_1}\delta_1 + \frac{\partial \Delta S}{\partial S_2}\delta_2 \tag{5.69}$$

$$\delta_{\Delta S} = (-1)\delta_1 + \delta_2 \tag{5.70}$$

$$\delta_{\Delta S} = 0.5\,\text{mm}$$

c. Calculate the total (combined systematic and random errors) on the measured distance *QR*.

Solution: Total error:

$$\sigma_T = \sqrt{\sigma_{\Delta S}^2 + \delta_{\Delta S}^2} = 9.0\,\text{mm} \tag{5.71}$$

d. The distance *QR* was later measured with the EDM with a specified standard deviation of $\sigma_S = 5\ \text{mm} + 2\,\text{ppm}$; the measured distance was 499.990 m. Is this distance significantly different (at 99% confidence level) from the derived value in (*a*) (assuming there was no correlation between the measurement and the other measurements)?

Solution: Significance of the measured and derived values of at 99% confidence level:

The difference between the derived and the measured distance is $d = 0.010\,\text{m}$.

The standard deviation calculated for the derived distance is $\sigma_{\Delta S} = 9.0\,\text{mm}$.

The standard deviation for the measured distance can be calculated from $\sigma_S = 5\ \text{mm} + 2\,\text{ppm}$, giving:

$$\sigma_s = \sqrt{(5\,\text{mm})^2 + (2 \times 10^{-6} \times 499{,}995\,\text{mm})^2} \text{ or } 5.1\,\text{mm} \tag{5.72}$$

Error propagation on the difference between the two distances gives

$$\sigma_d = \sqrt{\sigma_{\Delta s}^2 + \sigma_s^2} \tag{5.73}$$

Substituting the calculated values above into Equation (5.73) gives $\sigma_d = 10.3\,\text{mm}$. The significance of the difference can be tested using Equation (2.52) as follows:

$$|d| \le \sigma_d \sqrt{\chi^2_{\text{df}=1,\alpha}} \tag{5.74}$$

$$|10\,\text{mm}| \le 10.3\,\text{mm}\sqrt{6.635} \tag{5.75}$$

$$10\,\text{mm} \le 26.5\,\text{mm}$$

Since the expression is satisfied, it can be concluded that the derived distance and the measured distances are not significantly different at 99% confidence level.

Example 5.9 The recommended actual refractive index n_a for electro-optical instruments is determined using the following equation (IUGG, 1960):

$$n_a = 1 + \frac{(n_g - 1)273.15p}{(273.15 + t)1013.25} - \frac{(11.20 \times 10^{-6})e}{(273.15 + t)} \tag{5.76}$$

where p is the measured atmospheric pressure (mbar); t is the atmospheric temperature (°C); e is the measured partial water vapor pressure (mbar); and n_g is the group refractive index (for all frequencies making up the wave). Assuming the values of the variables in the equation are $t = 15\,°\text{C}$, $p = 1007$ mbar, $e = 13$ mbar and $n_g = 1.0003045$, answer the following:

a. What are the individual effect of systematic errors $dt = 1\,°\text{C}$, $dp = 1$ mbar, and $de = 1$ mbar on the derived quantity n_a (using their partial differentials)?

Solution: Individual effect of systematic errors on the derived quantity n_a
Equation (5.76) can be rewritten as

$$(n_a - 1) = \frac{(n_g - 1)273.15(273.15 + t)^{-1}p}{1013.25} - 11.27 \times 10^{-6}(273.15 + t)^{-1}e$$

Find the partial derivatives of the equation with respect to the variables p, t, and e as follows. For error in temperature measurement, finding the partial derivative with respect to temperature t gives

$$\frac{\partial n_a}{\partial t} = \left[\frac{\left(n_g - 1\right)273.15p}{1013.25} - 11.27 \times 10^{-6}e\right](273.15 + t)^{-2} \tag{5.77}$$

Using the principles of partial differentials, the following can be given:

$$dn_a = \frac{\partial n_a}{\partial t}dt \tag{5.78}$$

Substituting all the values $t = 15\,°\text{C}$, $p = 1007$ mbar, $e = 13$ mbar and $n_g = 1.0003045$ into Equations (5.77) and (5.78) gives

$$dn_a \times 10^6 = -1.00dt \tag{5.79}$$

For $dt = 1\,°\text{C}$: $dn_a \times 10^6 = -1.00$ or $dn_a = -1.00$ ppm.

For error in pressure measurement, finding the partial derivative with respect to pressure p gives:

$$\frac{\partial n_a}{\partial p} = \left[\frac{(n_g - 1)\,273.15}{1013.25}\right](273.15 + t)^{-1} \tag{5.80}$$

$$dn_a = \frac{\partial n_a}{\partial p}dp \tag{5.81}$$

Substituting all the values $t = 15\,°\text{C}$, $p = 1007$ mbar, $e = 13$ mbar, and $n_g = 1.0003045$ gives $dn_a \times 10^6 = 0.28dp$; for $dp = 1$ mbar, $dn_a \times 10^6 = 0.28$ or $dn_a = 0.28$ ppm.

For error in relative humidity measurement, find the partial derivative with respect to relative humidity e:

$$\frac{\partial n_a}{\partial e} = [-11.27 \times 10^{-6}](273.15 + t)^{-1} \tag{5.82}$$

$$dn_a = \frac{\partial n_a}{\partial e}de \tag{5.83}$$

Substituting all the values $t = 15\,°\text{C}$, $p = 1007$ mbar, $e = 13$ mbar, and $n_g = 1.0003045$ gives $dn_a \times 10^6 = -0.04de$; for $de = 1$ mbar, $dn_a \times 10^6 = -0.04$ or $dn_a = -0.04$ ppm.

b. What is the combined effect of the systematic errors on the derived quantity n_a?

Solution: For the combined effect on the derived refractive index, find the partial derivatives of the equation with respect to the variables (referring to Section 2.8.1), giving the following:

$$dn_a = \frac{\partial n_a}{\partial t}dt + \frac{\partial n_a}{\partial p}dp + \frac{\partial n_a}{\partial e}de \tag{5.84}$$

Substituting the partial differentials in Equations (5.77), (5.80), and (5.82) gives the following:

$$dn_a = \left[\frac{(n_g - 1)\,273.15p}{1013.25} - 11.27 \times 10^{-6}e\right](273.15 + t)^{-2}dt + \left[\frac{(n_g - 1)\,273.15}{1013.25}\right](273.15 + t)^{-1}dp + [-11.27 \times 10^{-6}](273.15 + t)^{-1}de \tag{5.85}$$

Substituting the values of the variables into the partial derivative gives

$$dn_a = -9.93789\text{E} - 7dt + 2.84874\text{E} - 7dp - 3.911157\text{E} - 8de \tag{5.86}$$

or

$$dn_a \times 10^6 = -1.00dt + 0.28dp - 0.04de \tag{5.87}$$

For error $dt = 1\,°\mathrm{C}$, $dp = 1$ mbar, and $de = 1$ mbar, the following result is obtained:

$$dn_{\mathrm{a}} \times 10^6 = -1.00 + 0.28 - 0.04 \rightarrow dn_{\mathrm{a}} = -0.76\,\mathrm{ppm}$$

The combined effect on $dn_{\mathrm{a}} = -0.76\,\mathrm{ppm}$; this value is the same as summing up the individual effect calculated in Solution (a) above.

c. What is the combined effect of the random errors $dt = 1\,°\mathrm{C}$, $dp = 1$ mbar, and $de = 1$ mbar on the derived quantity n_{a} (using the variance–covariance propagation laws, assuming no correlation amongst the variables)?

Solution: Use random error propagation laws (Section 2.8.2); and assume no correlation among the variables (so that the covariances among the variables can be set to zero). According to the concept of the general variance–covariance propagation law (Section 2.8.2), the variance of n_{a} can be determined as follows:

$$\sigma_{n_{\mathrm{a}}}^2 = \left(\frac{\partial n_{\mathrm{a}}}{\partial t}\sigma_t\right)^2 + \left(\frac{\partial n_{\mathrm{a}}}{\partial p}\sigma_p\right)^2 + \left(\frac{\partial n_{\mathrm{a}}}{\partial e}\sigma_e\right)^2 \tag{5.88}$$

The partial differentials in Equation (5.88) are the same as those in questions (a) and (b) above:

$$\frac{\partial n_{\mathrm{a}}}{\partial t} = 1.00\,\mathrm{ppm}; \; \frac{\partial n_{\mathrm{a}}}{\partial p} = 0.28\,\mathrm{ppm}; \; \frac{\partial n_{\mathrm{a}}}{\partial e} = -0.04\,\mathrm{ppm} \tag{5.89}$$

Substituting $\sigma_t = 1\,°\mathrm{C}$, $\sigma_p = 1\,\mathrm{mbar}$, and $\sigma_e = 1\,\mathrm{mbar}$ and Equations in (5.89) into Equation (5.88) gives

$$\sigma_{n_{\mathrm{a}}}^2 = (1.00)^2 + (0.28)^2 + (-0.04)^2 \rightarrow \sigma_{n_{\mathrm{a}}} = 1.039\,\mathrm{ppm}$$

Example 5.10 The standard deviation of measuring a baseline was 10 mm. The baseline was measured at two different epochs, using the identical instruments, as 1200.000 m and 1200.025 m. Are the baseline measurements significantly different at 90% confidence level?

Solution: This is a typical example of a case in which the measurements are from the same population and we want to test if the two sample means are the same. Referring to Section 2.9.2 (Equations (2.45) and (2.46)), the hypotheses can be stated as follows:

$$H_0 : \mu_1 - \mu_2 = 0 \quad H_{\mathrm{A}} : \mu_1 - \mu_2 \neq 0 \tag{5.90}$$

Since the standard deviations of measurements are well known, Equation (2.50) or (2.52) can be used, but Equation (2.52) will be used as an example as follows:

$$|\bar{x}_1 - \bar{x}_2| = (\mathrm{SE})\sqrt{\chi^2_{\alpha,\mathrm{df}=1}(\text{upper-tail area})} \tag{5.91}$$

where the misclosure $|\bar{x}_1 - \bar{x}_2| = 0.025\,\text{m}; \alpha = 0.1;$ $\sqrt{\chi^2_{\alpha=0.1,\text{df}=1}(\text{upper-tail area})} = \sqrt{2.706}$ (or 1.645) or $z_{1-0.1/2} = 1.645$; and the standard deviation of each distance, $\sigma_d = 10\,\text{mm}$ (0.010 m). The standard error (SE) of the misclosure can be propagated based on the difference between the two measured baselines as $\sigma_{\hat{\bar{x}}} = \sigma_d\sqrt{2}$ (or 14.1 mm). Substitute the corresponding values into the above equation and check if the condition of the equation is satisfied as follows:

$$0.025\ \text{m} \leq 0.014.1 \times 1.645 \quad \text{or} \quad 25\,\text{mm} \leq 23.26\,\text{m}$$

Since 25 mm is not less than nor equal to 23.26 mm, we can say at 95% confidence that the two baseline measurements are significantly different.

5.7 CALIBRATION AND TESTING PROCEDURES FOR EDM INSTRUMENTS

Calibration and testing procedures for geodetic EDM instruments are to provide the best achievable measure of precision (repeatability) of a particular electro-optical distance meters (EDM instruments) and their supporting equipment under field conditions. They are performed in order to determine the instrument (additive constant and cyclic error) and scale errors. The additive constant and scale error of the instrument tend to change due to usage, transportation, and aging of frequency oscillator in the instrument. The additive constant consists of two parts:

- Error due to uncertainty of the electronic origin of measurement with the EDM
- Error due to uncertainty of the reflected position of the EDM signal within the prism.

Note that the additive constant (or additive correction) is of equal magnitude but of opposite sign to the zero error. The EDM instruments should be calibrated whenever one or more of the following needs or requirements are to be satisfied:

1. Need to verify that the EDM equipment is working within the EDM manufacturer's stated specification for scale error and constant error.
2. A requirement before a survey control project for the establishment and/or maintenance of survey control markers.
3. A statutory requirement of the Surveys Act, which requires verification of all electronic linear measuring devices by comparison with calibration base lines established by the appropriate government agency for that purpose. EDM equipment must be calibrated over a certified baseline at intervals not exceeding 12 months or more frequently if conditions warrant it.
4. Need to check the quality of the EDM in situations where it has been damaged during regular surveying operations or when the EDM is old and may no longer be operating within the manufacturer's specifications.

The standard calibration approach for geodetic EDM instruments is to use distance measurements in all combinations on baselines of between 6 and 8 stations. The maximum baseline distance should correspond to the maximum range of the EDM to one prism under fair conditions. A calibration baseline consists of a set of forced-centering concrete filled steel pillars with an interpillar spacing of approximately 100 m to over 2 km. In the province of British Columbia in Canada, for example, there are currently six EDM basenets, located in Vernon, Prince George, Surrey, Victoria, Vancouver West, and Cranbrook. The Surrey basenet is located on the grounds of the Surrey Nursery and seed orchards and consists of 6-pier linear baseline with the 7th pier only visible with Piers 1–3. The Geodetic Survey Division (GSD) is responsible for determining the baseline lengths. All baselines are measured by GSD at regular intervals (epochs of observations) to verify the interpillar distances and pillar stability. In general, the remeasurement schedule has been between 1 and 3 years depending on the baseline. The current policy for reobservation of the baseline lengths is once every 5 years. Baseline lengths for all of the baselines are published on the Director of Surveys websites.

5.7.1 Observation and Data-Processing Methodology

It is the responsibility of the surveyor to ensure that all equipment used in a measurement will achieve a result in terms of the accuracy required. The following general information should be considered while conducting an EDM calibration survey:

(a) Users must fully understand the operating manual of the EDM being used.

(b) All equipment should be checked and confirmed to be in good adjustment and in good working order as specified by the manufacturer; appropriate steps should also be taken to ensure their proper use with appropriate tripods, forced-centering equipment, and recommended reflectors.

(c) All appropriate instrument self-checks as stated in the manufacturer's operating manual should be done.

(d) Calibration surveys should only be conducted within the allowable range of weather conditions as defined by the manufacturer.

(e) Each pillar-to-pillar slope distance should be measured at least twice with the average used as the observed value.

(f) Meteorological conditions such as variations in air temperature and barometric pressure should be measured at both the instrument station and the target (prism) station for all distances. The temperature and pressure values are to be recorded as the correct value(s) in the field regarding known standards. Measurements should also include wind speed, cloud cover, and visibility. Since inaccurate meteorological observations can contribute to as much as 1–2 ppm error in the scale determination of an EDM, it is important to verify the accuracy of the meteorological equipment by comparing it against a known standard.

(g) If least squares adjustment will be performed on the distance measurements in a network, the additive constant will be most important to be determined and be known, but the scale factor will be fixed by the network datum. In this case, the calibration of EDM (for scale error) may be considered not needed.

(h) When a linear array of markers (above what is required in unique determination) is used in the determination of the additive constant of an EDM, there are a number of advantages associated with this process:
   - There are redundant measurements so that least squares method of adjustment can be applied.
   - The linear array of markers provides a perfect geometry.
   - Redundant measurements will increase the reliability and precision of the estimated additive constant.

If desired, users of EDM equipment may submit their EDM calibration survey data to the Survey and Technical Services Section in order to determine the scale and constant error for their EDM. This service is offered free of charge, and it usually takes one business day to complete provided all the pertinent information for the EDM and the survey has been provided. Pertinent information will include the following:

1. EDM make, model, and serial number
2. Number and type of prisms employed in the calibration survey
3. The make, model, and serial number of barometer used
4. The make, model, and serial number of thermometer used
5. Carrier wavelength of the EDM
6. Modulation wavelength of the EDM
7. Modulation frequency of the EDM
8. Calibration survey data submitted in appropriate forms.

Items 5–7 are required in order to be able to derive the meteorological coefficient values needed within the evaluation software. The carrier wavelength, modulation wavelength, and the modulation frequency can be obtained by contacting the EDM equipment supplier or the manufacturer directly.

5.7.1.1 Temperature Sensor Types Some of the commonly used temperature sensor types are as follows:

- Precision Hygro-Thermometer with simultaneous display of humidity/temperature and humidity/wet bulb with an accuracy of 2% relative humidity
- Precision psychrometer with an accuracy of 2% relative humidity
- Precision psychrometer for simultaneous display of relative humidity (%), temperature and dew point or wet bulb with $\pm 0.1\,^{\circ}$C resolution of temperature and $\pm 1\%$ for relative humidity
- Thermistors with an accuracy of $\pm 0.1\,^{\circ}$C or $\pm 0.2\,^{\circ}$C
- Precision thermometer, which can measure with an accuracy of $\pm 0.01\,^{\circ}$C or $\pm 0.02\,^{\circ}$C.

5.7.1.2 ***Atmospheric Pressure and Relative Humidity Sensor Types*** Some of the common atmospheric pressure and relative humidity sensor types are as follows:

- Precision barometer (or digital barometer) with an accuracy of ±1.0 mbar
- Handheld multibarometers with an accuracy of ±5.0 mbar for pressure and ±1.0 °C for temperature
- Microbarometers with an accuracy of μbar
- Digiquartz pressure sensor with an accuracy of 0.01%.

A typical relative humidity sensor is Hygristor sensor, which measures humidity to an accuracy of 0.25% relative humidity.

5.7.2 EDM Baseline Designs

EDM baselines must be well designed in order to allow all the systematic errors in the EDM to be detected when used. The baseline facilities are provided by the government or its agencies. There are three basic EDM baseline design types (Hazelton, 2009):

1. *Aarau design*, which is named after a town in Switzerland where Kern instruments are made. In this baseline design, all baselines (which are straight) are measured as integral multiples of some numbers, such as 60 m; the baselines may consist of 4 or 9 points, depending on the range of the EDM. The design also requires that a separate cyclic error be done within the baselines.
2. *Hobart design*, which is named after the city in Australia where the authors of the design were faculty members of the University of Tasmania (Sprent and Zwart, 1978). This design requires the EDM instrument being calibrated to be set up at just two points (at the zero pillar and at a pillar that is half the unit length of the EDM from the zero pillar), where distances are measured to all the other points on the line. This has an advantage of making use of fewer measurements, but with a disadvantage that the design is restricted to a few instruments with certain unit lengths.
3. *Heerbrugg design*, which is named after the city in Switzerland where Wild (now Leica) is located. The baseline design allows combined zero and reflector offsets and cyclic error to be determined for a case where the baseline distances are known or unknown; and in addition, the design allows the scale error to be determined if the baseline distances are known. According to ISO standard 17123-4, an array of 7 collinear points (with 21 one-way distances being observable) is needed with spacing following one unit length ($U = \lambda/2$) of the electro-optical distance measuring instrument (EODMI) and the overall length of the array, which is usually at least as long as any intended use of the EODMI. On the basis of this design and considering the number of measurements and the number of unknown parameters, the least squares estimation is possible. With regard to Heerbrugg' design, the following setting out must be done in preparation for measurements with the EODMI (Rüeger, 1996):

- Design six distances (m_{12}, m_{23}, m_{34}, m_{45}, m_{56}, m_{67}) of the test line as shown in Figure 5.8 with the whole length (from points 1 to 7) being m_{17} using the following formulae:

$$m_{12} = \lambda + \beta + 3\gamma \quad (5.92)$$

$$m_{23} = \lambda + 3\beta + 7\gamma \quad (5.93)$$

$$m_{34} = \lambda + 5\beta + 11\gamma \quad (5.94)$$

$$m_{45} = \lambda + 4\beta + 9\gamma \quad (5.95)$$

$$m_{56} = \lambda + 2\beta + 5\gamma \quad (5.96)$$

$$m_{67} = \lambda + \gamma \quad (5.97)$$

where

$$\beta_0 = \frac{d - 6\lambda - U}{15} \quad (5.98)$$

d is the EODMI range to be tested;

U = unit length ($U = \lambda/2$) and λ is the modulation wavelength of the EODMI;

μ = integer(β_0/U), which means that μ is the integer value derived from β_0;

$\beta = \mu \times U$; and $\gamma = U/36$

- Set out collinear array of 7 points using a series of tribrachs on tripods for forced-centering interchange.

Based on the above, the information usually known prior to the EDM measurements in this case is as follows:

- Unit length of the instrument must be known in order to set up points with appropriate spacing (e.g., the unit length of Leica TPS 700 series is 1.5 m; it can measure 3000 m in average weather conditions using standard prism; and its standard deviation according to ISO 17123-4 EDM calibration and testing is 2 mm + 2 ppm for IR Fine).
- Points used shall be known to be stable during the test measurements

Using the Leica TPS 700 series as an example, $U = 1.5$ m, $\lambda = 3$ m and $d = 3000$ m (range of the EDM to be tested):

$$\beta_0 = \frac{3000 - 18 - 1.5}{15} = 198.7; \quad \mu = \text{integer}\left(\frac{198.7}{1.5}\right) = 132$$

$$\beta = 132 \times 1.5 = 198; \quad \gamma = \frac{1.5}{36}$$

The designed distances are $m_{12} = 201.125$ m; $m_{23} = 597.292$ m; $m_{34} = 993.458$ m; $m_{45} = 795.375$ m; $m_{56} = 399.208$ m; $m_{67} = 3.042$ m; sum of all the subsections gives $m_{17} = 2989.5$ m.

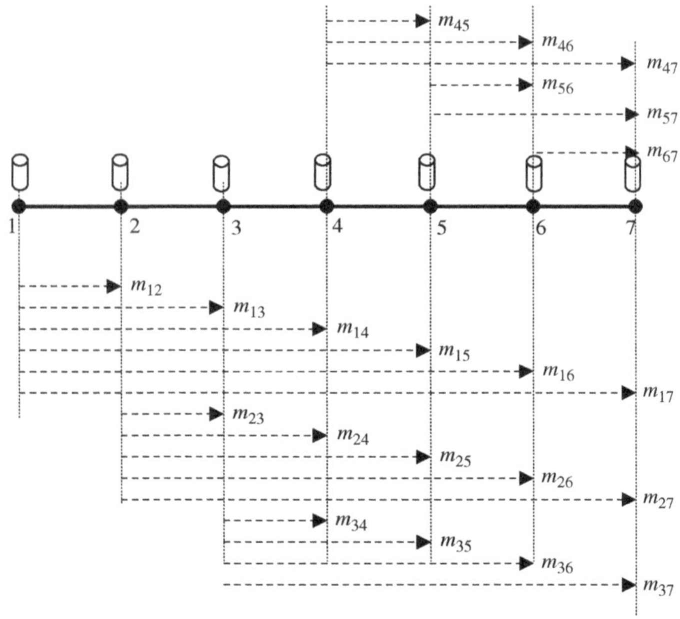

Figure 5.8 Baselines and measuring arrangement for EDM calibration.

5.7.3 EDM Calibration When Length of Baseline Is Known

The calibration of EDM should be done on the government-provided calibration baselines. A sample calibration baseline is shown in Figure 5.8. It consists of seven stations (labeled 1–7) established in a horizontal area. These stations, which should remain stable throughout the calibration measurements, are usually equipped with forced-centering devices.

The calibration measurement scheme in relation to the sample baseline in Figure 5.8 is as follows:

- Force-center the EDM instrument on the pillars with forced-centering devices in order to minimize centering errors.
- Use sufficient number of prisms in order to ensure that all the distances are measured with a good return signal.
- Measure all possible combinations of distances between the baseline pillars in the same day when visibility is good.
- Measure the air temperature and pressure and apply appropriate atmospheric corrections to the measured distances.

The distance measurements (the means of four measurements per section) made can be processed as follows:

1. Correct the slope distances for atmospheric conditions (calibration values of barometer and thermometer must be applied) by applying the first velocity

corrections; use the EDM manufacturer's provided formula for correcting the distances for metrological condition. This can be obtained from the instrument manual.

2. If the published distances (p) for the baselines are mark-to-mark distances, the corrected slope distances (SD) will have to be reduced to mark-to-mark distances (m). For example, the mark-to-mark calculated distance from pillar 1 to pillar 2 (Figure 5.8) can be given by

$$m_{12} = \sqrt{\mathrm{SD}^2 - [(H_1 + HI) - (H_2 + HT)]^2 + [H_1 - H_2]^2} \tag{5.99}$$

where m_{12} is the mark-to-mark distance from pillar 1 to pillar 2, SD is the slope distance from pillar 1 to pillar 2 (corrected for meteorological conditions), H_1 is the orthometric height of the instrument station (pillar 1), HI is the height of instrument at pillar 1, H_2 is the orthometric height of the target station (pillar 2), and HT is the target height at pillar 2.

3. Use the published distances (p) and the calculated mark-to-mark distances (m) in the following linear regression formula and perform the least squares adjustment to determine the calibration parameters (the system constant and the scale factor):

$$p = C + Sm \tag{5.100}$$

where p is a vector of published distances, m is a vector of measured mark-to-mark distances (corrected for meteorological conditions), C is the system (instrument/reflector) constant (which is expected to be close to zero value set in the instrument), and S is the scale factor (which is expected to be close to an ideal value of 1) with $1 - S$ as the scale error.

4. Since the standard deviations of the published distances are usually provided, and the standard deviations of the calculated distances can be propagated from the manufacturer's accuracy specification or from the repeated distance measurements, the least squares adjustment of general model approach will be appropriate. In this case, the regression formula can be rearranged as follows:

$$p - Sm - C = 0 \tag{5.101}$$

In relation to Figure 5.8, the least squares adjustment of the general model $f(x, \ell) = 0$ can be done with the vector of parameters (x) and the vector of observations (ℓ), respectively, as follows:

$$x = \begin{bmatrix} C \\ S \end{bmatrix} \tag{5.102}$$

$$\ell^{\mathrm{T}} = \begin{bmatrix} m_{12} & m_{13} & m_{14} & \cdots & m_{67} & p_{12} & p_{13} & p_{14} & \cdots & p_{67} \end{bmatrix} \tag{5.103}$$

The weight matrix (W) is a diagonal matrix with its elements corresponding to the weights of the measured and published distances in vector ℓ. For example, the weight (w_{ij}) of a distance measurement (m_{ij}) between points i and j can be derived from the manufacturer's specified accuracy for the EDM instrument and the errors of centering the instrument and targets as

$$w_{ij} = \frac{1}{[(a)^2 + (m_{ij} \times b \times 10^{-6})^2 + (\varepsilon_{\mathrm{cI}})^2 + (\varepsilon_{\mathrm{cT}})^2]} \tag{5.104}$$

where "a" is the EDM constant error (m), "b" is the distance-dependent error (ppm), and $\varepsilon_{\mathrm{cI}}$ and $\varepsilon_{\mathrm{cT}}$ are the centering errors (m) of the instrument and target, respectively. The weight of a published distance will be equal to the inverse of the variance of the distance.

5. Determine the adjusted values of the unknown parameters by the least squares method:

$$\hat{x} = \begin{bmatrix} C \\ S \end{bmatrix} = \begin{bmatrix} C^0 \\ S^0 \end{bmatrix} - (A^{\mathrm{T}} M^{-1} A)^{-1} A^{\mathrm{T}} M^{-1} w \tag{5.105}$$

where A is the first design matrix of the 21 general model equations formed in Equation (5.101) with respect to two unknown parameters C and S; B is the second design matrix of Equation (5.101) with respect to 42 observations given in Equation (5.103); w is a vector of misclosures obtained when approximate parameters $C^0 = 0.0$ and $S^0 = 1.0$ and the observations and published values are substituted into Equation (5.101); and

$$M = BW^{-1}B^{\mathrm{T}} \tag{5.106}$$

6. Calculate the standard factor of unit weight, $\hat{\sigma}_0$, for the EDM instrument:

$$\hat{\sigma}_0 = \sqrt{\frac{-k^{\mathrm{T}} w}{v}} \tag{5.107}$$

where $v = n - u$ is the number of degrees of freedom (in this case, $v = 19$), $n = 21$ is the number of general model equations, $u = 2$ is the number of unknown parameters, and k is the vector of correlates given as

$$k = -M^{-1}(A\delta + w) \tag{5.108}$$

with

$$\delta = -(A^{\mathrm{T}} M^{-1} A)^{-1} A^{\mathrm{T}} M^{-1} w \tag{5.109}$$

7. The standard deviation of the system constant ($\hat{\sigma}_{\mathrm{C}}$) and the scale factor standard deviation ($\hat{\sigma}_{\mathrm{S}}$) can be extracted from the covariance matrix of the adjusted parameters given as

$$C_{\hat{x}} = \hat{\sigma}_0^2 Q_{\hat{x}} \tag{5.110}$$

where

$$Q_{\hat{x}} = HQ_{\ell}H^{\mathrm{T}}$$

$$H = G^{-1}A^{\mathrm{T}}M^{-1}B$$

$$G = A^{\mathrm{T}}M^{-1}A$$

8. Perform statistical tests on the adjusted parameters (C, $1-S$). Use the Student's t statistical test in Section 2.9.2 (Equation (2.49)) to check if C and $(1-S)$ are statistically different from zero. In this case, $C = \overline{x}_1 - \overline{x}_2$ and $1 - S = \overline{x}_1 - \overline{x}_2$ are used in Equations (2.49)–(2.52) in Section 2.9.2. If the tests fail, then the system constant and the scale factor error are significantly different from zero. This is an indication that there may be problems with the modulation frequency of the EDM, reflector, or pointing of telescope, and further tests must be carried out on the instrument to confirm if actually the instrument is not working properly. There is a need to recalibrate the instrument on another baseline. If the same results are obtained, then it may be concluded that the instrument is not working properly and the service representatives should be consulted for possible repair of the instrument.
9. Perform Chi-squares statistical test (test of hypothesis for a population variance in Section 2.9.3, using Equation (2.56)) on the standard deviations ($\hat{\sigma}_{\mathrm{C}}$) and ($\hat{\sigma}_{\mathrm{S}}$) to check if they are significantly greater than those specified for the instrument by the manufacturer. For example, if the manufacturer specifies 2 mm ± 3 ppm for the instrument, 2 mm must be compared with the computed $\hat{\sigma}_{\mathrm{C}}$, and 3×10^{-6} with $\hat{\sigma}_{\mathrm{S}}$. If the two tests fail, then the manufacturer's constant error and scale error claims are too optimistic, meaning that the precision quoted by the manufacturer is different from what it is in reality.

It should be mentioned that using dedicated facilities (such as the government calibration baselines) for instrument calibration has a lot of advantages if the facilities are readily accessible at the time of need. Some of the advantages are as follows:

- Dedicated pillars are stable during measurements; no danger of movement when interchanging reflector and heavy instruments.
- Fast instrument and reflector setup resulting from forced-centering and preleveled centering plates.
- Constant height of instrument and reflectors resulting in standardized computations.
- High precision for additive constant (even if known distances are not available or are out of date).
- Distances are spread over the whole range of the instrument.

Example 5.11 Leica Distomat DI1600 EDM equipment was calibrated over Surrey EDM seven-point baseline in *BC*. The constant correction for the EDM is assumed to be equal to zero and the manufacturer's stated accuracy is 3 mm ± 2 ppm. Using the published mark-to-mark baseline distances (p) and the calculated mark-to-mark distances corrected for meteorological conditions (m) in the least squares adjustment based on the linear regression in Equation (5.100), the following least squares adjusted quantities were obtained with 13 degrees of freedom:

$$1 - S = -3.01\,\text{ppm} \quad \hat{\sigma}_{\mathrm{S}} = 1.05\,\text{ppm}$$

$$C = 0.70\,\text{mm} \quad \hat{\sigma}_{\mathrm{C}} = 0.77\,\text{mm}$$

Perform the following tasks:

a. Chi-squares test on $\hat{\sigma}_S$ at 95% and state if the manufacturer's claimed scale error is acceptable.

Solution:

Manufacturer's specification: $\sigma_S = 2.00\,\text{ppm}$
Computed: $\hat{\sigma}_S = 1.05\,\text{ppm}$
From Equation (2.56):

$$\hat{\sigma}_S \leq \sqrt{\frac{\chi^2_{\alpha=0.05,\text{df}=13}(\sigma_S)}{\text{df}}} \quad \text{or} \quad 1.05 \leq \sqrt{\frac{22.4(2.00)}{13}}$$

This gives $1.05\,\text{ppm} \leq 2.63\,\text{ppm}$?
Since the condition is satisfied, the manufacturer's scale error claim may be pessimistic but acceptable at 95% confidence level.

b. Chi-squares test on σ_C at 95% and state if the manufacturer's claimed constant error is acceptable.

Solution:

Manufacturer's specification: $\sigma_C = 3.00\,\text{mm}$
Computed: $\hat{\sigma}_C = 0.77\,\text{mm}$
From Equation (2.56)

$$\hat{\sigma}_C \leq \sqrt{\frac{\chi^2_{\alpha=0.05,\text{df}=13}(\sigma_C)}{\text{df}}} \quad \text{or} \quad 0.77 \leq \sqrt{\frac{22.4(3.00)}{13}}$$

This gives $0.77\,\text{mm} \leq 3.94\,\text{mm}$?
Since the condition is satisfied, the manufacturer's constant error claim may be pessimistic but acceptable at 95% confidence level.

c. Use the t-statistic to test if the system constant and the scale correction are significantly different from 0 at 95% confidence level.

Solution:

For the scale factor S: $1 - S = -3.01$ ppm and $\hat{\sigma}_S = 1.05\,\text{ppm}$
Using Equation (2.16):

$$|1 - S| \leq \hat{\sigma}_S t_{1-\alpha/2}(\text{df}) \quad t_{0.975}(13) = 2.160$$

At 95% confidence level:$3.01 \leq 1.05 \times 2.160$? or $3.01 \leq 2.27$?
Since the condition is not satisfied, the scale correction is considered to be significantly different from 0 at 95% confidence level.

For the constant correction C: $C = 0.70$ mm and $\hat{\sigma}_C = 0.77$ mm
Using Equation (2.16):

$$|C| \leq \hat{\sigma}_C t_{1-\alpha/2}(\text{df}) \quad t_{0.975}(13) = 2.160$$

At 95% confidence level:0.70 mm $\leq 0.77 \times 2.160$? or 0.70 mm $\leq$ 1.66 mm? Since the condition is satisfied, the constant correction is not significantly different from the expected value of 0 at 95% confidence level.

General Conclusion: The calibration of the instrument should be done again at another time and probably on a different baseline since the scale factor test failed; all the tests must pass in order to consider the calibration as successful.

5.7.4 EDM Calibration When Length of Baseline Is Unknown

The calibration of EDM can also be done in two steps: determining the system constant (C) on a baseline with unknown length and determining the scale factor (S) by calibrating the modulation frequency (EDM standardization). The procedure for determining the system constant will be discussed in this section, while the scale factor determination (EDM standardization) procedure will be given in the next section. In calibrating the EDM equipment when a known base line is not available, a collinear array of points represented by a series of tribrachs on tripods may be used. This type of calibration, however, cannot provide the scale errors of the EDM, but can only determine the system constant of the EDM. The system constant determination in this section is divided into three approaches: the standard approach, the modified standard approach, and the approximate approach.

5.7.4.1 System Constant Determination: Standard Approach The standard approach of EDM system constant determination is based on the ISO standards approach, which provides the procedure for testing the EDM equipment (as opposed to calibrating the EDM equipment). In the testing procedure, a straight line approximately 600 m long with seven points (designed based on unit length of the instrument) as shown in Figure 5.8 is to be measured in all possible combinations; for the seven-point baseline, 21 distance observations are measured. Forced-centering interchange should be used to eliminate centering errors and sufficient number of prisms must be used to ensure that all distances are measured with good return signals. The raw measurements (each distance measured three times and averaged) are corrected for systematic effects (atmospheric correction and slope reduction). Atmospheric corrections are to remove any scale bias in the distance observation due to change in velocity of propagation in the atmosphere. Slope reduction may require that the zenith angles be equally measured; there may also be a challenge in aligning all the seven points in the straight line as well as forced centering on the points. The corrected measurements are then evaluated by parametric least squares method with equal unit weights for all measurements to solve for seven unknown parameters, which are the six distances ($m_{12}, m_{23}, m_{34}, m_{45}, m_{56}, m_{67}$) and the system constant (C). For the 21 distance observations, the parametric least squares equations can be given as follows:

$$\begin{aligned}
\ell_{12} &= m_{12} + C \\
\ell_{13} &= m_{12} + m_{23} + C \\
\ell_{14} &= m_{12} + m_{23} + m_{34} + C \\
\ell_{15} &= m_{12} + m_{23} + m_{34} + m_{45} + C \\
\ell_{16} &= m_{12} + m_{23} + m_{34} + m_{45} + m_{56} + C \\
\ell_{17} &= m_{12} + m_{23} + m_{34} + m_{45} + m_{56} + m_{67} + C
\end{aligned} \tag{5.111}$$

$$\begin{aligned}
\ell_{23} &= m_{23} + C \\
\ell_{24} &= m_{23} + m_{34} + C \\
\ell_{25} &= m_{23} + m_{34} + m_{45} + C \\
\ell_{26} &= m_{23} + m_{34} + m_{45} + m_{56} + C \\
\ell_{27} &= m_{23} + m_{34} + m_{45} + m_{56} + m_{67} + C
\end{aligned} \tag{5.112}$$

$$\begin{aligned}
\ell_{34} &= m_{34} + C \\
\ell_{35} &= m_{34} + m_{45} + C \\
\ell_{36} &= m_{34} + m_{45} + m_{56} + C \\
\ell_{37} &= m_{34} + m_{45} + m_{56} + m_{67} + C
\end{aligned} \tag{5.113}$$

$$\begin{aligned}
\ell_{45} &= m_{45} + C \\
\ell_{46} &= m_{45} + m_{56} + C \\
\ell_{47} &= m_{45} + m_{56} + m_{67} + C
\end{aligned} \tag{5.114}$$

$$\begin{aligned}
\ell_{56} &= m_{56} + C \\
\ell_{57} &= m_{56} + m_{67} + C
\end{aligned} \tag{5.115}$$

$$\ell_{67} = m_{67} + C \tag{5.116}$$

where ℓ_{ij} is the adjusted distance observation from point i to point j; m_{ij} is the unknown distance from point i to point j to be determined; and C is the system constant also to be determined (depending on whether the manufacturer-supplied constant is zero or not). After the least squares adjustment of the observations, the experimental standard deviation (s) of a single measured distance is determined as

$$s = \sqrt{\frac{r^{\mathrm{T}}r}{\nu}} \tag{5.117}$$

where v is number of degrees of freedom (21-7 or 14); r is the vector of observation residuals determined after the least squares adjustment. The most important parameter calculated is the system constant C and its standard deviation. The constant (C) must be tested if it is significantly different from zero using t- or z-test, and the calculated standard deviation must be tested if it is compatible with the quoted value by the instrument manufacturer (Section 2.9). As it can be seen above that scale factor is not determined by the above procedure, but the experimental (or representative) standard deviation of a single distance measurement by the EDM under the same condition of testing is determined. Before testing the EDM equipment in this approach, the EDM must have been checked to be in an acceptable state of permanent adjustment with appropriate tripods, forced-centering equipment and reflectors used, and so on. The stability of the scale of the EDM (if suspected) is tested by another procedure (EDM standardization procedure discussed later). If the above setup is used to determine the system constant (C) of the EDM equipment, the following disadvantages may be experienced with the approach:

- It is very time-consuming, especially for centering tripods and measuring zenith angles every time a calibration is carried out.
- It is impossible to achieve the same level of accuracy in centering as in the case where dedicated pillars are used.
- There is a high degree of uncertainty about the stability of tripods due to the effects of the sun, interchange of reflectors and EDM instrument, and so on.
- It cannot provide the scale factor error for the EDM instrument.

5.7.4.2 System Constant Determination: Modified Standard Approach The modified standard approach of EDM system constant determination involves changing only the parametric equations in the standard approach as follows. The collinear array in Figure 5.8 is still used, but it is now assumed that the points are aligned in the x-axis direction with coordinate of point 1 fixed as $x_1 = 0$; the coordinates of the remaining points and the system constant (C) are then considered as the unknown parameters. In this case, the unknown parameters are $x^{\mathrm{T}} = [x_2, x_3, x_4, x_5, x_6, x_7, C]$ and the parametric least squares equations are as follows:

$$\begin{aligned}
m_{12} &= (x_2 - 0) + C \\
m_{13} &= (x_3 - 0) + C \\
m_{14} &= (x_4 - 0) + C \\
m_{15} &= (x_5 - 0) + C \\
m_{16} &= (x_6 - 0) + C \\
m_{17} &= (x_7 - 0) + C
\end{aligned} \tag{5.118}$$

$$m_{23} = (x_3 - x_2) + C$$
$$m_{24} = (x_4 - x_2) + C$$
$$m_{25} = (x_5 - x_2) + C$$
$$m_{26} = (x_6 - x_2) + C$$
$$m_{27} = (x_7 - x_2) + C \tag{5.119}$$

$$m_{34} = (x_4 - x_3) + C$$
$$m_{35} = (x_5 - x_3) + C$$
$$m_{36} = (x_6 - x_3) + C$$
$$m_{37} = (x_7 - x_3) + C \tag{5.120}$$

$$m_{45} = (x_5 - x_4) + C$$
$$m_{46} = (x_6 - x_4) + C$$
$$m_{47} = (x_7 - x_4) + C \tag{5.121}$$

$$m_{56} = (x_6 - x_5) + C$$
$$m_{57} = (x_7 - x_5) + C \tag{5.122}$$

$$m_{67} = (x_7 - x_6) + C \tag{5.123}$$

where m_{ij} is the distance observation from point i to point j. For the least squares adjustment of the observations, the standard deviation of each measurement can be taken as the precision of the measurement (σ_s) or estimated from the repeated measurements of the same distance. It should be mentioned that each measurement m_{ij} has the following uncertainties:

- Systematic error due to zero constant C not being accounted for
- Random error σ_s due to precision of instrument and measurements.

At the end of the least squares adjustment, the estimated distance (which will be equivalent to the coordinates of the stations, such as $x_2, x_3, x_4, x_5, x_6, x_7$), will have the following uncertainties:

- Random error σ_s due to precision of instrument and measurements.
- Random error due to uncertainty in removing the systematic error from the measurement through adjustment.
- Indicated random error will be higher than that of the direct measurement, but estimate is more certain since the systematic error is accounted for.

Generally, the least squares adjusted distances will be more precise and more accurate than the original measurements since the system constant is taken care of

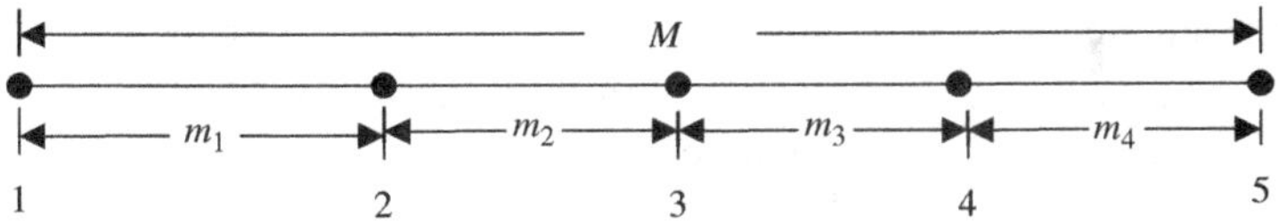

Figure 5.9 Approximate approach of EDM system constant determination.

already in the adjusted quantities, and the estimated system constant will be more precise than when uniquely determined from measurements.

5.7.4.3 System Constant Determination: Approximate Approach The approximate approach of EDM system constant determination involves measuring a line of unknown length in several sections and calculating the system constant by using a simple formula. For example, let the distance M in Figure 5.9 be divided into arbitrary four subsections with measured distances as m_1, m_2, m_3, and m_4 (not necessarily of the same length); there can be as many sections as needed for better accuracy, with the minimum for unique determination being two sections. Measure the total length M of the line with the EDM equipment to be calibrated and then measure the four sections separately. For electro-optical EDM equipment, the same reflector should be used throughout the measurement process and all distance measurements must be corrected for meteorological conditions and slope.

In Figure 5.9, let the system constant correction be C and the corrected measured total distance be $M + C$ and the corrected measured subsections be $m_1 + C$, $m_2 + C$, $m_3 + C$ and $m_4 + C$. The total measured distance $(M + C)$ can be expressed as

$$M + C = m_1 + C + m_2 + C + m_3 + C + m_4 + C \tag{5.124}$$

From Equation (5.124), the system constant correction can be determined as

$$\hat{C} = \frac{M - (m_1 + m_2 + m_3 + m_4)}{3} \tag{5.125}$$

Equation (5.125) can be generalized for n sections of a line, giving the computed system constant as

$$\hat{C} = \frac{M - (m_1 + m_2 + m_3 + m_4 + \cdots + m_n)}{n - 1} \tag{5.126}$$

Random error propagation laws can be applied to Equation (5.125) or generally to Equation (5.126) in order to determine the error $(\hat{\sigma}_C)$ of computing the system constant.

5.7.5 EDM Standardization

EDM standardization refers to a process of comparing the output of the EDM to a standard of length traceable to the National Standard. It is related to the EDM scale determination; the scale can be wrong due to some reasons, which include the following:

- The calculated refractive index (n_2) is incorrect or has been incorrectly applied.
- The reference frequency (f_{REF}) of the oscillator from which the reference wavelength (λ_{REF}) is derived has changed.

When EDM calibration is done on a known baseline, the calibration guidelines are usually given to assist users in verifying that their EDM equipment is working within the EDM manufacturer's stated specification for scale error and constant error. The Surveyors Board of each of the provinces in Canada sets requirements for calibration and standardization of survey equipment. The Surveyor General is responsible for issuing practical implementation advice and for providing certified calibration facilities.

Two different ways of standardizing an EDM instrument are as follows:

(a) Measuring the frequencies of the EDM, which is more precise, but has some major problem. The major problem is that the calibrated frequency counters sufficiently accurate for EDM standardization are normally not readily available to surveyors.

(b) Determining a scale factor from a baseline of known length (a baseline that has been previously known by invar taping, interferometric method, or a precision EDM instrument). The main problem with this approach is that the results may be affected by the instrumental (or constant) errors or errors in reduction of measurements.

5.7.5.1 EDM Standardization: Frequency Method If the actual frequency (f_a) is significantly different from the reference frequency (f_{REF}) for which the instrument is designed, the measured distance (S_{meas}) can be corrected for scale errors, giving the corrected distance (S_{corr}) as

$$S_{corr} = \frac{f_{REF}}{f_a} \times S_{meas} \tag{5.127}$$

Note that the scale error is not very critical; it may be assumed to be fairly constant during the period of observation in a project. However, in order to avoid large-scale errors in EDM, periodic standardization of EDM should be done. If the calibration of EDM is performed over a certified baseline to a prescribed level of precision, the EDM is also considered to be standardized. Note also that the adjustment process will automatically adjust the actual mean scale of the EDM to the grid scale defined by the two control points and their coordinates. Additive constant and cyclic errors are not eliminated, however, by adjustment and may cause systematic errors in the coordinates of traverse points if not accounted for.

5.7.6 Use of Calibration Parameters

After an EDM instrument calibration, the derived calibration parameters (instrument system constant, scale factor), if found to be statistically significant, must be applied to subsequent measurements made with the instrument. For example, if the calibration of an EDM is done on a calibration baseline with the system constant (*C*) and the scale factor (*S*) determined for it and the EDM is used to measure a distance *m*, the corrected distance measurement (*d*) will be given as $d = C + Sm$. In the case where only the system constant (*C*) is determined, the corrected distance will be $d = C + m$ (assuming the scale factor is good). Corrections should only be applied for statistically significant systematic errors in order to provide improvement in accuracy.

Example 5.12 The system constant, z_0, of an EDM is to be determined without using the calibration baseline, as shown in Figure 5.10. Answer the following:

a. Explain how z_0 can be uniquely determined.

Solution: Lay out three stations (A, B, C) separated by distances m_1 and m_2 as shown in Figure 5.10; set the EDM constant to zero and measure the distances m_1, m_2, and M. The distances, corrected for system constant (z_0), can be given as follows:

$$s_1 = m_1 + z_0 \quad (5.128)$$

$$s_2 = m_2 + z_0 \quad (5.129)$$

$$s_3 = M + z_0 \quad (5.130)$$

The corrected distances can be used to formulate the following:

$$s_3 = s_2 + s_1 \quad (5.131)$$

Substituting Equations (5.128)–(5.130) into Equation (5.131) gives the following:

$$M + z_0 = m_1 + z_0 + m_2 + z_0 \quad (5.132)$$

System constant is uniquely determined from Equation (5.132) as follows:

$$z_0 = M - (m_2 + m_1) \quad (5.133)$$

Alternatively, Equation (5.133) can be deduced from Equation (5.126) by substituting $n = 2$ (the number of sections measured in Figure 5.10) into the equation.

b. What is the uncertainty (standard deviation) of the uniquely determined system constant (z_0) if each distance involved is measured with uncertainty of ±0.003 m?

Solution: Apply error propagation law to Equation (5.133):

$$\sigma_{z_0}^2 = \sigma_M^2 + \sigma_{m_2}^2 + \sigma_{m_1}^2$$

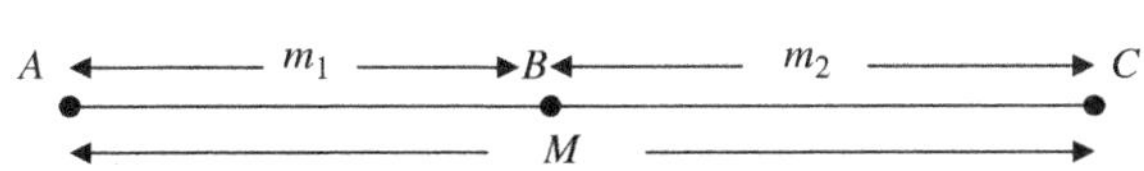

Figure 5.10 Determination of EDM system constant.

For uncertainty of ±0.003 m each:

$$\sigma_{z_0} = 0.003\sqrt{3} \rightarrow 0.0052\,\text{m}$$

c. Explain how you can improve the uncertainty of z_0.

Solution: A collinear array of more than three points would improve the uncertainty in the value of z_0. For example, for $n = 5$ sections, the following equation can be formulated from Equation (5.126):

$$z_0 = \frac{M - (m_1 + m_2 + m_3 + m_4 + m_5)}{4} \tag{5.134}$$

By applying random error propagation laws on Equation (5.134) and assuming the errors are equal ($\sigma_{m_1} = \sigma_{m_2} = \sigma_{m_3} = \sigma_{m_4} = \sigma_{m_5} = \sigma$) for all the section measurements, the variance of the system constant can be given as

$$\sigma_{z_0}^2 = \left(\frac{1}{4}\right)^2 (\sigma_M^2 + 5\sigma^2) \rightarrow \sigma_{z_0} = \left(\frac{1}{4}\right)\sqrt{(\sigma_M^2 + 5\sigma^2)} \tag{5.135}$$

For $\sigma_M = \sigma = 0.003\,\text{m}$, the error in computing the system constant will be $\sigma_{z_0} =$ 0.0018 m. Comparing this error with the value ($\sigma_{z_0} = 0.0052\,\text{m}$) computed in Question (b) for the unique determination of z_0, it can be seen that the size of the standard deviation in the case involving five sections (0.0018 m) is smaller. Thus, it can be concluded that increasing the number of sections will improve the uncertainty of determining z_0.

Example 5.13 In the calibration of an EDM instrument, the zero-point correction ($\widehat{x}$) to the baseline measurements is 1.3 mm and its standard deviation ($s_{\widehat{x}}$) is 0.7 mm. Evaluate if $\widehat{x}$ is equal to zero at 95% confidence level, assuming the number of degrees of freedom for the adjustment of the baseline measurements is 14.

Solution: The hypotheses to be tested are given from Table 2.7 as follows:

$$H_0: \mu = 0 \quad H_A: \mu \neq 0$$

Given $\overline{x} = 1.3\,\text{mm}$ and $\text{SE} = s_{\overrightarrow{x}} = 0.7\,\text{mm}$

Number of degrees of freedom, df = 14
Significant level, $\alpha = 0.05$

From Table 2.8, the H_0 is not rejected if the following condition is satisfied:

$$|\overline{x} - \mu| \leq (\text{SE})t_{\alpha/2,\text{df}=14} \tag{5.136}$$

$$|1.3\,\text{mm}| \leq (0.7\,\text{mm}) \times 2.14$$

$$1.3\,\text{mm} \leq 1.5\,\text{mm}?$$

Since the above condition is satisfied, the null hypothesis stating that the zero-point correction is zero is not rejected at the confidence level of 95%. Similar evaluation can be performed on the *vertical index error* of theodolite's zenith angle measurements.

Example 5.14 On the shelf in the company's survey stores, you have found a total station that has not been used for at least 20 years. The manufacturer's claim, following DIN 18723 (or ISO 17123, now), is a distance "accuracy" of ±2 mm ±2 ppm. Since there is no record of any testing or calibration of this particular instrument, explain the steps that you would recommend following to determine whether this total station is capable of behaving as the manufacturer claimed.

(Reproduced by permission of CBEPS.)

Suggested Solution: The ISO 17123-4 (EDM testing procedures) according to ISO17123-4(2001) determines only the representative standard deviation of distance measurement and the additive constant. Refer to Section 5.7.4.1 for the setup and measurement procedures, Section 5.7.2 for the Heerbrugg's design of the baseline used in the testing procedure, and Example 5.11 for typical statistical testing procedure.

Example 5.15 Rüeger (1996) offers solutions, using combinations of summations, for the additive constant and scale factor from calibration baseline observations and for the additive constant from linear arrays. For a calibration baseline, the solution uses linear regression in the form

$$\hat{y} = a + bx \tag{5.137}$$

(Reproduced by permission of CBEPS).

a. Explain which, of the known pillar distances and of the observed distances, is the independent variable and the dependent variable in the regression and why.

Suggested Solution: y represents the observed distances in all combinations; they can be used to determine the constant (a) without using a baseline with known distances; this variable is dependent on x to determine the scale.

x represents the known pillar distances corresponding to the observed distances (y), determined through a more precise procedure (such as using higher precision instrument and controlled procedure); this is needed to determine the scale factor of the EDM; this variable is independent.

b. Using the estimated values of "a" and "b", explain how the additive constant, z_0, and the scale factor, k, are calculated.

Suggested Solution: Rearrange the regression as follows:

$$x = \frac{-a}{b} + \frac{\hat{y}}{b} \tag{5.138}$$

$$\text{Additive constant, } z_0 = \frac{-a}{b}$$

$$\text{Scale factor, } k = \frac{1}{b}$$

For a calibration baseline, the linear regression solution is given in the form

$$\hat{y} = a + bx$$

For example, if the least squares adjusted quantities are given as $\hat{a} = +0.0057\,\text{m}$ and $\hat{b} = 1.000\,00218$, the additive constant and the scale correction (ppm) can be determined as follows:

$$\text{Additive constant, } z_0 = \frac{-0.0057}{1.00000218} = -5.7\,\text{mm}$$

$$\text{Scale factor, } k = \frac{1}{1.00000218} = 0.999\,99782$$

$$\text{Scale correction} = -2.18\,\text{ppm}$$

The amount of correction to be applied to any measurement D can be expressed as

$$\text{Instrument correction} = -5.7\,\text{mm} - 2.18\,\text{ppm}\ (D)$$

If the correction is applied to the measurement D, the corrected distance will be given as

$$\text{Corrected distance} = D - 5.7\,\text{mm} - 2.18\,\text{ppm}\ (D)$$

or

$$\text{Corrected distance} = -5.7\,\text{mm} + 0.999\,99782\ (D) \tag{5.139}$$

c. Explain whether these are rigorous least squares estimations and why.

Suggested Solution: The quantities a and b are determined through least squares process; the derived additive and scale factor are also rigorously determined from redundant measurements.

d. The standard deviations, σ_{z_0} and σ_k, can be estimated from the regression. Explain how this is done and why.

Suggested Solution: From the least squares adjustment procedure, the cofactor matrix of the adjusted parameters a and b (which are correlated) will be provided as Q:

$$z_0 = \frac{-a}{b} \tag{5.140}$$

$$k = \frac{1}{b} \tag{5.141}$$

The covariance matrix (C) of the additive constant (z_0) and the scale factor (k) is obtained through variance–covariance propagation:

$$C = J\left[s_0^2 Q\right] J^{\mathrm{T}} \tag{5.142}$$

where J is the Jacobian of Equations (5.140) and (5.141) with respect to a and b, given as

$$J = \begin{bmatrix} \frac{-1}{b} & \frac{a}{b^2} \\ 0 & \frac{-1}{b^2} \end{bmatrix} \tag{5.143}$$

and the a posteriori variance factor of unit weight (assuming unit weights for all the measurements and taking the true distances as errorless) can be given as

$$\hat{\sigma}_0^2 = s_0^2 = \frac{\sum (a + bx - y)^2}{n - 2} \tag{5.144}$$

If the correlations between parameters a and b are ignored, then their standard deviations can be given, respectively, as follows:

$$s_a = s_0 \sqrt{\frac{\sum x^2}{n\sum x^2 - \left(\sum x\right)^2}} \tag{5.145}$$

and

$$s_b = \frac{s_0}{\sqrt{\frac{\sum x^2 - \left(\sum x\right)^2}{n}}} \tag{5.146}$$

From Equation (5.142), the variance–covariance matrix can be given as

$$C = \begin{bmatrix} \frac{s_a^2}{b^2} + \frac{a^2}{b^4}s_b^2 & -\frac{a}{b^4}s_b^2 \\ -\frac{a}{b^4}s_b^2 & \frac{1}{b^4}s_b^2 \end{bmatrix} \tag{5.147}$$

so that

$$\sigma_{z_0} = \sqrt{\frac{s_a^2}{b^2} + \frac{a^2}{b^4}s_b^2} \tag{5.148}$$

$$\sigma_k = \sqrt{\frac{1}{b^4}s_b^2} \tag{5.149}$$

The standard deviations are determined in order to statistically test if z_0 and the scale correction $1 - k$ are significantly different from zero at 95% confidence level, and also to check if the standard deviation for z_0 is compatible with the manufacturer specification (a) for the additive constant and also if that of scale factor (k) is compatible with the value supplied by the manufacturer (b ppm).

e. For a linear array, the solution involves calculation of the adjusted distances and misclosures. This has been simplified for linear arrays with a particular number of points. For example, with seven points, the additive constant is

$$\begin{aligned} z_0 = &-\frac{1}{35}[5(s_{12} + s_{23} + s_{34} + s_{45} + s_{56} + s_{67} - s_{17})] \\ &-\frac{1}{35}[3(s_{13} + s_{35} + s_{57} - s_{27} + s_{24} + s_{46} - s_{16})] \\ &-\frac{1}{35}[(s_{14} + s_{47} - s_{37} + s_{36} - s_{26} + s_{25} - s_{15})] \end{aligned} \tag{5.150}$$

Explain why this is a rigorous least squares estimation of z_0 and what assumptions are the basis for that.

Suggested Solution: From the parametric Equations (5.111)–(5.116), form the design matrix, A, of size 21 by 7. This sized is based on 21 observations of all combinations of baselines in 7-point baseline with 6 distances between baseline points and the additive constant C forming the unknown parameters. From the least squares solution, $\hat{x} = -(A^{\mathrm{T}}A)^{-1}A^{\mathrm{T}}\ell$ (where ℓ is a vector of 21 observations, solve for the additive constant and rearrange to obtain Equation (5.150)). On the basis of satisfying the least squares criterion, the equation gives the least squares adjusted solution for the additive constant z_0. The basic assumptions in arriving at the above solution are listed as follows:

1. All the 7 points are colinear so that sums of measurements of sections of the baselines give the corresponding length of the whole section measured.
2. All measurements have equal weights (unit weights).
3. No initial values are assumed for the unknown 6 baseline distances between the markers and the unknown additive constant.
4. 21 distance observations in all combinations are measured to solve for the 7 unknowns, giving 14 redundancies (degrees of freedom).
5. Short periodic errors will not affect the additive constant.

f. From the 28 distances on an 8-point linear array, the additive constant, z_0, was estimated to be −0.91 mm with the standard deviation of an observation of unit weight being estimated to be ±1.14 mm. Explain whether the value z_0 is significant at 95%.

Suggested Solution: According to Rüeger (1996):

$$\hat{\sigma}_{z_0}^2 = \hat{\sigma}_0^2 \frac{6}{(N-1)(N-2)} \tag{5.151}$$

where N is the number of baseline stations, and

$$\hat{\sigma}_0^2 = \frac{\sum v^2}{n-u}.$$

The given quantities are $z_0 = -0.91$ mm and $s_0 = \pm 1.14$ mm; substituting these values into Equation (5.151) gives

$$\hat{\sigma}_{z_0} = s_0\sqrt{\frac{6}{(8-1)(8-2)}} = 1.14 \times \sqrt{0.143} = \pm 0.43\,\text{mm}$$

For the degrees of freedom, df = 28 − 8 = 20 (for 28 possible measurements and the 7 unknown distances between baseline points plus one additive constant forming the unknown parameters), the additive constant can be tested for significance by using Equation (2.16) and two-tailed test in Table 2.8:

$$|-0.91| \leq 0.431 \times t_{\text{df}=20,0.025}?$$

with $t_{\text{df}=20,0.025} = 2.086$, the following condition is obtained:

$$0.91 \leq 0.899?$$

Since the condition is not satisfied, the additive constant is (just) statistically significant at 95% confidence.

6

ACCURACY ANALYSIS AND EVALUATION OF ELEVATION AND COORDINATE DIFFERENCE MEASUREMENT SYSTEMS

OBJECTIVES

After studying this chapter, you should be able to

1. Analyze accuracy of elevation difference and coordinate difference measurements, including sources of errors and error propagation
2. Evaluate the precision of geodetic leveling equipment under field conditions
3. Evaluate the GPS equipment/software performance

6.1 INTRODUCTION

Precise elevation differences between accessible terrain points are precisely determined using geodetic leveling procedure. The internal errors with the procedure are due to the type of level instrument and level rods used. The major *internal errors* associated with a level instrument are given as follows:

- Pointing error
- Reading error
- Instrument leveling error

Precision Surveying: The Principles and Geomatics Practice, First Edition. John Olusegun Ogundare.

- Level collimation error, which is a systematic deviation of the line of sight from the horizontal plane that is perpendicular to the direction of gravity through the instrument.

The major *internal errors* associated with a level rod are given as follows:

- Rod scale error, which is due to the graduations on the rod not being uniform and not being consistent with the National standards of units.
- Rod index error, which is due to a possible constant offset of the zero mark on the rod from the base of the plate; even number of setups are usually required in geodetic leveling in order to eliminate its effect.

The major *external error sources* in geodetic leveling are given as follows:

- Vertical atmospheric refraction, which is due to vertical temperature gradient
- Sinking of instrument and turning points, which is due to the weight of the instrument or level rod on the points.
- Rebound of instrument and turning points, which is caused by the response of the spongy material of the locations of the instrument and turning points.
- Earth curvature, which is a consequence of the earth not being flat.
- Rod temperature, which tends to change the length of the leveling rod during the measurement process as a result of changing atmospheric temperature. This requires that the temperature of the rod be taken at regular intervals and the appropriate correction applied to the rod readings.
- Astronomic correction, which is due to the effects of Moon and Sun tides on the equipotential surfaces of the earth, may contribute up to 0.1 mm/km, accumulating along the north–south direction. These effects must be corrected for in a regional or continental leveling project.
- Orthometric correction, which is due to the effect of nonparallelism of equipotential surfaces (especially along the north–south direction).

Systematic errors commonly accounted for in geodetic leveling are the effects due to vertical collimation, rod scale, vertical atmospheric refraction, rod temperature, astronomic and orthometric corrections.

6.2 POINTING ERROR

Pointing error in geodetic leveling is due to prevailing atmospheric conditions and magnification of the instrument telescope. If the atmospheric condition is good (with clear visibility), the effect of pointing error on a leveling observation can be given as

$$\sigma_{\mathrm{P}} = \frac{C}{(206,265'')M}S \tag{6.1}$$

where C is a constant value ranging from $C = 30''$ to $C = 60''$, S is the length of the line of sight, and M is the magnification of the telescope of the level instrument.

6.3 READING/ROD PLUMBING ERROR

Reading and rod plumbing errors will be considered to be the same; if the rod is not plumb, a higher reading than the true reading will be read on the rod. Usually, reading error in geodetic leveling is due to the effects of nonverticality of the level rod and the imperfection in reading the rod. Geodetic level rods are usually equipped with level vial; the limited sensitivity of the level bubble, however, prevents the rod from being perfectly vertical. The effect of the nonverticality of a level rod on a level measurement is demonstrated in Figure 6.1. In the figure, v_r'' is the sensitivity of the level bubble on the rod in arc-seconds and ℓ is the length O-R1 of the rod. (The sensitivity of the level bubble of most of the geodetic level rods is $10'$.) The actual reading on the rod is R1 (affected by nonverticality of the rod) while R2 is desired. The effect of nonverticality of the level rod increases the rod reading by an amount that can be given as

$$\sigma_r \approx \frac{\ell}{2}\left(\frac{v_r''}{206,265}\right)^2 \tag{6.2}$$

In geodetic leveling, some of the errors in reading the rod are averaged out by reading and averaging the three stadia hairs read on the rod.

6.4 LEVELING ERROR

Leveling error (σ_L) of a level instrument is due to the sensitivity of the level bubble on the instrument. Any misleveling of the instrument will cause the line of sight to deviate from the horizon as illustrated in Figure 6.2.

The effect of the instrument leveling error on a level measurement can be estimated by

$$\sigma_L = \left(\frac{\sigma_v}{206,265''}\right) S \tag{6.3}$$

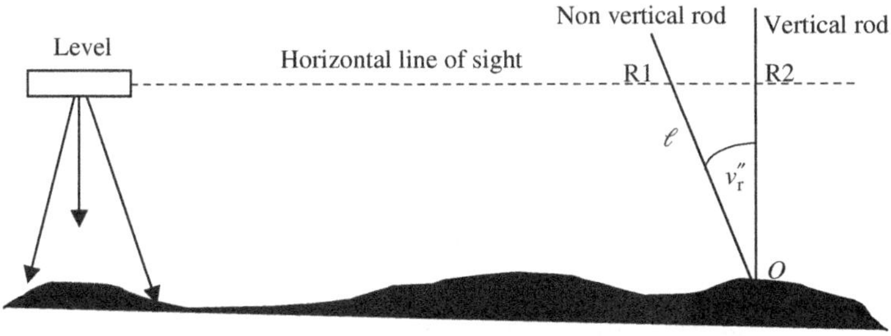

Figure 6.1 Relationship between nonverticality of level rod and rod readings.

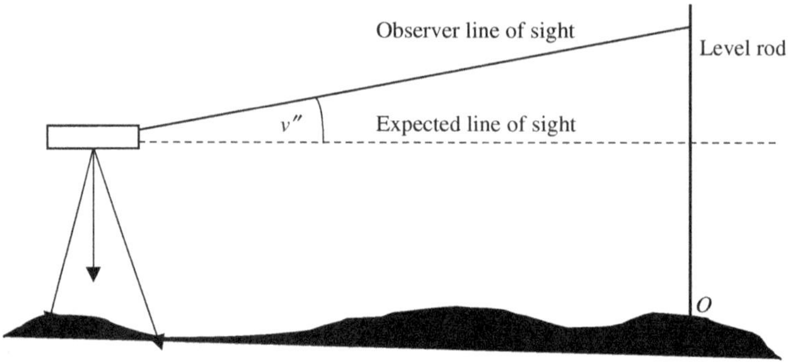

Figure 6.2 Relationship between instrument leveling error and rod readings.

where $\sigma_v = 0.2v''$ is the error in leveling the instrument and S is the horizontal distance between the level instrument and the level rod.

6.5 COLLIMATION, ROD SCALE, AND ROD INDEX ERRORS

Level collimation error results from a systematic deviation of the line of sight from the horizontal plane. In precision leveling, a compensating device is used to orient the line of sight in a horizontal plane after approximately leveling the level instrument. Depending on the type of leveling job, the collimation error factor (or C-factor) of a precision level instrument is not to exceed 0.05 mm/m for a single line of sight. If the instrument uses a reversible compensator, C-factor of the mean of two lines of sight should not exceed 0.02 mm/m. In order to further control the effects of collimation errors on leveling, it is usually specified that the imbalance between backsight and foresight distances at each instrument setup and their total for each section being leveled must not exceed a specified tolerance; for example, for special-order geodetic leveling, it must not exceed 5 m per setup or accumulate algebraically to more than 5 m in a section (NRC, 1978).

The effect of *collimation error* on a level measurement is the same as that of the instrument leveling error, except that collimation error is systematic and its effect can be removed by balancing the lengths of the backsights (BS) and foresights (FS) or by instrument calibration. By calibrating the instrument, the C-factor for the level is determined and the following correction can be added to the elevation difference in a leveled section:

$$\text{Correction} = C \times \left(\sum_{1}^{n} d_{\text{BS}_i} - \sum_{1}^{n} d_{\text{FS}_i} \right) \tag{6.4}$$

where C is the C-factor in mm/m and n is the number of instrument setups in the leveled section, d_{BS_i} and d_{FS_i} are the backsight (BS) and foresight (FS) distances at a given setup number i. The collimation error per setup can be given as

$$\varepsilon_{\mathrm{col}} = -C \times \Delta s \tag{6.5}$$

where Δs is the difference in the length of the backsight and foresight in (m) given as $(d_{\mathrm{BS}_i} - d_{\mathrm{FS}_i})$. If $C = 0.05$ mm/m (the maximum collimation error for single line of sight for first order) and $\Delta s = 5$ m, the error over a setup is 0.25 mm. This correction is applied for every setup in a section when instruments other than those having a reversible compensator are used. When double-compensator instruments are used, the C is not expected to reach more than 0.02 mm/m or 0.08 mm/section length when the maximum discrepancy of Δs does not exceed 5 m. The collimation check of instrument must be performed daily to keep instrument within 0.02 mm/m; if this amount is exceeded, the instrument must be readjusted.

The effects of *rod scale* and *rod index errors* are also systematic. The *rod scale error* can be removed or reduced to a negligible amount by calibrating the rod just before a leveling campaign. By using an even number of setups or using the same rod for both the backsight and foresight readings in each setup, the effect of *rod index error* can be completely removed.

6.6 EFFECTS OF VERTICAL ATMOSPHERIC REFRACTION AND EARTH CURVATURE

The effect of vertical refraction in reading a rod at a backsight distance S_{B} can be given as $\delta Z_{\mathrm{B}}(S_{\mathrm{B}})$ and for the foresight distance S_{F} as $\delta Z_{\mathrm{F}}(S_{\mathrm{F}})$. The error (δh_{r}) in the difference between the backsight and foresight rod readings as a result of *vertical atmospheric refraction* can be given as

$$\delta h_{\mathrm{r}} = \delta Z_{\mathrm{F}}(S_{\mathrm{F}}) - \delta Z_{\mathrm{B}}(S_{\mathrm{B}})\delta h_{\mathrm{r}} \tag{6.6}$$

or

$$\delta h_{\mathrm{r}} = \frac{k_{\mathrm{vF}} S_{\mathrm{F}}^2 - k_{\mathrm{vB}} S_{\mathrm{B}}^2}{2R} \tag{6.7}$$

where k_{vF} and k_{vB} are the coefficients of vertical refraction for the foresight and backsight readings, respectively; δZ_{F} and δZ_{B} are the atmospheric refraction effects in the vertical directions (in radians) to the foresight and backsight, respectively; and R is the mean radius of the earth. This effect is systematic and can be applied to the measurement; the residual error due to the inaccuracy in determining the coefficient of vertical refraction could be negligible if short distances are involved in the leveling and if the foresight and backsight distances are balanced during the survey.

The effect of *earth curvature* on a leveling measurement can be given as follows:

$$\delta h_{\mathrm{e}} = \frac{S_{\mathrm{F}}^2 - S_{\mathrm{B}}^2}{2R} \tag{6.8}$$

This effect is removed from leveling by balancing the foresight and backsight distances. The effects of the *sinking and rebound of the instrument and the turning points* can be minimized by setting up the instrument and the level rods on stable locations and also by alternating the backsight and foresight readings at every other setup.

6.7 RANDOM ERROR PROPAGATION FOR ELEVATION DIFFERENCE MEASUREMENTS

The standard deviation of leveled elevation difference will vary depending on the leveling procedure adopted. One of the geodetic leveling procedures is known as double simultaneous observation with invar double-scale (high and low scales) rods and a geodetic level. The usual reading procedure of double-scale rod can be summarized (cf. FGCC, 1984) as follows:

First setup:

- Take backsight – read low-scale stadia
- Take foresight – read low-scale stadia
- Off-level/relevel or reverse compensator
- Take foresight – read high-scale stadia
- Take backsight – read high-scale stadia.

Second setup:

- Take foresight – read low-scale stadia
- Take backsight – read low-scale stadia
- Off-level/relevel or reverse compensator
- Take backsight – read high-scale stadia
- Take foresight – read high-scale stadia.

As can be seen in the aforementioned steps, four readings are taken at each setup, and there is releveling between every pair of measurements. The elevation difference at each setup can be determined from the following:

$$\Delta h_i = \frac{(\mathrm{BS}_{\mathrm{L}i} - \mathrm{FS}_{\mathrm{L}i}) + (\mathrm{BS}_{\mathrm{H}i} - \mathrm{FS}_{\mathrm{H}i})}{2} \tag{6.9}$$

where $\mathrm{BS}_{\mathrm{L}i}$ and $\mathrm{FS}_{\mathrm{L}i}$ are the backsight and foresight readings on the low scale; $\mathrm{BS}_{\mathrm{H}i}$ and $\mathrm{FS}_{\mathrm{H}i}$ are the backsight and foresight readings on the high scale. From Equation (6.9), assuming the pointing error for each sighting is the same and the reading error for each sighting is the same, the variance of Δh_i in Equation (6.9) can be determined by the law of error propagation as follows (remember that leveling is done twice in each setup):

$$\sigma^2_{\Delta h_i} = \sigma^2_{\mathrm{P}} + \sigma^2_{\mathrm{r}} + \sigma^2_{\mathrm{L}} \tag{6.10}$$

where $\sigma_{\rm p}$, $\sigma_{\rm r}$, and $\sigma_{\rm L}$ are the pointing, reading, and leveling errors as expressed in Equations (6.1)–(6.3), respectively. Remember that in Equation (6.10) the effect of refraction is ignored since necessary precautions are taken in minimizing its effect on geodetic leveling; the effects of residual systematic errors are also not considered. The variance of a leveled section with m number of setups can be deduced from Equation (6.10) as

$$\sigma^2_{\rm section} = m \cdot \sigma^2_{\Delta h_i} \tag{6.11}$$

To achieve higher accuracy in geodetic leveling surveys, field crews, instruments, and sections to be leveled must be chosen randomly; the temperature is to be measured by using temperature probes at chosen different heights above the terrain (e.g., at 0.3, 0.7, 1.2, 1.8, and 3.0 m) for calculating the vertical temperature gradients along the leveling line. Since the temperature probes are capable of storing temperature readings for a whole-day probes, the stored data can be downloaded and the memory cleared daily.

Example 6.1 Consider a differential leveling with the Leica NA2 automatic level with the telescope magnification of 32× and a compensator setting accuracy of $\sigma_{\rm v} = 0.3''$. Determine the standard deviation of elevation differences over 1 km (for single and double leveling runs) and the section closure and the loop closure over $L = 3$ km.

Solution: From Equation (6.1), the pointing error over a sight distance, $S = 50$ m, can be determined as follows:

$$\sigma_{\rm P} = \frac{C}{(206,265'')M} S \quad (\text{Taking } C = 45'')$$

$$\sigma_{\rm P} = \frac{45''}{(206,265'') \times 32} 50{,}000 \text{ mm} \quad (\text{or } 0.4 \text{ mm})$$

From Equation (6.2), the reading error can be calculated as follows:

$$\sigma_{\rm r} \approx \frac{\ell}{2}\left(\frac{v''_{\rm r}}{206,265}\right)^2$$

Taking the rod length (ℓ) as 3000 mm and sensitivity of the rod level bubble as 600″, the reading error can be given as follows:

$$\sigma_{\rm r} \approx \frac{3000 \text{ mm}}{2}\left(\frac{600''}{206,265''}\right)^2 = 0.01 \text{ mm}$$

From Equation (6.3), the leveling error (over 50 m sight distance) can be calculated as follows:

$$\sigma_{\rm L} = \left(\frac{\sigma_v}{206,265''}\right) S$$

With $S = 50{,}000$ mm and $\sigma_{\text{v}} = 0.3''$, the leveling error is determined as follows:

$$\sigma_{\text{L}} = \left(\frac{0.3''}{206,265''}\right) 50{,}000 \text{ mm (or } 0.07 \text{ mm)}$$

The total standard deviation of the elevation difference in one setup based on the procedure expressed in Equation (6.9) can be given as

$$\sigma^2_{\Delta h(100)} = \sigma^2_{\text{P}} + \sigma^2_{\text{r}} + \sigma^2_{\text{L}}$$

or

$$\sigma_{\Delta h(100)} = \sqrt{0.4^2 + 0.01^2 + (0.07)^2} = 0.41 \text{ mm}$$

where the total standard deviation of the elevation difference ($\sigma_{\Delta h(100)}$) is over the distance of 100 m. The total standard deviation of elevation differences over 1000 m is given using Equation (6.11):

$$\text{No of setups, } m = 1000/100 \text{ (or 10)}$$

$$\sigma_{\Delta h(1000)} = \sqrt{(0.41)^2 \times 10}/1 \text{ km (or } 1.\ 30 \text{ mm/km)}$$

For single run: $\sigma_{\text{ran}} = 1.30$ mm/km: For double leveling run with the average of forward and backward runs considered, the standard deviation is 1.30 mm/$\sqrt{2}$ (or 1.0 mm/km).

Section closure: For section closure, which is the discrepancy (at 95% confidence level) between the measured forward and backward elevation differences, the following can be obtained from Equation (3.6):

$$\Delta = 1.96 \times \sigma_{\text{ran}} \sqrt{2L} \tag{6.12}$$

$$\sigma_{\text{ran}} = 1.30 \text{ mm/km}; L = 3 \text{ km}; \Delta = 1.96 \times 1.3\sqrt{2 \times 3} \text{ mm (or } 6.2 \text{ mm)}$$

Loop closure: Loop closure is the discrepancy (at 95% confidence level) from zero of the sum of elevation differences over a total leveling loop length *L*, which can be expressed from Equation (3.6) as follows:

$$\Delta = 1.96 \times \sigma_{\text{ran}} \sqrt{L} \tag{6.13}$$

$$\sigma_{\text{ran}} = 1.30 \text{ mm/km}; L = 3 \text{ km}; \Delta = 1.96 \times 1.3\sqrt{3} \text{ mm (or 4.4 mm)}$$

Example 6.2 A geodetic leveling survey is to be carried out such that it satisfies the requirement that the difference between backsight and foresight distances at each setup and their total for each section is not to exceed 5 m with a maximum length of sight of 50 m. What is the standard deviation expected for each length of sight if "not to exceed" is considered the expected error at 99%.

Solution: Using Equation (2.52):

$$|\Delta s| \le \sigma_{\Delta s}\sqrt{\chi^2_{\mathrm{df}=1,\alpha}} \tag{6.14}$$

where $\Delta s = s_\mathrm{b} - s_\mathrm{f}$; $\sigma_{\Delta s}$ is the standard deviation of Δs; s_f and s_b are the foresight and backsight distances, respectively; $\chi^2_{\mathrm{df}=1,\alpha}$ is the upper area Chi-square distribution value at the significance level of $\alpha = 0.01$; and $\mathrm{df} = 1$ is the number of degrees of freedom, which is 1 (for one-dimensional cases). In this problem, the maximum discrepancy between the backsight and foresight distances ($\Delta s = 5$ m) can be given as being equivalent to the 99% confidence interval as follows:

$$\Delta s = \sigma_{\Delta s}\sqrt{6.63}$$
$$5 = \sigma_{\Delta s} \times 2.575 \rightarrow \sigma_{\Delta s} = 1.9417 \tag{6.15}$$

Applying the error propagation laws on the discrepancy $\Delta s = s_\mathrm{b} - s_\mathrm{f}$ (with equal contribution from the backsight and foresight distances, s_b and s_f, respectively) gives

$$\sigma_{\Delta s} = \sigma_\mathrm{s}\sqrt{2} \tag{6.16}$$

From Equations (6.15) and (6.16):

$$\sigma_\mathrm{s} = \pm\frac{1.9417}{\sqrt{2}} \rightarrow \pm 1.373\ \mathrm{m}$$

The standard deviation of sight distance measurement should be less than 1.4 m.

6.8 TESTING PROCEDURES FOR LEVELING EQUIPMENT

The optical-mechanical levels are calibrated by using a set of collimators, precise invar rods, and sighting distances of 30 m; the mean error of difference in height is determined as well as setting accuracy of compensator, collimation errors, and so on. The testing procedure for leveling equipment is determining the best achievable measure of precision of a particular precision level and its supporting equipment under field conditions. The measure of precision of the leveling equipment is usually expressed in terms of the experimental standard deviation of a 1-km double-run leveling. At the end of the testing procedure, statistical tests should be applied to determine whether the calculated standard deviation ($\widehat{\sigma}$) obtained compares with the manufacturer's claimed standard deviation (σ) and whether the difference (dz_0) of the zero-points of the leveling staffs used is equal to zero. The testing procedure, however, does not check collimation error of the instrument.

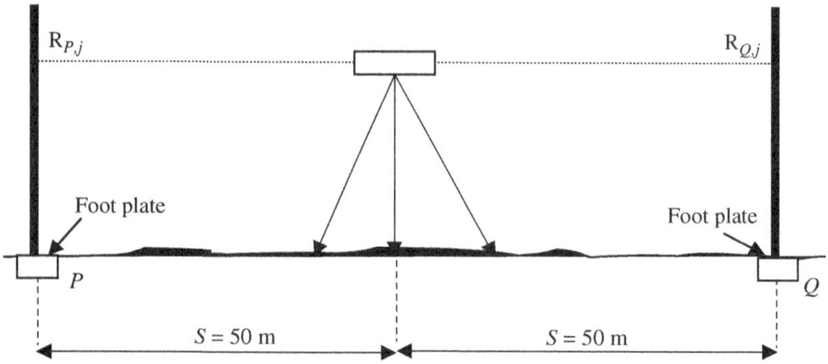

Figure 6.3 A typical setup of level on a test line.

The testing procedure to be discussed in this section is to illustrate how the precision of leveling equipment can be evaluated. The ISO standards (ISO17123-2, 2001) are the internationally recommended standards for testing procedures and should be consulted in practice. The testing procedure in this section (Figure 6.3) consists of setting up two leveling points P and Q at approximately 100 m apart (based on special-order specification for precision leveling; the ISO standards procedures use 60 m instead); the leveling staffs are to be set up on positions that will remain stable during the measurements. The ground must be compact and uniform (or fairly horizontal) in order to keep the influence of refraction as minimal as possible. Since refraction effects can be more troublesome on the roads covered with asphalt or concrete, such roads are considered unsuitable as test lines.

From Figure 6.3, the leveling instrument is to be set up approximately at an equal sight distance of about $S = 50$ m from the leveling points P and Q. This is to reduce the influence of refraction and the displacement of the collimation axis. The leveling instrument must be shaded with umbrella from any direct sunlight on the instrument during data acquisition. The collimation error of the instrument should also be checked before taking the measurements. It is important that the leveling instrument be allowed to acclimatize to the ambient temperature for about 2 min/°C temperature difference before taking the measurements. The data acquisition procedure can be illustrated as follows. For an example, let 10 pairs of rod readings be made using the setup in Figure 6.3. Each pair of readings shall comprise of one backward reading, R_{Pj}, to the leveling staff at point P and one forward reading, $R_{Q,j}$, to the leveling staff at point Q with $j = 1, \ldots, 10$. After each pair of readings, the instrument must be lifted and placed at a slightly different location and leveled in order to randomize the measurement errors. It should be mentioned that the ISO standards testing procedure recommends 20 pairs of readings (with the distance between the rods being 60 m) for each of the two sets of measurements (ISO17123-2, 2001). Refer to Example 6.5 for further analysis of ISO standards testing procedure for leveling equipment. The procedure stated in the following section, however, is found by the author to be consistent with the ISO standards procedure for leveling equipment.

6.8.1 Precision Determination of Leveling Equipment

The steps for determining the precision (experimental standard deviation) of leveling equipment are as follows:

1. Determine the height differences (dh_j) between the backward readings, $R_{P,j}$, and the forward readings, $R_{Q,j}$, of the 10 pairs of readings:

$$dh_j = R_{P,j} - R_{Q,j} \quad j = 1, \dots, 10 \tag{6.17}$$

2. If the elevation of point P is assumed to be $h_P = 0.0\ m$ (known) and the elevation of point Q (h_Q) is unknown, the least squares parametric equations can be formulated as follows:

$$dh_i = h_Q - h_P + dz_0 \quad i = 1, \dots, 10 \tag{6.18}$$

   where h_Q and dz_0 are the unknown parameters, and dz_0 is the difference in the zero-point offsets of the two leveling staffs used. Assume that another series of measurements consisting of 10 pairs of rod readings are added to the original measurements. If for the new series of measurement the two leveling staffs at the points P and Q are interchanged for randomizing the errors, the following parametric equations can be added to Equation (6.18):

$$dh_j = h_Q - h_P - dz_0 \quad j = 1, \dots, 10 \tag{6.19}$$

   where Equation (6.18) is for the first 10 pairs of elevation difference measurements and Equation (6.19) is for the second 10 pairs of elevation difference measurements. From Equations (6.18) and (6.19) there will be $n = 20$ parametric equations formulated with the $u = 2$ unknown values of the parameters (h_Q and dz_0) to be solved for.
3. The calculated values for the unknown parameters are given from the least squares solution as

$$\hat{\delta} = \begin{bmatrix} \hat{h}_Q \\ d\hat{z}_0 \end{bmatrix} = \begin{bmatrix} h'_Q \\ dz'_0 \end{bmatrix} + \delta \tag{6.20}$$

   where h'_Q, dz'_0 are the approximate elevation of point Q and the approximate value of the difference in the zero-point offsets of the two leveling staffs used (there values can be set to zeroes); the corrections to be applied to the approximate values can be given from least squares adjustment as

$$\delta = -(A^{\mathrm{T}}A)^{-1}A^{\mathrm{T}}w; \tag{6.21}$$

   A is the Jacobian matrix of Equations (6.18) and (6.19) with respect to the $u = 2$ unknown parameters (h_Q and dz_0) and w is the misclosure vector given as follows:

$$w = \begin{bmatrix} w_i \\ w_j \end{bmatrix}_{20} \tag{6.22}$$

with

$$w_i = h_Q - h'_P + dz'_0 - dh_i \quad i = 1, \ldots, 10 \tag{6.23}$$

$$w_j = h_Q - h'_P - dz'_0 - dh_j \quad j = 1, \ldots, 10 \tag{6.24}$$

and h'_Q and dz'_0 are the approximate values of the parameters.

4. Calculate the experimental standard deviation ($\hat{\sigma}$) of the height difference for a distance of 100 m:

$$\hat{\sigma} = \sqrt{\frac{r^{\mathrm{T}} r}{\mathrm{df}}} \tag{6.25}$$

where df $= n - u$ is the number of degrees of freedom (in this case, df $= 18$) and r is the residual vector given as follows:

$$r = A\hat{\delta} + w \tag{6.26}$$

5. Calculate the experimental standard deviation for 1-km double-run leveling (usual form of expressing precision of level equipment):

$$\hat{\sigma}_{\mathrm{LEV}} = \frac{\hat{\sigma}}{\sqrt{2}} \times \sqrt{\frac{1000\ \mathrm{m}}{100\ \mathrm{m}}} \tag{6.27}$$

or

$$\hat{\sigma}_{\mathrm{LEV}} = \hat{\sigma} \times 2.24 \tag{6.28}$$

6. The experimental standard deviation ($\hat{\sigma}_{dz_0}$) of the zero-point offsets of the two leveling staffs can be extracted from the covariance matrix of the adjusted parameters, given as

$$C_{\hat{\delta}} = \hat{\sigma}^2 (A^{\mathrm{T}} A)^{-1} \tag{6.29}$$

7. In order to interpret the results, appropriate statistical tests must be performed. The tests will determine if the experimental standard deviation ($\hat{\sigma}$) of a height difference measured on the test line is statistically the same as that claimed by the manufacturer and will also determine if the difference (dz_0) in the zero-point offsets of the two leveling staffs with its experimental standard deviation ($\hat{\sigma}_{dz_0}$) is statistically equal to zero. The calculated zero-point offset of any two leveling staffs can be tested if it is statistically different from zero by using the concept of test of hypothesis about difference of two population means given in Section 2.9.2, in this case $dz_0 = \bar{x}_1 - \bar{x}_2$ is used in Equations (2.49)–(2.52) in the section; or using the two-tailed tests in Tables 2.7 and 2.8.

Example 6.3 In order to investigate that the precision in use of leveling equipment is appropriate to the intended measuring task, two experiments were carried out with two samples of measurements by the same instrument but different observers. The results of the experiments are as follows:

Experiment 1: Computed standard deviation of instrument (s_1) = 2.0 mm, number of degrees of freedom (df_1) = 38.

Experiment 2: Computed standard deviation of instrument (s_2) = 2.5 mm, number of degrees of freedom (df_2) = 38.

Do the two experimental standard deviations, s_1 and s_2, as determined from the two different samples of measurements belong to the same population at 95% confidence level?

Solution: This example deals with comparing two sample standard deviations discussed in Section 2.9.4. In this case, two experimental standard deviations, s_1 and s_2, determined from two different samples of measurements belonging to the same population (σ) are to be compared at the confidence level, $1-\alpha$. The statistical tests can be expressed as follows:

$$H_0: s_1^2 = s_2^2 \quad H_A: s_1^2 \neq s_2^2$$

Given $s_1 = 2.0$ mm, $s_2 = 2.5$ mm, $\upsilon_1 = 38$ and $\upsilon_2 = 38$.
The confidence interval expression to be used (at $\alpha = 0.05$) is given in Section 2.9.4, Equation (2.58), as follows:

$$F_{1-\alpha/2,\mathrm{df}_1,\mathrm{df}_2}(\text{upper-tail area}) \leq \frac{s_1^2}{s_2^2} \leq F_{\alpha/2,\mathrm{df}_1,\mathrm{df}_2}(\text{upper-tail area}) \tag{6.30}$$

$$\frac{1}{F_{\alpha/2,\mathrm{df}_1,\mathrm{df}_2}(\text{upper-tail area})} \leq \frac{s_1^2}{s_2^2} \leq F_{\alpha/2,\mathrm{df}_1,\mathrm{df}_2}(\text{upper-tail area})$$

$$F_{0.975,38,38}(\text{upper-tail area}) \leq \frac{2.0^2}{2.5^2} \leq F_{0.025,38,38}(\text{upper-tail area}) \tag{6.31}$$

$$0.52 \leq \frac{4.0}{6.25} \leq 1.91 \quad \text{or} \quad 0.52 \leq 1.56 \leq 1.91$$

Since the condition is fulfilled, the null hypothesis stating that the experimental standard deviations $s_1 = 2.0$ mm and $s_2 = 2.5$ mm belong to the same population is not rejected at the confidence level of 95%.

Example 6.4 In the calibration of some geodetic leveling equipment, two leveling staffs were used. After the least squares adjustment of the measurements, the difference in the zero-point offsets of the two leveling staffs and its experimental standard deviation were calculated as −0.3 mm and 0.2 mm, respectively. Evaluate if the difference in the zero-point offsets of the two leveling staffs is equal to zero at 95% confidence level, assuming the number of degrees of freedom for the adjustment is 38.

Solution:
The hypotheses to be tested (from Table 2.7 in Chapter 2):

$$H_0 : \mu = 0 \quad H_A : \mu \neq 0$$

Difference in the zero-point offsets of two leveling staffs, $dz_0 = -0.3$ mm
Standard deviation of the difference, SE $= s_{dz_0} = 0.2$ mm.
Number of degrees of freedom, df = 38; significant level, $\alpha = 0.05$.
The H_0 is not rejected if the following condition (Table 2.8 in Chapter 2) is satisfied:

$$|dz_0| \leq s_{dz_0} t_{\alpha/2,\text{df}} \tag{6.32}$$

$$|dz_0| \leq s_{dz_0} t_{\alpha/2=0.025,\ \text{df}=38}$$

$$0.3 \text{ mm} \leq 0.2 \text{ mm} \times 2.02 \quad \text{or} \quad 0.3 \leq 0.4 \text{ mm}$$

Since the aforementioned condition is satisfied, the null hypothesis stating that the zero-point offset of the leveling staffs is zero is not rejected at the confidence level of 95%.

Example 6.5 A leveling instrument that has not been used for over 20 years is to be used for a survey project. The manufacturer claims, following DIN 18723 (or ISO 17123, now), that the equipment has a standard deviation of ±0.2 mm over 1-km double-run leveling. Since there is no record of any testing or calibration of this particular instrument, explain (with reasons) all of the necessary setting out, some important quality assurance (QA)/quality control (QC) measures, and the field procedure (including the number and types of measurements made in the field) that you would recommend following in order to determine whether the level is capable of behaving as the manufacturer claimed. Explain four quantities that will be determined from the measurements and fully discuss the statistical tests that will be performed on some of the quantities in order to determine whether the level is capable of behaving as the manufacturer claimed.

Suggested Answer: Perform the equipment test satisfying the ISO17123-2 according to the following procedures (ISO17123-2, 2001):

Setting out:

(i) Setting up 2 leveling points *P* and *Q* at approximately 60 m apart on fairly horizontal test area.

(ii) Set the instrument approximately equidistant between points *P* and *Q*.

QA/QC measures:

(i) Check the collimation error of the leveling instrument to check that it is within acceptable limits.

(ii) Avoid direct sunlight on the instrument by shading the instrument to avoid differential heating of instrument that may affect internal working of the instrument.

(iii) Allow the instrument to acclimatize to ambient temperature for about 2 min/°C temperature difference to avoid initial blunders in measurements.
(iv) Avoid roads covered with asphalt or concrete as test site to reduce refraction effects.
(v) Choose stable points (P and Q) for your leveling staffs to avoid sinking of rod between measurements.
(vi) Set the instrument approximately equidistant between points P and Q (to reduce effects of refraction and collimation error).

Field measurements and procedure:

(i) Take and record the temperature reading of the barometer before and after the test.
(ii) Take and record 20 pairs of rod readings (a pair consists of one back reading to P and one forward reading to Q); before each pair of readings, lift up the instrument, place it at a slightly different location, and relevel it.
(iii) Switch the two leveling staffs at points P and Q (to randomize difference in zero-points errors of staffs) and repeat step (ii).
(iv) At the end of the data collection, you should have a total of 40 pairs of rod readings (or 80 readings in all).

Quantities calculated:

- Experimental standard deviation over your leveling distance of 60 m
- Adjusted value of difference in the zero-point offsets of the two leveling staffs used
- The propagated standard for the adjusted value of the difference in the zero-point offsets
- Propagated standard deviation over 60 m to mm/km double run.

Statistical tests:

- Perform t-test (two-tailed test in Table 2.8 in Chapter 2) to check if your adjusted value of the difference in the zero-point offsets is statistically different from zero at 95% confidence level.
- Perform the Chi-square test using Equation (2.56) in Chapter 2 to check if the experimental standard deviation for 1-km double-run is the same as that specified by the manufacturer at 95% confidence level.

6.9 CALIBRATION OF COORDINATE DIFFERENCE MEASUREMENT SYSTEM (GNSS EQUIPMENT)

Since the basic observables of Global Navigation Satellite Systems (GNSS) surveys are baseline vectors (coordinate differences), the GNSS equipment is considered in this chapter as a coordinate difference measurement system. Basic instrumentation for GNSS network survey includes multiple sets of geodetic receivers, antennas, fixed-height tripods, and meteorological instruments. The equipment must be maintained according to manufacturer specifications and calibrated on a regular basis. Equipment calibrations should be performed at the start and end of a project, before and after any maintenance, and at sufficient intervals to maintain data integrity. Any data not supported by successful calibrations are suspects. To prevent the invalidation of good data, frequent calibrations are recommended.

Field calibration is necessary to control systematic errors that may be critical to GNSS satellite surveys. This will verify the adequacy of the GNSS survey equipment, observation procedures, processing software, and steps implemented in the data analysis, thereby determining whether significant biases exist. Examples of systematic errors in relative position determination in the static mode of GNSS surveys include the following:

(a) Errors in satellite positions
(b) Atmospheric refraction (ionospheric and tropospheric) model errors
(c) Receiver timing bias
(d) Field procedural errors
(e) Antenna setup (plumbing, centering, measurement of height of antenna phase center above the station mark)
(f) Antenna phase center stability
(g) Signal multipath.

The following systematic errors are possible in GNSS derived orthometric heights:

- Bias in GNSS ellipsoidal height differences
- Bias in orthometric heights for the vertical control
- Bias in geoid undulation differences.

Development of models, methods, and techniques to bring these error sources under control will enhance survey capability in terms of accuracy, logistics, and economy.

Calibration tests are performed for a number of reasons, such as testing the total system with the purpose of determining the overall characteristics of GNSS performance (e.g., GNSS measurement validation) and testing in order to isolate as many of the external errors/biases as possible (e.g., GNSS zero-baseline and GNSS antenna phase center variations tests). Some tests are conducted once, either in the laboratory (by the manufacturer or an independent organization) or in the field; others are conducted on a continuous basis.

6.9.1 GNSS Measurement Validation

Field calibration or *GNSS measurement validation* consists of testing the GNSS equipment performance (measurement techniques) and the associated baseline processing software on an approved *GNSS three-dimensional test network.* The entire system of GNSS equipment and processing procedures are proved with a *validation survey* as a final check to ensure all components interact properly.

The approved GNSS 3D test networks or validation networks, which usually include existing electromagnetic distance measurement (EDM) baselines, are coordinated three dimensionally in the local coordinate system. By holding the coordinate values of at least one of these network points fixed, the coordinates for all the other known points are derived independently using the GNSS observations. The differences between the derived coordinates and those provided for the network points are used to determine whether or not the validation is acceptable. The validation networks may be used to check the full range of GNSS equipment from hand-held C/A code receivers to geodetic quality dual-frequency receivers. Generally, they can be used in evaluating the following:

- Results obtained from a specific combination of GNSS equipment, software, and observation procedures
- The proposed GNSS equipment, procedures for data collection, software and procedures used for the data processing and adjustment, and determining with confidence whether they can meet contract accuracy requirements.

GNSS measurement validation is usually repeated if any significant modifications or upgrades are made to the GNSS receiver or the postprocessing software. However, in order to avoid additional fieldwork for every software upgrade, it is recommended that the original validation raw data be reprocessed with the new software version so that any changes in the results are only evaluated.

The establishment of GNSS validation networks (also known as basenets) across Canada is due to the need for a physical standard for evaluating GNSS positioning accuracy and precision, GNSS equipment and software, and positioning methodologies. Currently, there are two GNSS validation networks in the Province of British Columbia (BC) in Canada: the Greater Vancouver basenet and the Okanagan basenet. The maintenance responsibility of the basenets is shared by the Geodetic Survey Division (GSD), Natural Resources Canada (NRCan), and the provincial survey agencies. In sharing the maintenance responsibility, for example, the Geographic Data BC (the provincial survey agency) located, designed, and installed the Okanagan and Greater Vancouver networks, while the GSD and NRCan established the validation coordinates for the networks through precise GNSS measurements.

6.9.1.1 Basic Configuration of GNSS Validation Networks The configuration of a GNSS validation network is composed of between 5 and 10 forced centering pillars or piers. Usually, some of the pillars are also parts of an EDM calibration baseline, which forms the core of the network. The network design provides GNSS baselines of varying lengths, usually ranging between 1 and 100 km; the forced-centering devices on the pillars are to help in minimizing centering errors of GNSS antennas; and the pillars are located where they are easily accessible with clear visibility above 10° from the horizon. For stability and longevity of the pillars, the pillars are built to the same specifications as the EDM calibration baseline pillars.

One of the examples of GNSS validation networks in British Columbia, Canada, is the Greater Vancouver GNSS validation network, which is comprised of seven concrete forced centering pillars (GSD and NRCan, 1997). The network, which is centered in Surrey, spans the entire lower Fraser Valley from Mission to West Vancouver with one of the network pillars (Pier 3) relating to the West Vancouver EDM baseline and another two (Pier 1 and Pier 6) belonging to the City of Surrey EDM baseline. The baseline lengths range from 800 m to 74 km; and all the pillars in the network are positioned three dimensionally using GNSS with the orthometric heights of the pillars established through first-order leveling. More details on the Greater Vancouver validation network can be found in GSD and NRCan (1997).

6.9.2 GNSS Zero-Baseline Test

GNSS receivers must be calibrated to ensure that they contain the latest manufacturer's firmware upgrades. A *zero-baseline test* can measure receiver internal noise if the performance is a suspect. Some of the basic items evaluated in the test include the following:

- Receiver hardware variations (which can be up to 10 m), which are primarily due to temperature effects
- Receiver characteristics, such as its correct operation, its measuring precision, and its data processing software.

A zero baseline test consists of hooking up two (or more) receivers to the same antenna (using antenna splitter) and observing the differences between the measurements made by the two receivers, which would ideally be zero. The two receivers connected to the same antenna may be running from a common clock or from separate clocks. When two receivers share the same antenna, biases such as those depending on the satellite (clock and ephemeris) and the atmospheric path (troposphere and ionosphere), as well as errors due to multipath, will cancel out during data processing. The quality of the resulting zero-baseline is a function of the random observation error (or noise) and the propagation of any receiver biases that do not cancel in data differencing. The impact of residual bias effects on the baseline solutions is a function of baseline length, such as satellite ephemeris bias, handling of observation time-tags and atmospheric delay, and cannot be evaluated by zero-baseline test.

Some of the important advantages of zero-baseline test are as follows:

- It is comparatively simple to administer since no specialized software or *ground truth* data is required, and the location of the antenna is immaterial.
- GNSS surveying receiver manufacturers use this test procedure to perform final product testing of all receiver units before they leave the factory.
- No significant time-dependency to the quality of the zero baseline results should be evident, apart from a very small effect that is due to the daily variation in receiver-satellite geometry.

One important disadvantage of zero-baseline test is that it cannot be applied to integrated antenna/receiver systems such as some of Leica, Trimble and Ashtech GNSS instruments.

6.9.3 GNSS Antennas Phase Center Variations

Differential GNSS solutions are used routinely to provide geodetic positions with precisions that are often as good as a few millimeters. A GNSS geodetic solution for a baseline provides the vector between the phase centers of the antennas at either end of the baseline. The phase center of a GNSS antenna, however, is neither a physical point that can be accessed with a tape measurement by a user nor a stable point. For any given GNSS antenna, the phase center is a function of the changing direction of the signal from a satellite. Ideally, most of these phase center variations depend on the satellite elevation. Azimuthal effects are only introduced by the local environment around each individual antenna site. If the phase center variation is ignored, the measured baseline will be between the averages of all the individual phase centers for each of the measurements included in the solution. When the antennas at opposite ends of relatively short baselines are identical, these variations should cancel out and no effect should be seen. Different antenna types, however, exhibit different phase variations so that baselines with different antenna types will show increasing sensitivity to things such as elevation cut-off angle and the distribution of observations within a solution. GNSS antennas, therefore, must be calibrated in order to determine the *antennas phase center variations*. Since the phase center variations affect the antenna offsets that are needed to connect GNSS measurements to physical monuments, ignoring them can lead to serious (up to 10 cm) vertical errors.

An antenna calibration is an essential part of making the most precise GNSS surveying possible. Since all antennas have an average phase center offset and a phase center variation with respect to an antenna reference point, an antenna calibration by itself cannot be considered as a statement about the relative merits of any particular models of antennas. The most significant contribution of antenna calibrations, however, is ensuring interoperability within the growing community of GNSS antenna types.

6.9.4 Supplementary GNSS Equipment Calibration

The tripods to be used in GNSS survey must be calibrated; they must be examined for stability and ensured that hinges, clamps, and feet are secure and in good repair. Fixed-height tripods must be tested for stability, plumb alignment, and height verification at the start and end of each project. The tribrachs to be used for the antennas must be calibrated to ensure that the optical plummet alignment is correct. The meteorological equipment, which includes wet-bulb and dry-bulb thermometers to measure temperatures and a barometer or altimeter to measure atmospheric pressure, should also be calibrated at the beginning and end of a project.

6.9.5 General Concerns on GNSS Equipment Calibration

The following concerns may arise from some or all of the testing procedures for GNSS surveying systems:

(a) Test may not be conclusive if it is carried out only once since GNSS errors/biases are a function of time.

(b) Test may not be conclusive if it is carried out in only one location. Some of the GNSS errors/biases are a function of geographic location.

(c) There is usually a general question whether all the baseline lengths be sampled or whether all the possible satellite geometries be sampled, and so on. Propagation of most GNSS errors/biases into the baseline solution is a complex combination of factors, such as time, location, baseline length, and satellite geometry.

(d) There is usually confusion about the operational procedures to be adopted for data collection and data processing. The quality of GNSS baselines is a function of length of observation session, type of carrier phase solution, and the processing techniques. The quality is also influenced by data editing and preprocessing procedures that are used. Some may adopt automatic data processing procedures and some may not.

(e) The other concern may be on how often one should conduct the testing of procedures for GNSS surveying systems.

7

SURVEY DESIGN AND ANALYSIS

OBJECTIVES

At the end of this chapter, you should be able to

1. Discuss the elements and problems of network design
2. Carry out simple two-dimensional network design by simulation
3. Perform simple two-dimensional network analysis
4. Design deformation monitoring scheme
5. Discuss the purpose of simulating survey measurements
6. Carry out simple simulation of survey measurements to check if tolerance limits of measurements can be met

7.1 INTRODUCTION

The survey design considered in this chapter is essentially *network design* and the accompanying simulations. Network design is to estimate the confidence of future survey before it is actually carried out. It allows one to experiment with different surveying variables so as to meet or exceed the desired survey accuracy requirements. After an initial design, it may be discovered that the desired accuracy requirements are not met; in this case, there may be a need to iteratively change the surveying

Precision Surveying: The Principles and Geomatics Practice, First Edition. John Olusegun Ogundare.

variables until the accuracy requirements are satisfied. This is usually done through a process of computer simulation. *Simulation* of survey measurements is an imitation process (before the measurements are made) to see how those measurements would be made under different conditions and also to analyze component measurements of new design. It includes planning and laying out of a project and proper selection of equipment, measurement methods, and procedures. It provides a basis for evaluating the accuracies of the survey measurements and for meeting tolerances that may have been imposed on these measurements. Currently, there is a great demand for more accurate survey measurements, which requires that the surveyor chooses an appropriate survey instrument out of several models of surveying instruments and an appropriate survey technique out of possible survey techniques. For an illustration, consider Figure 7.1, in which the coordinates (X_C, Y_C) of point C are to be determined, given that the coordinates of points A (X_A, Y_A) and B (X_B, Y_B) are known.

The surveyor can solve the above problem using a number of measuring techniques, such as the following:

- Triangulation – measuring three angles α_A, α_B, and α_C
- Trilateration – measuring two distances ℓ_a and ℓ_b
- Triangulateration – measuring the three angles α_A, α_B, and α_C and the two distances ℓ_a and ℓ_b.

In addition, the surveyor will have to decide on which type of instrument to use out of a number of models of surveying instruments to choose from. The surveyor's choice of survey techniques and instruments must be based on a thorough simulation of the project so that the selected techniques and instruments would satisfy the accuracy requirements of the client in an economical way. The simulation is used in order to predict (or design) what type of instrumentation and what procedure of measurements should be used in order to satisfy the specifications of the client. It should also be mentioned that when a tolerance to be achieved is provided by the

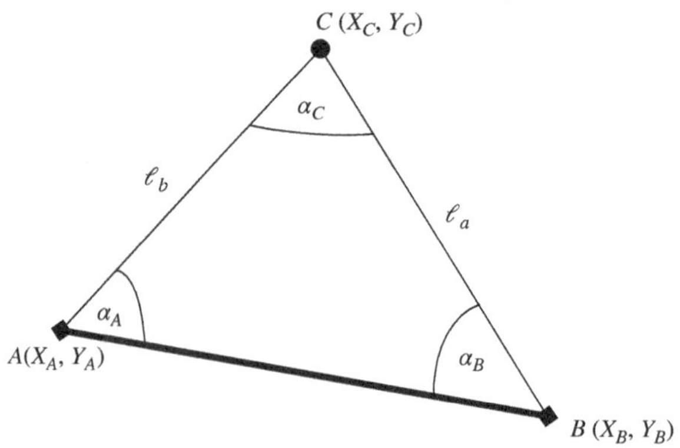

Figure 7.1 A simple surveying problem.

client, it is customary for the surveyor to reserve half of the error budget for systematic errors (such as refraction effects that cannot be completely eliminated) and to reserve the other half for random errors, which will be used in determining the type of measurement and instruments to use for the project.

After the design and simulation processes have been completed, a blueprint for the field crew is usually created, such as where the network stations would be located, types of observables to measure at each station, level of accuracy needed for the observations.

7.2 NETWORK DESIGN

The level of accuracy of geodetic positioning has increased in the past few years, requiring that the geodetic surveyors should shift their focus away from just being able to make quick and accurate observations; they should also be able to perform *survey design*, *data processing*, and *analysis* in order to produce reliable results. Although the commercial computer software packages for network design and adjustment have made the work of design and analysis easy, a good understanding of the terminology, concepts, and procedures involved are still needed in properly interpreting the results and successfully designing and/or analyzing a survey.

Network design, according to Kuang (1996), will help in identifying and eliminating blunders in network measurements; it will also ensure that the effects of the blunders that are not detected and eliminated are minimal on the network solution. Some of the other benefits of network design can be summarized as follows:

- It helps in reducing the amount of time and effort required in carrying out a field project, which will also result in the reduction of the cost of the project.
- It provides a measure of confidence that the project can or cannot be completed as specified by the client.
- It will afford the surveyor the opportunity to experiment with different design variables such as
  - (i) network geometry (number and physical location of survey points);
  - (ii) network accuracy (or precision of measurements);
  - (iii) reliability (ensuring enough redundant observations in order to be able to assess the accuracy of network); and
  - (iv) cost of survey (if number of measurement to be made is reduced and if the observation procedure is made simple).

A network must be designed to satisfy the preset precision, reliability, and cost criteria. In order to achieve the network quality set by a client, the network design essentially involves the following:

- Deciding on the best configuration (or geometry) of a geodetic network or deciding on the location of station points

- Choosing the measuring techniques and the types of geodetic observables to be measured
- Making decisions on which instruments to use among hundreds of available models of various geodetic instruments
- Computing the optimal distribution of required observational precisions among heterogeneous observables.

7.2.1 Geodetic Network Design

Much can be done to design a network to ensure that it will achieve its desired aim before any measurements are made. From the least squares adjustment of network survey measurements, the adjusted coordinates of the network points ($\hat{x}$) can be expressed as

$$\hat{x} = x^0 + \delta \tag{7.1}$$

where x^0 is the vector of approximate coordinates of the network points (taken from a large-scale map or an aerial photograph); and δ is the vector of unknown corrections to the approximate coordinates of the network points. The least squares solution for δ can be given as

$$\delta = -(A^{\mathrm{T}}PA)^{-1}A^{\mathrm{T}}Pw \tag{7.2}$$

where A is the first design matrix also known as the configuration matrix, P is the weight matrix (inverse of the covariance matrix of the observations), and w is the vector of the observations computed by using approximate coordinates minus the original observations. From the variance–covariance propagation laws, it can be shown from Equation (7.1) that the covariance matrix ($C_{\hat{x}}$) of the estimated parameters (unknown coordinates of the network points) is as follows:

$$C_{\hat{x}} = (A^{\mathrm{T}}PA)^{-1} \tag{7.3}$$

The basis on which a design can be carried out is seen from the covariance matrix of the estimated parameters in Equation (7.3). It can be seen from the equation that the covariance of the estimated parameters ($C_{\hat{x}}$) can be determined before making the actual field measurements if the approximate coordinates of the network stations are known. This situation represents the most usual design problem, which is to decide where to position observation stations and which measurements to make in order to satisfy the defined (precision) criteria. The problem of network design has been divided into the following (Grafarend, 1974; Grafarend et al., 1979): zero-order design (ZOD), first-order design (FOD), second-order design (SOD), third-order design (THOD), and combined-order design (COMD). The characteristics of the different classes of design problems are given in Table 7.1.

The design problem is not only limited to solving the problem of meeting precision criteria, but it also includes providing the minimum-cost solution. When a design satisfies both the precision and the minimum-cost criteria, it is often referred to as the *optimum design*. The cost element can be very difficult to quantify so that designs

TABLE 7.1 Problem of Network Design.

Problem Order	Unknown (To Be Determined)	Known (or Provided A Priori)
Zero-order design (ZOD) – also known as *datum problem*	Reference coordinate system is unknown so that the optimal values of unknown parameters (x) and their covariance matrix ($C_{\hat{x}}$) are unknown	Matrix indicating the configuration or geometry of network (A) and weight matrix of observations (P)
First-order design (FOD) – also known as *configuration problem*	Optimal locations (or configuration) of network stations (A) and observation technique or plan (how network points are to be connected)	Weight matrix of observations (P) and covariance matrix of parameters ($C_{\hat{x}}$) are known
Second-order design (SOD) – also known as *weight problem*	What type of observations to make and their precisions or weight matrix (P)	Network geometry (A matrix) and covariance matrix of parameters ($C_{\hat{x}}$)
Third-order design (THOD) – also known as *improvement problem*	Improvement of existing network configuration (A) and weight or precision of observations to be made (P). Combining modified FOD and modified SOD	Covariance matrix ($C_{\hat{x}}$) of parameters are known
Combined design (COMD)	Solves combined FOD and SOD simultaneously (A and P are not known)	Covariance matrix ($C_{\hat{x}}$) of parameters are known

are usually assessed subjectively by considering the previous experience of the surveyor.

7.2.2 Design of GNSS Survey

One of the significant advantages of the Global Navigation Satellite System (GNSS) survey technique over conventional techniques is that survey stations may be placed where they are required, irrespective of whether intervisibility between stations is preserved, provided there are no obstructions between the stations and the satellites to be tracked. It should be remembered that the GNSS technology is continually evolving and the following are continually changing:

- Requirements for classification of geodetic control surveys by GNSS techniques
- Definitions for GNSS accuracy standards
- Experience in performing GNSS surveys
- GNSS surveying equipment improvement
- GNSS field procedures
- Refinements to processing software.

The design of GNSS surveys is a component of the GNSS specifications. The specifications are for control surveys performed by relative positioning techniques where two or more receivers are collecting carrier phase measurement data simultaneously; and they include network design and geometry, instrumentation, calibration procedures, field procedures, and office reduction (processing) procedures. Some of the guidelines for GNSS network design, geometry, and connections are given for the highest orders of surveys by FGCC (1989) as shown in Table 7.2.

7.2.3 Design of Deformation Monitoring Scheme

Deformation monitoring schemes of objects are designed to help in accurately determining the expected deformation model and deformation parameters (which must be known a priori) of the object. In practice (Chen, 1983; Secord, 1985), the initial design is usually based on single-point displacement models with x and y displacements set to zero for each reference network point and each object network point is given constant x and y displacements whose variances and covariances are later solved for. With some modifications based on the different purpose of the geodetic network (which is to provide absolute positions of network points) compared to that of the monitoring network (which is to determine displacements of network points between epochs), the design of deformation monitoring networks can be classified based on the design orders for geodetic networks given in section 7.2.1, as follows (Kuang, 1996):

TABLE 7.2 Guidelines for GNSS Network Design, Geometry and Connections.

Geometric Accuracy Standards	
1. Minimum number of stations of the horizontal network control of a reference system to be connected to	4 to 3
2. Minimum number of stations of the vertical network control of a reference system to be connected to	5
3. For continuously tracking stations (master or fiducials): minimum number of stations to be connected to	4 to 2
4. Station spacing (km) between old network control and center of project should not be more than (and 50% not less than $\sqrt{5d}$): *(where d is the maximum distance (km) between the center of the project area and any station of the project)*	$100d$ to $7d$
5. Station spacing (km) between old network control located outside of the project's outer boundary and edge of the boundary, not more than	3000 to 100
6. Location of network control (relative to center of project): number of "quadrants," not less than	4 to 3
7. Direct connections should be performed, if practical, between any adjacent stations (new or old, GNSS or non-GNSS) located near or within the project area, when spacing is less than (km)	30 to 5

1. The Zero-order design (ZOD) problem is about confirming the stability (in position and orientation) of reference network points between monitoring epochs; the reference network that remains stable over several epochs is considered optimal.
2. The First-order design (FOD) problem is about locating monitoring points where maximum deformations are expected and ensuring that the reference network points are located in stable regions. The locations of points with expected maximum deformations can be determined by modeling the deformations using finite element method.
3. The Second-order design (SOD) problem is to determine the types of observables and their accuracies that will provide accurate deformation parameters. The Third-order design (THOD) is about improving the accuracies of the deformation parameters.

The important requirements to be satisfied in the solution for the design parameters of the monitoring scheme are accuracy, reliability, separability (or discriminability), and cost-effectiveness.

7.2.3.1 Accuracy Requirement The sources of errors causing uncertainty in engineering survey measurements are numerous and diverse. The main concerns include factors such as physical instability of observation stations; atmospheric refractions along the line of observation, thermal effects on the mechanical, electronic, and optical components of the instrument used; instrument malfunction and human ability to fail. Since on many occasions, particularly precision observations are required, special attention must also be paid to matters such as centering, targeting, and leveling of instruments.

Generally, the main problem of deformation monitoring scheme is not to define an optimum datum (or reference system) for the initial epoch of measurements but to confirm the stability of the datum. For example, if a set of reference points (serving as reference datum) used to constrain the network adjustment are erroneously assumed stable while they are not, a biased displacement pattern that can be misinterpreted as monitoring results could be obtained. Unstable reference points must be identified prior to data acquisition stage based on the knowledge of boundaries of deformation zone or during data processing by using appropriate algorithm.

There are several ways of attempting to reduce the errors due to the effects of systematic errors on surveying measurements. It may be possible to calibrate the instrument concerned, quantify the error, and apply corrections to subsequent measurements. The effects of random errors on measurements, which are represented by the precision (or internal accuracy) of the measurements, cannot be completely eliminated.

As a general rule, the accuracy of monitoring ground displacements at 95% confidence level should be at least three times smaller than the expected (or predicted) average displacements over the observation time span. The frequency of observations will then depend on the expected rates and magnitudes of the practically detectable deformations. In this case, the standard deviation of monitoring displacements can be taken as the predicted displacement reduced by a factor of 3×2.48 for horizontal displacements and reduced by a factor of 3×1.96 for vertical displacements. This requires that the predicted maximum ground displacement over the total period of the deformation activities be available in order to be able to determine the accuracy of monitoring surveys. For example, from the predicted maximum deformation over a period in which gas will be withdrawn from an underground reservoir, one can determine the annual rate of subsidence and the accuracy of monitoring surveys. Since the predicted values, however, may be different from the actual values, the accuracy requirement for the surveys will have to be revisited time to time depending on the depth and geometry of the mine and also on the mining method being adopted.

The ability of a monitoring scheme (configuration of stations and object points and observables) to reveal relative movement is related to the relative positioning error (relative 95% confidence error ellipse) between any pair of stations (Chrzanowski and Secord, 1985). In this case, relative movement must be greater than $a_{95} \times \sqrt{2}$ (where a_{95} is the semi-major axis value of the relative 95% confidence error ellipse) in the case of two-dimensional networks in order to be able to detect the movement.

Relative confidence regions provide the accuracy of coordinate differences among monitoring network stations and are a measure of the internal accuracy of the network. The relative error ellipses between the object points and selected reference stations become indicators of the ability of the scheme, configuration, and observables to monitor the behavior of the structure represented by the object points with respect to the reference created by the network stations. For example, assuming a network is to be designed such that it is capable of detecting the minimum horizontal displacement of $d_{min} = \pm 3.0$ mm and assuming no correlation between a pair of measurement epochs, it can be expressed that

$$d_{min} \geq \sqrt{(a_{I}^{2} + a_{II}^{2})} \tag{7.4}$$

where a_I and a_{II} are the semi-major axis values of the standard confidence ellipse for epochs I and II, respectively. Assuming the same value (a_{std}) for the semi-major axis values for the two epochs, then $d_{min} \geq a_{std}\sqrt{2}$. Thus, for the detection of ± 3.0 mm horizontal displacement at the 95% confidence level, the tolerance for the semi-major axis of the standard confidence ellipses becomes

$$a_{std} = \pm \frac{3.0}{\sqrt{\chi^{2}_{0.95,df=2}}\sqrt{2}} \tag{7.5}$$

for the positional accuracy in a single campaign, where $\sqrt{\chi^{2}_{0.95,df=2}} = 2.4484$ is the Chi-distribution value at 95% confidence level for the degrees of freedom of $df = 2$ (for two-dimensional cases). Similarly, for the detection of ± 3.0 mm vertical displacement at the 95% confidence level, the tolerance for the standard confidence level becomes

$$a_{\text{std}} = \pm \frac{3.0}{Z_{0.975}\sqrt{2}} \tag{7.6}$$

for the positional accuracy in a single campaign, where $Z_{0.975} = 1.9600$ is the standard normal distribution value at 95% confidence level (for one-dimensional cases). The amount of displacement to be detected (e.g., ±3.0 mm) can be predicted based on the annual rate of displacement determined initially from the geotechnical measurements.

7.2.3.2 ***Reliability Requirement*** Reliability of a monitoring network is a measure of the ability of a measuring scheme to detect and eliminate blunders from observations. It is a function of both the network geometry and the precision of observations. An observation that is reliable is unlikely to contain an undetected blunder, and, conversely, a blunder is unlikely to be detected in an unreliable observation. The reliability of the monitoring network will generally improve if the design is capable of producing redundant observations and if the sources of errors are well understood and well taken care of. An unreliable and poorly designed monitoring system will lead to false conclusions and misinterpretation of deformation of the monitored object. The design of monitoring network should also make sure that at least a monitoring point is located at the point of expected maximum movement; otherwise, the deformation analysis based on this design may become inconclusive.

7.2.3.3 ***Separability or Discriminability Requirement*** In a specific case of a deformation monitoring network, the design may not only be required to meet precision (e.g., variances of point positions or derived quantities) and reliability criteria, but also to be *sensitive* to the deformation pattern that is expected to take place. If such a pattern of a deformation can be formulated in a mathematical model, then network designs can be quantitatively assessed as to their capability to identify the true deformation. Such ability is sometimes referred to as *separability* (*or discriminability*) of the network. When considering design for deformation analysis, it is important to consider the separability of the resulting network to the particular deformation expected. This is because the purpose of such a network is usually not only to detect possible movements but also to try and establish the general mechanism of the motion taking place.

Separability is to indicate if the network is sensitive enough to detect and discriminate possible causative factors and the mechanisms postulated as responsible for the deformation of the object. This concept was extended into what is known as *discriminability* of the monitoring network (Ogundare, 1995). Discriminability incorporates all possible causative factors and the postulated deformation mechanisms in its formulation while separability considers only one possible causative factor and a postulated mechanism at a time.

7.2.3.4 ***Cost-Effectiveness Requirement*** The network design must satisfy the required accuracy, reliability, and three-dimensional monitoring criteria in the most economical way. The choice of monitoring technology or monitoring system, however, must be chosen according to the accuracy requirement, reliability of the network design, and cost requirements.

7.3 SOLUTION APPROACHES TO DESIGN PROBLEMS

Once the design problem has been formulated, there are two basic approaches to its solution:

- Computer simulation or trial and error methods
- Analytical methods, which attempt to mathematically formulate the design problem in terms of equations or inequalities and then explicitly solve for the optimal solution.

The *trial and error* method uses personal judgment at every step of the design. It requires repeated postulation of solution until a satisfactory (unlikely to be optimal) network is found. With the development of modern computers, the *trial and error method* is now referred to as the *computer simulation method.*

7.3.1 Simulation Steps for Network Design

In simulation, the problem of propagation of errors is reversed in order to determine the accuracy of measurements that will satisfy a specified tolerance limit for the unknown quantities to be determined. A simulation may tell us that the requirements for the accuracies of measurements are within or beyond our capabilities. If the requirements are beyond our capabilities, the client must be told that the tolerance limits specified are beyond what can be satisfied. The common steps for carrying out a network design by computer simulation method are as follows (cf. Cross, 1985):

1. Specify precision and reliability desired for the new network.
2. Choose a measurement scheme, such as station locations, types of observations, and weights (from precisions) of observations.
3. Select preliminary locations of control stations on an existing map or on an existing photographs based on the specific needs of the survey project control required by the client.
4. Perform a preliminary field reconnaissance, and based on the available instrumentation, determine the possible interconnection of stations by geodetic observables.
5. The proposed station locations and geodetic observables in steps 1–4 constitute an initial design of the network. According to this initial design, hypothetical precisions or weights of observables are then used to simulate the quality of the network. This is done by computing the covariance matrices of the desired least squares estimates and deriving the values of the quantities specified as precision and reliability criteria (such as standard deviation, standard error ellipse, and relative error ellipse or redundancy number).
6. If the values derived in step 5 are close enough to those specified in step 1, go to step 7; otherwise, alter the observation scheme in step 2 (by removing observations or decreasing weights if the selected network is too good, or by

adding more observations or increasing weights if it is not good enough) and return to step 5.

7. Compute the cost of the network and consider the possibility of returning to steps 2 and 3 and restarting the process with a completely different type of network (e.g., using traverse instead of using triangulation, etc.). Stop when it is believed that the optimum (minimum cost) network has been achieved.
8. Perform a field reconnaissance to examine the physical possibilities of the simulated network. Control stations are temporarily marked on the ground. If conventional terrestrial geodetic observables are proposed, intervisibility of control stations must be ensured. If the GNSS technique is to be used, the station site should be wide open with no obstructions to block the GNSS satellite signal between the stations and the satellites (within 10–15° above the horizon).
9. If step 8 is successfully done, the network will be monumented and surveyed.

This method has been used for decades, and a number of commercial software packages, such as *Microsearch GeoLab* and *Star*Net Pro*, are available for the network simulations. The main advantage of this method over analytical methods is that there is no need of evaluating any complex mathematical formulation, unlike in the analytical methods. The main disadvantage of the method is that an optimum (minimum-cost) solution may never be achieved.

Some of the properties of a well-designed control network should include the following:

(a) Stations must be as evenly spaced as possible; ratio of the longest length to the shortest should never be greater than 5.
(b) Adjacent pairs of stations should be connected by direct measurements.
(c) There should be reasonable number of redundant measurements in the network.
(d) Good a priori estimates of the accuracies of various instruments used with various techniques must be available in order to design a network that will achieve required accuracies. Accuracy of a horizontal control can be assessed properly from the results of a rigorous least squares adjustment of the measurements.

Typical examples of simulation problems are given as follows. Consider a case where you are to design a survey scheme (i.e., deciding on the best choice of equipment and procedures) for horizontal positioning by the process of trial and error or simulation assuming the following:

- Two types of observables (angles and distances) are to be measured.
- Standard deviation of each angle is σ_{β_i}, and of each distance is σ_{s_i}.
- Potential geometry is expressed as approximate coordinates, x^0.
- Ninety-five percent relative positioning tolerance (relative ellipses at 95% confidence) is to be achieved in the survey, that is, the semi-major axis of the relative error ellipses should be a_{95}.

The steps for carrying out the network design by the computer simulation method can be given as follows:

1. Specify precision and reliability desired of the new network:
   - Semi-major axis of the relative error ellipses expected is a_{95}.
2. Choose a measurement scheme, such as station locations, types of observations, and weights (from precisions) of observations.
   - Standard deviations of observations are provided ($\sigma_{\beta_i}, \sigma_{s_i}$).
   - Use them to form the cofactor matrix Q (matrix of variances of observations, assuming observations are uncorrelated).
   - Form the weight matrix (P) from Q, assuming $\sigma_0 = 1$, giving $P = Q^{-1}$.
3. Select preliminary locations of control stations on an existing map or on an existing photographs based on the specific needs of the survey project control required by the client.
   - Use approximate coordinates (x^0) of network points to create the first design matrix (A) based on the observation equations of distances and angles as functions of unknown coordinates of network points.
4. Compute from the available data in steps 2–3, the achievable semi-major axis of the relative error ellipses a'_{95} as follows:
   (i) Create the covariance matrix ($C_{\hat{x}}$) of the adjusted coordinates of the network points from Equation (7.3).
   (ii) Determine the relative standard deviations and covariances of pairs of points (x_1, y_1) and (x_2, y_2) involved in the network from the $C_{\hat{x}}$. Refer to Equations (2.32)–(2.40) for a typical covariance matrix $C_{\hat{x}}$ and for the relative variances and covariances of a pair of points.
   (iii) Obtain the standard relative error ellipses from Equations (2.41)–(2.44) as $a_s = \sqrt{\lambda_1}, b_s = \sqrt{\lambda_2}$ and θ, where a_s is the semi-major axis of the standard relative error ellipse, b_s is the semi-minor axis of the standard relative error ellipse, θ is the orientation of the semi-major axis of the standard relative error ellipse with λ_1 and λ_2 as the maximum and the minimum eigenvalues of the relative covariance matrix.
   (iv) Obtain the 95% relative error ellipses from Equations (2.26)–(2.28), which can also be expressed as follows:

$$a'_{95} = k_{95} a_s \tag{7.7}$$

$$b'_{95} = k_{95} b_s \tag{7.8}$$

$$\beta = \frac{1}{2} \arctan \left[\frac{2 s_{\Delta x \Delta y}}{s^2_{\Delta y} - s^2_{\Delta x}} \right] \tag{7.9}$$

$$k_{95} = \sqrt{\chi^2_{\alpha=0.05, df=2} (\text{upper-tail area})} \tag{7.10}$$

where a'_{95} is the computed semi-major axis of the 95% relative error ellipse, b'_{95} is the semi-minor axis of the 95% relative error ellipse, β is the orientation of the semi-major axis of the 95% relative error ellipse, and k_{95} is obtained from the Chi-square statistical table. Note that Equations (7.7)–(7.10) give identical results as those given in Equations (2.26)–(2.28); the above equations are just to show the variations in formulas commonly used.

(v) The semi-major axis of the relative error ellipses computed is a'_{95} in Equation (7.7).

5. Compare the obtained a'_{95} in step 4 with the limit on the relative ellipses (a_{95}) from step 1; if a'_{95} is less than the tolerance a_{95}, then the precision of potential observables and potential geometry are considered acceptable. If a'_{95} in step 4, however, is greater than the tolerance a_{95}, modify the precision and the geometry of observables or one of them and repeat the simulation until a'_{95} is less than the tolerance (a_{95}).

Referring to the above illustration, one can ensure that the intended standard deviations σ_{β_i} and σ_{s_i} are realized during the observations by taking the following steps:

1. Confirm that the instruments are well calibrated and their standard deviations quoted are correct.
2. Avoid sources of systematic errors such as
   - refractions (avoid temperature variations or make measurement at different atmospheric conditions and average the results);
   - consider leveling and tilting axis error (use dual-axis compensators and ensure that instrument is in good adjustment);
   - design allowable discrepancy for testing the acceptability of the set of measurements and implement this during the data acquisition stage.
3. Use face left and face right positions of instrument for direction measurements.
4. Targets must be well designed and appropriate for the project.
5. Targets must be well illuminated and visible from the instrument stations.
6. Use appropriate centering devices; well-adjusted optical/laser plummet or use forced centering procedure.
7. Minimize pointing error by using experienced instrument persons.
8. Use instruments with slightly better precision than those designed.

Simulation of 3D Traverse: Modern precision total stations can be used in three-dimensional traversing to result in Easting (*E*), Northing (*N*), and orthometric height (*H*) simultaneously, provided that points with known *E*, *N*, and *H* are available. The 3D design is accompanied using appropriate computer software simulation. The design (especially the network configuration) may be modified based on the outcome of the reconnaissance survey. The following steps may be taken:

1. In the simulation process, the approximate coordinates (N, E, H) of network points are to be used to determine the standard deviations of the horizontal and zenith angles to be measured.
2. For simulation purpose, the typical height of instrument (HI), height of reflector or target (HR or HT) can be assumed to be 1.600 m; and heights above pillar plates used as reference can be assumed to be 0.300 m.
3. In simulation and later in the least squares adjustment of measurements, two measurements between two stations taken at both ends will not necessarily be of the same observable since the HIs, HRs, and, possibly, the meteorological conditions at the time of measurements may be different. If this is the case, the average measurements corrected for meteorological conditions and reduced to mark to mark can be used or the two measurements involved can be treated as separate observables.
4. Use the input from the design in the simulation software and generate station (or relative) error ellipses and confidence intervals at 95%. Take note of the following with regard to relative error ellipses:
   - They are unaffected by the choice of origin of a network in the minimally constrained adjustment.
   - They are the precisions of relative positions of two points; they represent the relative precision of each station pair.
   - They can be smaller than the absolute (station) error ellipse on each end, that is, the coordinates for each station could be completely wrong (e.g., based on incorrectly used fixed coordinates), but the relative errors between stations give the best estimate of the precision of the survey regardless of the coordinates. For example, in terms of GPS measurements, the station coordinates determined using GPS may be off by meters, but the vector (the difference between these coordinates) can be accurate to centimeter level or better. The error in this vector is the best indicator as to the quality of the measurement.
5. Impose minimal constraints on the traverse by fixing (E, N, H) one of the control points available and an azimuth to another control point (assuming distances have been measured).
6. Often the quality of a traverse depends on its not exceeding a maximum allowable linear misclosure or the ratio of misclosure. It should be mentioned that the ratio of misclosure only implies general quality of relative precision of a closed traverse; it does not evaluate scale, rotational errors, blunders, and positional errors of the traverse. The product of a simulation suggesting that the expected quality would meet that ratio of misclosure criterion is the 95% confidence relative error ellipse between a pair of points or the relative distance accuracy estimates between the points in the network. The relative distance accuracy estimate is determined by error propagation using the positional standard errors at each end of the given line. If only approximate adjustments are being performed, the relative distance accuracies may be taken as a function of position misclosure.

7. The generated relative error ellipses are to suggest how well the 3D connection between every pair of points should be determined by the scheme that has been designed.

7.4 NETWORK ADJUSTMENT AND ANALYSIS

According to Kuang (1996), network analysis in surveying is about processing and analyzing survey data and reporting the outcome with its quality to the client. The steps involved in network analysis are given as follows (cf. Kuang, 1996):

1. Accuracy analysis of observations
2. Observation data preprocessing
3. Preadjustment data screening
4. Least squares network adjustment
5. Postadjustment data screening
6. Quality analysis of the results
7. Reporting network results and their quality to the user.

In general terms, items 1–3 can be considered as *preanalysis of measurements*; and items 5 and 6 as *postanalysis of measurements* and results.

Basic problem in surveying is to determine coordinates of a network of points using various types of measurements that establish a known geometrical relationship between them. Points with unknown spatial coordinates are connected to the network by the measurements. Network adjustment permits all of the available survey measurements to be processed together to determine a weighted mean value for the coordinates. Coordinate accuracy is determined by the application of error propagation to the observation equations. A predetermined uncertainty (standard deviation) is assigned to each measurement, which then propagates to the coordinates during the adjustment. The probable error in the coordinates (or positioning accuracy) is reported by the absolute confidence ellipse for each point or by the relative confidence ellipse between two points. It is essential to determine the positioning accuracy; without an adequate knowledge of the positioning accuracy, the survey (and the network adjustment) should be considered incomplete.

7.5 ANGULAR MEASUREMENT DESIGN EXAMPLE

In monitoring a dyke system along a certain coast, direction, distance, and height difference measurements were made to a network of stations on the dyke system. Directions were observed using a Kern DKM3 optical-mechanical theodolite (with adapter to fit onto a Wild tribrach) and Wild traversing targets, both onto Wild trivets; the distances were measured using a Tellurometer MA-100 infrared distance meter; and the height differences were measured using a Wild N3 tilting level and invar

TABLE 7.3 Approximate Coordinates, Heights of Instrument and Pillar Plate Elevations.

Station	*x*/*E* (m)	*y*/*N* (m)	*H* (m above msl)	HI (m)
TJ-8	2037.384	1197.560	1.464	0.300
TJ-8A	2050.536	1138.241	1.249	0.300
TJ-8B	2051.170	1075.133	0.646	1.080

staves. The approximate coordinates, heights of instrument (DKM3, tilting axis above pillar plate), and pillar plate elevations are given in Table 7.3 for three of the stations involved in the network. TJ-8B required tripod setup (using Wild tribrach with optical plummet) since the pillar top was too close to the ground to be used directly.

If the directions at station TJ-8A are to be observed to each of the two other stations, what would you expect to be the standard deviation of each of the directions, measured in one set using DKM3?

Solution: Some of the specifications of Kern DKM3 are as follows: $M = 45\times$; micrometer $= 0.5''$; plate vial sensitivity $= 10''/2$ mm. Figure 7.2 and Equations (7.11)–(7.13) can be used in computing the corresponding horizontal distance (HD), change in elevation (Warning: pl. change to entity$H$), and slope distance (s):

$$\text{HD} = \sqrt{(\Delta E)^2 + (\Delta N)^2} \tag{7.11}$$

$$\Delta H = H_B + \text{HT} - H_A - \text{HI} \tag{7.12}$$

$$s = \sqrt{(\text{HD})^2 + (\Delta H)^2} \tag{7.13}$$

where HI is the height of instrument, HT is the height of target, and s is the slope distance.

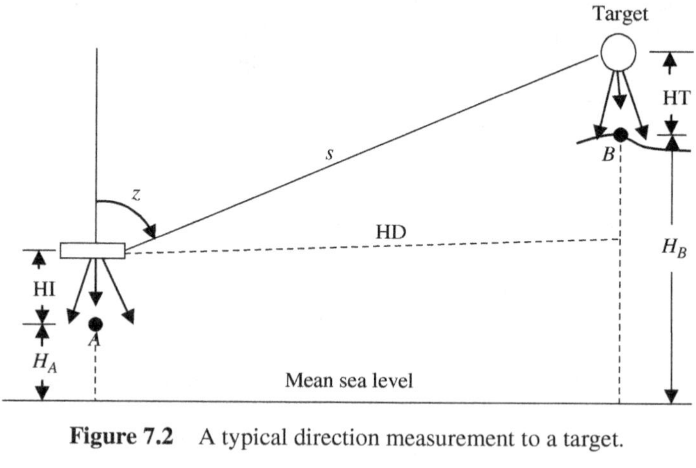

Figure 7.2 A typical direction measurement to a target.

Using Equations (7.11)–(7.13), the following are calculated:

$$\text{Direction TJ-8A} \rightarrow \text{TJ-8} : \text{HD} = 60.760\,\text{m}; \Delta H = 0.215\,\text{m}; s = 60.760\,\text{m}$$

$$\text{Direction TJ-8A} \rightarrow \text{TJ-8B} : \text{HD} = 63.111\,\text{m}; \Delta H = 0.177\,\text{m}; s = 63.111\,\text{m}$$

Compute the total random error in a single direction due to centering, pointing, reading, leveling as follows:

Stations TJ-8 and TJ-8A: Use forced-centering device, $\sigma_c = \pm 0.0001$ m;
Stations TJ-8B: Use optical plummet, $\sigma_c = \pm 0.0005 \times \text{HI}$ (m).
Direction TJ-8A to TJ-8B: For direction TJ-8A to TJ-8B, the chosen centering error for each station are as follows:

$$\sigma_{\text{cTJ-8A}} = \pm 0.0001\,\text{m}; \sigma_{\text{cTJ-8B}} = 0.0005 \times 1.080\,\text{m} \;\; (\text{or } \; 0.00054\,\text{m});$$
$$\text{Distance} = 63.111\,\text{m}$$

Using Equation (4.48), the centering for a direction measurement can be given as

$$\sigma_c = \frac{206,265}{63.111}\sqrt{(0.0001)^2 + (0.00054)^2} = 1.795'' \tag{7.14}$$

From Equation (4.21) and using $C = 45''$, the pointing error for direction measurement for $n = 1$ set can be calculated as

$$\sigma_P = \frac{45}{45\sqrt{2}} = 0.707'' \tag{7.15}$$

From Equation (4.32), the reading error for direction measurement for $n = 1$ set can be given as

$$\sigma_R = \frac{2.5(0.5)}{\sqrt{2}} = 0.884'' \tag{7.16}$$

From Equations (4.34) and (4.36), the leveling error for direction measurement in one set can be given as

$$\sigma_l = 0.2(10) \times \text{cotan}(Z) \tag{7.17}$$

where $\text{cotan}(Z) = (\Delta H/HD)$ and the level bubble sensitivity per division is 10"/div. This gives the leveling error of direction as $\sigma_l = 0.0056''$. The total error due to centering, pointing, reading, and leveling is calculated to be $2.12''$.

Direction TJ-8A to TJ-8: For direction TJ-8A to TJ-8, the chosen centering error for each station is ±0.0001 m; Distance = 60.760 m.

Using Equation (4.48), the centering for a direction measurement can be given as

$$\sigma_c = \frac{206,265}{60.760}\sqrt{(0.0001)^2 + (0.0001)^2} = 0.48'' \tag{7.18}$$

Using the same approach as in the case of direction TJ-8A to TJ-8B, the calculated pointing error is 0.707″; the reading error is 0.884″; the leveling error is 0.0071″; and the total error is 1.23″.

7.6 DISTANCE MEASUREMENT DESIGN EXAMPLE

For visible and near infrared radiation and neglecting the effects of water vapor pressure, the refractive correction, ΔN, can be determined by

$$\Delta N_i = N_D - N_i = 281.8 - \frac{0.29065p}{1 + 0.00366086t} \tag{7.19}$$

The meteorological correction is in the sense that $s = s' + C_{met}$, with $C_{met} = \Delta N_i s'$.

Temperature and pressure are to be measured at each end of an 1800 m distance, the refractivity correction at each end will be calculated, and the average value of ΔN_i will be used to determine the meteorological correction, C_{met}. The instrument being used has a design $n_D = 1.0002818$ (so that $N_D = 281.8$) and the average temperature and pressure during the measurements are expected to be +35 °C and 1000 mb. What would be the largest values of σ_t and σ_p that, together with equal contribution to $\sigma_{\Delta N}$, would result in a meteorological correction that would contribute uncertainty of no more than 2 ppm to the corrected distance? (Reproduced by permission of CBEPS.)

Solution: From Equation (5.40), the first velocity correction (or meteorological correction) can be given as $C_{met} = \Delta\overline{N}s'$ (where $\Delta\overline{N}$ is in ppm). Uncertainty in meteorological correction (m) by error propagation of C_{met} can be given as $\sigma_{C_{met}} = s'\sigma_{\Delta\overline{N}}$. The error propagation of the average value of refractive correction ($\Delta\overline{N} = (\Delta N_1 + \Delta N_2)/2$ with ΔN_1 and ΔN_2 as the refractive corrections at the two ends of the measured line) can be given as

$$\sigma^2_{\Delta\overline{N}} = \frac{1}{4}(\sigma^2_{\Delta N_1} + \sigma^2_{\Delta N_2}) \tag{7.20}$$

Assuming equal contribution of error with $\sigma^2_{\Delta N} = \sigma^2_{\Delta N_1} = \sigma^2_{\Delta N_2}$, then

$$\sigma^2_{\Delta\overline{N}} = \frac{1}{2}(\sigma^2_{\Delta N}) \rightarrow \sigma_{\Delta N} = \sigma_{\Delta\overline{N}}\sqrt{2} \tag{7.21}$$

Applying the laws of variance–covariance propagation on Equation (7.19) with pressure (p) and temperature (t) as variables gives the following:

$$\sigma^2_{\Delta N_i} = \left(\frac{0.29065}{1 + 0.00366086t}\right)^2 \sigma_p^2 + \left(\frac{0.29065 \times 0.00366086}{(1 + 0.00366086t)^2}P\right)^2 \sigma_t^2 \tag{7.22}$$

In order to solve for the unknown quantities σ_p^2 and σ_t^2 in Equation (7.22), it will be assumed that each term in Equation (7.22) contributes equally to $\sigma_{\Delta N_i}^2$, resulting in the following relationships:

$$\left(\frac{0.29065}{1+0.00366086t}\right)^2 \sigma_p^2 = \left(\frac{0.29065 \times 0.00366086}{(1+0.00366086t)^2}P\right)^2 \sigma_t^2 = \frac{\sigma_{\Delta N_i}^2}{2} \tag{7.23}$$

Using the given uncertainty in meteorological correction ($\sigma_{\Delta \overline{N}} = 2\,\text{ppm}$) in Equation (7.21) and substituting the value (ppm) for $\sigma_{\Delta N_i}^2$ in Equation (7.23) give the following:

$$\left(\frac{0.29065}{1+0.00366086t}\right)^2 \sigma_p^2 = 4 \text{ or } \left(\frac{0.29065}{1+0.00366086 \times 35}\right) \sigma_p = 2 \tag{7.24}$$

$$0.257639\sigma_p = 2 \text{ or } \sigma_p = 7.76\,\text{mbar} \tag{7.25}$$

Similarly for the other term in Equation (7.23):

$$\left(\frac{0.29065 \times 0.00366086}{(1+0.00366086t)^2}P\right) \sigma_t = 2 \tag{7.26}$$

$$0.83605543\sigma_t = 2 \rightarrow \sigma_t = 2.39°\text{C} \tag{7.27}$$

The values $\sigma_t = 2.39°\text{C}$ and $\sigma_p = 7.76$ mbar are the largest errors expected.

7.7 TRAVERSE MEASUREMENT DESIGN EXAMPLES

Example 7.1 A closed-loop traverse of 5 points is to be run in a fairly flat and homogeneous terrain. Assume that the traverse legs will be approximately equal to 300 m and the specified allowable maximum misclosure of the five angles is to be 15″. Design the measurement scheme and the type of theodolite to be used for this traverse.

Solution: Let $3\sigma = 15''$; the permissible standard deviation of closure of the traverse will be 5″. Assuming the same precision (σ_θ) of angular measurements at each station (with five stations), from error propagation rule:

$$5\sigma_\theta^2 = (5'')^2 = 25 \tag{7.28}$$

The permissible standard deviation of the angle measurements at each station will be

$$\sigma_\theta = \sqrt{\frac{25}{5}} = \sqrt{5} \tag{7.29}$$

The permissible standard deviation of the angle measurements at each station will be due to reading error ($\sigma_{\theta\text{r}}$), pointing error ($\sigma_{\theta\text{p}}$), and centering error ($\sigma_{\theta\text{c}}$) (assuming the instruments are in good adjustment and the targets will be well designed). The leveling error will be ignored since the terrain is fairly flat. The permissible standard deviation for each angular measurement becomes

$$\sigma_\theta = \sqrt{\sigma_{\theta\text{r}}^2 + \sigma_{\theta\text{p}}^2 + \sigma_{\theta\text{c}}^2} \tag{7.30}$$

Let us assume (for the sake of preanalysis) that the error components will have equal contribution, so that $\sigma_{\theta\text{r}} = \sigma_{\theta\text{p}} = \sigma_{\theta\text{c}} = \sigma$. In this case, each error will be equal to $\sigma_\theta/\sqrt{3}$ (or $\sqrt{5/3}$):

$$\sigma_{\theta\text{r}} = \sigma_{\theta\text{p}} = \sigma_{\theta\text{c}} = \sqrt{\frac{5}{3}} \tag{7.31}$$

From Chrzanowski (1977) and Section 4.5.2.2, the reading error for an angle measured (based on directional method) in n sets for theodolites with optical micrometers and with the smallest division of 1″ or 0.5″ is given as

$$\sigma_{\theta\text{r}} = \frac{2.5d''}{\sqrt{n}} \tag{7.32}$$

where d is the nominal value of the smallest division of the instrument (in arc seconds).

In the current problem, the least count of the instrument to be used can be estimated as follows:

$$\sigma_{\theta\text{r}} = \sqrt{\frac{5}{3}} = \frac{2.5d''}{\sqrt{n}} \tag{7.33}$$

or

$$d'' = \frac{\sqrt{5n}}{2.5\sqrt{3}} \tag{7.34}$$

From Equation (7.34), it can be deduced that for $n = 1$ *set*, $d'' = \sqrt{5}/(2.5\sqrt{3})$ or 0.5″, meaning that 0.5″ theodolite should be used; for $n = 4$ *sets*, $d\varepsilon = \sqrt{20}/(2.5\sqrt{3})$ or 1″, meaning that 1″ theodolite should be used.

Considering a case of an average atmospheric condition (average visibility and thermal turbulence over short traverse legs) and the use of well-designed targets, it is understood from Chrzanowski (1977) and that provided in Section 4.5.1 that the pointing error for angular measurement can be expressed by

$$\sigma_{\theta\text{p}} = \frac{45''}{M\sqrt{n}} \tag{7.35}$$

Given in Equation (7.31) that $\sigma_{\theta r} = \sigma_{\theta p} = \sigma_{\theta c} = \sqrt{5/3}$, the magnification of instrument telescope can be computed as follows:

$$\sigma_{\theta p} = \sqrt{\frac{5}{3}} = \frac{45''}{M\sqrt{n}} \tag{7.36}$$

or

$$M = \frac{45''\sqrt{3}}{\sqrt{5n}} \tag{7.37}$$

From Equation (7.37), it can be deduced that for $n = 1$ set, $M = 45''\sqrt{3/5}$ or 35, meaning that a theodolite with a magnification of 35× should be used; for $n = 4$ sets, $M = 45''\sqrt{3/20}$ or 18, meaning that a theodolite with a magnification of 18× should be used.

The influence of centering errors (σ_c) on an angle measurement is given by Chrzanowski (1977) and can be deduced from Equation (4.46) from Section 4.5.4 by assuming the angle measurement $\theta = 180°$; centering errors of target and instrument are the same; and the distances (D) are the same. From the appropriate substitution into Equation (4.46), the instrument centering error on angle measurement is deduced as

$$\sigma_{\theta c_i} = \frac{2\sigma_c}{D}(206, 265) \tag{7.38}$$

and similarly, the target centering error on angle measurement can be deduced from Equation (4.44) as

$$\sigma_{\theta c_t} = \frac{\sigma_c}{D}(206, 265)\sqrt{2} \tag{7.39}$$

Given in Equation (7.31) that $\sigma_{\theta r} = \sigma_{\theta p} = \sigma_{\theta c} = \sqrt{5/3}$ and assuming the centering errors of target and instrument are the same, the error due to each component will be $\sigma_{\theta c}/\sqrt{2}$ or $\sqrt{5/6}$. The type of instrument centering device to be used can be determined from the estimated centering error (σ_c) as follows:

$$\sqrt{\frac{5}{6}} = \frac{2\sigma_c}{D}(206, 265) \tag{7.40}$$

so that by rearranging Equation (7.40):

$$\sigma_c = \frac{D}{2(206, 265)}\sqrt{\frac{5}{6}} \tag{7.41}$$

Substituting $D = 300{,}000$ mm into Equation (7.41) gives $\sigma_c = 0.7$ mm or 0.4 mm/m if the height of instrument is taken as 1.6 m. The centering device (such as forced centering with tripod) that will give a centering error of 0.7 mm at the height of the instrument (about 1.6 m) above the ground will be suitable. Similarly, solving

TABLE 7.4 Summary of Traverse Design.

Option	Magnification (M)	Least Count (d)	Number of Sets (n)	Type of Suitable Theodolite
1.	35	0.5″	1	Kern DKM3 ($M=45$, $d=0.5''$)
2.	18	1″	4	Wild T2 ($M=28$, $d=1''$)

Equation (7.39) gives target centering error $\sigma_c = 0.9$ mm or better than 0.6 mm/m for the height of target of 1.6 m; a plumb line will be suitable as a centering device if it is not windy. The summary of the design is given in Table 7.4.

In each option, optical plummet, laser plummet, or plumbing rods (0.5 mm/m) can be used as a centering device for the target and a forced-centering device with tripod for the instrument. If there will be recentering of the instrument between sets, then the optical plummet, laser plummet, or plumbing rods can be used as the centering device for the instrument.

Example 7.2 The maximum allowable angular misclosure (m_θ) in a traverse of 10 angles is 50″ at 99% confidence level, what is the expected standard deviation of measuring each of the angles of the traverse, assuming equal error for each angle?

Solution: Equation (2.15) or (2.16) can be used, but Equation (2.15) will be used as an example as follows:

$$|m_\theta| \le \sigma_{m_\theta} z_{\alpha/2} \tag{7.42}$$

where the misclosure $m_\theta = \bar{x} - \mu$ or 50″, $\alpha = 0.01$, $z_{\alpha/2} = 2.576$ and σ_{m_θ} is the standard deviation (which should be considered as the SE) of the misclosure; note also from Equation (2.51) that $\sqrt{\chi^2_{\text{df}=1,0.01}} = \sqrt{6.635}$ (or 2.576). Determine the unknown σ_{m_θ} from the equation as follows:

$$\sigma_{m_\theta} = \frac{|m_\theta|}{z_{1-\alpha/2}} \rightarrow \sigma_{m_\theta} = \frac{50''}{2.576} \quad (\text{or } 19.4'')$$

By error propagation, the standard deviation of the misclosure can be expressed in terms of the standard deviation of individual measured angle as follows:

$$\sigma_{m_\theta} = \sqrt{\sigma^2_{\theta_1} + \sigma^2_{\theta_2} + \sigma^2_{\theta_3} + \cdots + \sigma^2_{\theta_{10}}} \tag{7.43}$$

For this problem, $\sigma^2_{\theta_1} = \sigma^2_{\theta_2} = \sigma^2_{\theta_3} = \cdots = \sigma^2_{\theta_{10}} = \sigma^2_\theta$; the propagated error of misclosure can be given as $\sigma_{m_\theta} = \sqrt{10}\sigma_\theta$ so that

$$\sigma_{m_\theta} = \sqrt{10}\sigma_\theta = 19.4'' \rightarrow \sigma_\theta = 6.1''$$

Each of the traverse angles should be measured with a precision of not more than 6.1″.

Example 7.3 A traverse is to be measured around a rectangular city block, which is 100 m by 210 m as shown in Figure 7.3. For subsequent use, there has to be an intermediate point along each 210 m side so that there would be six angles (one, ~90°, at each corner and one, ~180°, in the middle of each long side) with approximate horizontal "lengths of sight" of ~100 or ~105 m. The two 100 m sides are relatively flat while the other two have slopes of +18% and of −18%. The equipment (theodolite or targets) would be set up on tripods with HI or HT of 1.755 m. Since the survey may extend over one session, forced centering cannot be assumed, but this would be inappropriate anyway since only ground mark points (monumented in the concrete of the sidewalk by either brass plates or finely cut crosses) will be occupied. At least two sets would be observed. Using theodolites with lower precision may require more sets to make the mean values compliant with the misclosure limit or compatible with the result from two sets using the highest precision instrument. Offer one choice of equipment and the associated procedures for observing the angles associated with the given situation, with consideration for the effects of centering, leveling, pointing, and reading. Determine the maximum misclosure in the loop of six angles.

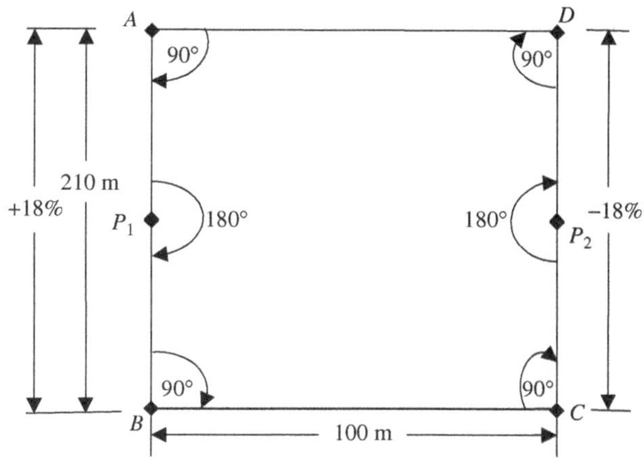

Figure 7.3 A sketch of a traverse around a rectangular city block.

Solution: Information supplied: HI or HT = 1.755 m; at least two sets observed; and forced centering is not assumed since the survey may extend beyond one session. Horizontal distance of 210 m side is given, at a slope of +18% and −18%. The calculated slope distances AB is 213.375 m; AP_1, P_1B, CP_2, and P_2D are all equal to 106.687 m. Assume the choice of Leica TC2003 with a standard deviation of angle

measurement of 0.5" (ISO 17123-3) and electronic dual-axis compensator with a setting accuracy of 0.3".
Centering error of instrument and targets on horizontal angles can be determined from Equation (4.47):

$$\sigma_{\theta c} = (206,265'')\sqrt{\frac{\sigma^2_{\text{cns}}}{S^2_{\text{bs}}} + \frac{\sigma^2_{\text{cfs}}}{S^2_{\text{fs}}} + \frac{\sigma^2_{\text{cst}}}{S^2_{\text{bs}}S^2_{\text{fs}}}(S^2_{\text{bs}} + S^2_{\text{fs}} - 2S_{\text{bs}}S_{\text{fs}} \cos\theta)} \qquad (7.44)$$

Assuming all backsight (bs), foresight (fs), and setup (st) points are all centered and leveled using the same methods ($\sigma_{C\text{bs}} = \sigma_{C\text{fs}} = \sigma_{C\text{st}} = \sigma_c$). If optical plummet will be used, then

$$\sigma_c = (0.0005) \times (\text{HI}) \text{ or } \sigma_c = (0.0005) \times (1.755)\,\text{m} = 0.0008775\,\text{m}$$

Centering error at stations A, B, C, and D: Substituting $S_{\text{bs}} = 106.687$ m, $S_{\text{fs}} = 100.000$ m, $\theta = 90°$, and $\sigma_c = 0.0008775$ m in Equation (7.44) gives the centering error as $\sigma_{\theta c} = 3.51''$. For recentering between two sets, the centering error will be 2.48".
Centering error at stations P_1 and P_2: Substituting $S_{\text{bs}} = 106.687$ m, $S_{\text{fs}} = 106.687$ m, $\theta = 180°$, and $\sigma_c = 0.0008775$ m in Equation (7.44) gives the centering error as $\sigma_{\theta c} = 4.16''$. For recentering between two sets, the centering error will be 2.94".
Leveling error on angle measurement is determined from Equation (4.38) assuming electronic dual-axis compensator with a setting accuracy (σ_v) of ±0.3" (for Leica TC2003) will be used:

$$\sigma_{\theta L} = 0.3''\sqrt{[\cot(Z_{\text{bs}})]^2 + [\cot(Z_{\text{fs}})]^2} \qquad (7.45)$$

Leveling errors at A, B, C, D: For 18% slope, the backsight zenith angle at each station will be 79.796° and the foresight zenith angle will be 90°; substituting these values into Equation (7.45) gives $\sigma_{\theta L} = 0.05''$. For releveling between two sets, the leveling error will be 0.04".
Leveling errors at P_1 and P_2: For 18% slope, the backsight and foresight zenith angles at each station will be 79.796°; substituting these values into Equation (7.45) gives $\sigma_{\theta L} = 0.08''$. For releveling between two sets, the leveling error will be 0.06".
Pointing and reading errors at each station using the chosen Leica TC2003 is 0.5" for angle measurement in two sets:

Total error (for two sets) at each of stations A, B, C, D: $\sigma_\beta = 2.53''$.
Total error (for two sets) at each of stations P_1, P_2: $\sigma_\beta = 2.98''$.

The total error for the six stations is $\sigma_m = 6.59''$; the expected maximum error (at 99% confidence level) for the loop traverse will be determined from Equation (2.15) as $\sigma_m \times 2.57$ or 16.9".

Example 7.4 As part of a special traverse of "n" angles around a city block, a total station is to be set up along one side of the block, at one station with sight distances of 50 and 200 m with the angle being very close to 180°. The 50 m sight is nearly horizontal, but the 200 m sight is at a slope of 15%. These are the extreme values for this situation. Accounting for the effects of centering, leveling, pointing, and reading, recommend an instrument that would be capable of meeting the requirement that the block angular misclosure is not to exceed $10''\sqrt{n}$. "Not to exceed" is to be regarded as being at 99%. The values taken in the calculation of the misclosure would be averages from at least two sets (a set being the average of face left and face right sightings)

(reproduced by permission of CBEPS).

Solution: Standard deviation of misclosure of "n" angles: $\sigma_{\mathrm{m}} = \sigma_\theta\sqrt{n}$.
Using Equation (2.52), the maximum misclosure (at 99%) can be given as

$$|m| = \sigma_{\mathrm{m}}\sqrt{\chi^2_{\alpha=0.01,\mathrm{df}=1}} \quad \text{or} \quad |m| = \sigma_\theta\sqrt{n}\sqrt{6.63} \tag{7.46}$$

Equate the maximum misclosure to the angular misclosure and solve for σ_θ:

$$\sigma_\theta\sqrt{n}\sqrt{6.63} = 10''\sqrt{n} \to \sigma_\theta = \frac{10''}{\sqrt{6.63}} \to \sigma_\theta = \frac{10''}{2.575}(\text{or } 3.88'') \tag{7.47}$$

Error propagation for each angle θ due to centering, leveling, pointing, and reading errors:

$$\sigma_\theta^2 = \sigma_{\mathrm{c}}^2 + \sigma_{\mathrm{L}}^2 + \sigma_{\mathrm{P}}^2 + \sigma_{\mathrm{R}}^2 \tag{7.48}$$

Assuming equal contribution (σ) of all the errors: $\sigma_\theta = \sigma\sqrt{4}$ so that each error will contribute $\sigma = \sigma_\theta/2$ (or 1.94″), and $\sigma_{\theta\mathrm{p}} = \sigma_{\theta\mathrm{r}} = \sigma_{\theta\mathrm{c}} = \sigma_{\theta\mathrm{L}} = 1.94''$.

- From Equation (4.22), the pointing error (two sets):

$$\sigma_{\theta\mathrm{p}} = \frac{\sigma_{\mathrm{p}}}{\sqrt{2}} \quad \sigma_{\mathrm{p}} = \frac{60''}{M} \tag{7.49}$$

  Solve for M in Equation (7.49):

$$M = \frac{60''}{\sigma_{\theta\mathrm{p}}\sqrt{2}} \quad (\text{or } M = 21.9\times) \tag{7.50}$$

- From Equation (4.31), the reading error (two sets):

$$\sigma_{\theta\mathrm{r}} = \frac{\sigma_{\mathrm{r}}}{\sqrt{n}} \quad \sigma_{\mathrm{r}} = 2.5\mathrm{div} \tag{7.51}$$

Solve for div (given $\sigma_{\theta r} = 1.94''$):

$$\text{div} = \frac{\sigma_{\theta r}\sqrt{2}}{2.5}(\text{or div} = 1.1'') \tag{7.52}$$

- Centering error (due to target and instrument): Since the question is specific about the distances, we cannot make any other assumptions about them but use Equation (4.47) and assume that the worst angle θ will be 180° and the centering error (σ_c) for target and instrument are equal. The centering error σ_c is then solved for as follows:

$$\sigma_{\theta c} = \sqrt{\frac{\sigma^2}{(200)^2} + \frac{\sigma^2}{(50)^2} + \frac{\sigma^2}{(200)^2(50)^2}(200^2 + 50^2 - 2(200)(50)\cos 180)} \tag{7.53}$$

or

$$\sigma_{\theta c}^2 = 0.000025\sigma^2 + 0.0004\sigma^2 + 0.000625\sigma^2 \text{ or } \sigma_{\theta c}^2 = 0.00105\sigma^2 \tag{7.54}$$

For recentering between two sets, the centering error on one set of angle will be $1.94''\sqrt{2}$ or 2.74". The error in centering the instrument and the target can be calculated from Equation (7.54) as

$$0.00105\sigma^2 = \left(\frac{2.74}{206,265}\right)^2 \text{ or } \sigma^2 = 1.680586011\text{E–7} \tag{7.55}$$

with $\sigma = 4.1\text{E–4}$ m (or 0.41 mm or 0.00041 m) as the expected centering error of the instrument and the target, which requires forced centering device of ±0.0001 m.

- From Equation (4.40), the leveling error (releveling between two sets) can be given as

$$\sigma_{\theta L} = \frac{\sigma_v}{\sqrt{2}}\sqrt{\cot^2(Z_b) + \cot^2(Z_f)} \quad \text{cotan } Z_b = 0 \tag{7.56}$$

Since the slope angle (tangent of the vertical angle $(90° - Z_f)$) is 15%, $\text{cotan}(Z_f) = 0.15$, so that the following is obtained:

$$\sigma_{\theta L} = \frac{\sigma_v}{\sqrt{2}}(0.15) = 1.94'' \text{ or } \sigma_v = 18.3'' \tag{7.57}$$

From Equation (7.57), $\sigma_v = 0.2\text{div}$ or 18.3" so that the sensitivity of plate bubble div is equal to 92"/2 mm. On the basis of this design, the recommended instrument will have the following features: $M = 22\times$; least count $= 1''$; forced centering with target and instrument interchange on tripods (with equivalent centering error of 0.0001 × height of instrument); and bubble sensitivity better than 92"/2 mm.

7.8 ELEVATION DIFFERENCE MEASUREMENT DESIGN EXAMPLE

Determine, by the propagation of variance, whether a Wild N3 could be used, with double-scale invar staves, for Canadian special-order leveling. If not, suggest the order for which it would be suitable and why.

Solution: Some of the specifications for Wild N3 level are as follows: standard deviation for 1 km double run leveling is 0.2 mm; setting accuracy (split bubble) is 0.25″; parallel-plate micrometer (with a range of 10 mm, interval of graduation of 0.1 mm with estimation to 0.01 mm possible); magnification of telescope, $M = 42\times$; and tubular level sensitivity per 2 mm is 10″.
The sources of error are pointing, reading, and leveling of the instrument; the magnitude of each error is estimated as follows:

Pointing error can be calculated from Equation (6.1) as

$$\sigma_{\mathrm{P}} = \frac{45}{206,265 \times M} S \tag{7.58}$$

For the given sight distance $S = 50$ m and magnification $M = 42$, the calculated pointing is $\sigma_P = 0.2597$ mm. The reading/plumbing error is estimated using Equation (6.2):

$$\sigma_{\mathrm{r}} = \frac{\ell}{2}\left(\frac{v_{\mathrm{r}}}{206,265}\right)^2 \tag{7.59}$$

For the given length of rod $\ell = 3000$ mm and sensitivity of leveling rod $v_{\mathrm{r}} = 600''$, the calculated reading error is $\sigma_{\mathrm{r}} = 0.0127$ mm. The instrument leveling error is calculated from Equation (6.3):

$$\sigma_{\mathrm{L}} = \left(\frac{\sigma_{\mathrm{v}}}{206,265}\right) S \tag{7.60}$$

Given the error in leveling the instrument as $\sigma_{\mathrm{v}} = 0.25''$ and the sight distance as $S = 50$ m, the leveling error is calculated as $\sigma_{\mathrm{L}} = 0.0606$ mm. The total in an elevation difference measurement in a setup is calculated as

$$\sigma_{\Delta h} = \sqrt{(0.2597)^2 + (0.0127)^2 + (0.0606)^2} \ \text{ or } \ 0.267 \text{ mm} \tag{7.61}$$

The value calculated in Equation (7.61) is for the average of two leveling runs in one setup as the double-scale invar rod readings suggest. It can be concluded from the above calculations that Wild N3 level with double-scale invar rods will yield a standard deviation of elevation difference of 0.267 mm per setup with 50 m sight lengths. The following relationship can be established for special-order leveling using Equation (3.6) and the specification in Table 3.1:

$$\sigma_{\mathrm{ran}} \sqrt{2} \times 1.96\sqrt{L} = 3\sqrt{L} \text{ mm} \tag{7.62}$$

where σ_{ran} is the standard deviation of elevation difference over 1 km, L is the total length of leveling section, and $3\sqrt{L}$ mm is the allowable discrepancy between independent forward and backward leveling runs between benchmarks (at 95% confidence) for special-order leveling. The value of σ_{ran} from Equation (7.62) is 1.0823 mm/$\sqrt{\text{km}}$. The expected standard deviation of elevation difference ($\sigma_{\Delta h1}$) at every instrument setup (with a total backsight and foresight distance of 100 m per setup or 10 setups in 1 km leveling section) can be calculated as $\sigma_{\Delta h1} = 1.0823/\sqrt{10}$ or 0.342 mm. This is considered as the value for leveling done twice in a setup; leveling with double-scale invar staves involves leveling twice per setup with the average of the two elevation differences determined and used as the elevation difference for that setup. On this basis, 0.342 mm is considered as the standard deviation of the average of two elevation differences at a setup. Since the standard deviation (±0.267 mm per setup) achievable with N3 with double-scale rods is less than ±0.342 mm per setup required by special order, then Wild N3 can be used for the Canadian special-order leveling.

8

THREE-DIMENSIONAL COORDINATING SYSTEMS

OBJECTIVES

At the end of this chapter, you should be able to

1. Describe the commonly used three-dimensional coordinate reference systems
2. Discuss the needs and the common models for three-dimensional coordinating systems
3. Explain the concepts and principle of electronic coordinating systems
4. Describe the features of three-dimensional coordination with Global Navigation Satellite System (GNSS)
5. Discuss the features and applications of three-dimensional coordination with electronic theodolites
6. Analyze the accuracy limitations of three-dimensional coordination with electronic theodolite, including three-dimensional traverse surveys
7. Describe the features and accuracy limitations of airborne laser scanning system as coordinating system
8. Describe the features and accuracy limitations of terrestrial laser scanning system as coordinating system

Precision Surveying: The Principles and Geomatics Practice, First Edition. John Olusegun Ogundare.

8.1 INTRODUCTION

A three-dimensional coordinating system is considered in this book as a system of hardware and software that allows three-dimensional (x, y, z) coordinates of any targeted or nontargeted point to be determined through direct measurements or through internal transformation of the measured quantities. Certain types of three-dimensional coordinating systems are becoming increasingly important for industrial metrology (such as measurement of antennas and measurements on aircraft for its dimensional control) and deformation monitoring applications, since they provide a portable, noncontact, and real-time method of acquiring three-dimensional coordinates about objects smaller than a meter in size to objects of several meters in size.

Three types of coordinating systems are discussed in this chapter: Coordinating with Global Navigation Satellite System (GNSS), the electronic coordinating system, and the terrestrial laser scanning system. Each method offers complementary advantages to surveyors and engineers; however, since the solution for real-time three-dimensional coordinates of remote points is central to most coordinating system applications, more emphasis will be placed on the electronic coordinating system and the terrestrial laser scanning system.

An integral part of any coordinating system is a *reference coordinate system*, which must be well understood in order to properly use survey measurements for calculating positions (coordinates of points) and for solving difficult problems in geomatics. *Coordinates* can be simply defined as separations from a given origin, in certain directions or ordered values (x, y, z) in a given *coordinate system*. A coordinate system is a methodology or an idealized abstraction for defining the coordinates (or location) of a feature in space. In order for a coordinate system to be usable in locating a point in space, it must have an origin as well as properly defined reference directions for its axes. The coordinate system, therefore, specifies how coordinates are assigned to points (or locations) on the earth and its environment. When the origin and orientation of axes of the coordinate system are specified with regard to the earth, the coordinate system is known as a *datum* or a *coordinate reference system*. A *datum*, however, may be associated with a reference ellipsoid (on which measurements can be reduced for further computations) in addition to a coordinate system or the geoid (in the case of height system). There are three types of coordinate reference systems: *one-dimensional coordinate reference systems*, *two-dimensional coordinate reference systems*, and *three-dimensional coordinate reference systems*.

The one-dimensional coordinate system is basically about height determination for points on the earth surface or near the earth surface. The determined heights, however, are only useful as one-dimensional coordinates if they are referenced to a well-defined origin or datum (e.g., the geoid) and if they have well-defined unit of measurement in a geometrical sense. In surveying, precise heights are determined from measured elevation differences.

8.1.1 Two-Dimensional Coordinate Reference Systems

Two-dimensional coordinate reference systems can be divided into two types: coordinate reference systems on reference ellipsoid and coordinate systems on the plane. The coordinate reference systems on reference ellipsoid tend to locate positions in angular units (as latitude and longitude) on the surface of the three-dimensional model of the earth (the ellipsoid). The latitude and longitude values are known as geodetic coordinates, and they are considered as two-dimensional coordinates on the ellipsoid. An example of such a system is the North American Datum of 1983 (NAD83).

The coordinate reference systems on the plane tend to locate positions in linear units (Easting and Northing) on the two-dimensional model of the earth, when the portion of the earth being mapped is considered to be so small that it can be represented by a plane as in the case of plane surveying or in the case where the three-dimensional model of the earth is transformed into two dimensions through a process of map projections. A map projection system consists of conventions that prescribe how geodetic coordinates (latitude and longitude) are transformed to and from grid coordinates by means of map projections. Commonly used map projection coordinate systems (also known as grid coordinate systems) are Universal Transverse Mercator (UTM) and the stereographic double projection for project areas of circular shape, such as super collider ring in Texas, USA. They are both conformal map projections that preserve local shapes by preserving local angles at each point of the areas being projected. The shape and angle preservation properties of conformal projections are very attractive to geomatics professionals.

On the basis of the definition of a coordinate reference system, it can be said that a topographic map will have two types of coordinate systems defined for it as follows:

- *Two-dimensional coordinate system*: origin is at the center of the map projection; x-axis direction in the East–West direction; and the y-axis direction in the North–South direction. For example, the UTM projection will have its origin on the surface of the reference ellipsoid at the intersection of the equator and the central meridian; x-direction is in the East–West direction and y-axis is in the North–South direction.
- *One-dimensional coordinate system:* (for representing elevations or orthometric heights on the map): origin is on the surface of the geoid (or mean sea level) and z-axis in the direction of gravity, perpendicular to the geoid.

Some of the important advantages of using two-dimensional coordinate system as a computation model are as follows:

(a) It allows positional accuracy of 10 ppm or better to be achieved by separating horizontal control survey projects from vertical control survey projects. For example, horizontal traverse surveys are separated from the leveling surveys so that the effects of atmospheric conditions are reduced in the process.

(b) Orthometric (mean sea level) height (H) produced through leveling is practically meaningful in engineering; points having the same orthometric heights are at the same geometric height above the geoid (mean sea level).

(c) The (Northing, Easting, Orthometric height) coordinates are easy to manipulate in survey computations, for example, using plane geometry and plane trigonometry in computations. Analyses of some engineering projects are better done in map projection coordinate system since (Northing, Easting, Orthometric height) coordinates are easy to manipulate, and they are easily used to produce more practical results.

One important disadvantage of using the two-dimensional coordinate system as a computation model is that measurements must first be rigorously reduced to the reference ellipsoid and then to the map projection plane before using them to calculate the two-dimensional (Northing, Easting) coordinates. Many reductions must also be applied to leveled heights in order to obtain the orthometric heights.

8.1.2 Three-Dimensional Coordinate Reference Systems

Three-dimensional coordinate reference systems can be of three different types, such as Conventional Terrestrial Reference System (CTRS) (now known as International Terrestrial Reference System (ITRS)), local geodetic (LG) system, and local astronomic (LA) system. These systems tend to locate positions in three linear dimensions (X, Y, Z) with respect to their origins. The properties of the systems are described in Table 8.1.

The ITRS is a reference system that can be accessed for practical use as a coordinate system through its realization called International Terrestrial Reference Frame (ITRF). The ITRF is a dynamic datum, which is regularly updated by the International Earth Rotation and Reference System Service (IERS) to account for the dynamics of the earth. It is a global network of control stations (with known coordinates and velocity components) that binds an earth-centered, earth-fixed three-dimensional coordinate system to the earth. One of the important properties of ITRS is that it is the closest approximation of the geocentric natural coordinate system, whose coordinate axes are defined by the directions of gravity and the spin axis of the earth. The natural coordinate system, whose natural coordinates are the astronomic latitude and longitude and gravity potential, can be determined by measurements: the latitude and longitude values are determined by astronomic positioning (such as observation of star positions) and a particular gravity potential is derived from leveling and gravity measurements with reference to a selected level surface.

As can be seen in Table 8.1, the LA system is a coordinate system in which observations can be considered as a natural coordinate system. The LG system is close to the LA system (both are left-handed systems) except that the LG system is in relation to the reference ellipsoid while the LA system deals with the natural earth surface or the geoid and its gravity field.

One important advantage of using the three-dimensional coordinate system as a computation model is that in using measurements to compute the X, Y, Z coordinates of a point, one does not need to reduce the measurements (distances, angles, and azimuth) to reference ellipsoid, but only needs to correct for atmospheric and instrumental errors. This model is commonly applied in the positioning and orientation of

TABLE 8.1 Properties of the Three Common 3D Coordinate Systems.

	ITRS System	LG System	LA System
Origin	Center of mass of the earth	A chosen point on the reference ellipsoid or the instrument setup station projected to the reference ellipsoid	A point on the earth surface where an observation is made (instrument setup point)
Primary (or z-) axis	A line from the origin pointing in the direction of the conventional terrestrial pole (CTP)	An orthogonal line passing through the origin on the reference ellipsoid	A line that is orthogonal at the origin to the geoid (direction of gravity or zenith at the setup point – direction of plumb line when the survey instrument is level). This is usually referred to as *Up* direction
Secondary (or x-) axis	A line from the origin corresponding to the intersection of the mean equator plane and the mean meridian plane of Greenwich	A line tangent at the origin and aligned along the geodetic meridian, pointing toward the geodetic North	A line tangent at the origin and aligned along the astronomical meridian, pointing toward the true North. This is referred to as *Northing* direction
Tertiary (or y-) axis	A line from the origin that is orthogonal to the z–x plane in a right-handed system	A line from the origin that is orthogonal to the z–x plane in a left-handed system	A line from the origin that is orthogonal to the up-North plane in a left-handed system. This is referred to as *Easting* direction

nuclear accelerator and in the alignment of radio telescope aerial arrays over a very long distance, relative to the center of mass of the earth. GNSS is an example of a system that provides coordinates in this model. The important disadvantage of using this model is that the ellipsoidal height (h) derived in this process is not practically useful in engineering; orthometric (or mean sea level) height is commonly used.

8.1.2.1 Topographic Coordinate System Topographic coordinate system is helpful for application where the area being mapped is sufficiently small as to allow the curvature of the earth to be ignored, thereby rendering map projections in the area unnecessary. Topographic surveying is a special type of three-dimensional surveying for determining the three-dimensional (x, y, elevation) coordinates of selected natural and artificial features on the earth surface. It ranges from aerial mapping to ground and underground surveys. Some of the projects requiring topographic survey include the following:

- Locating invert elevations of structures
- Determining the horizontal location of building corners and road centerlines
- Determining the positions of trees and identifying the sizes of the trees
- Locating all the high points and low points among ridges and valleys
- Providing cross sections at specified intervals
- Locating all buildings and dwellings at the wall or footer lines
- Identifying structure addresses (house or box numbers)
- Locating all the government benchmarks
- Locating utility items above (utility poles, manholes, fire hydrants, etc.) and underground (sewage disposal and water supply).

The main deliverable of a topographic survey is usually a topographic plan or map. A typical topographic plan may include all or some of the following:

(a) Location of surrounding structures and services (above and below ground).
(b) Some spot heights.
(c) Contours with appropriate intervals (constant elevation difference between two adjacent contour lines) with steep slopes having more contour interval to make map more legible; flat areas will decrease the contour interval to a limit that will not interfere with planimetric details located on the topographic map. In essence, contour intervals must be selected to allow good interpretation of the character of the terrain.
(d) Plan scale, which is well chosen so that the plan can serve as base map over which subsequent project drawings can be drawn at the same scale. The detailed topographic plan will then serve as a base upon which to prepare utility maps.

Errors in map plotting and scaling should be checked to ensure that appropriate map accuracy standards are complied with.

8.2 COORDINATE SYSTEM FOR THREE-DIMENSIONAL COORDINATING SYSTEMS

The ITRS is commonly used in space or extraterrestrial techniques in three-dimensional positioning and orientation of nuclear accelerator and in the alignment of radio telescope aerial arrays over very long distances, relative to the center of mass of the earth. The LG system is commonly used in engineering projects, such as local deformation monitoring and alignment of machine components in industrial metrology, where the earth's curvature can be ignored. The LG coordinate system is illustrated in Figure 8.1, where ϕ, λ are the geodetic coordinates of the origin P and the coordinate axes are represented by x^{LG}, y^{LG}, and z^{LG}.

Based on the right-hand convention of the coordinate system in North America, it is necessary to have a right-handed local geodetic (rLG) coordinate system. The difference between the LG system and the rLG will just be that (x, y, z) in LG is switched to (y, x, z) in rLG system; as can be seen, the x and y values are switched around so as to satisfy their right-handed representation in a diagram. If the terms *Northing*, *Easting*, and *UP* are used to represent the coordinate values, the meanings of the terms are not changed. For example, if the coordinates of point P in LG are given as (1000, 2000, 10), the coordinates of the point in rLG will be (2000, 1000, 10). Since rLG system is only different from LG system in the order in which coordinate values are presented, the LG and rLG systems will be used to mean the same thing. Note that the geodetic North is not an observable that can be measured since it is not a physical quantity like astronomic North.

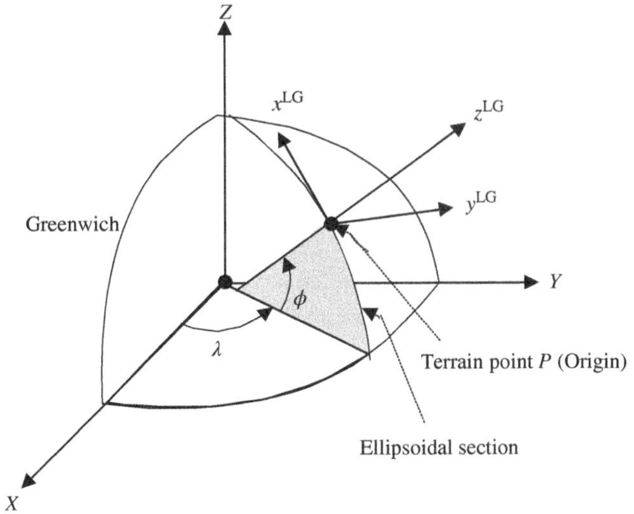

Figure 8.1 Representation of local geodetic (LG) coordinate system.

8.3 THREE-DIMENSIONAL COORDINATION WITH GLOBAL NAVIGATION SATELLITE SYSTEM

GNSS, for example, the Global Positioning System (GPS), uses messages received from space-based satellites and the positions of the satellites to compute positions of earth-based antennas by using some navigation equations. The positions of the antennas are given as latitude, longitude, and ellipsoidal height coordinates or as three-dimensional Cartesian (X, Y, Z) coordinates, which are based on World geodetic system of 1984 (WGS84) geodetic datum (coordinate reference frame). The current version of WGS84 as of Year 2002 is WGS84 (G1150), which is closely related to the ITRF2000 (epoch 2001.0) (Department of Rural Development and Land Reform, 2013). In this case, the three-dimensional (X, Y, Z) coordinates produced by GPS is closely related to the ITRS parameters given in Table 8.1, knowing that ITRS is the most precise earth-centered, earth-fixed datum currently available. Further information and details on the use of GNSS as a coordinating system can be found in any modern Geodesy book.

Accurate three-dimensional data are also possible from extraterrestrial positioning techniques, such as very long baseline interferometry (VLBI) and satellite laser ranging (SLR). They produce a three-dimensional global coordinate system based on ITRF, while the terrestrial positioning is done in the LA system.

8.4 THREE-DIMENSIONAL COORDINATION WITH ELECTRONIC THEODOLITES

8.4.1 Coordinating Techniques

The electronic coordinating system usually consists of two or more high-magnification, short-focus model electronic theodolites linked to a microcomputer for real-time calculations of three-dimensional coordinates of target points. The system is commonly used for the highest precision positioning of targets and deformation monitoring surveys over small areas. Sokkia NET2100 and Leica (Wild) tunnel measurement system (TMS) are examples of such systems.

The principle of coordinate determination used in the electronic coordinating system is based on the surveying technique known as three-dimensional intersection. This principle involves the simultaneous measurement, using two theodolites (T1 and T2), of horizontal angles θ_1 and θ_2 from either end of a precisely measured baseline (b) and the zenith angles z_1 and z_2 as shown in Figure 8.2. Before the measurements are made, the optical lines of sight through the telescopes of the two theodolites are first made to coincide with each other, and the telescopes are then turned to measure the given targets. Standard surveying computational techniques can then be applied to derive the three-dimensional coordinates of any unknown target point P.

An alternative approach to coordinate determination with the electronic coordinating system may be adopted if the simultaneous use of two theodolites providing real-time coordinates is not required. In such a case, data acquisition may be accomplished by using a single theodolite and there will be no need to align the theodolite

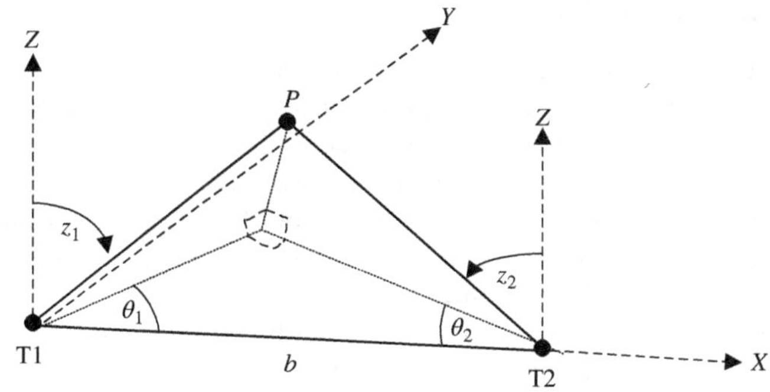

Figure 8.2 Three-dimensional intersection problem.

in relation to another instrument before obtaining measurements. More importantly, the acquired data can be rigorously processed using least-squares techniques to fully exploit any redundant data. The observation equations may be formulated for the problem and then solved by the method of least squares adjustment in order to determine the three-dimensional coordinates of the unknown points.

From Figure 8.2, if the baseline length b, the zenith angles (z_1 and z_2), and the horizontal angles (θ_1 and θ_2) are known, the coordinates of point P can be determined by using trigonometric functions. The baseline length b, however, must be accurately known in order to accurately determine the coordinates of point P. Two ways of determining the baseline length are by directly measuring the baseline or by introducing into the measuring scheme a different scaling mechanism. Since short baselines are commonly involved in the applications of the coordinating system, direct measurement of the baselines may be imprecise. The most common technique is to introduce into the measuring scheme a scaling mechanism, which involves setting up short invar scaling bars of known lengths at suitable locations as part of the micronetwork to be measured. The invar scaling bar is used because of its low coefficient of linear expansion, which ensures that no systematic change in its length occurs while it is being used; its use generally reduces the measuring time, and if well calibrated, it will allow the baseline to be determined to a high level of accuracy with relative ease. The calibration of the invar bar is essential if systematic errors are to be avoided.

The type of network usually established for electronic theodolite coordination is known as metrological micronetwork. A metrological micronetwork is typically a triangulation network with the observables being the horizontal directions and zenith angles; distances are not measured but the invar scale bars with known lengths (marked with targets on the bar) are to provide scales for the networks. In some cases, wall targets are used as control points with directions and zenith angles measured (in both faces) to the wall targets, the targets on the invar scale bars and the targeted points on the already positioned components in the work area. A single simultaneous least squares adjustment is then performed to all the measurements to obtain the adjusted coordinates of all the targets in the design coordinate system.

8.4.2 Field Data Reductions

The typical field data acquired in the electronic coordinating system are azimuth, horizontal directions, or horizontal angles, zenith angles, and slope distances. Before these field data are used in calculating the three-dimensional (x, y, z) coordinates of points, they must first be corrected for instrumental errors, meteorological effects, and gravity effects (such as deflection of the vertical). In industrial metrology, spatial distances are much short. As a consequence of this, the effects of refraction on distance measurements are reduced, while the precision of measuring such distances are reduced. In order to improve the accuracy of three-dimensional positioning, horizontal angles and zenith angles, which can be measured more accurately, are measured instead of distances.

Since the vertical axes of surveyor's instruments are aligned in the direction of local plumb lines (directions of gravity), survey observations must be corrected for deflection of the vertical to reduce them to a reference ellipsoid, along the normal to the ellipsoid. The deflection of the vertical, which may vary from several seconds in flat areas to up to 60″ in the mountains, causes angular traverse loop misclosures as in the case of instrument leveling errors. The components (ξ, η) of the deflection of the vertical at a given point can be given as follows:

$$\xi = \Phi - \phi \tag{8.1}$$

$$\eta = (\Lambda - \lambda)\cos\phi \tag{8.2}$$

or

$$\eta = (A - \alpha)\cot\phi \tag{8.3}$$

where

ξ is the deflection of the vertical in the North–South direction;
η is the deflection of the vertical in the East–West direction;
Φ is the astronomic latitude; ϕ is the geodetic latitude;
λ is the geodetic longitude; Λ is the astronomic longitude;
A is the astronomic azimuth of a line from the given point to another point; and
α is the geodetic azimuth of the line from the given point to another point.

The *slope distance* measurements are not affected by gravity effects and are not to be corrected for gravity effects. The measured *zenith angle* (Z'_{ij}) from point "i" to "j" is corrected for gravity (or deflection of the vertical) effects as follows:

$$Z_{ij} = Z'_{ij} + \varepsilon_{ij} \tag{8.4}$$

where

$$\varepsilon_{ij} = \xi_i \cos\alpha_{ij} + \eta_i \sin\alpha_{ij} \tag{8.5}$$

or ε_{ij} is the deflection of the vertical in the direction of the geodetic azimuth α_{ij};

ξ_i, η_i are the North–South and East–West components of the deflection of the vertical at the instrument station "i," respectively; and α_{ij} is the geodetic azimuth of line "i" to "j."

The astronomic *azimuth (or direction) measurement* (A_{ij}) is corrected for gravity effects as follows:

$$\alpha_{ij} = A_{ij} - \eta_i \; tan \; \phi_i + C_{ij} \tag{8.6}$$

where

$$C_{ij} = -(\xi_i \sin \alpha_{ij} - \eta_i \cos \alpha_{ij}) \cot Z_{ij} \tag{8.7}$$

or C_{ij} is the correction to be applied to the observed astronomical azimuth A_{ij} to relate it to the same ellipsoidal normal as the geodetic azimuth α_{ij} (the offset due to the deflection of the vertical);

$-\eta_i \tan \phi_i$ is the Laplace correction or the azimuth correction to line up the x^{LG} and the x^{LA} axes (with x as the direction of the North as shown in Figure 8.1);

A_{ij} is the measured astronomic azimuth (or the total station direction measurement) from i to j;

ξ_i, η_i are the North–South and East–West components of the deflection of the vertical at the instrument station i, respectively; and

ϕ_i is the geodetic latitude of point i.

If a horizontal angle θ' is measured at station 2, backsighting to station 1 and foresighting to station 3, the corrected angle θ (reduced to the reference ellipsoid) can be formulated from Equation (8.6) as follows:

$$\theta = (A_{23} - A_{21}) - (\xi_2 \sin \alpha'_{23} - \eta_2 \cos \alpha'_{23}) \cot Z_{23} + (\xi_2 \sin \alpha'_{21} - \eta_2 \cos \alpha'_{21}) \cot Z_{21} \tag{8.8}$$

where

$\theta' = (A_{23} - A_{21})$ is the measured horizontal angle;

α'_{23} and α'_{21} are the geodetic azimuths in the directions 2-3 and 2-1, respectively; Z_{23} and Z_{21} are the measured zenith angles;

ξ_2 and η_2 are the components of the deflection of the vertical measured at setup point 2.

In horizontal angle measurements, the influence of deflection of the vertical is identical to the influence of misleveling the theodolite. Corrections to horizontal angle measurements for almost all practical situations are insignificantly small, except if the lines of sights have large zenith angles. If the terrain is relatively flat, where zenith angles Z_{23} and Z_{21} are likely to be 90° (in horizontal sightings), Equation (8.8) will be reduced to $\theta = (A_{23} - A_{21})$, meaning that the effect of the deflection of the vertical will be zero in the relatively flat terrains.

8.4.3 Three-Dimensional Coordinate Determination

As can be seen in Table 8.1, the origin of the LA system is the instrument setup station, meaning that every instrument setup station in a micronetwork must have its own separate LA system. If this is the case, and understanding that the directions of gravity from one setup station to another are usually not parallel (due to earth curvature), the LA coordinate system axes established for one setup station will not be parallel to the corresponding LA coordinate system axes at another setup point. Also, the azimuth of a line in one system will be different in another system for the same line due to the convergence of meridian. Since distance measurements are usually avoided (because of precision problem) in industrial metrology, targets in a geodetic micronetwork must be sighted and intersected from at least two stations in triangulation techniques usually adopted. In this case, every micronetwork station that is introduced due to setting up of theodolite will have its own LA coordinate system with its origin at the station, which will be different from each other, and the orientation of each coordinate system will be defined by the astronomic latitude (ϕ) and longitude (λ) and a tangent to the local gravity vector at each station.

The traditional approach in micronetwork establishment is to simply fix one of the several LA coordinate systems as a reference and relate the rest to it by solving for translation components (station coordinates) with respect to this fixed coordinate system, for each of the other systems. Solving for the translation components, however, will not make all the corresponding axes of all the different LA coordinate systems parallel, although some of the effects will be absorbed in the estimated translation components. The nonparallelism property is due to both gravimetric and geometric causes. The gravimetric cause is the effect of deflection of the vertical components in the North–South direction (ξ) and in the East–West direction (η) and the difference between the direction of astronomic and geodetic North, which can be given from Equation (8.3) as

$$\Delta A = A - \alpha = \eta \tan \phi \tag{8.9}$$

where A is the astronomic azimuth of a line, α is the geodetic azimuth of the same line, and ΔA is the change that only exists because of the initial condition enforced when determining the biaxial ellipsoid's position with respect to the ITRS system. If (ξ, η) are the same in the area, they will affect the transformation from LA to LG identically for all stations in the network, except that the different directions of astronomic and geodetic North will not be affected identically. The transformation of coordinates (x_j, y_j, z_j) of point "j" in LA system to LG coordinate system with the origin at the instrument's setup station "i" can be given mathematically as (Vanicek and Krakiwsky, 1986)

$$\begin{bmatrix} x_j \\ y_j \\ z_j \end{bmatrix}^{\text{LG}_i} = \begin{bmatrix} 1 & \eta_i \ tan \ \phi_i & \xi_i \\ -\eta_i \ tan \ \phi_i & 1 & \eta_i \\ -\xi_i & -\eta_i & 1 \end{bmatrix} \begin{bmatrix} x_j \\ y_j \\ z_j \end{bmatrix}^{\text{LA}_i} \tag{8.10}$$

where ξ_i and η_i are the deflection of the vertical components (radians) in the North–South and East–West directions at point "i," respectively. If the coordinates of the setup station at "i" are (0, 0, 0), the coordinates of point "j" in the LA_i system can be given as follows:

$$\begin{bmatrix} x_j \\ y_j \\ z_j \end{bmatrix}^{LA_i} = s_{ij} \begin{bmatrix} \sin Z'_{ij} \cos A_{ij} \\ \sin Z'_{ij} \sin A_{ij} \\ \cos Z'_{ij} \end{bmatrix} \tag{8.11}$$

where s_{ij}, A_{ij}, and Z'_{ij} represent the measured slope distance, astronomic azimuth, and zenith angle, respectively, from setup station "i" to target point "j"; and (x_j, y_j, z_j) are the LA_i coordinates of target point "j." As can be seen in Equation (8.11), the slope distance (s_{ij}) and zenith angle (Z'_{ij}) measurements provide the necessary link between the (x, y) horizontal coordinates and the z coordinate. The vertical information obtained from the spatial distance, however, is only useful when the lines of sight are steeply inclined. Alternatively, the coordinates in the LG_i coordinate system, in Equation (8.10), can be obtained directly by substituting into Equation (8.11), the corrected zenith angle Z_{ij} from Equation (8.4) and the geodetic azimuth α_{ij} from Equation (8.6).

After the effects of gravity have been taken care of as shown in Equation (8.10), the LA coordinate system axes at each theodolite station will become transformed to their corresponding LG coordinate system axes at that theodolite station; but the corresponding axes of the LG coordinate systems for all the different theodolite stations are not likely to be parallel to each other.

In order to align all the axes of all the different LG coordinate systems in the micronetwork, the geometric effects must be taken care of. In this case, the effects of curvature are effectively taken care of by performing corresponding rotations of each coordinate system to line them up with the fixed LG_k reference coordinate system (with the setup station at point "k"). For example, the transformation of (x_j, y_j, z_j) coordinates of point "j" from LG_i system (for setup station "i") into the fixed LG_k system of another setup station "k" can be given mathematically as follows (Wilkins, 1989):

$$\begin{bmatrix} x_j \\ y_j \\ z_j \end{bmatrix}^{LG_k} = \begin{bmatrix} 1 & \Delta\lambda & \Delta\phi \\ -\Delta\lambda & 1 & \omega \\ -\Delta\phi & -\omega & 1 \end{bmatrix} \begin{bmatrix} x_j \\ y_j \\ z_j \end{bmatrix}^{LG_i} + \begin{bmatrix} x_0 \\ y_0 \\ z_0 \end{bmatrix}^{LG_k} \tag{8.12}$$

where $\Delta\lambda$ and $\Delta\phi$ are small changes (radians) in longitude and latitude of the origins of the LG_i and LG_k systems; (x_0, y_0, z_0) are the coordinates of the origin of LG_i system in the LG_k coordinate system; and ω is a small angle (radians) given as

$$\omega = \arctan\left(\frac{d}{R+h}\right) \tag{8.13}$$

with d as the distance from the origin of the fixed LG_k system to the z axis of the LG_i system in the horizontal plane of the fixed LG_k system; R is the radius of the earth; h is the ellipsoidal (or spherical) height of the fixed LG_k system origin.

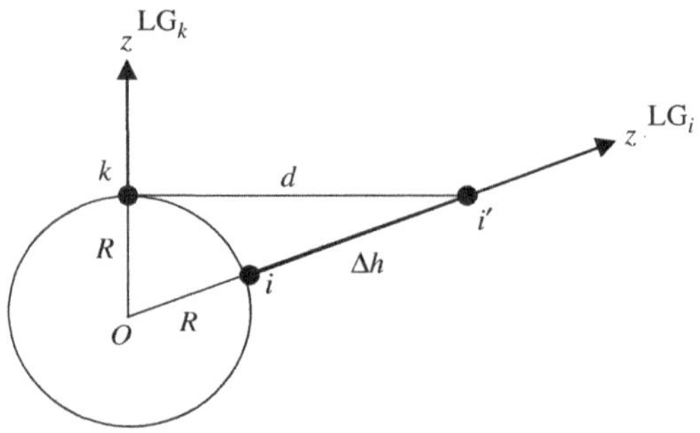

Figure 8.3 Relationship between a plane and a level surface.

In geodetic micronetworks, the transformation Equation (8.12) is usually applied implicitly in the adjustment equations. In this case, the earth is assumed flat and the datum is defined by fixing an LG$_k$ system and the translation coordinates (x_0, y_0, z_0) of other origins with respect to it, and taking the small changes $(\Delta\lambda, \Delta\phi, \omega)$ as negligible. The assumption of a flat earth, while acceptable for horizontal positional applications, may not be acceptable for finding elevations, as the geoid or the reference ellipsoid may deviate from the tangent plane by about several millimeters at 1 km from the point of contact.

Since engineering projects are usually limited to small areas, the reference surface may be considered as a plane to allow the use of simple plane trigonometry for coordinate computation. There is usually a limit on the length of sight that will allow plane trigonometry to be used, beyond which the curvature of the earth would have to be considered, especially with regard to the height system. This is illustrated by Figure 8.3. For simplicity, consider the earth as a sphere of radius R centered at point O; points "k" and "i" are the setup stations with the z component of the LG coordinate systems passing through the setup stations as shown in Figure 8.3; curve k-i is a level surface with stations k and i being at the same elevation. Areas around station "k" (the fixed origin) can be considered as a plane, but as one moves toward point "i," the plane k-i' (of length d) deviates by length i-i' (or Δh) from the level surface k-i. The length i-i' (or Δh) can be derived using Pythagoras theorem on the right-angle triangle k-O-i' as follows:

From Figure 8.3, the following relationship can be established:

$$R^2 + d^2 = (R + \Delta h)^2 \tag{8.14}$$

which reduces to the following:

$$d^2 = 2R(\Delta h) + \Delta h^2 \tag{8.15}$$

We can set $\Delta h^2 \approx 0$ in Equation (8.15) since adding it to $2R(\Delta h)$ will not significantly change d^2 (knowing that the radius of the earth, R, is very large). With this,

Equation (8.15) can be reduced to the following:

$$\Delta h = \frac{d^2}{2R} \tag{8.16}$$

where Δh is the error in height if the curvature of the earth is neglected over a distance d from a tangent plane. In most metrology applications, however, Δh in Equation (8.16) can be neglected since the project area involved is usually small and relatively smooth. For instance, the micronetworks in industrial metrology applications would rarely exceed a distance of 100 m so that Δh is likely to be less than 0.8 mm.

From Equation (8.10), it can be seen that if there is no deflection of the vertical (i.e., the direction of normal to the reference ellipsoid and the direction of gravity line up), the coordinates of the same point in both LA and LG systems will be the same, since the two coordinate systems will be the same. If there is no curvature of the earth, Equation (8.12) will be simplified to a simple problem of translating coordinates from one system to another by fixed amounts (i.e., x_0, y_0, z_0 coordinates), without any need for the rotation of the axes. In this case, all the corresponding coordinate axes of all the setup stations in the network will be parallel to each other.

In practice, the only time that the corresponding axes of all the LG systems constituting a micronetwork can become parallel is when the projections of all the setup stations are on the same plane surface, assuming the reference ellipsoid is a plane. These can be assumed to be the case in practice since the size of a micronetwork in industrial metrology applications rarely exceeds a distance of 100 m in one direction. With the area of application of this size, a plane surface can be assumed with respect to the reference ellipsoid.

Generally, the majority of engineering surveys are carried out in areas of limited extent, in which case the reference surface may be taken as a tangent plane to the geoid and the principles of plane surveying applied. The plane surveying principles ignore the curvature of the earth and take all points on the physical surface as orthogonally projected onto a flat plane. For areas less than 10 km^2, the assumption of a flat earth is acceptable when one considers that in a triangle of approximately 200 km^2, the difference between the sum of the spherical angles and the plane angles would be 1 arcsec, or when one considers that the difference in length of an arc of approximately 20 km on the earth surface and its equivalent chord length is just 8 mm.

8.4.4 Factors Influencing the Accuracy of Electronic Coordinating Systems

The metrological micronetworks established for electronic coordinating systems are essentially three-dimensional networks requiring very high precision and some critical considerations when designing and measuring the micronetworks. Apart from the possible effects of the sources of systematic errors discussed in Sections 8.4.2 and 8.4.3, the overall accuracy of the coordinates determined by the coordinating system will also depend on other factors. Some of these factors are the equipment and target design used for measurements, the geometry of the measurement scheme, and the influence of the environment (due to vibration, wind, temperature fluctuations, refraction and varying lighting conditions, etc.), discussed as follows (Wilkins, 1989).

8.4.4.1 ***Effect of Equipment and Target Design*** The use of precision electronic theodolites that are able to resolve to a fraction of a second is essential if high-accuracy measurements are to be obtained. The effects of any instrumental errors must be eliminated and the pointing of the telescope to well-designed targets on the structure must be precise. The design of the targets should enable precise centering of the telescope crosshairs over a wide angular range (60–120°). Other consideration is about how the scale is determined for the network adjustment.

In metrology networks of few tens of meters distances, short distances cannot be measured accurately enough (0.05 mm) to satisfy the scale requirements. The real solution to this is to use calibrated invar scale bars, which are well positioned in the work area to provide the needed scale for the network. The two targets that define the scale bar length can be tied to the network through triangulation (observed in the same way as the wall targets), with the known distance between them added as spatial distance observable in an adjustment procedure. The three-dimensional coordinates (x, y, z) of the scale bars' target locations are estimated in the adjustment process also; and through calibration procedure using interferometric comparator, the absolute lengths of the bars can be determined to an accuracy of 0.01 mm.

The other critical consideration with regard to equipment is the centering error of the coordinating system. In order to reduce the contribution of the centering error in geodetic measurements to ±0.1 mm, forced centering procedure is commonly used. Forced centering procedure for achieving this level of accuracy, however, will require that permanent and stable pillars be constructed at network stations. Because of restricted space and the special requirement that instrument locations be close to the structures being set out or monitored, it is impossible to establish permanent pillars on the project sites. Generally, if one does not need to center on any specific survey marker or if one does not need to set on the marker in the next session of measurement, the location of the instrument during measurement will not matter, and the instrument can be located in any convenient location. In this case, the location of the instrument whenever it is set up can be determined through resection by observing to distant control points (i.e., by free-stationing method). The wall targets (established from previous surveys) are used as control points, and observations between instrument locations are not necessary if there are enough wall targets to create enough redundancy. This makes reference wall target coordination a primary concern in industrial metrology, with the instrument locations serving only as a link between the different wall targets. The instrument can be located anywhere in the project area; the coordinates of the instrument locations are determined by resection, then the resected coordinates of the instrument locations are used to obtain the intersected coordinates of targets (i.e., the object points) located on the structures to be aligned or monitored. The intersected coordinates of the object points can be used to determine corrections, offsets, or calibration values for the structures being aligned or monitored. Some polynomial functions can also be fitted to the coordinates of the object points in order to determine discrepancies of the surface of the structures. Since the coordinates of the object points are computed using the coordinates of the instrument locations, any error in these computed instrument coordinates (x, y, z) will be reproduced in the object point coordinates (X, Y, Z).

8.4.4.2 Effect of Geometry of Measurement Scheme The geometrical relationship between the theodolites and the points on the object to be measured should be in such a way that the length of the baseline is restricted to between 5 and 10 m. The intersection angle range suggested for high-accuracy projects is between 78° and 142°. The position and orientation of the scaling bar are also very important. The bar should be located so that the angle of intersection at the target points on the bar is close to 90°; the bar should also be oriented in such a way as to enable a clear view of the targets on the bar.

8.4.4.3 Effect of the Environment The bulk of the stations in metrology networks are wall targets in addition to other targets that may be attached to the structures being setup or aligned using the coordinating system. Horizontal and vertical refractions must be considered. The only real solution to the refraction is to try to keep the effects to a minimum through the design of the network. Most refraction effects can be eliminated by keeping the sight distances very short (<20 m) and by keeping the lines of sight away from a large bulky apparatus (e.g., large machinery, hanging fixtures, jutting walls), which may be a source of heat or processes that release large irregular quantities of heat energy into air. Other solution is to keep the temperature distribution constant within the work area (i.e., keeping doors and windows closed, switching off machinery).

8.4.5 Analysis of Three-Dimensional Traverse Surveys

Three-dimensional traverse surveys are different from the usual two-dimensional type in which (x, y) coordinates are determined. Apart from determining the (x, y) coordinates of points in three-dimensional traverse surveys, the elevations of those points are also determined. The total station equipment or the EDM and theodolite combination is commonly used in the modern three-dimensional traverse, which is a combination of trigonometric leveling and two-dimensional traverse methods.

In addition to predicting the quality of positioning in simulation or preanalysis, it is necessary to suggest quality assurance (QA)/quality control (QC) measures before embarking on three-dimensional traverse surveys. This entails determining discrepancies (between sets of angles, directions, zenith angles, and distances) used in field assessment and specifying what field "reductions" should be done before ending a station occupation (e.g., mark-to-mark for comparison of sets). Usually, the steps involved in three-dimensional traverse surveys are:

1. *Reconnaissance*: Identifying subject points; confirming locations of possible control points (from map and sketches, etc.); making final choice of intermediate or temporary points; and providing station sketches.

2. *Design and simulation*: Designing measurement processes (equipment, techniques, specifications for QA/QC); predicting the precision and accuracy of the expected results using appropriate network adjustment software, forming the basis for QA/QC.
3. *Equipment testing*: Testing the optical plummets and additive constants of total station equipment (a value for additive constant should be determined for each combination of total station and reflector); testing collimation errors of levels; rod constants; different offsets (for two rods used).
4. *Field observation*: Data gathering; QA/QC evidenced in field notes. The field notes are the only record of what activities have taken place in the field and are very important to those who would be interested in those field activities. Because of this, the field notes must be organized and formatted (demonstrating due care in QA/QC during the observations) so that those who may not have been in the field can understand the field notes.
5. *Data processing*: Postanalysis and verification; "reduction" or preprocessing of data; estimation of coordinates and elevations using appropriate software; statistical assessment of results; comparison of results with other methods.
6. *Reporting and presenting the deliverables.*

8.4.5.1 Observables in Three-Dimensional Traverse Surveys The observables commonly measured at each setup of total station equipment for three-dimensional traverse surveys are as follows:

- Heights of instruments and reflectors/targets above traverse stations
- Horizontal angles between backsight and foresight stations (if only two rays at a station are being measured) or horizontal directions (if three or more rays at a station are being measured)
- Zenith angles to targets
- Slope distances to target points, including the following:
  - Measurement of meteorological data to allow correcting the slope distances for systematic errors caused by the atmospheric conditions
  - Reduction of corrected slope distances to horizontal using average zenith angles or using elevation differences if available.

In a three-dimensional traverse, each successive station is occupied so that observations from "B" to "A" are done as well as those from "A" to "B" (Figure 8.4). Observations of directions and zenith angles are always done in at least two sets, with field checks through limits on the discrepancies between individual sets or between a set and the mean. The averages calculated from the acceptable sets are then used in subsequent calculations. In order to randomize certain errors, a fresh setup should be done before each set. Further, even with keeping the same tripod and tribrach setups at "A" and at "B," the heights of instrument and of reflector/target are not necessarily the same, especially with Wild or Leica style three foot-screw tribrachs; and

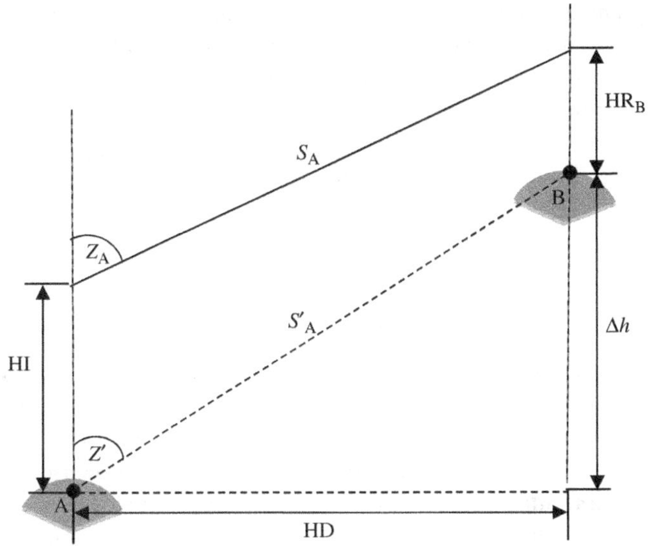

Figure 8.4 Example of three-dimensional traverse survey.

the meteorological conditions may be different at the different occupations of "A" and "B."

The slope distance, zenith angle, and elevation differences measured in an occupation from station A to B (forward) will be similar to corresponding values from B to A (backward) if these values are reduced to mark to mark. Otherwise, because of different heights of instrument and targets and different meteorological conditions, their corresponding values will be different and would have to be treated as two separate observables. The forward and backward mark-to-mark values can be used to check the accuracy of measurements and the average of the values will have its error reduced by a square root of two. If the distance is appreciably longer than 200 m, the effect of vertical refraction on zenith angle will become significant and will have to be considered.

For example, from Figure 8.4, while occupying station A sighting to B, z_{A} is the average zenith angle from $z_{\mathrm{A}i}$, each from one set of which there are n_{sA} sets; the standard deviation of z_{A} can be given as $\pm\sigma_{z_{\mathrm{A}}} = \pm\sigma_{z_{\mathrm{A}i}}/\sqrt{n_{\mathrm{sA}}}$. The limit on the discrepancy between any two sets will be $\sigma_{z_{\mathrm{A}i}}\sqrt{2}$. At the same time, $S_{\mathrm{A}i}$ is the slope distance, which is the average from several face left and face right observations associated with each set. The average of the $S_{\mathrm{A}i}$ is S_{A} with a standard deviation of $\pm\sigma_{S_{\mathrm{A}}}$. It should be noted, however, that this value will not be made smaller in the averaging, except if the atmospheric conditions are considerably different at the time of data acquisition for each set.

For each set, the height of instrument is $\mathrm{HI}_{\mathrm{A}i}$ and the height of reflector/target is $H\mathrm{R}_{\mathrm{B}i}$. In order to compare sets during an occupation, it is necessary to account for the variation in the HIs and HRs by "reducing" the $z_{\mathrm{A}i}$ and $S_{\mathrm{A}i}$ to "mark-to-mark" values, that is, to z'_{Ai} and S'_{Ai} in Figure 8.4 with the assumption that the slope distance and height difference, between the ground marks A and B, remain the same for each set. The mark-to-mark distance and the mark-to-mark zenith angle can be derived from Figure 8.4, as follows:

$$S' \sqrt{(S_A \sin z_A)^2 + (HI_A + S_A \times \cos z_A - HR_B)^2} \tag{8.17}$$

and

$$z'_A = 90 - \tan^{-1}\left(\frac{HI_A + (S_A \cos z_A) - HR_B}{S_A \sin z_A}\right) \tag{8.18}$$

In the occupation of B sighting to A, the circumstances are similar with $z_B \pm \sigma_{z_B}$ from n_{sB} sets and with $S_B \pm \sigma_{S_B}$ and with HI_{Bi} and HR_{Ai} based on the assumption that the slope distance and height difference between A and B remains the same as in the occupation of station A. The mark-to-mark values can be used in appropriate map projection formulas in order to derive the Easting and Northing coordinates for the horizontal component of the traverse.

8.4.5.2 Data Processing and Analysis With regard to network least squares adjustment of three-dimensional traverse surveys, the overconstrained adjustment is usually misleading, showing more error of observations since control points considered errorless are actually not errorless, so that positions are overprecise. In simulation, when two control points are fixed, more errors show up in measurements; and outlier detection is more difficult. The following are therefore possible in overconstrained adjustment:

- More errors show up in the residuals, making the measurement appears less precise.
- Computed coordinates are more precise.
- In adjustment, there will be more false outliers and high variance factor.

In minimal constrained adjustment, the following are possible:

- Errors in measurements are unbiased.
- Coordinates are less accurate (more errors in computations of coordinates due to errors in fixed point); because there are uncontrolled errors in measurements, there will be more errors in positions, making the positions less precise.
- In adjustment, measurement outliers will be appropriate.

In traverse surveys, horizontal angles, zenith angles, directions, and horizontal distances may need to be measured in order to determine coordinates of traverse points. Since errors are involved in each component measurement, there is usually a need to analyze the accuracy of the traverse surveys. In the design of expected standard deviation of measuring angles in a traverse, the expected maximum (at specific confidence level) misclosure of the traverse may be given. Equations (2.49), (2.50), and (2.52) can also be interpreted to mean the maximum allowable errors. In this case, $|\overline{x}_1 - \overline{x}_2|$ will be considered the maximum allowable error at the given confidence level $(1 - \alpha)$. Equation (2.50) or (2.52) can also be used to check misclosures of traverses in which angles and distances are measured. Let the coordinates of the last point k of the traverse be given as (x_k, y_k); and let the coordinates of this last point k calculated with the unadjusted measured angles and distances in the traverse be $(x_{k'}, y_{k'})$ with their propagated standard deviations as $(\sigma_{x_{k'}}, \sigma_{y_{k'}})$, respectively; using Equation (2.50), the following are obtained:

$$|x_k - x_{k'}| \leq \sigma_{x_{k'}} z_{\alpha/2} \tag{8.19}$$

$$|y_k - y_{k'}| \leq \sigma_{y_{k'}} z_{\alpha/2} \tag{8.20}$$

If Equations (8.19) and (8.20) are both satisfied, then the linear misclosures of the traverse are not significant at $(1 - \alpha)$ 100% confidence level. The above tests can be applied to traverses that close on the same point (loop traverse) or at both ends to different points of a higher order control network, for example, as in connecting traverses. If the traverse closes at both ends to different points of a higher order control network, the expected linear misclosure will be larger than that of a loop traverse because of the additional effect of relative errors of coordinates of the terminal points of the higher order control.

8.4.5.3 Effect of Correlation on Traverse Closure Sometimes it may be required to check the significance of a group of parameters that are likely to be correlated; for example, it may be required to check if two sets of coordinates (x and $\widehat{x}$) are statistically the same (where the elements of the vector $\widehat{x}$ are correlated or related to each other). In this case, it could be that one is testing for compatibility of estimated parameters ($\widehat{x}$) with existing independent estimates (x). This can be stated in another way: testing whether independently determined values x lie within a given confidence region about adjusted values $\widehat{x}$, which can be expressed mathematically as

$$y = (x - \widehat{x})^{\mathrm{T}} C_{\widehat{x}}^{-1} (x - \widehat{x}) \tag{8.21}$$

where $C_{\widehat{x}}$ is a fully populated covariance matrix of the adjusted coordinates of the network point(s) considered. For a given significance level α, x and $\widehat{x}$ may be assumed compatible (using modified version of Equation (2.52)) if $y < \chi^2_{u,\alpha}$ (for upper-tail areas in the case where the variance factor is known and u is the number of parameters being tested); for the case where the variance factor is unknown, the estimated a posteriori variance factor may be used. In this case, x and $\widehat{x}$ may be assumed compatible if $y < uF_{u,\mathrm{df}_2,\alpha}$ (for upper-tail areas, where df_2 is the number of degrees of freedom for determining the u unknown parameters).

For example, given the covariance matrix of the adjusted coordinates of a point as

$$C_{\widehat{x}} = \begin{bmatrix} \sigma_{\mathrm{E}}^2 & \sigma_{\mathrm{EN}} \\ \sigma_{\mathrm{NE}} & \sigma_{\mathrm{N}}^2 \end{bmatrix} \tag{8.22}$$

and letting the vector of coordinate differences be given as

$$x - \hat{x} = \begin{bmatrix} \Delta E \\ \Delta N \end{bmatrix} \tag{8.23}$$

Equation (8.21) can be rewritten as follows:

$$y = \begin{bmatrix} \Delta E & \Delta N \end{bmatrix} \begin{bmatrix} \sigma_{\mathrm{E}}^2 & \sigma_{\mathrm{EN}} \\ \sigma_{\mathrm{NE}} & \sigma_{\mathrm{N}}^2 \end{bmatrix}^{-1} \begin{bmatrix} \Delta E \\ \Delta N \end{bmatrix} \tag{8.24}$$

or

$$y = \begin{bmatrix} \Delta E & \Delta N \end{bmatrix} \frac{1}{(\sigma_{\mathrm{E}}^2\sigma_{\mathrm{N}}^2 - \sigma_{\mathrm{EN}}^2)} \begin{bmatrix} \sigma_{\mathrm{N}}^2 & -\sigma_{\mathrm{EN}} \\ -\sigma_{\mathrm{NE}} & \sigma_{\mathrm{E}}^2 \end{bmatrix} \begin{bmatrix} \Delta E \\ \Delta N \end{bmatrix} \tag{8.25}$$

or

$$y = \frac{\Delta N^2\sigma_{\mathrm{E}}^2 + \Delta E^2\sigma_{\mathrm{N}}^2 - 2\Delta E\Delta N\sigma_{\mathrm{EN}}}{(\sigma_{\mathrm{E}}^2\sigma_{\mathrm{N}}^2 - \sigma_{\mathrm{EN}}^2)} \tag{8.26}$$

Using Equation (8.24), or (8.25) or (8.26), for a given significance level α, x and $\hat{x}$ may be assumed compatible at 95% confidence level if y is less than $\chi^2_{2,0.05}$ (for upper-tail areas in the case where the variance factor is known and $u = 2$, the number of coordinate differences tested). These equations can be applied to each of the network points to check the compatibility of the two independent determinations of the coordinates of each point. Equation (8.24) or (8.25) or (8.26) can also be used to check if the misclosure of a traverse is acceptable at a particular confidence level by considering ΔE and ΔN as the misclosure of the traverse in Northing and Easting with $C_{\hat{x}}$ as the covariance matrix of the adjusted coordinates of the unclosed traverse point. If y is less than $\chi^2_{2,0.05}$, the misclosure is acceptable at the 95% confidence level.

8.5 THREE-DIMENSIONAL COORDINATION WITH LASER SYSTEMS

Two types of laser systems can be identified as three-dimensional coordinating systems: *airborne laser scanning system* and *terrestrial laser scanning system*. The characteristics of these systems as coordinating systems are discussed in this section. Further details on some aspects of terrestrial laser scanning system are discussed in Chapter 10.

8.5.1 Coordination with Airborne Laser Scanning System

The operational principle of airborne laser scanning system is based on that of *laser profiler*, which is a system that uses phase comparison and pulse echo methods of measuring distance from the airborne platform to the ground. The laser profiler, however, can only acquire elevation data over a single line crossing the terrain during an individual flight. Airborne laser scanning system is an upgrade of laser profiler with a scanning mechanism (rotating mirror or prism) added so that it can measure and map the topographic features of an area in detail instead of simply determining elevation values along a line in the terrain. The airborne laser scanning systems operate over ranges of several hundred meters to several kilometers from helicopters or fixed-wing airplanes. The system uses laser mounted beneath an airplane or helicopter to scan the ground by emitting tens of thousands of pulses per second as the airplane or helicopter follows a predetermined path, producing the LiDAR three-dimensional point cloud. In order to get measurements for the horizontal coordinates (x, y) and elevation (z) of the objects scanned, the aircraft position is determined with GNSS measurements and the distance measurements from the aircraft to the ground.

The airborne laser scanning systems, also known as LiDAR systems, consist of an airborne and ground segment. The airborne segment consists of *airborne platform*, *laser unit*, and *position and orientation system* (POS). The laser unit is to provide range (distance) information from the laser beam firing point to the ground point. The POS component consists of GNSS system to provide positional information and an inertial measurement unit (IMU) for attitude determination with the ground segment consisting of GPS reference stations, processing hardware and software for synchronization and registration, which is done off-line. During a laser scanning process, the time it takes each laser pulse to travel to the target and return to the aircraft is recorded along with the angle from nadir at which each pulse is emitted to produce the line-of-sight slant ranges referenced to the laser unit coordinate system. The POS will then store, for the entire session, the airborne GPS data (including carrier phase information) recorded at a rate of 1 Hz and the IMU attitude data of the aircraft at a rate of 50 Hz for the entire session. Each calculated slant distance is corrected for atmospheric conditions, and for roll, pitch, and yaw of the aircraft using the IMU data. GPS data is processed separately and imported into the LiDAR solution, and each corrected slant distance is transformed to a ground surface elevation.

The laser unit and the POS will sample the data independently; at the same time, on-ground GPS stations gather GPS data and GPS carrier phase data at known earth-fixed positions for later off-line computing of differential Global Positioning System (DGPS) positions of the airborne platform. Using DGPS and inertial data, the position of the laser scanner can be computed with centimeter to decimeter accuracy, and its orientation can be determined to better than about $40''$. The position and orientation data are stored as a function of the GPS time. As the laser scanner data are also stored with timestamps generated from the received GPS signal, the scanner and POS data sets can be synchronized. After synchronization, the laser vector for each sampled ground point can be directly transformed into an earth-fixed coordinate system, producing geocoded laser data. The modern LiDAR systems can also capture intensity images over the mapped area. Currently, registered laser scanner data with accuracy better than 10 cm in three-dimensional space are possible and the accuracy is primarily determined by the accuracy of POS.

8.5.1.1 Accuracy Analysis of Airborne Laser Scanning System Airborne laser scanning systems or airborne LiDAR systems are accepted for the acquisition of dense and accurate surface models over extended areas. Derived footprints from this system are not based on *redundant measurements*, making the LiDAR data and, consequently, the final products, less reliable. Moreover, the quality of surfaces derived from LiDAR data depends on the accuracy of the involved subsystems (laser, GNSS, and IMU) and the calibration parameters relating these components. The calibration process of LiDAR systems, however, is still not transparent and remains restricted to the system's manufacturer, so that the systems are usually viewed as black boxes (Brinkman and O'Neil, n.d.). In general, the LiDAR system manufacturers usually provide a range of expected accuracy of the derived point cloud. A typical horizontal accuracy is usually 1/2000th of the flying height and the vertical accuracy is between 15 and 35 cm depending on the flying height (Brinkman and O'Neil, n.d.). In this case, lower flying heights will provide a smaller laser spot size or footprint, allowing for more accurate data. Operating altitudes of LiDAR projects are generally 400–1200 m or up to 3000 m. The other rules of thumb relating to accuracy are that the slower the aircraft, the denser the spot spacing; the denser the spot spacing, the more reliable the digital terrain model (DTM); and the laser spots at nadir are more accurate than the spots at the outside edge of the swath or field of view.

Generally, when discussing the accuracy of airborne LiDAR data, the following should be considered (Brinkman and O'Neill, n.d.):

1. Total error for a LiDAR system is the contributing error budgets from each subsystem of LiDAR, such as laser ranger, GPS, IMU, and so on. Final accuracy of LiDAR data are, therefore, significantly affected by variation in quality of these subsystems. The laser ranger errors may be due to the distortion of the radiation path by the varying atmospheric conditions (introducing error of the laser pulse), pointing error of the laser, error in recording the scanner angle at the moment of each laser pulse; the GPS sources of error include satellite geometry, orbital biases, multipath, antenna phase center variations, integer resolution and atmospheric errors, and the effects of the operational distance from the ground GPS stations; and the IMU sources of error include typical small angular misalignments between the laser reference frame and the IMU reference frame, such as errors of pitch, roll and heading.
2. Since rigorous theoretical error analysis of LiDAR system is difficult or impossible to do, there is a possibility of wrongly interpreting what is meant by the accuracy of the LiDAR data.
3. The current method of accuracy analysis of LiDAR data tends to focus on vertical accuracy (z), and details on how planimetric accuracy (x, y) is verified are usually not clear.
4. Accuracies of LiDAR data and products will vary under different conditions across a project, such as in the areas of steep slope from the maximum angle of the scan to the minimum.
5. Geoid height model errors will impact final accuracy. Any vertical GPS error, such as geoid height modeling, will directly influence the accuracy of any LiDAR product.
6. Skill of personnel in project planning and execution will have impact on data accuracy and quality.

8.5.2 Coordination with Terrestrial Laser Scanning System

Terrestrial laser scanners are neither automated total stations nor digital cameras, but they are currently being accepted as surveying tools in surveying profession. Their acceptance may be due to the current development in the design of modern terrestrial laser scanners in which some of them now comply with the standards required of the geodetic surveying total station instruments. For example, some of the scanners are now equipped with typical geodetic devices such as leveling, centering, and orienting devices. However, the scanners do not place crosshairs on specific ground features in order to measure them; instead, they allow automated measurement and location of tens or hundreds or thousands of nonspecific points in the area surrounding the positions where the instruments are set up all within a very short time frame. When a reflective target is used, a terrestrial scanner only returns a cluster of responses from the target with a need to reduce the responses to a position estimate for the center of the target.

Compared with the surveyor's total station equipment, the terrestrial laser scanning techniques require only a relatively short time for data acquisition, which may be very important if there is a need to reduce the interruption time in the workplace to a minimum during the survey. In addition, scanning system will provide a permanent historical record of the raw data when saved digitally in computer disks. With this, remeasurement process can take place by using the record at a later stage if required.

Three raw observables that are measured by terrestrial laser scanners are slant range (based on pulse or phase-shift method as discussed in Chapters 5 and 10) and the two associated angles taken by the angular encoders in the horizontal and vertical planes (horizontal and vertical angles) passing through the center of the instrument. Some scanners, however, are capable of recording the intensity of the reflected laser beam at each object point as the fourth observable. These raw observables are simultaneously measured in a highly automated manner using a predetermined scan pattern often at a measuring rate of 1 kHz or more. The range measurements are usually made in uniform angular increments in both horizontal and vertical planes with their accuracy depending on the method of measurement, such as pulse or phase method.

The measured ranges and vertical (or zenith) and horizontal angles by the scanners are used to calculate positions of each returned laser signal in the scanner's internally defined coordinate system. This coordinate system (Figure 8.5) is defined (Lichti et al., 2002; Balis et al., 2004) as follows:

Origin: The electro-optical center of the scanner or the point of intersection of the horizontal and vertical rotation axes of scanner or the zero distance measurement point of the scanner.

z-axis: From the origin along the instrument vertical (rotation) axis

x-axis: From the origin along the instrument optical axis based on some arbitrary horizontal angle or a built-in magnetic compass direction

y-axis: Orthogonal to x–z plane in a right-handed system.

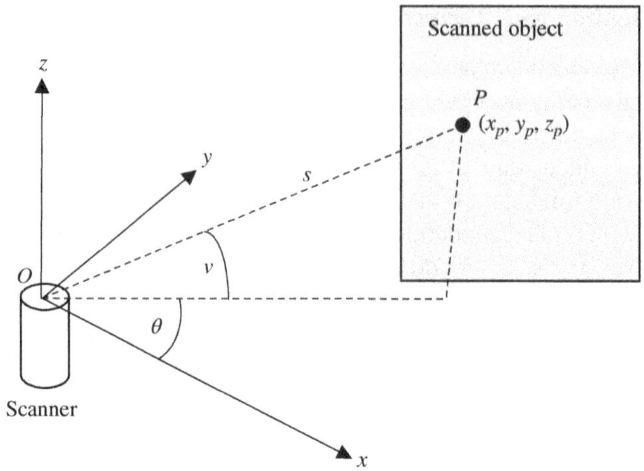

Figure 8.5 Coordinate system of a terrestrial laser scanner.

The relationship between the range (s), horizontal direction (θ) and vertical angle (v) and the coordinates of an object point P (x_p, y_p, z_p) in the instrument's internally defined coordinate system can be given as

$$\begin{aligned} x_p &= s \cos \theta \cos v \\ y_p &= s \sin \theta \cos v \\ z_p &= s \sin v \end{aligned} \tag{8.27}$$

The x, y, z output coordinates of several points constituting what is known as *point cloud* (or scan) are all referenced to the instrument's internally defined coordinate system. These Cartesian coordinates (x, y, z) in the scanner coordinate system are the quantities usually provided as output from most of the scanner software packages, and these coordinates are usually treated as observables instead of the measured quantities such as distances (s), vertical angles (v), and horizontal directions (θ). More details on the operation principle of terrestrial laser scanners can be found in Chapter 10 and in Lichti et al. (2002) and Lichti and Gordon (2004).

8.5.2.1 Georeferencing Problem Since coordinates in a point cloud (scan) are all referenced to the instrument's internally defined coordinate system, there is a need to georeference the coordinates to the ground coordinate system (X, Y, Z). The georeferencing problem involves transforming the point clouds (or scan data) from the scanner's internally defined coordinate system to the ground coordinate system. Two methods of georeferencing the scan data are *direct method* and *indirect direct* (Gordon and Lichti, 2004). The techniques used in direct georeferencing method are well known to surveyors, who are now able to integrate them with their traditional survey practice. The direct method is discussed further in this section while the indirect method is discussed later in Chapter 10.

In direct georeferencing method, a scanner is set up over a known point centered and leveled; its height over the point is measured; and its telescope is oriented toward another target (backsight) like a total station. In this case, the method requires that the scanner be equipped with leveling bubble, dual-axis compensator for precise leveling, optical plummet, a mark to which the instrument height is measured, and a telescope for backsighting to targets. The position and orientation information as well as the instrument height may be entered into the software before scanning or used later during the data processing (Gordon, 2005). This method is similar to reflectorless total station survey method with similar limitations, such as the uncertainty in the angular location of a range measurement due to finite diameter of propagated laser beam and the uncertainty due to the model for centroid-derived target pointing. Unlike with a total station, however, it is not possible to optically orient the telescopes of some terrestrial laser scanners toward known target points. With such scanners, the centers of structured targets are usually estimated using high-resolution scanning and centroid-estimation algorithm (Gordon and Lichti, 2004). In this case, the pointing error is governed by the uniform angular sampling interval (Δ), which is assumed equal in both the horizontal and vertical angle measurements.

A typical relationship between a vector of directly georeferenced ground coordinates of a point (P) and a vector of the corresponding scanner space coordinates of the same point can be given as follows:

$$\begin{bmatrix} X \\ Y \\ Z \end{bmatrix}_P^{\text{ground}} = \begin{bmatrix} X \\ Y \\ Z \end{bmatrix}_{\text{O}}^{\text{ground}} + \begin{bmatrix} \cos k & \sin k & 0 \\ -\sin k & \cos k & 0 \\ 0 & 0 & 1 \end{bmatrix} \begin{bmatrix} x \\ y \\ z \end{bmatrix}_P^{\text{scanner}} \tag{8.28}$$

where

$\begin{bmatrix} X \\ Y \\ Z \end{bmatrix}_P^{\text{ground}}$ is a vector of georeferenced object space coordinates of point P;

$\begin{bmatrix} X \\ Y \\ Z \end{bmatrix}_{\text{O}}^{\text{ground}}$ is a vector of object space coordinates of setup station O;

$\begin{bmatrix} x \\ y \\ z \end{bmatrix}_P^{\text{scanner}}$ is a vector of scanner space coordinates of point P; and

k is the derived azimuth from the setup station to the backsight station.

8.5.2.2 Accuracy Analysis of Terrestrial Laser Scanning System In order to determine the propagated variance–covariance matrix of a vector of directly georeferenced coordinates of any point P, variance–covariance propagation laws can be applied to Equation (8.28) with respect to the measured quantities, such as ranges, horizontal directions, vertical angles, derived azimuth, instrument setup coordinates, and their corresponding variance–covariance matrices. The possible sources of errors for each of the measured quantities are discussed as follows.

In determining the error budget for points in a scanned point cloud, the contributions of random errors due to *internal sources* (noise in the observations and beam width uncertainty) and *external sources* (instrument setup errors and errors due to survey points used for georeferencing) are considered. These error sources associated with direct georeferencing method are considered here since direct georeferencing is a more familiar approach to surveyors. Detailed descriptions of error budgeting for direct georeferencing, which can be found in Lichti and Gordon (2004), are summarized as follows:

1. Random errors in coordinates of the electro-optical center of the scanner are due to the variance–covariance of the coordinates of the scanner setup point with the variance in the z-component increased by the variance of measuring the instrument height with a tape.
2. Random errors in coordinates of the center of the backsight target are due to the variance–covariance of the coordinates of the target setup point with the variance in the z-component increased by the variance of measuring the target height with a tape.
3. Random error in the azimuth from the scanner setup point to the backsight target is obtained by the error propagation of the azimuth based on the coordinates of the scanner setup point and the backsight setup point.
4. Errors in horizontal direction measurement from the scanner setup point to the backsight target are propagated from the following:
   - Leveling errors of the scanner and target according to Section 4.5.3.
   - Centering errors of the scanner and target according to Section 4.5.4.
   - Pointing error to the backsight target with a telescope according to Section 4.5.1; if the centroid of the backsight target is determined by scanning the target at dense sampling interval (Δ), the pointing error can be replaced in this case by the error in determining the centroid (assumed to be equal in the horizontal and vertical directions) given as (Lichti and Gordon, 2004):

$$\sigma = \pm\frac{\Delta}{2\sqrt{3}} \tag{8.29}$$

5. Errors in vertical angle measurement from the scanner setup point to the backsight target are propagated from the following:
   - Leveling errors of the scanner and target according to Section 4.5.3 with the error in the vertical angle measurement being equal to the error in leveling the bubble given as σ_v in Section 4.5.3 (a fraction of the bubble sensitivity) or the compensator setting accuracy; Equation (8.29) can be used instead, if the center of the backsight target is determined by the scanning technique. For example, Leica ScanStation P20 has a dual-axis compensator setting accuracy (σ_v) of $\pm 1.5''$.

6. Errors in scanner measurement of the three observables (range, horizontal direction, and vertical angle), which are usually provided by the scanner manufacturer, are due to the following:
   - Errors in range measurements are similar to errors in distance measurements with EDM according to Section 5.6; the scanner manufacturer may provide the specifications for the error propagation. For example, in Leica ScanStation P20, the specified standard deviation for the range measurement up to 100 m is ±1.5 mm.
   - Errors in horizontal direction measurements are due to the manufacturer-specified variance for the horizontal angle measurement plus the effect of beam width uncertainty. The standard deviation of beam width uncertainty is given (Lichti and Gordon, 2004) as

     $$\sigma_{\mathrm{b}} = \pm\frac{\delta}{4} \tag{8.30}$$

     where δ is the diameter (angular units) of laser beam of circular cross section. The laser beam width is known to strongly influence both point cloud resolution and positional uncertainty since it determines both the uncertainty in the angular location of the point to which the range measurement is made and the spot size at different ranges from the instrument. For example, for Leica ScanStation P20, the beam divergence (δ) is quoted as 0.2 mrad (or $42''$), giving the uncertainty (σ_{b}) of $\pm 11''$; and the standard deviation of horizontal angle measurement is $\pm 8''$. The combined error in horizontal direction measurement will be $\pm 13.6''$.
   - Errors in vertical angle measurements are due to the manufacturer-specified variance for the vertical angle measurement plus the effect of beam width uncertainty given in Equation (8.30). For example, the specified error in vertical angle measurement for Leica ScanStation P20 is $\pm 8''$.

9

DEFORMATION MONITORING AND ANALYSIS: GEODETIC TECHNIQUES

OBJECTIVES

At the end of this chapter, you should be able to

1. Discuss the role of geodetic deformation monitoring and analysis
2. Discuss the characteristics of geodetic deformation monitoring techniques in contrast with other similar techniques
3. Discuss the important differences between absolute and relative geodetic networks and the importance of datum definition
4. Discuss the differences between deformation monitoring and control surveys
5. Use the design elements of deformation monitoring schemes to carry out deformation monitoring surveys
6. Describe the various monumentation and targeting requirements for deformation monitoring networks
7. Perform geodetic deformation monitoring surveys for hydroelectric dam structures and for subsidence areas
8. Reduce deformation monitoring data for input into least squares network adjustment software package for further processing
9. Explain the importance of single-point movement in absolute geodetic deformation monitoring networks

Precision Surveying: The Principles and Geomatics Practice, First Edition. John Olusegun Ogundare.

10. Explain the concept of the iterative weighted similarity transformation (IWST) and use it to solve the problem of datum instability
11. Discuss the differences between the observation-difference and coordinate-difference approaches in deformation analysis
12. Perform statistical and graphical trend analyses of deformations
13. Discuss the new developments in the automation of geodetic deformation monitoring of slope walls in open-pit mining
14. Discuss the geodetic techniques for deformation monitoring of tunnels during their construction
15. Discuss the use of geodetic leveling procedure in deducing tilt, strain, and curvature resulting from ground subsidence

9.1 INTRODUCTION

Deformation refers to changes in shape, dimension, and position of a deformable object. Deformations of objects are essentially three dimensional, but it is common to measure the horizontal and vertical deformations separately for better accuracy. In this book, vertical deformation of ground surface will be referred to as *ground subsidence*.

The most common parameters of a deformable object commonly monitored are deformation, strain, load, stress, ground water pressure, and so on; among them, surveyors are mainly interested in the deformation parameter. The goal of geodetic *deformation monitoring* is to determine changes in positions (or displacements) of points constituting the object being monitored. The techniques, although becoming less attractive, are still important since they produce absolute data and allow localized measuring devices, such as geotechnical instrumentation, to be connected together in a complementary way. Deformation monitoring involves periodic and probably automatic measurement of reference and object points in or around the active area in order to determine the deformation of those objects; in most cases, deformation is a continuous process affecting the whole object. *Deformation analysis* is about detecting, localizing, and modeling monitoring network point movements based on deformation measurements. Over the last several years, the role of deformation monitoring and analysis has significantly increased to include the following (Chrzanowski and Bazanowski, 2011; Chen, 2011):

1. Provide safety assurance against possible failure of the monitored object. This requires determining the deformations of the object and comparing the deformations with given tolerances. Since engineering companies are now held liable for the health of structures they create and maintain, it is important that they have accurate and timely information on the actual status of the structure for evaluating the safety of the structure so as to initiate necessary amendments to their initial designs.

2. Gaining better understanding of the mechanism of rock deformation through scientific experimentation and research. This requires correlating observed deformations with their causative factors in order to provide further knowledge for the future design of safer structures and in the case of mining areas, to provide better planning and safer operation in mines.
3. Verify behaviors of rock masses against their predicted patterns in order to refine the prediction models or validate design assumptions made with regard to the monitored object. Some parameters such as properties of soil or rock of a cut slope are often assumed at the design stage based on some limited field investigations. Results of monitoring during or after construction of the structures can help in validating such assumptions so as to be able to do remedial work if needed or to constitute the basis for future design.
4. Deriving information in order to resolve dispute on how the effects of mining impact on surface infrastructure and to help protect the infrastructure.
5. Deriving information for the purpose of identifying and separating various causes of deformation.

Deformation monitoring is one of the most important activities in engineering surveying. The number of objects requiring monitoring, such as dams, tunnels, high-rise buildings, bridges, industrial complexes, slopes, glaciers, and areas of landslide, subsidence, and recent crustal motion, in highly populated areas is increasing every day. These objects are subject to deformation as a result of many factors such as tidal effect, changing ground water level, mining activities, tectonic phenomena, landslide. The structures of a dam may undergo deformation due to a number of factors, including alkaline aggregate reaction expansion of concrete, instability of surrounding bedrock, changeable water load on the dam structures, seasonal thermal-induced deformations, and possible seismic events. A dam failure is also possible if an embankment dam is overflown beyond its spillway, requiring a high safety measure for the spillway to ensure that it is capable of containing a maximum flood stage. Vertical deformation of ground surface or ground subsidence may be due to a number of factors, such as mining activities; withdrawal of oil, gas, sulfur or other minerals, and excessive ground water withdrawal; effects of tectonic movements; long-term trend of permafrost compaction or frost heave; change in the sedimentation loading; earthquakes and other movements; and the instability of reference points. These movements may be difficult to detect over short time periods, but as they accumulate over long time periods, their effects may become significant enough to cause serious concerns. In densely populated areas, ground subsidence due to mining activities and the withdrawal of oil, gas, and salt or other minerals, and excessive ground water withdrawal are usually a major concern. In some areas, seasonal deformation of the active layer and long-term settlement of permafrost will be the main problem in separating ground subsidence due to gas withdrawal from the total surface deformation resulting from a combination of many factors (Chrzanowski and Szostak-Chrzanowski, 2010).

9.1.1 Characteristics of Geodetic Monitoring Techniques

In comparison with other monitoring techniques, such as high-definition surveying and remote sensing and geotechnical instrumentation techniques, the geodetic deformation monitoring techniques have the following characteristics:

1. They are based on a ground surface network of points interconnected by angle and/or distance measurements; they measure only the ground surface deformations.
2. They are usually conducted so as to provide sufficient redundant measurements for statistical evaluation of the quality of the measurements and for detection of errors in the measurements. This makes the techniques more reliable than the geotechnical/structural techniques. In data collection procedure, a campaign involves several locations as the stations in a network and requires a campaign "adjustment" to obtain least squares estimates and statistical assessment of the observations.
3. They provide overall picture of deformation trend of the whole object being monitored and of the surroundings, with respect to some stable reference points.
4. They require skillful observers and data analysts and are labor-intensive and are not done frequently, except when they are operated in a fully automated mode.
5. Instruments involved (i.e., robotic total station (RTS) equipment and GPS) can be automated to provide continuous information of behavior of the monitored structures, but can be more expensive to install and operate compared with geotechnical/structural instruments.
6. They require intervisibility between observing stations and are affected by the environment such as atmospheric refractions, effect of thermal expansion on the measuring equipment and on the monitored structure, possible influence of the changeable water level of the reservoir of a dam, tropospheric delay, and instability of geodetic reference network stations.

Geodetic deformation monitoring starts with the establishment of monitoring networks. The geodetic monitoring networks can be put into two classes (Chrzanowski et al., 1986): *absolute geodetic networks* and *relative geodetic networks*. An absolute geodetic network has some of its network points that are not likely to move over time and some that are subject to movement or are within the deformable object. Those network points that are not subject to movement are usually outside the area of influence of deformation and those points constitute the *reference network* or *reference datum* for least squares adjustment and deformation analysis. The points that are subject to movement are usually the points that are being monitored and are known as *object points*. Absolute deformation is described with respect to the reference datum that is assumed to be stable. Since the stability is not so easy to determine a priori, it should be ensured that there are sufficient number and suitable distribution of reference points that their relative stability can be assessed as part of the monitoring process. The relative network, however, has all its network points located within

TABLE 9.1 Summary of the Traditional Geodetic Technologies Used in Deformation Monitoring.

Technology	Accuracy	Advantages	Limitations
1. Robotic total stations (RTS)	• Angle measurements can be better than 1″ • Precision of single pointing at distances 400–1500 m in harsh conditions: 3″ horizontal, and 4″ vertical • Distance measurements can be better than 1 mm ± 1 ppm	• Provides 3D positions in almost real time • Used as automatic deformation monitoring system (can operate continuously in time domain and can communicate data to remote station)	• Can only measure discrete points • Affected by atmospheric refraction • Limited by the ATR resolution and range when used in automatic mode for direction measurements
2. Precise leveling with precision level or automatic levels with parallel-plate micrometer (Wild N3 precision level)	• For special-order: ±3 mm $\sqrt{L}$ where L (in km) is one-way distance between benchmarks; leveling accuracy of ±0.2 mm/1 km double run	• Capable of high precision • Reliable	• Affected by atmospheric refraction • Slow (maximum of 5 km/day) and labor-intensive with survey crew of 3 • Provides only 1D information
3. GNSS positioning (GPS, GLONASS, Galileo, Compass, etc.) with or without pseudolites	• Can provide millimeter accuracy in relative positioning (2 mm horizontal and 4 mm vertical) • Can monitor slow deformation in campaign mode and fast or dynamic deformation in RTK mode	• Provides 3D positions in almost real time • Line of sight between ground points not required	• Limited by satellite visibility • Major source of error is multipath • More tropospheric delay with elevation difference greater than 100 m between antenna locations • Uneconomical if large number of points are monitored • Requires up to 12 h per session for vertical component

the area of influence of deformation, and all the network points are subject to movement with no stable points to be used as reference datum. The relative monitoring network will allow the detection of strain components derived from relative displacements, differential rotation, and relative rigid body movements, while the absolute network can detect absolute movements of object points relative to the stable reference points, in addition to what can be detected in relative monitoring networks. A list of some of the advanced geodetic technologies used in deformation monitoring is given in Table 9.1 (Chrzanowski, 2009; Chrzanowski and Chrzanowski, 2012; Leica, 2006).

In Table 9.1, pseudolites (or pseudo-satellites) are ground-based emitters of GPS signals, which can be used to complement GPS measurements where there is limited visibility of satellites. With regard to GNSS positioning technique, it should be noted that the longer the length of sessions, the better the solution; 12 h of observations will give reasonable accuracies in the horizontal and the vertical with the error in vertical component being about twice that of the horizontal component (Chrzanowski and Chrzanowski, 2012). It can also be understood from the table that all of the geodetic technologies are affected by atmospheric refraction and/or tropospheric delay and not all are suitable for fully automated and continuous monitoring.

9.1.2 Deformation Monitoring and Control Surveys

Geodetic deformation monitoring must be distinguished from *geodetic control surveys*. In geodetic control surveys, the absolute positions (coordinates) of points are of interest and common systematic errors due to the effects of constant refraction, calibration error, scale error, and configuration defects are physically removed or randomized while they are expected to cancel out in deformation surveys if they are the same in all epochs of observations. The requirement for absolute scale of the geodetic control network is not necessary in networks established for the monitoring of deformations; what is more important is the ability to detect and control a change in scale between measurement epochs. Configuration defects such as eccentricities of instruments with respect to targets, triangular misclosures are permitted in deformation surveys, but not in geodetic positioning surveys. Generally, geodetic deformation monitoring encourages large correlation between repeated observations of the same observable, while geodetic positioning survey does not, but instead attempts to randomize the effects of all sources of errors. In order to obtain a strong correlation and thus the highest possible accuracy in the displacement calculation, observations should be made in the same environment and observation conditions, and the same observables, observer, and instruments should be used in all epochs of observations.

9.1.3 Geodetic Monitoring Measurements and Error Sources

Monitoring tasks and deformation analysis present some of the most important challenges in the surveying industry today because they require higher accuracy of measurements, maximum reliability of measuring instruments, ability to automate measuring system, and high flexibility of computation and analysis tools. Geodetic measurements are usually considered as contaminated with the following effects (Chrzanowski and Secord, 1987):

- Observation random errors
- Systematic errors due to inconsistencies in the instrument construction (axial) errors and the atmospheric conditions such as atmospheric refraction or tropospheric delay in the case where GPS technology is used
- Seasonal (thermal) cyclic expansions of the measured objects
- Other systematic errors arising from lack of proper calibration of instruments (especially, distance measuring equipment).

The observation random errors are caused due to reading, pointing, centering, and leveling of the instrument. The reading error is nonexistent in electronic instruments except for residual graduation errors with the suggested reading error, for example, for Kern and Leica instruments being 0.5″ based on laboratory tests. Pointing of instrument is very critical in distance measurements since changes in the return signal strength may introduce a bias in phase measurements for distance determination. Refer to Chapters 2, 4–6 for full discussion on the sources of systematic and random errors and their treatment. It should be mentioned that the index error in electronic theodolites equipped with two-axis liquid compensators is directly affected by temperature variations, necessitating that observations be made in different telescope positions in the shortest possible time interval. Moreover, in order to increase the usefulness of geodetic surveys for the detection of systematic deformations, calculated displacements must be corrected for thermal expansion of the structures (after correcting for the atmospheric effects on the measurements) and for a possible influence of the changeable water level, in the case where dam structures are monitored. In this case, the cyclic effect due to those effects must be separated from the systematic deformations, which are of main interest.

9.2 GEODETIC DEFORMATION MONITORING SCHEMES AND THE DESIGN APPROACH

Deformation monitoring scheme is an elaborate and systematic plan of action to be followed in monitoring deformation of an object. The scheme systematically identifies and arranges all of the interrelated elements needed in successfully detecting deformations. These elements include making choices about the type and locations of observables; timing of measurement campaign, determining the stability of reference points; selecting monitoring techniques, suitable instrumentation and type of monumentation and targeting (for geodetic monitoring); identifying the data processing and analysis techniques; and determining the actual deformations. Since deformations that are to be detected are usually within the margin of measurement errors, it is required that the scheme be carefully designed.

Some of the main criteria for the design of deformation monitoring schemes are accuracy, reliability, temporal and spatial continuity, stability of reference points, cost-effectiveness, and choice of monitoring technology. The criteria of accuracy, reliability, cost-effectiveness, and choice of monitoring technology were discussed in Chapter 7. *Temporal and spatial continuity* both depend on the type of rock materials in the area being investigated. Temporal continuity has to do with the frequency of monitoring an object and the spatial continuity is about whether sufficient number and location of discrete monitoring points are achieved. For example, the process of ground subsidence in viscous rock (such as salt and potash) is slow so that temporal continuity of the monitoring surveys above the salt and potash extraction will not be critical.

Spatial continuity design criterion is a requirement for appropriate distribution of reference and object points. This criterion requires that the network of discrete points be as dense as possible and the sensors or the monitoring points be located where maximum or critical deformations are expected. The selection of site points are also based on field reconnaissance with the criteria of accessibility to the sites and good visibility to GPS satellites (if GPS monitoring survey is being considered). A reference datum erroneously assumed stable will give a biased displacement pattern that can be misinterpreted as monitoring results. Unstable reference points must be identified prior to data acquisition stage based on the knowledge of boundaries of deformation zone or during data processing using appropriate algorithm.

In the total effort of deformation monitoring, the quality of the analysis of the behavior of the object being monitored depends on the location, frequency, type, and reliability of the data gathered. Since the object of interest is being monitored at discrete object points, a campaign of observations must be done within a short time to ensure that all of the points are being observed while in the same state; the points must be known to remain unchanged in position during the campaign. In both the horizontal and vertical observation schemes, the duration of the campaign must not exceed the interval within which the observations would all remain among the same points; the typical duration should not be more than 1 week, depending on the rate of movement taking place. The amount of movement to be detected must be predicted and the desired accuracy of measurements must be better with the measuring instruments to be chosen to satisfy the accuracy requirements. It should be mentioned that with regard to dam monitoring, there are no universally accepted standards and specifications for the choice of monitoring schemes; monitoring schemes are usually designed based on individual guidelines or those provided by the International Commission of Large Dams (ICOLD) (Avella, 1993). According to Chrzanowski et al. (1992), the accuracy of monitoring both horizontal and vertical displacements in concrete dams should be around 1–2 mm; and for embankment dams, the accuracy should be approximately 10 mm for horizontal displacements, 5–10 mm for settlements during construction, and 5 mm for horizontal and 3–5 mm for vertical displacements, during normal operation of the dams. Sample geodetic specifications for dam monitoring by the New Zealand Electric Corporation (ECNZ) are given (Avella, 1993) in Table 9.2.

After the initial design of the monitoring schemes, they must be revisited and enhanced time to time with regard to the following:

TABLE 9.2 Geodetic Observables and Their Specifications for Dam Monitoring.

Observables	Recommended Accuracy
Horizontal observation	Standard deviation of mean direction/angle measurement should be $\leq \pm 1.5''$
Vertical angles	Standard deviation of mean angle should be $\leq \pm 2.0''$
Height by vertical angle	Accuracy of final height should be $\leq \pm 5.0$ mm
Distances	All distances are to be accurate to within ± 3.0 mm
Precise leveling	• Maximum difference between pairs of reading between two consecutive marks should be $\leq \pm 0.7$ mm • Maximum difference between forward and backward runs between benchmarks should be $\leq \pm 3\,\text{mm}\sqrt{\text{km}}$ • For concrete structures, maximum difference between two consecutive marks should be $\leq \pm 0.3$ mm • Precise leveling is carried out to $\pm 1.0\,\text{mm}\sqrt{\text{km}}$
Optical plumbing	• Accuracy of final results should be $\leq \pm 3$ mm
Crack or joint movement	• For cracks or joints with markers <500 mm apart, a measurement accuracy of $\leq \pm 0.2$ mm is required
Offsets	• Accuracy of observations should be $\leq \pm 2.0$ mm

- Configuration (or geometry) of reference network stations and the object points
- Types of observable, depending on the angular and linear relationships among the reference network stations and object points
- Timing of campaigns, including duration and appropriate sequencing of observations
- Accuracy of measurements, which depends on the situation of the reference network stations about the structure and the choice of observables and their measurement accuracies
- Economy, which depends on the choice of observables and the procedures and instrumentation necessary to ensure the measurement accuracies.

The original design of the monitoring schemes may be altered if network stations or points are damaged and lost between observation epochs, new stations or points are to be installed to strengthen the network, new instruments with better accuracy are available, or there is a need for observation schemes to be reduced to cut costs without compromising the integrity of monitoring. As the monitoring schemes and network points increase with time, it may become necessary to perform optimization analysis of the existing network in order to improve overall accuracy in the detection of displacements of object points. This may require adding the measurement of different types of observables (e.g., measuring direction observable with high precision if the existing network is a trilateration network) and deleting possible redundant measurements from the existing scheme. It may also become necessary to modify the frequency of measuring the networks or abandon some aspects of the monitoring schemes, depending on the situations of the monitored objects. Surveys, however, must be repeated at intervals necessary to detect short periodic or long-term variations in the rate of deformation.

A typical geodetic deformation monitoring scheme for a hydroelectric dam will consist of the following elements:

1. Horizontal angle or horizontal direction measurements.
2. Distance measurements (short distances of 1–2 km in length are usually involved), which are reduced to mark to mark and all meteorological and calibration reductions applied. The reduced distances are then projected onto a horizontal plane using station elevations obtained in the area leveling before they are adjusted by the method of least squares.
3. Zenith angle measurements (usually around 90–95°); it should be noted that accuracy of reducing slope distances using zenith angles is inferior to using elevations.
4. Orthometric height difference measurements (usually from differential leveling).
5. Forced-centering monuments (with only few tripod setups) are commonly used.
6. Network points are in such a way that the reference points are in stable regions and are close to the dam, both upstream and downstream, outside the influence zone of the structure; the points must be intervisible and accessible all year round. The minimum reference points should be four. The reference points are used to intersect the object points.
7. A reference point should be a deeply anchored round double-walled concrete pillar with a forced-centering device and must be well protected from the sun and vandals. The stability of the points (observation pillars) must be checked by resection from close (relocation points) and distant targets; each reference point should have clear sights to at least four relocation marks that are at a close range for checking the reference point movements.
8. The object points are spread on the crest, inside the galleries, and on the base of the upstream and downstream parts of the dam structure, and they are usually targets with concentric circles, typically installed in grid pattern (in rows (horizontally) and columns (vertically)). The object points can be pillars (with forced-centering system), brackets (with forced-centering system), and bolts (with forced-centering system). Electromagnetic distance measurement (EDM) reflectors must be able to be fitted into the object points. Object points located inside the galleries (of concrete dams) can be connected to the exterior geodetic network to provide absolute deformation information, except if the points have been referenced to some stable points in the foundation or in the abutments.
9. Settlement of the dam is either monitored by leveling runs across the crest and along the base of the dam, or less often, by zenith angles from the reference points (pillars) to the object points (targets on the dam).

10. Optical alignment on the crest is not considered suitable for highest precision because of huge refraction problems usually experienced when measuring along the crest. Refraction problems on the crest are due to the lines of sight being close to the ground or structures and the effects of the blending of the upstream or downstream winds over the crest and the strong temperature gradients associated with it.
11. The *X*, *Y* plane coordinates determined from the adjustment are based on a local reference coordinate system with the axes usually defined as follows:
    - The origin is assigned assumed coordinates, such as 1000 m N, 1000 m E.
    - *X*-axis is parallel to the longitudinal centerline of the dam units with the *X*-axis increasing easterly or to the right-looking upstream.
    - Positive *Y*-axis is directed upstream through the dam units.
    - The *Z*-axis is along the direction of gravity with the origin as the mean sea level; the elevations above mean sea level are taken as the *Z*-coordinates without concern for any geodetic complications of curvature or nonorthogonality.
12. Instead of the local reference coordinate system defined above, the map projection coordinate system with a translated origin can also be used to create a local coordinate system.

With regard to distance measurements, two-way distances among the network points should be measured with about 15 measurements each way. The standard deviations associated with these measurements are computed and those measurements that are different from their mean value by twice the standard deviation are eliminated; the remaining measurements are corrected for the effects of the atmosphere and height differences in order to obtain mark-to-mark reductions. Although the computed standard deviations are used to eliminate inconsistent measurements, they are not used in the least squares adjustment; the manufacturer-specified standard deviation for the equipment is used for each distance observation. Each angle or round of directions must be done in several sets, usually at least three sets. The circle readings and consequent micrometer readings are to be sampled at various positions of the horizontal circle with the instrument releveled before each set. A standard deviation is associated with each mean of the sets and used in least squares adjustment of the overall network. Direction measurements to every visible reference station and object point and distances to every other reference station should be observed. In order to obtain height differences from differential geometric leveling, precision levels and invar rods are used among the points of interest. The inverse of the number of setups can be used in relative weighting of the measured height differences.

In order to account for seasonal variations in the behavior of a monitored object, the geodetic surveys should be performed more often throughout the year. For example, overall geodetic network (including precision leveling of reference network) of an hydroelectric generating station should be measured once per year; subnetwork for the Powerhouse/Intake structures of the station should be measured four times a year with the leveling of various levels of the Powerhouse done as often as activity permits; and the main dam/slope indicator stations should be measured twice a year. The campaign should be repeated using the same observation scheme and procedures and at the same times each year, especially in concurrence with the activity of the reservoir. This should continue for at least 2 years so as to provide sufficient number of repeated campaigns for evaluating the consistency of the geotechnical measurements (if available) and of their compatibility with the geodetic surveys. After data analysis from the 2 years of campaigns and a possible advancement in instrumentation and data processing, the overall monitoring scheme can be assessed and revised for further enhancement.

The current trend in dam deformation monitoring, however, is to integrate various geotechnical/structural and geodetic surveys techniques into integrated monitoring scheme (Chrzanowski and Secord, 1987). The initial network configuration for the dam monitoring described above may need some modifications with time; network stations and points may have to be located so that they can be interconnected among themselves; some stations (such as the slope indicator stations) may not require setting up instruments on them; and inverted pendulums may be included as stations of the network rather than concrete pillars. If inverted pendulums are included, the pendulums may need to be anchored to 30 m within the bedrock, at geometrically suitable locations within the network (Chrzanowski and Secord, 1985). If the inverted pendulum is situated within the structure being monitored, then the pendulum can also be used to determine the horizontal movement of the structure apart from serving as a stable reference point.

9.3 MONUMENTATION AND TARGETING

In deformation studies, types of monuments and targets to be used will depend on the level of accuracy of the monitoring survey and the location of the monitored object. Two different monument design philosophies commonly followed are as follows:

- Points forming the reference network must be designed to minimize the effects of local movements; the points must be durable and stable.
- Object points must be designed to be able to reveal what local movements are actually taking place.

These philosophies require that one be familiar with the characteristics of the site or the object being monitored, including construction constraints such as location, rock and soil types, and other information that may help to determine the durability and stability of the reference network points. In this case, the expertise of soil and geotechnical specialists is required prior to constructing monitoring survey monuments.

Consistent repeatability of centering is very important in monumentation and targeting of a monitored structure. The location of the object points on the monitored structure must be in such a way as to allow a better accuracy for centering and easier connection to the geotechnical observables, if available. The reference network and object points can be designed to allow instrument setup, target setup, or both. These instrument and target setup points are usually pillars made of concrete materials that are installed into exposed bedrock to a certain depth (usually between 1 and 2 m depending on the nature of the site) below the surface of the ground. After installation, dry shrinkage of the pillars (affecting only the heights of the pillars) will take place for a long period of time with the rate of shrinkage decreasing with time. According to P.R. Zwart (unpublished), a dry shrinkage in the order of less than 1.0 mm is possible if high-quality aggregates such as quartz, limestone, or granite are used with the water–cement ratio kept to 0.5 or less. Pillar installation is expected to be completed in approximately 60–90 days before the first measurement campaign is done in order to allow for curing and the initial dry shrinkage of the concrete (Rohde, 1991).

The instrument and target centering devices on monitoring survey monuments are usually of forced-centering types so that setup errors can be eliminated. A typical example of such devices is Wild tribrach centering system with the quoted accuracy of the ball and socket arrangement being ±0.1 mm or less (Deumlich, 1980). The selection of an appropriate centering mechanism, however, will depend on survey accuracy specifications, available instrumentation, and the type of survey control being established. A typical reference control pillar for geodetic monitoring of a dam is shown in Figure 9.1, where the extensometer anchor on the pillar is for stability test of the pillar. The schematic design of a typical dam monitoring instrument pillar installed in a bedrock, which is based on the US Army Corps of Engineers (2012), Rohde (1991), and the personal investigation of the author, is shown in Figure 9.2. In the figure, the white polyurethane foam pads, usually 50 mm thick, are cut to fit and wrapped around the pillar to reduce the effects of thermal expansion and contraction during a survey campaign according to J.H. Chrzanowski (personal communication).

(a) (b) (c)

Figure 9.1 Typical reference control pillar (showing extensometer anchor) for geodetic monitoring: (a) GPS unit setup, (b) top of survey pillar, and (c) whole length of survey pillar.

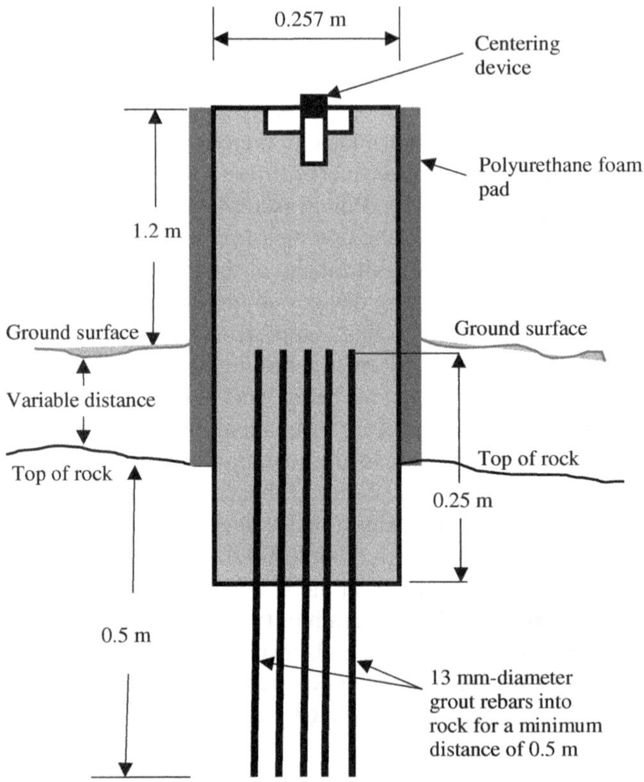

Figure 9.2 Typical dam monitoring instrument pillar design.

It is believed that the centering accuracy of better than ±0.3 mm can be maintained with this approach.

It should be mentioned that a thermal expansion of pillar material due to differential heating of the pillar can cause lateral shift of the top of the pillar depending on the coefficient of expansion of concrete and the temperature difference between the two sides of the pillar. Generally, the movements of some of the reference pillars will not be considered dangerous as long as, at least, two pillars in the reference network are identified as stable during the process of determining the displacements of the object points on the monitored structures. One of the means of clarifying local movements (i.e., movements of pillars with respect to bedrock or movements of pillars caused by bedrock) is illustrated in Figure 9.3(a). In the figure, the extensometer anchor point on a reference control pillar and the survey markers on the other two monitoring pillars (Monitor 1 and Monitor 2) are measured with an extensometer; a typical monitoring pillar with a survey marker is shown in Figure 9.3(b). The extensometer measurement of the network is used to perform the control pillar stability test.

With regard to Figure 9.3(a), the following steps are usually taken (J.H. Chrzanowski, personal communication) in performing the control pillar stability test:

(a)

(b)

Figure 9.3 (a) Two monitoring pillars (Monitor 1 and Monitor 2) for stability test of another pillar (control pillar). (b) A monitoring pillar with a survey marker (e.g., Monitor 1).

- Establish a minimum of two monitoring concrete monuments (Monitor 1 and Monitor 2 in Figure 9.3(a)) within a few meters (e.g., 10 m away) from the reference control pillar and test their relative stability by measuring the short distances and angles from the reference pillar to the two monitoring monuments. Note that the monitoring concrete monuments shown in Figure 9.3(a), in practice, are to be flush with the ground.
- Measure the distances among the reference pillar and the two monuments either with Kern invar wire distometer (with an accuracy of ±0.05 mm) or with tape extensometer.
- Measure the vertical and horizontal angles from the reference pillar to the two monitoring monuments using precision theodolite.
- Perform the measurement procedure every month for 1 year and determine the possible relative movements by performing least squares adjustment of the measurements for the determination of positions of the reference pillar and the monitoring monuments. Iterative weighted similarity transformation (IWST) described in Section 9.4.3.5 can be performed to determine the relative displacements of the pillars.
- Once the unstable pillars are identified, they are not included in the final determination of displacements of the object points.

In the case where a reference monument is to be installed in the soil (rather than in the rock), the US Army Corps of Engineers (2012) suggest that 1.2 m by 1.2 m concrete footing with 0.6 m thickness be constructed below the frost line so that a 10-cm-diameter concrete monument (similar to that in Figure 9.2) is attached to it with five 13-mm-diameter rebars. It is suggested that at least 50 cm of the length of each rebar be embedded both in the footing and the monument.

9.3.1 Dam Slope and Crest Monuments and Targets

The design of dam slope and crest monuments is difficult due to the steepness and composition of the downstream rockfill shell. The shell is composed of placed rock ranging in size from large cobble to small boulders with sizable voids. Survey requirement is to monitor the surface and near-surface movements (1.5–1.8 m depth) of the downstream slope. The observed movement must relate to the actual motion of the downstream slope, and the monuments must be designed to be able to accommodate all standard targets and prism holders. Accurate repeatability of centering in all three (X, Y, Z) coordinates must also be ensured for targets. Since targets are likely to be rotated to be viewed from several different pillar locations, eccentricity about the vertical axis must be minimized by using a specially designed circular spirit leveling device.

The concept behind designing a dam slope monument is to bond the rebar and pipe with the sides of the core hole and surrounding rock with concrete, thus providing good stability and a more representative indication of local surface and near-surface movements. The crest monument design is essentially the same as the slope monument. However, because the crest of the dam also serves as part of the access road to the dam, it is necessary to set the monuments flush with the surface of the road. A typical dam crest monument is shown in Figure 9.4. In this monument, a brass disk embedded in concrete with centering mark may be used as reference point by occupying the site with a heavy-duty tripod, thereby using the tripod as survey marker. With a well-adjusted optical plummet, survey instrument set on the tripod can be centered with an error smaller than 1 mm.

Suitable target plates are to be used for slope and crest monuments. Wild concentric circle insert target can be used for the ball and socket centering device installed in the instrument pillars. These targets are designed for line-of-sight distances up to about 300 m; this design assembly appears to favor the horizontal pointing accuracy over the vertical by a factor of 2 (Rohde, 1991). Specially designed targets are required for instrument pillars, the types that are omnidirectional (360° target) so as to be able to accommodate all line-of-sight distances ranging from 70 to 400 m.

9.3.2 Monuments for Subsidence Monitoring in Mining Area

Some of the monuments used in mining subsidence monitoring (Figure 9.5) are drilled to between 1.5 and 3.4 m depth depending on the nature of the soil (Chrzanowski and Bazanowski, 2011). Typical monuments are made of drilled-in 4″

Figure 9.4 A typical dam crest monument installation.

Figure 9.5 Typical leveling markers used in subsidence monitoring surveys.

pipes with survey markers welded to the inner surfaces of the pipes (Figure 9.5(d)). The design depth of the installation is usually below the freezing level, which is about 1.4 m.

9.4 HORIZONTAL DEFORMATION MONITORING AND ANALYSIS

9.4.1 Monitoring Techniques

The traditional geodetic techniques for monitoring horizontal deformation of an object are based on the use of terrestrial positioning with total stations, theodolites, and EDM, and the use of space-borne GPS survey techniques augmented with GLobal Orbiting NAvigation Satellite System (GLONASS). A typical GPS survey of a mining area may require simultaneous use of up to six or more geodetic grade receivers/antennas in static relative positioning mode with the data rate set at 10 s. Figure 9.6 shows the typical GPS antenna setups on monitoring points in a mining area. A reference monument, preferably a high-precision GPS control network point (Figure 9.6), will be considered fixed for the network adjustment, and the horizontal coordinates are commonly provided in the appropriate map projection grid coordinate system, such as Universal Transverse Mercator (UTM).

In GPS survey, the measurement procedure is a repetitive one in which the tripod/tribrach and GPS antenna are centered on a monitoring point and the slant antenna height from the survey point to the marked edge of the antenna is measured; in the case of the reference pillar, the antenna height is measured from the top of the pillar to the marked edge of the antenna. If the visibility to GPS satellites is poor in the area where the GPS survey is being conducted, a total station traverse

(a) (b)

Figure 9.6 Geodetic grade GPS unit setup to monitor subsidence-induced horizontal displacements in a mining area: GPS unit setup on a (a) tripod over a monitoring point and (b) high-precision pillar.

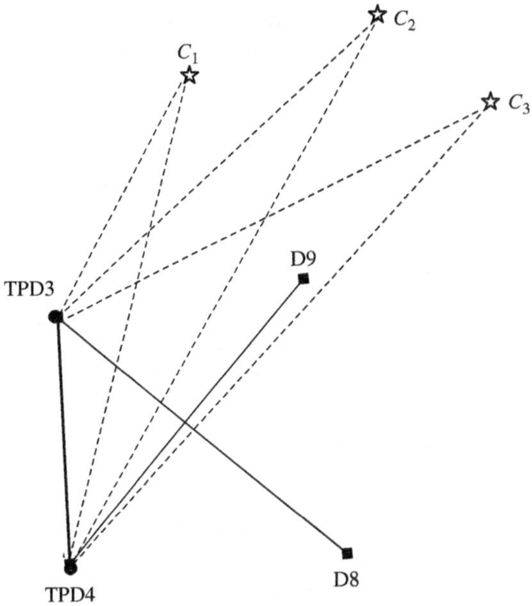

Figure 9.7 Simple total station subnetwork traverse controlled by GPS control points C_1, C_2, and C_3 in three-baseline surveys.

survey subnet connecting to the main project network can be created; and if the high-precision pillar being used is stable, it can be considered as a fixed reference point for GPS three-baseline surveys. An example of total station subnetwork traverse controlled by GPS control points C_1, C_2, and C_3 in three-baseline surveys with dotted lines as measured GPS baselines is shown in Figure 9.7.

A typical procedure for total station traverse and the GPS three-baseline surveys of a subnetwork illustrated in Figure 9.7 is as follows (A. Chrzanowski and M. Bazanowski, personal communication):

- Set up two local GPS points TPD3 and TPD4 in suitable locations and determine the positions of the points in relation to three GPS reference points C_1, C_2, and C_3 containing geodetic grade GPS antennas, which are continuously running in the overall project network. The coordinates of points TPD3 and TPD4 will be used to control the traverse subnetwork (providing azimuth and translation for the subnetwork).
- Measure the included angles at the local GPS points TPD3 and TPD4 and the distance TPD3-TPD4 and perform an open traverse to connect the points to other points (e.g., D8 and D9) using RTS with automatic target recognition (ATR) capability.
- In Figure 9.7, it is assumed that points TPD3 and TPD4 are temporary points located where there is visibility to the sky to allow for GPS positioning of the

points; and points D8 and D9 are traverse points located in some vegetated areas with no visibility to the sky for GPS measurements.

- Forced-centering procedure with tribrachs left on tripods each time is to be followed during the traverse.

In order to determine the positions of D8 and D9 in Figure 9.7, for example, the following steps can be taken:

1. Set up the RTS on TPD3 (with reflectors set up on points TPD4, D8, and D9) and make two series of direction and distance measurements. Each series, consisting of Group A and Group B measurements, must be completed in order to randomize measurement errors. The measurement steps are given as follows:

Series I, Group A measurement steps:

(i) While the total station is set up on TPD3, train the ATR device to automatically locate points TPD4, D8, D9 in that order in the first half of set 1 and complete the set measurements; at the end of set 1 measurements, input temperature, pressure, and humidity (to be associated with the set measurements at the processing stage) into the automatic data recorder; the observables that are automatically measured and recorded by the RTS data recorder are the horizontal direction (HZ), zenith angle (Z), and slope distance (SD).

(ii) Add two more sets of measurements to the data file (while inputting to the data recorder, the temperature, pressure, and humidity at the end of each set) to complete three sets of measurements to each of the three points; this completes series I, Group A measurements.

(iii) Compute the standard deviations of measurements for the Group.

(iv) Start series I, Group B measurement steps as follows.

Series I, Group B measurement steps:

(i) Repeat steps (i)–(iii) for another three sets of measurements, giving a total of six sets of measurements to complete series I, Group B measurements.

(ii) Compare the standard deviations of measurements in Group A and Group B in this series for consistency; if the computed corresponding standard deviations of measurements for the two groups in this series are not consistent, another group of measurements must be made until consistency is achieved.

(iii) Start series II as follows.

Series II

- While the instrument is still set up on TPD3, repeat series I but now change the order of ATR pointing and measurements to TPD4, D9, and D8. At the end of this series, the total number of sets of measurements should be 12 (combining 6 sets of measurements from series I with 6 sets from series II).

2. Switch the RTS from TPD3 and the reflector from TPD4 (without moving the tribrachs from their tripods for forced-centering procedure) and repeat step 1 above.
3. After completing the total station measurements in steps 1 and 2, set the GPS antennas on TPD3 and TPD4 and leave the antennas to make continuous measurements for at least 6 h relative to the three GPS reference points (C_1, C_2, and C_3) in the overall project network. Postprocess the measurements and determine the positions of TPD3 and TPD4.
4. Use the computed coordinates of TPD3 and TPD4 in step 3 with their covariance matrices and the total station measurements in steps 1 and 2 and their estimated standard deviations, in a combined least squares adjustment to determine the adjusted coordinates of D8 and D9 in the overall project network.

9.4.2 Observables and Data Preprocessing

The typical geodetic observables in deformation monitoring are slope distances, horizontal angles, directions, zenith (or vertical) angles, and height differences. After the measurement of the observables, the measurements must be preprocessed before they are used in network adjustment by the method of least squares. In monitoring dam structures for horizontal displacements, *distance observables* are usually measured in direct and reverse directions in trilateration networks using precise EDM, such as DM502 (with manufacturer specified accuracy of 0.005 m $\pm$ 5 ppm), Kern Mekometer ME3000 (with manufacturer-specified accuracy of 0.0003 m $\pm$ 3 ppm), Tellurometer MA200 (with manufacturer-specified accuracy of $\pm$0.3 mm $\pm$ 2 ppm), or ComRad Geomensor 204DME (with manufacturer-specified accuracy of 0.0001 m $\pm$ 0.1 ppm). With the high precision of EDM, they can also be used as electronic extensometers to complement the geotechnical extensometers commonly used in monitoring deformations of dam structures. In using EDM for deformation monitoring of dam structures, the following procedures are recommended (Chrzanowski and Secord, 1985):

- Calibrating the equipment on the calibration baseline established near the structures.
- Measuring two-way distance observables with 5 measurements for each of the three electronic pointings taken each way, giving a total of 15 measurements each way.
- Computing the mean and the standard deviation of the 15 distance measurements and prescreening the measurement using $\pm 2\sigma$ (with σ as the standard deviation of measurement) as the tolerance limit between any two measurements.
- Using height differences based on the values deduced from differential leveling for distance reduction.

- Correcting reduced distances for atmospheric conditions at the time of measurements and for the effect of differences in elevation between station pairs involved and producing *corrected mark-to-mark* distances. The mark-to-mark distance is a useful distance that is independent of instrument and reflector heights and is usually the output of GPS processing; a corrected EDM distance can be reduced to mark to mark using the following:

$$d_{\mathrm{m}} = \frac{d}{\sqrt{1 + \frac{(\mathrm{HI}-\mathrm{HR})^2}{d_{\mathrm{m}}^2} - \frac{2(\mathrm{HI}-\mathrm{HR})\cos Z_{\mathrm{m}}}{d_{\mathrm{m}}}}} \tag{9.1}$$

  where d_{m} is the mark-to-mark distance, d is a curved EDM distance between the instrument setup point and the reflector (after velocity corrections), HI is the height of instrument, HR is the height of reflector, and Z_{m} is the mark-to-mark zenith angle. Since d_{m} is on both sides of Equation (9.1) and is also needed in computing Z_{m} later, there will be a need for rigorous solution for d_{m} and Z_{m} by performing several iterations. In the case where elevations are available instead of zenith distances, a simple correction to be applied to measured distance (d) in order to obtain the mark-to-mark distance can be given as follows (assuming the EDM is collinear with the theodolite and the reflector and target are the same):

$$C_{\mathrm{d}} = \frac{(H_2 - H_1) \times (h_2 - h_1)}{d} + \frac{d(h_1 + h_2)}{2R} \tag{9.2}$$

  where H_2 is the elevation of the high-point marker; H_1 is the elevation of the low-point marker; h_2 is the height of the reflector/instrument at the high point; h_1 is the height of the reflector/instrument at the low point; and R is the radius of the spherical earth.
- Performing further reduction of corrected mark-to-mark-distances to a reference ellipsoid and to a mapping plane if computations are to be done on a mapping plane.
- Assigning to each reduced distance a variance based on the manufacturer-specified standard deviation of the equipment used in the measurement process.

In measuring *angle and direction observables* for deformation purpose, the commonly recommended instruments are the optical theodolites such as DKM2-A (for angles), DKM3 (for angles and directions); or the electronic theodolites, such as Kern E2 electronic theodolite (capable of direction measurement to an accuracy of 0.7″). If angle or direction observables are measured, then each angle or round of directions must be measured in at least three sets, with releveling between sets, and sampling the circle readings and consequent micrometer readings at various positions of the horizontal circle (Chrzanowski and Secord, 1985).

When observing horizontal directions in multiple sets, if one or more sets show a systematic trend in the computed discrepancies between the reduced directions and the averaged values, then the effect of horizontal refraction may be considered as a suspect. The only solution to minimizing the effect of refraction is by selecting lines of sight away from heat sources and to randomize the effect by performing observations at different times under varying atmospheric conditions.

It is reported (Vanicek and Krakiwsky, 1986) that the effect of refraction in the vertical direction is at least one order of magnitude larger than in the horizontal direction. Vertical temperature and thus density gradients are substantially greater than horizontal gradients; therefore, the curvature of the ray path is much greater in the vertical than in the horizontal direction. In correcting for the effect of vertical refraction on zenith angle measurements, the observed zenith angle is smaller than the actual value and the computed height difference is too large since the telescope direction is sighted to a point higher than the point actually being observed. The air is warmer than the ground and the coefficient of refraction k is positive (for positive gradient); the coefficient of refraction (k) can be included in the parametric model as unknown at each station. The variation of coefficient of refraction (k), however, from station to station is complex, depending on the following:

- The azimuth and length of the observed line
- The time of day when the observation is acquired
- Topography or terrain profile of the line of sight
- Variation of vegetation along the line of sight
- The height of the line of sight above the ground
- Atmospheric conditions at the time of observation.

On the basis of this list, it is commonly suggested that angular observations be made quickly over a short period of time to avoid rapidly changing refraction field, which may result in inconsistent data.

In trigonometric leveling or three-dimensional triangulation networks, it may be necessary to reduce zenith angles to their mark-to-mark equivalent due to differences in the heights of instruments and targets. This is necessary since the theodolite height could not be reset to the same height with each epoch, so that this effect is not cancelled out by comparing two epochs of measurements. The associated correction is also known as the *eye-to-object correction* (Clark, 1973). The mark-to-mark reduction (C_{zm}) for zenith angle can be approximated as

$$C_{zm} = \sin^{-1}\left(\frac{\Delta h \sin Z}{d}\right) \tag{9.3}$$

or

$$C_{zm} = \sin^{-1}\left(\frac{\Delta h \sin^2 Z}{D}\right) \tag{9.4}$$

or

$$C_{zm} = \frac{\Delta h \sin Z}{d_m} \tag{9.5}$$

where Δh is the difference between the height of instrument (HI) and height of target (HT) given as (HT − HI); Z is the observed zenith angle; d is the measured slope distance corrected for meteorological effects; D is the reduced horizontal distance; and d_{m} is the mark-to-mark distance that can be determined iteratively with Equation (9.1). Equation (9.5) is approximate, but valid if the height difference of target and instrument is less than 0.5 m. The reduced zenith angle (Z') can be expressed as

$$Z' = Z + C_{\mathrm{zm}} \tag{9.6}$$

Generally, before zenith angle observations are used in least squares adjustment method, it is necessary that they be reduced first to their mark-to-mark equivalent. For the purpose of network adjustment, a vertical datum is established by fixing Z coordinate of a station, which serves as the origin. In this case, the vertical datum is a horizontal plane tangent to the equipotential surface with a local vertical passing through the origin. At other stations, the local verticals are not parallel to the vertical at the origin point and the zenith angle measurements at those stations will be smaller than what would be expected if all the verticals were parallel. In order to account for the nonparallelism of the local verticals at the observing stations, appropriate corrections must be applied to the zenith angle measurements. The correction (C_{z}) to be added to the corresponding zenith angle measurements can be calculated (Rohde, 1991) as

$$C_{\mathrm{z}} = \frac{\Delta x \sin\alpha + \Delta y \cos\alpha}{R} \tag{9.7}$$

where Δx and Δy are the coordinate differences between the observation network station and the fixed station considered as the origin, and α is the azimuth of the line connecting the observation station to the origin. Since the ellipsoid of revolution is not being used as a mathematical approximation of the earth, deflections of the vertical are not required. If the vertical datum is not being imposed by arbitrarily fixing the Z coordinate at the origin, the resulting heights will be orthometric heights based on the geoid as the datum.

If zenith (vertical) angles are measured for determining height differences, the zenith (vertical) angles must first be corrected for the effects of *earth curvature*, instrument and target eccentricities (or mark-to-mark correction), and the effects of refraction before they are used in computing the height differences. *Earth curvature correction* is applied since in the absence of refraction, the line of sight follows a horizontal line and not the expected curved line of the level surface. The height difference between forward and backward points separated by distance d is corrected for earth curvature by adding the following earth curvature correction C_{c} to it to cancel the effect:

$$C_{\mathrm{c}} = \frac{d^2}{2R} \tag{9.8}$$

where R is the mean radius of the earth; and d and R do not need to be precisely known. It is important that data collector be used during geodetic observations, as the data collector is used to automatically perform several quality checks that aid in avoiding blunders and obtaining reliable data. ATR, if available in the survey instrument, should also be used to free the operator from the time-consuming and repetitive task of accurately pointing the instrument to the survey target. ATR is better than bisecting using crosshairs with the human eye, so that the angular measuring accuracy of the instrument is not compromised.

9.4.3 Monitoring-Data Processing Techniques

In determining the deformation of an object, the geodetic monitoring data (distances, angles, directions, etc.) of an object are collected at discrete points of the object over certain epochs of time. These data, however, must be transformed to horizontal displacements of those points between epochs of time, which are more useful as a measure of deformation of the object. Two ways of doing this transformation are by *two-epoch* (or *coordinate differencing*) approach and *observation differencing* approach. In each approach, the concepts of least squares parametric model adjustment are employed. The two-epoch approach consists of least squares adjustment of *single-epoch measurements* performed in two separate epochs (one adjustment for each epoch) with their results compared later to determine possible deformation between the two epochs.

9.4.3.1 Least Squares Adjustment of Single-Epoch Measurements Adjustment of single-epoch monitoring data is performed with the purpose of determining the coordinates of points representing the monitored object at a given time epoch. The monitoring data (observations) for that epoch can be expressed in the form of least squares parametric model as follows:

$$\hat{\ell} = f(\hat{x}) \tag{9.9}$$

where $\hat{\ell}$ is a vector of adjusted monitoring data (observations) and $\hat{x}$ is a vector of adjusted coordinates of network points. Since coordinates of network points are of interest, a geodetic datum (or Cartesian reference frame) must also be defined in order to solve for the coordinates. This is done by specifying the values for the datum elements, such as origin, orientation, and scale of the Cartesian reference frame for the network. From the defined reference frame (computational base), the approximate coordinates (x^0) of all the other network points are computed using suitably selected measurements from the first or reference epoch. This same set of approximate coordinates is then used for the least squares estimation of coordinates for each of the subsequent epochs of measurements. These approximate coordinates define the coordinate system and serve as the Taylor point for the linearization of Equation (9.9) as follows:

$$\ell + V = \ell^0 + A\delta \tag{9.10}$$

or

$$V = w + A\delta \tag{9.11}$$

where V is a vector of observation residuals; ℓ is a vector of monitoring data (observations) for the given epoch; ℓ^0 is a vector of approximate values of monitoring data for the given epoch, calculated from the approximate coordinates (x^0) of the network points; A is the first design matrix or the configuration matrix; δ is a vector of corrections to be applied to the approximate network coordinates; and $w = (\ell^0 - \ell)$ is a vector of misclosures. The configuration matrix A depends on the geodetic observations (such as distances, angles, directions), which define the *internal network geometry* and the coordinates and elevations of the network points, which constitute the *external network geometry*. Equations (9.9)–(9.11) are formulated for pairs of epochs to be evaluated. The linearized Equation (9.11) for each epoch is adjusted separately by the method of least squares and analyzed in pairs in the two-epoch approach. The least squares adjustment solution of the linearized Equation (9.11) gives the vector of unknown corrections (δ) to the approximate values of the unknown coordinates (x^0) for each epoch as

$$\delta = -(A^{\mathrm{T}}PA)^{-1}A^{\mathrm{T}}Pw \tag{9.12}$$

where P is a weight matrix formed by inversing the variances of the observations (usually a diagonal matrix if the observations are assumed uncorrelated). The use of weight matrix is an indication that the quality of monitoring data is important and must be known in order to avoid misinterpreting possible systematic errors or outliers in the observations as deformation issue (Chen, 1983). The adjusted coordinates ($\hat{x}$) for each epoch can be given as

$$\hat{x} = x^0 + \delta \tag{9.13}$$

and the covariance matrix of the adjusted coordinates can be given for each epoch as

$$C_{\hat{x}} = s_0^2 Q_{\hat{x}} \tag{9.14}$$

where $Q_{\hat{x}} = (A^{\mathrm{T}}PA)^{-1}$ is the cofactor matrix of the adjusted coordinates; s_0^2 is the variance factor of observation of unit weight, which can be calculated for each epoch from

$$s_0^2 = \frac{V^{\mathrm{T}}PV}{n-u} \tag{9.15}$$

V is the vector of residuals (corrections to observations); n is the number of equations, including parametric and constraint equations; and u is the number of unknown parameters, including the number of unknown coordinates of the network points to be determined and the number of nuisance parameters, such as orientation parameters and scale factor changes. More details about the least squares adjustment procedure and statistical analysis steps used in this book can be found in Ogundare (2012).

It is assumed in the foregoing adjustment procedure that the datum defect has been effectively taken care of. In deformation networks, however, the least squares estimate of unknown parameters cannot be obtained from the solution given in Equation (9.12) without a datum being defined since the normal equation coefficient matrix ($A^{\mathrm{T}}PA$)

will be singular or rank deficient. The datum (or rank) will be deficient because the coordinate datum is not completely defined by the observations. The classical solution to overcoming the rank deficiencies is to adequately define the network datum through the addition of absolute or weight constraints on the unknown coordinates of the network points. When the coordinates of a network point are fixed (assumed errorless), the point is said to be *absolutely constrained*; when the coordinates of the point are assigned some precisions, the point is said to have *weight constraint*. Overcoming rank deficiencies, however, depends on the number of datum elements to be defined in an adjustment; the datum elements are those parameters that constitute what is known as the *datum of the adjustment model*. Generally, a datum will be defined, for example, for two-dimensional geodetic networks, if four datum elements are known, such as two coordinates, one scale, and one orientation (or azimuth of a line). For three-dimensional geodetic networks, seven datum elements must be determined, such as three coordinates, three rotations, and one scale. The number of datum elements to be defined in an adjustment also depends on the type of measurements available since certain types of measurement will implicitly define datum elements. For example, a distance measurement included in a network will provide scale to that network; gyrotheodolite azimuth will provide orientation; and a point-positioning GPS measurement will provide position in terms of needed coordinates. If more datum elements are added than are necessary to remove the rank deficiency, then the network is said to be *overconstrained*.

Network adjustment that incorporates a minimal amount of information necessary to define a datum so that a unique coordinate solution is obtained is called *minimal constraint* (*minimal datum constraints*) or *free network* adjustment (Leick, 1982). Such a network is considered free in the sense that its *geometrical size and shape* is determined while remaining essentially independent of the reference coordinate system (or datum). Since a minimal constraint network adjustment must have the coordinates of one of the network points fixed, there is usually a problem of how to choose the point to fix. This is particularly important in deformation analysis since arbitrarily fixing a point will lead to arbitrary estimates of the solution and the associated precisions. If the coordinates of the network points are fixed in defining the datum, then the network is said to be *externally constrained*. Constraint equations (or datum elements) can also be added to remove the rank deficiency by the so called *free-network* adjustment, usually associated with minimal constraints where the center of gravity of the network is fixed; this is referred to as *inner constraint* adjustment. In this study, free network adjustment will be used to mean the same thing as inner constraint adjustment.

9.4.3.2 Free Network Adjustment Model Free network (inner constraint) adjustment model provides a means of solving rank-deficient systems through the imposition of particular set of minimal constraints that do not limit the freedom of the network to translate, rotate, or change size. This method of adjustment has been widely used and is reputed to remove the problem of datum definition. Without the tool of free network adjustment, movements of the datum points could not be detected or might lead to erroneous conclusions. This adjustment method imposes the following constraints on the adjustment:

- There will be no change in the coordinates of the centroid after the adjustment (the translations in the x and y coordinate axes are zero, for example, translations $\delta X_G = 0$ and $\delta Y_G = 0$). This means that the center of gravity (G) of the network (the centroid) is fixed.
- The average bearing from the centroid to each other point remains unchanged, that is, no differential change in rotation of the network.
- Average distance from the centroid to each other point remains unchanged, that is, no differential change in scale of the network.

Generally, the inner constraints state that the initial coordinate values assigned to each of the network points at the start of the iterative least square solution define the datum and as such, the solution is affected by any change in those initial coordinates.

In free network adjustment, two types of models are created: the model that relates observations to coordinates (parametric model) and the model that constrains the parameters to allow the solution of the parametric model (constraint model). The constraint model equations are most of the time based on the datum definition. For free two-dimensional network adjustment (inner constraint case), the maximum number of datum defects possible will be four; this will result in creating four constraint equations to define the datum. The constraint model will consists of the following four general constraint equations:

1. Equations that will define the origin (or translations from this origin) of the coordinate system. The usually chosen equations for *inner constraint adjustment* impose the positional (or translational) constraints on the centroid (with coordinates X_G, Y_G) of the network by specifying that changes in those coordinates of the centroid remain zero after the adjustment, that is, $\delta X_G = 0$ and $\delta Y_G = 0$. The centroid of the network can be determined as

$$X_G = \frac{\sum_{i=1}^{m} X_i}{m} \quad Y_G = \frac{\sum_{i=1}^{m} Y_i}{m} \tag{9.16}$$

where (X_i, Y_i) [for $i = 1, 2, \ldots, m$] are the approximate coordinates of the network points. In order to satisfy the constraint condition that $\delta X_G = 0$ and $\delta Y_G = 0$, the partial derivatives of equations in Equation (9.16) are done with respect to the coordinates of the network points to be used for defining the datum (this could be a subset of the network points or the whole network points) and the derivatives set equal to zero. The following two constraint equations (9.17) and (9.18) are then obtained:

$$\sum_{i=1}^{n} \delta x_i = 0 \quad \text{(for network } x\text{-translation)} \tag{9.17}$$

$$\sum_{i=1}^{n} \delta y_i = 0 \quad \text{(for network } y\text{-translation)} \tag{9.18}$$

2. Equation that will define the orientations (or rotations) of the coordinate system. The equation imposes rotational constraint by specifying that the average bearing from the centroid to each of the network datum point must not change after the adjustment. This will require that the sum of all the changes in bearings from all the datum network points to the centroid must be equal to zero. This is done by calculating bearings from the centroid to the datum network points (using the tangent function and the coordinates of the centroid and the corresponding points), finding their partial derivatives, and setting the sum of those derivatives to zero. The following constraint equation is then obtained.

$$\sum_{i=1}^{n} (y_i \delta x_i - x_i \delta y_i) = 0 \quad \text{(for network rotation)} \tag{9.19}$$

3. Equation that will define the scale (or provide the idea of distance) of the coordinate system.

   The equation imposes scalar constraint by specifying that the average distance from the centroid to each of the network datum point must not change after the adjustment. This requires the sum of changes in distances from the centroid to the datum network points must be equal to zero. In this case, the distance equations from the datum network points to the centroid are formed, and the sum of partial derivatives of those distances is set equal to zero. The following constraint equation is then obtained.

$$\sum_{i=1}^{n} (x_i \delta x_i + y_i \delta y_i) = 0 \quad \text{(for network scale)} \tag{9.20}$$

Equations (9.17)–(9.20) are the constraints mathematically expressing that the network is invariant in its shape with respect to small differential translation, rotation, and scale change. This means that the adjustment preserves the shape of the network defined by the observations. Equations (9.17)–(9.20) can be expanded and presented in matrix form as follows for horizontal network where there are no fixed point, no azimuth, and no scale (with the number of datum deficiencies equal to four):

$$G^{\mathrm{T}}\delta = \begin{bmatrix} 1 & 0 & 1 & 0 & \ldots & 1 & 0 \\ 0 & 1 & 0 & 1 & \ldots & 0 & 1 \\ y_1 & -x_1 & y_2 & -x_2 & \ldots & y_n & -x_n \\ x_1 & y_1 & x_2 & y_2 & \ldots & x_n & y_n \end{bmatrix} \begin{bmatrix} \delta x_1 \\ \delta y_1 \\ \delta x_2 \\ \delta y_2 \\ \vdots \\ \delta x_n \\ \delta y_n \end{bmatrix} = 0 \tag{9.21}$$

where

$$G^{\mathrm{T}} = \begin{bmatrix} 1 & 0 & 1 & 0 & \dots & 1 & 0 \\ 0 & 1 & 0 & 1 & \dots & 0 & 1 \\ y_1 & -x_1 & y_2 & -x_2 & \dots & y_n & -x_n \\ x_1 & y_1 & x_2 & y_2 & \dots & x_n & y_n \end{bmatrix}; \tag{9.22}$$

$$\delta = \begin{bmatrix} \delta x_1 & \delta y_1 & \delta x_2 & \delta y_2 & \cdots & \delta x_n & \delta y_n \end{bmatrix}^{\mathrm{T}}; \tag{9.23}$$

$$x_i = X_i - X_G \quad y_i = Y_i - Y_G \quad (\text{for } i = 1, 2, \ \dots \ , m) \tag{9.24}$$

The coordinates used in the G^{T} matrix in Equation (9.22) can be reduced to the centroid as shown in Equation (9.24) to reduce rounding error. However, it can be shown that the original network coordinates (unreduced to the centroid) can also be used directly. The solutions obtained in both cases will be identical.

In free-network adjustment, the following properties are also satisfied:

$$|\delta| = \delta^{\mathrm{T}}\delta = \min \tag{9.25}$$

and

$$\operatorname{trace}(Q_{\hat{x}}) = \min \tag{9.26}$$

where δ is the vector of corrections to the approximate values of the unknown parameters (as given in Equation (9.23)), and $Q_{\hat{x}}$ is cofactor matrix of the adjusted parameters. Equation (9.26) is satisfied by the free network adjustment since the adjustment minimizes the trace or a subtrace of the cofactor matrix of the adjusted parameters in order to arrive at a "best" precision. The resulting estimated parameter vector $\hat{x}$ from the adjustment based on the above constraints is referred to as the best linear unbiased estimates (BLUE) for the unknown parameter x.

The solution vector for the free network adjustment model can be expressed as

$$\delta = -Q_{\hat{x}}(A^{\mathrm{T}}Pw) \tag{9.27}$$

where A is the first design matrix from the parametric model equations relating observations and unknown parameters, P is the weight matrix of observations, w is the misclosure vector, and $Q_{\hat{x}}$ is the pseudo-inverse, which can be given (Ogundare, 2012) as

$$Q_{\hat{x}} = (N + GG^{\mathrm{T}})^{-1}N(N + GG^{\mathrm{T}})^{-1} \tag{9.28}$$

with $N = A^{\mathrm{T}}PA$, and G is given in Equation (9.22). The solution vector in Equation (9.27) can be modified if nuisance parameters (δ_2) are involved; in this case, we want to eliminate the effect of the nuisance parameters before the inversion is done. The solution vector can be modified as follows:

$$\text{Let}: \begin{bmatrix} \delta_1 \\ \delta_2 \end{bmatrix} \quad A_1 = \frac{\partial f}{\partial x}, \quad A_2 = \frac{\partial f}{\partial z}$$

where δ_1 is a vector of coordinate corrections, δ_2 is a vector of nuisance parameters, A_1 and A_2 are the first design matrices with respect to the unknown coordinates (x) and the nuisance parameters (z), respectively. The pseudo-inverse (Equation (9.28)) is modified as follows (Ogundare, 2012):

$$Q_{\hat{x}}^* = (N^* + GG^{\mathrm{T}})^{-1} N^* (N^* + GG^{\mathrm{T}})^{-1} \tag{9.29}$$

where

$$N^* = A_1^{\mathrm{T}} P^* A_1 \tag{9.30}$$

$$P^* = P\left[I - A_2\left(A_2^{\mathrm{T}} P A_2\right)^{-1} A_2^{\mathrm{T}} P\right] \tag{9.31}$$

The modified solution vector of corrections to the approximate coordinates (x_1^0) is given as

$$\delta_1 = -Q_{\hat{x}}^*(A_1^{\mathrm{T}} P^* w) \tag{9.32}$$

where w is the misclosure vector. The nuisance parameters can be calculated as follows:

$$\delta_2 = -(A_2^{\mathrm{T}} P A_2)^{-1}(A_2^{\mathrm{T}} P A_1 \delta_1 + A_2^{\mathrm{T}} P w) \tag{9.33}$$

The residual vector of the observations can be given as

$$v = A_1\delta_1 + A_2\delta_2 + w \tag{9.34}$$

The covariance matrix of the solution vector can be given as

$$C_{\hat{x}} = s_0^2 Q_{\hat{x}}^* \tag{9.35}$$

where s_0^2 is the a posteriori variance factor of unit weight expressed as

$$s_0^2 = \frac{v^{\mathrm{T}} P v}{n + d - u} \tag{9.36}$$

where n is the number of observations, d is the number of datum deficiencies (number of parameters to fix in order to define the datum), and u is the number of unknown parameters in the network adjustment. The a posteriori variance factor provides a global measure of the precision of the observations.

9.4.3.3 Statistical Analysis of Single-Epoch Measurements Since the magnitude of deformation to be detected is sometimes in the order of observation accuracy, the statistical analysis of each epoch of measurements is critical. The statistical analysis plays several important roles in the processing and analysis of deformation survey data, such as the following:

- Assessment of observation quality to decide whether to include the observation or not in the adjustment. This is done through the least squares adjustment blunder detection algorithm.
- Assignment of relative weights in deformation analysis; ensuring that variance factors computed for pairs of epochs are compatible so that the comparison of epochs is not biased. If appropriate relative weights are not assigned to the measurements, they will affect the estimated deformation parameters.

Before determining the adjusted coordinates of network points, the covariance matrix of the adjusted coordinates, and the a posteriori variance factor of unit weight for the monitored points, the appropriate statistical tests must be performed to detect and eliminate possible outliers from the measurements. Detection of outliers in each of the single-epoch adjustments is very important since any undetected errors will likely be assessed as deformations later in the analysis. Generally, in order to detect outlier (gross) error (due to systematic components) in geodetic measurements, the number of observations must be approximately twice the number of unknown coordinates.

The outlier detection is based on the outcomes of *global* and *local* tests and must be based on minimum constraint least squares adjustment. In global (or Chi-square) test, the computed (a posteriori) variance factor of unit weight, s_0^2, is tested statistically against the given (a priori) variance factor, $\sigma_0^2 = 1$ as follows:

$$\frac{(\mathrm{df}_i) \times s_{0i}^2}{\chi^2_{\text{larger}}} < \sigma_0^2 < \frac{(\mathrm{df}_i) \times s_{0i}^2}{\chi^2_{\text{smaller}}} \quad \text{or} \quad \frac{(\mathrm{df}_i) \times s_{0i}^2}{\chi^2_{\frac{\alpha}{2},\mathrm{df}_i}} < \sigma_0^2 < \frac{(\mathrm{df}_i) \times s_{0i}^2}{\chi^2_{1-\frac{\alpha}{2},\mathrm{df}_i}} \tag{9.37}$$

where subscript i represents an epoch i, α is the chosen significance level for the test (usually, $\alpha = 0.05$, using lower tail areas of Chi-square table), $\mathrm{df}_i = n_i - u_i$ is the number of degrees of freedom for the adjustment at epoch i, s_{0i}^2 is the computed variance factor being tested at epoch i, χ^2_{larger} is the larger (upper) value of the Chi-square value extracted from the statistical table, n_i is the number of observations (or least squares equations), u_i is the number of unknown parameters (including nuisance parameters), and $\chi^2_{\frac{\alpha}{2},\mathrm{df}_i}$, $\chi^2_{1-\frac{\alpha}{2},\mathrm{df}_i}$ are the values from the Chi-square table where α is the upper area value.

The local test can be based on the tau method by Pope (1976) or on the data snooping technique by Baarda (1968). In the Pope's method, the a priori variance factor, σ_0^2, is assumed unknown so that the ratio of the residual of observation to its standard error (standardized residual) is distributed as *Tau*, τ, which is expressed as follows:

$$\tau_i = \frac{|V_i|}{s_{V_i}} \tag{9.38}$$

where $C_V = C_\ell - AC_{\hat{x}}A^{\mathrm{T}}$ is the covariance matrix (with the diagonal elements as $s^2_{V_i}$) of the residual vector V, C_ℓ is the covariance matrix of the observation, A is the first design matrix, C_ℓ is the covariance matrix of the observations, and $C_{\hat{x}}$ is the covariance matrix of the adjusted coordinates. The statistics τ_i (for i =1, … , n) is then compared in a one-dimensional test with a value, $\tau(n, \mathrm{df}, \alpha_0)$, computed from the *Tau* table for the given number of observations (n), degrees of freedom (df) and the level of significance, α_0. The value of α_0 is related to the α used in the global test (Equation (9.37)) as follows (Pope, 1976):

$$\alpha_0 = 1 - (1 - \alpha)^{1/n} \text{ or } \alpha_0 \approx \frac{\alpha}{n} \tag{9.39}$$

If $\tau_i > \tau(n, \mathrm{df}, \alpha_0)$, the associated observation i in Equation (9.38) is considered a potential outlier. This is a form of in-context testing of each observation for the outlier. Using α directly in the test will result in out-of-context testing. An observation that does not pass the test is then subjected to further investigation and possible rejection. The Baarda's approach, which assumes that the a priori variance factor is known, will not be discussed any further.

9.4.3.4 Deformation Estimation from Two-Epoch Measurements Three important prerequisites for the comparison of least squares adjustment results of two epochs deformation monitoring are given as follows:

- Observation models for the two epochs of measurements must be based on the same geodetic datum (fixed points, network scale, and orientation of network must remain stable).
- Appropriate standard deviations of observations are available for weighting the observations.
- The same approximate coordinates for the common stations have been used for linearization purpose.

If the above-listed conditions are satisfied, the approach has some advantages, which include the following:

- No need of measuring the same observable in each epoch.
- It provides a means of assessing the consistency of the observations together in a network with the observations checking each other for blunder detection.

Some of the disadvantages of this approach include the following:

- Usual problem of datum definition and the stability of reference datum between epochs.

- The approach may not be able to handle any contamination of observations due to systematic errors relating to choice of instruments and observers. In this case, if there are systematic errors in each epoch measurement, these errors will impact the corresponding variance factor of unit weight, thereby affecting the accuracy of computed displacements.

The main objectives of two-epoch deformation analysis are as follows:

- To confirm the stability of reference points and to detect single-point movements. Single-point movements are considered as discontinuities with respect to time and locality, thus not conforming to continuous deformation models.
- To provide a plot of deformation vectors for assisting in developing a suitable deformation model. This geometrical aid is particularly useful for the detection of trends, which may not be detectable by statistical tests.
- To inform about the most recent deformations, which may be important for quick decisions, which cannot wait until the entire material is analyzed.

The displacements between two epochs of measurements can be determined from two separate single-epoch outlier-free least squares adjustments of coordinates using external constraint (fixing the stable reference points as datum) or inner constraint (fixing the center of gravity of the network). The results of the two single-epoch adjustments are as follows:

$$\hat{x}_1, \hat{x}_2; P_1, P_2; Q_{\hat{x}_1}, Q_{\hat{x}_2}; s_{01}^2, s_{02}^2 \tag{9.40}$$

where $\hat{x}_1, \hat{x}_2$ are the estimated coordinates of the network points; P_1, P_2 are the weight matrices; $Q_{\hat{x}_1}, Q_{\hat{x}_2}$ are the absolute cofactor matrices; s_{01}^2, s_{02}^2 are the variance factors of unit weight; and the subscripts 1 and 2 representing *epoch 1* and *epoch 2*, respectively. Absolute cofactor matrices provide a measure of the absolute accuracy of the station coordinate determination with respect to the origin of the network and in the case of free networks, with respect to the selected minimal constraints for the adjustment. Alternatively, the two epochs of measurements can be combined and adjusted with each object point in the network represented as two different points in the adjustment. This type of adjustment is only necessary in the case where correlations between epochs exist. Many practical reasons as well as experience support the presumption of correlations between epochs. However, no proven method is known to provide reasonable estimation of correlations. If actual combined adjustment is not carried out, some calculations are required to make the epochs comparable. In combined adjustments such as this, one variance factor $\hat{\sigma}_0^2$ common to both epochs is calculated. The estimate of this variance factor can also be computed from the two single-epoch adjustments provided that the same a priori factor σ_0^2 has been used or, more precisely, the estimates have the same expectation $E(s_{01}^2) = E(s_{02}^2) = \sigma_0^2$ and verified with F-test as follows. The estimates have the same expectation if the following is true (for α being the area in the upper tail of F-distribution):

$$F_{1-\frac{\alpha}{2},\mathrm{df}_1,\mathrm{df}_2} \leq \frac{s_{01}^2}{s_{02}^2} \leq F_{\frac{\alpha}{2},\mathrm{df}_1,\mathrm{df}_2} \tag{9.41}$$

Note that the F values refer to the upper-tail areas of the *F-distribution*, meaning that $F_{1-\frac{\alpha}{2},\mathrm{df}_1,\mathrm{df}_2}$ value should be smaller than $F_{\frac{\alpha}{2},\mathrm{df}_1,\mathrm{df}_2}$. If Equation (9.41) is satisfied, the combined (or pooled) variance factor $\hat{\sigma}_0^2$ is calculated from

$$\hat{\sigma}_0^2 = \frac{\mathrm{df}_1 s_{01}^2 + \mathrm{df}_2 s_{02}^2}{\mathrm{df}_1 + \mathrm{df}_2} \tag{9.42}$$

where df_1 and df_2 are the degrees of freedom for least squares adjustment of the measurements from the first and second epochs, respectively; and $\mathrm{df}_\mathrm{p} = \mathrm{df}_1 + \mathrm{df}_2$ is the pooled degrees of freedom.

The simplest method of showing deformations of a deformable object from repeated geodetic surveys is in form of displacements of the observed points with respect to selected reference points (or datum). From the results of the two single-epoch adjustments, it is possible to calculate the displacements $\hat{d}$ and the associated cofactor matrix $Q_{\hat{d}}$ from

$$\hat{d} = \hat{x}_2 - \hat{x}_1 \tag{9.43}$$

$$Q_{\hat{d}} = Q_{\hat{x}_2} + Q_{\hat{x}_1} \tag{9.44}$$

where $\hat{x}_1, \hat{x}_2$ and $Q_{\hat{x}_1}, Q_{\hat{x}_2}$ are obtained from two separate single-epoch adjustments of the network using minimum constraints (assuming $\hat{x}_1, \hat{x}_2$ are uncorrelated). The general process resulting in Equations (9.43) and (9.44) is known as *coordinate differencing* or *two-epoch approach*. The direct subtraction and addition given in Equations (9.43) and (9.44) are simplistic; these cannot be done directly as given, for a number of reasons, which include (A. Chrzanowski, unpublished) the following:

- The reference points used in providing minimum constraints for least squares adjustment for the two epochs may not be stable from one epoch to the other. This means that the two quantities being compared do not have the same reference.
- The values of the quantities $\hat{x}_1, \hat{x}_2$, $Q_{\hat{x}_1}, Q_{\hat{x}_2}$, and $\hat{d}$ depend on the choice of the minimal datum constraints; different minimum constraints will produce different values of the quantities. The quantities are incomparable except they are based on the same datum constraints. The datum constraints may be different in two epochs if some reference points used in epoch 1 were damaged so that different points have to be used in the following epoch.

- There may be different types of datum defects in the two epochs considered, for example, trilateration network may be measured in one epoch while triangulation network is considered in the other epoch.
- Inner constraint solutions in both epochs might have been defined by different sets of points, for example, if new points are added in the following epoch.

Due to the above-listed reasons, the vector of displacements will not provide the real picture of displacements and may likely contain false displacements caused by different datum parameters. The above conditions require that the two epochs concerned be evaluated for the detection of unstable reference stations first. The assumption can be made that $\hat{d}$ has been created with the reference stations remaining stable between epochs, that is, that the reference stations have not moved between the two epochs of measurements. However, to ensure that the assumption is valid, it is necessary to transform $\hat{x}_1, \hat{x}_2$ and their cofactor matrices $Q_{\hat{x}_1}, Q_{\hat{x}_2}$ into forms that are independent of the choice of datum or into a common datum. This is done by the IWST (Chen, 1983; Secord, 1985; Chrzanowski et al., 1986; A. Chrzanowski, unpublished) discussed in the following section.

9.4.3.5 Iterative Weighted Similarity Transformation The corrections to approximate values of unknown parameters (δ) and the cofactor matrix of adjusted parameters ($Q_{\hat{x}}$) for each epoch of measurements will vary depending on the choice of datum constraints. However, the so-called similarity (Helmert's) transformation or *S*-transformation can be used to transforms values (δ_i, Q_i) of a certain set of constraints *i* into another values (δ_j, Q_j) of a different set of constraints *j*. This transformation process preserves the network geometry by translating, rotating, and scaling the given network differentially into the other datum using a well-chosen transformation matrix *S* as follows:

$$\delta_j = S_j \delta_i \tag{9.45}$$

$$Q_j = S_j Q_i S_j^{\mathrm{T}} \tag{9.46}$$

$$S_j = I - G(D_j^{\mathrm{T}} G)^{-1} D_j^{\mathrm{T}} \quad \text{with} \quad D_j^{\mathrm{T}} = G^{\mathrm{T}} P_j \tag{9.47}$$

where *G* matrix is defined in Equation (9.22); *I* is the usual identity matrix (a matrix with all of the main diagonal elements equal to one and all of the off-diagonal elements equal to zero) with the matrix size equal to the size of the matrix product being subtracted from it in Equation (9.47); P_j is a special type of identity matrix whose diagonal elements are equal to unity for the corresponding coordinate parameters defining the datum with all other elements being zero. When all points of a control network constitute the datum, like in the free-networks (inner constraint) adjustment, matrix *P* is purely an identity matrix. The idea of IWST is to find such a "weight" matrix *P* for appropriate definition of the datum so that $D^{\mathrm{T}} = G^{\mathrm{T}} P$ and the first norm of the vector of displacement will be minimized, that is, $\|\hat{d}\|_1 = \min$. This optimization problem is generally nonlinear, requiring that iterations be performed, hence, the IWST. This procedure helps in finding the "best" datum in such a sense that such a datum will have the minimal distorting influence on the vector of displacements. The steps for the IWST can be described as follows (Chen, 1983; Secord, 1985; Chrzanowski et al., 1986; A. Chrzanowski, unpublished):

1. Perform the minimal constraint least squares adjustment of the reference network with the object points first treated as nuisance parameters with the accompanying quality assessment of the network.
2. Determine the vector of displacements $\hat{d}$ and its cofactor $Q_{\hat{d}}$ from Equations (9.43) and (9.44), respectively.
3. Confirm the compatibility of the a posteriori variance factors of unit weight (s_{01}^2, s_{02}^2) of the least squares adjustments of the epoch 1 and epoch 2 measurements using the statistical F-distribution test in Equation (9.41). If the test fails, then the weight matrices used in each epoch adjustment must be reevaluated and the adjustments repeated until the test passes. Then determine the pooled variance factor and its degrees of freedom from Equation (9.42).
4. In the first iteration, the weight matrix, P is taken as identity matrix ($P = I$) and used in the following transformation:

$$\hat{d}^{(1)} = S\hat{d} \quad \text{where} \quad S = (I - G(G^{\mathrm{T}}PG)^{-1}G^{\mathrm{T}}P) \tag{9.48}$$

5. Perform subsequent iterations until convergence criterion is achieved using the following:

$$\hat{d}^{(k+1)} = (I - G(G^{\mathrm{T}}PG)^{-1}G^{\mathrm{T}}P)\hat{d}^{(k)} = S_k\hat{d}^{(k)} \tag{9.49}$$

where G is the transformation matrix (scale, rotation, translations) given in Equation (9.22);

$$P = \mathrm{diag}\left[\frac{1}{\hat{d}_i^{(k)} + c}\right] \tag{9.50}$$

$\hat{d}_i^{(k)}$ is the ith component of the vector of displacements $\hat{d}^{(k)}$ computed after the kth iteration; and c is a small constant value chosen to avoid having zero denominators. The convergence criterion can be given that the maximum difference between two corresponding values of displacements of the two consecutive iterations is less than a predetermined small tolerance value Δ:

$$\max(|\hat{d}_i^{(k+1)} - \hat{d}_i^{(k)}|) < \Delta \tag{9.51}$$

6. If the process converged in step (5), compute the transformed cofactor matrix of the final vector of displacements $(\hat{d}_i^{(k+1)})$ as follows (using P determined in step (5)):

$$Q_{\hat{d}^{(k+1)}} = SQ_{\hat{d}}S^{\mathrm{T}} \tag{9.52}$$

7. From now on the superscript and subscript on the transformed vector of displacements can be dropped for the sake of simplicity for $\hat{\hat{d}}$ and its cofactor matrix as $Q_{\hat{\hat{d}}}$.

9.4.4 Observation Differencing Adjustment Approach

The model for the observation differencing approach is obtained by subtracting the parametric model equations for epoch 1 from those for epoch 2. For example, the observation equations for the two epochs of observations, for subsequent least squares adjustment, can be expressed in linearized form (after Equation (9.10)) as follows:

$$\ell_1 + V_1 = \ell_1^0 + A_1\delta_1 \tag{9.53}$$

$$\ell_2 + V_2 = \ell_2^0 + A_2\delta_2 \tag{9.54}$$

where ℓ_i is a vector of observations; V_i is a vector of observation residuals; ℓ_i^0 is a vector of approximate observations calculated from approximate coordinates (x^0) of network points; A_i is the first design matrix or the configuration matrix; δ_i is a vector of corrections to be applied to the approximate network coordinates; and $i = 1, 2$ for epochs 1 and 2. The model for the observation differencing approach is obtained by subtracting Equation (9.53) from Equation (9.54), giving the following:

$$\ell_2 - \ell_1 + V = \ell_2^0 - \ell_1^0 + A_2\delta_2 - A_1\delta_1 \tag{9.55}$$

where $V = (V_2 - V_1)$ is the vector of residuals of the observation differences. Assume that the following conditions are satisfied for the two epochs of monitoring: the same approximate coordinates (datum for the adjustment remains the same between epochs); the same observables (measured by the same observers using the same instrumentation); and the same network geometry. On the basis of the above assumptions, $\ell_2^0 = \ell_1^0$, $A_2 = A_1 = A$, and $d = \delta_2 - \delta_1$, so that Equation (9.55) can be simplified to give the following:

$$\ell_2 - \ell_1 + V = Ad \tag{9.56}$$

where d is the vector of displacements of network points. Equation (9.56) can be solved for "d" by the method of least squares, provided appropriate constraints (or datum) exist for the adjustment. The concept of constraining the network for least squares adjustment was discussed in Section 9.4.3.1. The advantages of using observation-differencing approach include the following:

- Considering the conditions satisfied before using Equation (9.56), common systematic errors (e.g., due to instrumentation, observer, and atmosphere) would be removed, except those due to seasonal variations of the atmospheric conditions.
- If reference points used as constraints are unstable, strain analysis (which is independent of datum) could still be done for a deformable object.
- This approach can easily accommodate geotechnical data, such as extensometer and tiltmeter measurements, in an integrated least squares adjustment method. For example, the change in dimension of a homogeneous structure derived from the measurements of multipoint extensometers anchored to the structure at two

points can be considered as the change in distances measured in two epochs. This observation difference can then be transformed into an observation equation, which is a function of displacements similar to Equation (9.56).

- Geometrical defects are permitted, for example, the defects caused by eccentric targets at the observing stations or by points established by only observing a single distance. This results in geometrical misclosures but still allows for the calculation of displacements.

There are also some disadvantages of using observation differencing approach, which include the following:

- Need to measure the same observables, use the same instrumentation and the same observers in all the epochs, which may be practically impossible.
- Outlier detection procedure for individual epoch of measurements cannot be done, making the approach more susceptible to blunders.

9.4.5 Geometrical Analysis of Deformation Measurements

Deformation of an object is now widely accepted in geodetic engineering surveying as a consequence of a *dynamic process* or *dynamic system* (Welsch and Heunecke, 2001), which is composed of the following three integrated elements:

- Factors causing the deformation (causative forces, internal and external loads) as input
- Physical properties of the monitored object (material properties, material response behavior, extension coefficients, geometry, etc.)
- Response of the object in form of deformation, as output. Being one of the elements of a dynamic process or dynamic system, deformation can be considered as an aspect of a dynamic system (or dynamic process).

The dynamic processes (or systems) are completely described and explained by *dynamic models*, which are to help in studying the effects of the different elements listed above and to make predictions. The models are considered the most general and complete with one of the simplifications being the deformation model of an object in space and time. The dynamic models consider monitored object points as continuously moving with time as a result of variable acting forces (loads). The model is called *deterministic* (Chrzanowski et al., 1990a, 1990b) if the factor causing the deformation of the object and the physical properties of the object are known and the deformations of the object are only to be predicted (e.g., by finite element method). If the factors causing the deformation and the physical properties of the object are known with deformation measurements available from geodetic monitoring, the dynamic model will be referred to as *integrated model* (Welsch and Heunecke, 2001). The complexity of dynamic modeling makes interdisciplinary cooperation a necessity. The typical dynamic models applied to deformation analysis are empirical models or experimental models, which require that permanent and automatic observation procedures be available for deformation monitoring.

The dynamic models can be broken into three simpler types of models: *kinematic, static,* and *geometrical*. The kinematic models consider the monitored object points as moving continuously with movements as functions of time only, and no acting forces or loads are involved. On this basis, a kinematic model will describe and explain deformation (in form of velocity and acceleration of the object points) using time functions with no regard for the factors causing the deformation. In the case of static models, time is not involved and the monitored objects are not in continuous motion (i.e., not moving at least during the time of monitoring), but are at equilibrium under the acting forces (loads). The deformations of an object, in this case, are functions of only the acting forces (loads) and not of time. Those elements that need to be known in static modeling are the physical and geometrical structures of the object, the material properties, and other characteristic quantities of the object. The geometric models of deformation processes in space and time, however, model an object (which is a continuum) as a set of discrete points in space; these discrete object points, which are supposed to be in continuous motion, are modeled as moving only within certain time intervals. The models do not explicitly consider time factors; they do not consider the acting forces (loads) that are responsible for deformation; and the monitored object points are not considered as moving continuously, but are taken as being at equilibrium under the acting forces (loads). In general term, geometric models are about modeling monitoring network point movements or changes in the geometry of the monitored object in space and time. These models are then used in what is called *geometrical deformation analysis*.

Geometrical deformation analysis is about detecting, localizing, and modeling monitoring network point movements based on deformation monitoring. In the case of deformation monitoring based on absolute networks, the usual problem (or main task) of geometrical deformation analysis is to confirm the stability of the reference points and to identify the possible single-point movement that may be due to local phenomena or wrong monumentation of survey markers. If the unstable reference points are not identified, the object points and the other reference points (that are stable) will show movements even when, in reality, they are truly stable. The subsequent analysis and interpretation will then be distorted and biased. Once the stable reference points have been determined, the deformation of the object can easily be determined. In relative networks, geometrical deformation analysis is not that easy; in addition to possible single-point movement, all the network points also may undergo relative movements due to strains in the materials of a deformable object. Moreover, if there is a discontinuity (as in the case of tectonic faults) in the object, relative rigid translations and rotations of a block of the object with respect to other blocks may occur. The main problem in this type of network, therefore, is how to identify, on the basis of repeated geodetic observations, the deformations caused by strains, relative rigid body translations, and single-point movements. In analyzing a relative network, however, the first step is usually to establish whether any group of points in the network has retained its shape between the two epochs of measurements by using the *IWST* with an appropriate statistical testing. If such a group of points can be identified, then the points may be used as a datum, thus providing an absolute network for the analysis of the other points. If no group of stable points can be identified, then the resulting relative network can be assessed in terms of datum invariant criteria, using *IWST*.

The overall task of deformation analysis is to obtain a displacement function (deformation model), which characterizes the deformation in space and time. The best deformation model produces the overall geometrical deformation trend. Once the displacement function is determined, all the basic deformation parameters such as strain components, rotations, and rigid body movements can be calculated at any desired point of the monitored object.

9.4.5.1 Statistical Trend Analysis of Deformations After performing an appropriate least squares adjustment and a transformation using IWST, the geometrical analysis of deformation monitoring networks usually consists of detecting deformation using two-epoch analysis, which will include determining stable and unstable reference points by localization of deformation through single-point statistical test or trend analysis. The steps for statistical trend analysis of deformations can be given as follows (Chen, 1983):

1. Perform least squares estimation of the coordinates of the points and their variances and covariances from each campaign separately, using minimal constraints (holding a reference station and the direction from one station to another fixed and errorless).
2. Determine the datum-dependent displacements from the estimated coordinates.
3. Perform IWST of the displacements to obtain datum-independent relative displacements and identify the stable reference points.
4. After the IWST, determine the statistical significance of the displacements by testing the subvector of displacements at each point ($\hat{\hat{d}}_j$) against the corresponding cofactor submatrix $Q_{\hat{\hat{d}}_j}$ at certain confidence region as follows:

$$F_{\mathrm{c}} = \frac{\hat{\hat{d}}_j^{\mathrm{T}} Q_{\hat{\hat{d}}_j}^{-1} \hat{\hat{d}}_j}{\hat{\sigma}_0^2 u_{\mathrm{d}}} < F(\alpha_0, u_{\mathrm{d}}, \mathrm{df}_{\mathrm{p}}) \tag{9.57}$$

where α_0 is the chosen significance level (area in the upper tail of F-distribution), u_{d} is the dimension of the confidence region, which could be $u_{\mathrm{d}} = 1$ (for one-dimensional cases), $u_{\mathrm{d}} = 2$ (for two-dimensional cases), and $u_{\mathrm{d}} = 3$ (for three-dimensional cases); $\hat{\sigma}_0^2$ is the pooled variance factor; and df_{p} is the number of degrees of freedom for the pooled variance factor. A point that does not pass the above F-test (if $F_{\mathrm{c}} > F$) is flagged as unstable and can be grouped with the unstable object points in further analysis. The points that are found to be stable are used to define the datum used in determining the final quantities $\hat{x}_1, \hat{x}_2$ and their cofactor matrices $Q_{\hat{x}_1}, Q_{\hat{x}_2}$. These final quantities are then used in Equations (9.43) and (9.44). According to Caspary (1987), the significance level (α_0) to be used in Equation (9.57) can be given as follows:

$$\alpha_0 = 1 - (1 - \alpha)^{1/m} \quad \text{or} \quad \alpha_0 \approx \frac{\alpha}{m} \tag{9.58}$$

where m is the number of points (not coordinates) being considered in the network. If the a priori variance factor σ_0^2 is known, the statistic given in Equation (9.57) can be tested against the Chi-square value $\chi^2_{\alpha_0,\text{df}=u_\text{d}}/u_\text{d}$ for α_0 being the area in the upper tail of Chi-square distribution. In this case, the displacement $\hat{d}_j$ will be considered significant if the computed F is greater than $\chi^2_{\alpha_0,\text{df}=u_\text{d}}/u_\text{d}$.

5. Model the stable points as a fixed reference block and determine the displacements of the object points.

The whole process stated in step 4 constitutes what is referred to as statistical analysis of deformation trend or the *localization of deformation*; in the process, the differences in coordinates at different epochs are compared. Generally, trend analysis is the intermediate link between deformation measurements and the deformation modeling referred to in step 5. More details on deformation modeling can be found in Chen (1983) and Ogundare (1990).

9.4.5.2 Graphical Trend Analysis of Deformations Graphical trend analysis of deformation consists of plotting network points displacement vectors along with their corresponding error ellipses as a graphical representation of the significance of any movement of the network points. This plot shows the *spatial trend* over time interval between the given two epochs. If a displacement vector extends outside the error ellipse, the movement can be seen as being significant at the specified level of significance and the associated point will be considered to be unstable. Any unstable reference points are segregated from the rest of the reference points and considered as object points during *deformation modeling*. The point displacement error ellipse can be computed by using the appropriate displacement cofactor submatrix of $Q_{\hat{d}}$ (i.e., submatrix $Q_{\hat{d}_i}$ for a given network point i) which can be given as

$$Q_{\hat{d}_i} = \begin{bmatrix} q_{xx}^2 & q_{xy} \\ q_{yx} & q_{yy}^2 \end{bmatrix} \tag{9.59}$$

Using the elements of the submatrix, $Q_{\hat{d}_i}$, the parameters of the point displacement error ellipse can be given as follows, in the case where the a priori variance factor of unit weight (σ_0^2) is known for the least squares adjustment of the deformation measurements with the pooled a posteriori standard factor of unit as $\hat{\sigma}_0$ computed from Equation (9.42) (with α or α_0 being the area in the upper tail of the distribution):

$$a = \hat{\sigma}_0 \sqrt{\lambda_1 \chi^2_{\alpha_0,\text{df}=2}} \tag{9.60}$$

$$b = \hat{\sigma}_0 \sqrt{\lambda_2 \chi^2_{\alpha_0,\mathrm{df}=2}} \tag{9.61}$$

$$\theta = \arctan\left(\frac{\lambda_1 - q^2_{yy}}{q_{xy}}\right) \tag{9.62}$$

where

$$\lambda_1 = \frac{1}{2}(q^2_{xx} + q^2_{yy} + z) \tag{9.63}$$

$$\lambda_2 = \frac{1}{2}(q^2_{xx} + q^2_{yy} - z) \tag{9.64}$$

$$z = \left[\left(q^2_{xx} - q^2_{yy}\right) + 4q^2_{xy}\right]^{1/2} \tag{9.65}$$

λ_1 and λ_2 are the maximum and minimum eigenvalues of the cofactor matrix $Q_{\hat{d}_i}$, respectively; $\chi^2_{\alpha_0,\mathrm{df}=2}$ is the Chi-square distribution value at α_0 significance level (using the in-context value) and df = 2 degrees of freedom (for 2D or x–y problem); and θ is the orientation of the semi-major axis of the point displacement ellipse. In the case where the a priori variance factor of unit weight (σ_0^2) is unknown for the least squares adjustment of the deformation measurements, the following are used:

$$a = \hat{\sigma}_0 \sqrt{2\lambda_1 F_{\alpha_0,\mathrm{df}_1=2,\mathrm{df}_2=\mathrm{df}_\mathrm{p}}} \tag{9.66}$$

$$b = \hat{\sigma}_0 \sqrt{2\lambda_2 F_{\alpha_0,\mathrm{df}_1=2,\mathrm{df}_2=\mathrm{df}_\mathrm{p}}} \tag{9.67}$$

$$\theta = \arctan\left(\frac{\lambda_1 - q^2_{yy}}{q_{xy}}\right) \tag{9.68}$$

where $\hat{\sigma}_0$ is the pooled a posteriori standard factor of unit weight computed from Equation (9.42), $F_{\alpha_0,\mathrm{df}_1=2,\mathrm{df}_2=\mathrm{df}_\mathrm{p}}$ is the Fisher distribution value at $\alpha_0 = 1 - (1 - \alpha)^{1/m}$, α_0 is the significance level (in-context value), m is the number of free points (if one of the coordinates of the point is fixed, the point can still be considered as free) whose coordinates were determined, and $\mathrm{df}_1 = 2$ (for 2D or x–y problem) and $\mathrm{df}_2 = \mathrm{df}_\mathrm{p}$ is the pooled degrees of freedom (the sum of degrees of freedom from the two epochs of measurements).

9.4.6 Examples: Deformation Monitoring and Analysis of Hydroelectric Dams

There are several thousands of registered dams operating all over the world. The purposes of those dams range from hydroelectric to domestic supply. A typical hydroelectric generating station consisting of an embankment dam (rock-filled structure with watertight clay core), concrete diversion sluiceway, concrete main spillway, and Intake/Powerhouse structures is shown in Figure 9.8. The embankment dam shown in

Figure 9.8 Main features of a typical hydroelectric generating station. *Source:* Background image is reproduced by permission of NB Power.

the figure is 42.367 m high and 518.16 m long and can be considered as a large dam according to the International Commission of Large Dams (ICOLD, 1977). Water flows through the *Water Channel* into the *Intake* structure by the opening below the *head gate* and falls down a chute called the *Penstock* to spin the turbines that drive the electric generators in the *Powerhouse*; the water then comes out downstream from the *tailrace*. The *Main Spillway* and the *Diversion Sluiceway* structures of the generating station are to provide controlled release of floods from the dam into the downstream area so that the water does not overflow and damage the dam.

Dams must be carefully and precisely monitored as required by law since they are considered to be inherently dangerous to lives and properties if they fail, and also to keep the use of the structures of the dams longer than are usually expected. The general norm is that dams should be measured during the first filling and emptying to test if the measured deformations agree with the expected deformations. Some of the factors that may cause dam structures to deform include the following:

- Alkaline aggregate reaction expansion of concrete
- Instability of surrounding bedrock
- Changeable water load on the dam structures, with the reservoir behind the dam placing new weight on the floor and sides of the valley of the reservoir with the water pushing against the upstream face of the dam

- Seasonal thermal-induced deformations
- Possible seismic events.

In order to be able to discriminate among these possible causes of deformations, a long-term pattern, usually based on a minimum of 2 years of data, must be analyzed (Chrzanowski and Secord, 1987). It should also be mentioned that a dam failure is also possible if an embankment dam is overflown beyond its spillway. In this case, it will be required that a high safety measure for the spillway be provided to ensure that it will be capable of containing a maximum flood stage.

9.4.6.1 Simulated Dam Deformation Monitoring and Analysis A simulated monitoring network presented in Figure 9.9 is to illustrate the steps of the two-epoch approach of deformation analysis. The monitoring network is in a local (X, Y) coordinate system with all the measurements and deformations simulated. Figure 9.9 shows the monitoring scheme with the network point P (assumed to be located on the crest of the dam) serving as an object point that is unstable. Points A, B, and C constitute the reference network, which is assumed stable during the two epochs of measurement. The approximate coordinates of the network points are given in Table 9.3 and the simulated horizontal angle and horizontal distance measurements for two epochs are given in Table 9.4. It is assumed that TCR705 total station instrument with the distance accuracy specification of 2 mm + 2 ppm and the accuracy specification for direction measurement (according to ISO 17123-3) as 5″ was used for the measurements. It is further assumed that each angle is measured in two sets with the propagated standard deviation for the average of the two sets

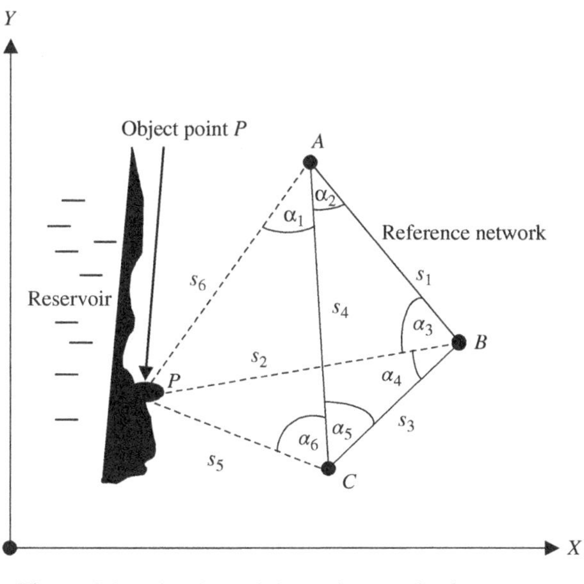

Figure 9.9 Simulated deformation monitoring scheme.

TABLE 9.3 Approximate Coordinates of Points.

Point	Y (m)	X (m)
1	800.000	600.000
2	700.000	900.000
3	100.000	600.000
P	400.000	200.000

TABLE 9.4 Simulated Field Measurements for Both Epoch 1 and Epoch 2.

At	From	To	Epoch 1 Observation	Epoch 2 Observation	Comments
A	C	P	45°00′10″	45°00′15″	Angle α_1
A	B	C	71°33′50″	71°33′50″	Angle α_2
B	P	A	41°37′55″	41°37′50″	Angle α_3
B	C	P	40°14′10″	40°14′30″	Angle α_4
C	A	B	26°33′55″	26°34′00″	Angle α_5
C	P	A	53°07′50″	53°07′45″	Angle α_6
	A	B	316.228 m	316.230 m	Distance s_1
	B	P	761.577 m	761.575 m	Distance s_2
	B	C	670.820 m	670.815 m	Distance s_3
	C	A	700.000 m	699.990 m	Distance s_4
	C	P	500.000 m	500.025 m	Distance s_5
	A	P	565.675 m	565.670 m	Distance s_6

as 5″; and the centering errors of the instrument and targets are 0.2 mm each (with forced-centering pillars used).

The required tasks to be carried out can be given as follows:

- Perform statistical analysis of the deformation trend at 95% confidence level according to Section 9.4.5.1 with point A and azimuth A-B (108°26′05″) held fixed (with no errors) as minimum constraints.
- Represent the deformation trend graphically based on 95% confidence level according to Section 9.4.5.2.

In this problem, the following two constraints are imposed on the given deformation monitoring network before determining displacements (which are to be analyzed statistically and graphically):

1. External minimal constraints
2. Iterative weighted transformation constraints.

Results of External Minimal Constraint Adjustment From the minimal constraint least squares adjustment of each epoch of measurements in which point A and azimuth

TABLE 9.5 Horizontal Displacements Based on External Minimal Constraints.

Displacement (Epoch 2 – Epoch 1)			95% Confidence Error Ellipses		
Point	*dx* (mm)	*dy* (mm)	*a* (mm)	*b* (mm)	Orientation
A	0	0	–	–	–
B	0	0	6.9	0.1	161°34′
C	+5	+7	19.9	6.1	167°48′
P	−12	+24	19.4	6.4	126°18′

A–B are held fixed with no errors, and based on the two-epoch approach, the horizontal displacements given in Table 9.5 and the following results were determined.

- Variance factor of unit weight for epoch 1 adjustment is $\hat{\sigma}_{01}^2 = 1.1706$, $\text{df}_1 = 7$.
- Variance factor of unit weight for epoch 2 adjustment is $\hat{\sigma}_{02}^2 = 1.0492$, $\text{df}_2 = 7$.
- According to Fisher test for compatibility of variance factors (Equation (9.41)) for the two epochs, the two variance factors are compatible since the following is true: $0.200 < 1.245 < 4.995$.
- From Equation (9.57), assuming the a priori variance factor is known, $F(\alpha_0 = 0.05, u_\text{d} = 2, \text{df}_\text{p} = 14) = 3.739$ and the pooled a posteriori standard factor of unit weight, $\hat{\sigma}_0 = 1.1115$
- From Equation (9.57), the following were obtained: $F_\text{c} = 0.000$ for point *B*; $F_\text{c} = 3.510$ for point *C*; and $F_\text{c} = 11.442$ for point *P*.

According to Equation (9.57), if F_c value is greater than the critical *F* value, the network point concerned is suspected to be unstable. Based on this statistical test procedure, we can consider the reference points *B* and *C* to be stable since their computed F_c values are less than the critical value ($F = 3.739$); and the object point *P* seems to have significantly moved at 95% confidence level since its F_c value is greater than the *F* value.

Assuming the a priori variance factor of unit weight is known, Equations (9.60) to (9.65) are used in computing point displacement 95% confidence error ellipses whose parameters are shown in Table 9.5. The plot of the displacement field based on the external minimal constraint solutions is given in Figure 9.10; the graphical analysis is according to Section 9.4.5.2. From the figure, it can be seen that the plotted displacement for point *P* extends outside the point displacement ellipse, indicating that the point has significantly moved at 95% confidence level; for point *C*, it can be seen that there is a small displacement, which is not clearly significant at 95% confidence level. It can also be seen that the analysis of the plot is in agreement with the result of the statistical analysis done.

Iterative Weighted Transformation Results The results of the external minimally constrained displacements are subject to IWST according to Section 9.4.3.5. The

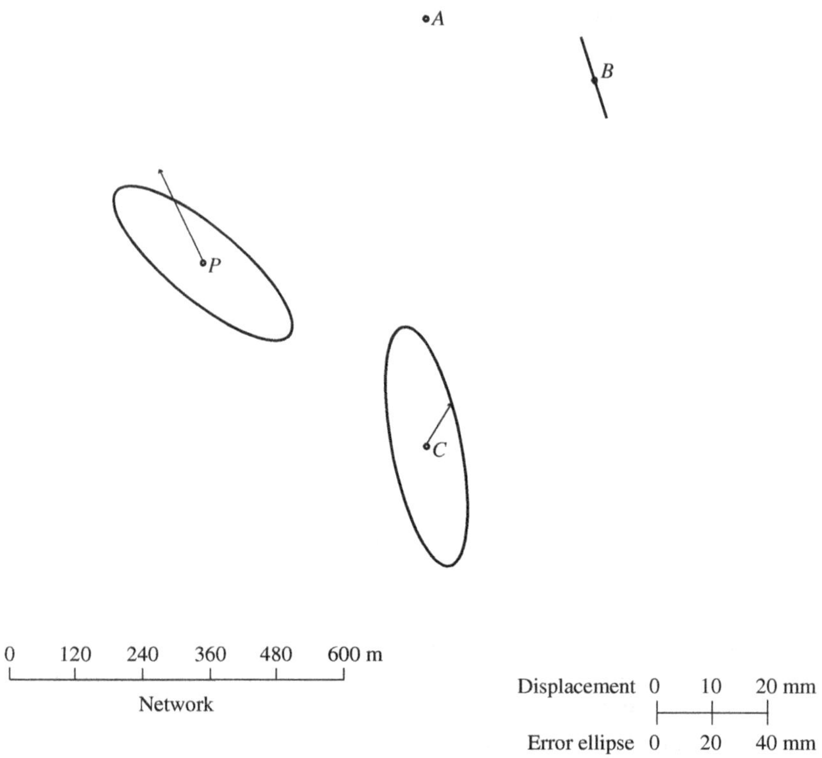

Figure 9.10 External minimally constrained displacements with point *A* and azimuth *A-B* held fixed (error ellipses at 95% confidence level).

TABLE 9.6 Horizontal Displacements Based on IWST.

	Displacement (Epoch 2 – Epoch 1)		95% Confidence Error Ellipses		
Point	*dx* (mm)	*dy* (mm)	*a* (mm)	*b* (mm)	Orientation
A	−2.0	+0.1	8.4	0.7	86°30′
B	+0.2	−2.2	10.7	1.2	05°00′
C	0.0	0.0	0.1	0.0	164°24′
P	−19.5	+21.7	11.7	10.3	59°48′

results of the transformation and the parameters of the point displacement 95% confidence error ellipses are given in Table 9.6. In computing the displacement 95% confidence error ellipses, the a priori variance factor of unit weight is assumed known so that Equations (9.60) to (9.65) are used.

The plot of the horizontal displacements after IWST has been performed is given in Figure 9.11; this is done according to Section 9.4.5.2. From the figure, it can be

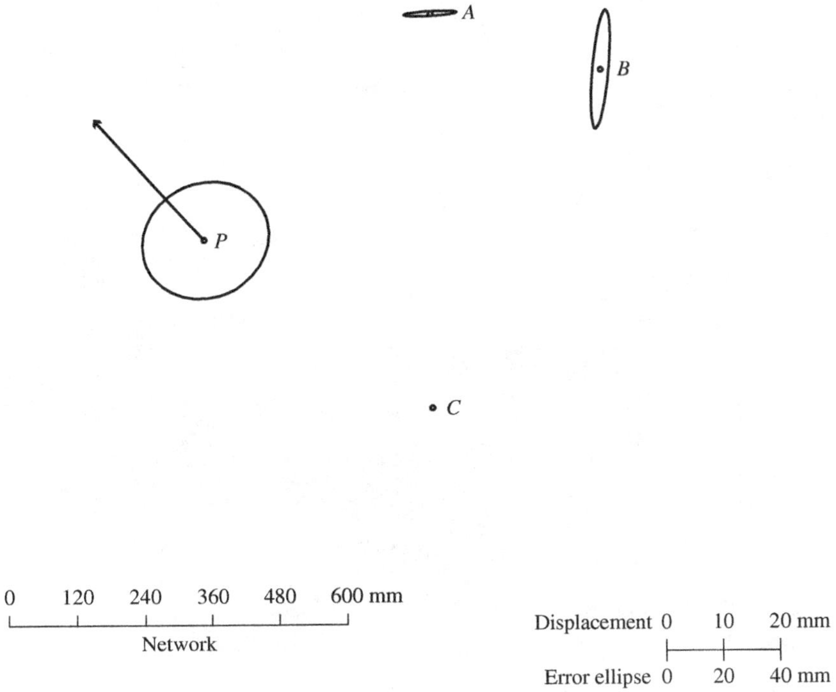

Figure 9.11 Displacement field after IWST (error ellipses at 95% confidence level).

seen that the plotted displacement for point *P* extends outside the displacement error ellipse, indicating that the point has significantly moved at 95% confidence level. It can also be seen that all the reference points *A*, *B*, and *C* are clearly shown to be stable with no possibility of misinterpreting any movement as deformation, unlike in the case of external minimally constrained displacements shown in Figure 9.10, in which there is uncertainty in the interpretation of the movement of point *C*.

9.4.6.2 Dam Deformation Monitoring and Analysis in Practice The traditional geodetic monitoring schemes for the purpose of monitoring structures (Powerhouse, Intake, main dam, and sluiceways) of a generating station, in practice, are based on both precise vertical and horizontal control networks. The horizontal reference network points (usually not less than five points) are established at stable areas around the structures while the horizontal object network points are established mainly on the Powerhouse rooftop, tailrace deck, Intake, crest of main dam, slope indicator areas, diversion sluiceway/spillway decks, riverbank, abutment, and available access roads. Figure 9.12 shows a typical absolute geodetic network for deformation monitoring of a hydro dam, which is a trilateration network consisting of a reference network of Pillars REF100–REF800 with the object points located on the structures of the generating station, within the reference network.

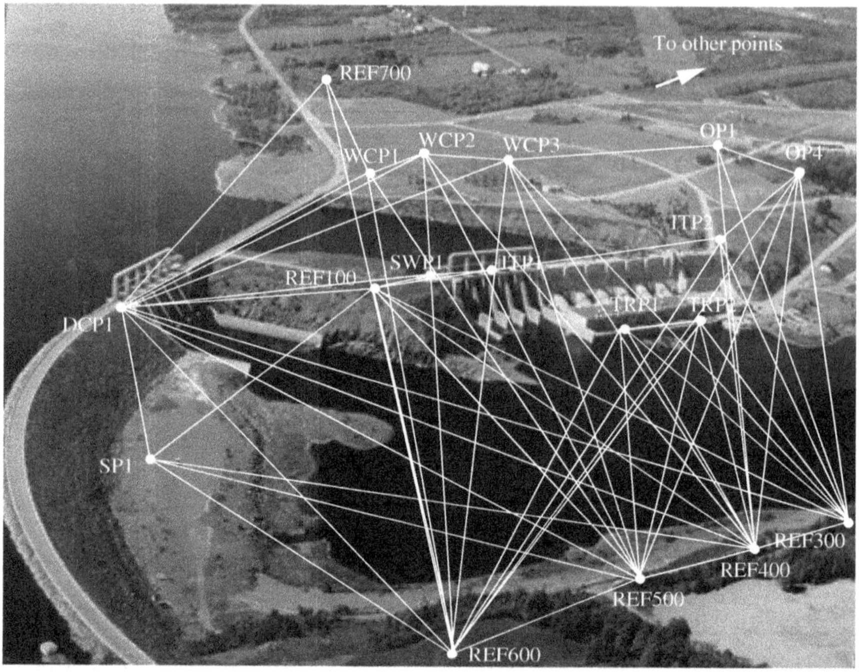

Figure 9.12 Typical trilateration network for deformation monitoring of an hydroelectric dam (not to scale). *Source:* Background image is reproduced by permission of NB Power.

In an absolute horizontal geodetic monitoring network, some of the object and reference points may be adapted for GPS surveys as can be seen in Figure 9.13, where GPS antennas are installed on brackets. In Figure 9.13(a), a GPS antenna is installed eccentrically on a bracket, away from a geodetic pillar, for deformation monitoring. This type of GPS setup is usually designed for automatic data collection and data transfer to remote station. In order to achieve subcentimeter accuracy in horizontal positioning at 95% confidence level, the duration of the observation sessions has to be up to 6 h (Chrzanowski and Szostak-Chrzanowski, 2010; Chrzanowski and Bazanowski, 2011).

The precise vertical control network measurements are based on precise differential leveling procedure connecting the following important regions of the generating station:

- Points on the generator floor, tailrace, and turbine floor of the Powerhouse
- Points on the galleries of the Intake structure
- Points on the diversion sluiceways/spillways
- Possible suspended and inverted plumbline points and borehole extensometer collar points, in the Powerhouse, Intake structure, and diversion sluiceways/spillways.

(a) (b)

Figure 9.13 (a) GPS unit installed eccentrically from a geodetic pillar on the Intake structure of a generating station. (b) GPS unit installed on the crest of the gravity dam/diversion sluiceway structure of a generating station.

Deformation Trend Analysis Two-epoch deformation analysis can be performed on the trilateration network in order to determine the deformation trend of the structure. This network is based on a local coordinate system with reference network points as REF200 to REF800. The displacement field based on external minimally constrained (point REF800 and azimuth REF800 to REF600 held fixed) adjustment of deformation measurements taken in two epochs can be processed according to Equation (9.12); the parameters of the 95% confidence error ellipses can be computed using Equations (9.66) to (9.68), assuming the a priori variance factor of unit weight is unknown. A figure similar to Figure 9.10, showing the displacements of points, will then be plotted. The point displacements that are outside their displacement error ellipses are considered to have significantly moved at 95% confidence level. By applying the IWST to the displacements and their corresponding cofactor matrices, the new displacement field, similar to that shown in Figure 9.11, can be plotted. Points that are now outside their displacement ellipses will be confirmed to have significantly moved; points that have been wrongly identified based on the results of external constraint adjustments will also be revealed.

9.4.7 Deformation Monitoring of Slope Walls

Developments in total station design and construction have led to the development of fully automated monitoring systems. The current direction in monitoring and deformation surveys of slope walls in open-pit mines includes the following:

- Creating fully automated monitoring scheme based on RTSs, active GPS, and an assortment of geotechnical instrumentation
- Integrating a number of monitoring techniques, including GPS, total stations, reflectorless EDM, and differential leveling.

The automated real-time deformation monitoring system is currently being applied in deformation monitoring of slope walls formed in open-pit mining; the slope walls can be a few hundred meters deep and 1 or 2 km long and wide. Since the inclination angles of a slope wall are directly related to profit-to-cost ratio of a mining operation, with steeper angles requiring less waste removal, which also results in less cost of operation, it has become advantageous to design very steep mine slopes to reduce cost of mining. The consequence of this is the very high frequency of slope failures that often brings about significant safety and financial problems. As a result of this, monitoring of slope walls has become an integral part of a mining operation, but with geotechnical instruments being commonly used since they are easily automated.

The conventional surveying procedure to monitor the stability of slope walls involves measuring spatial displacements of selected object points (located on the slope) from reference points that are fixed in position. The spatial displacements, which are considered as the coordinate differences of the same network of reference and object points from repeated independent surveys, can be determined from the terrestrial and satellite-based measurements as discussed in Section 9.4. The advantages and disadvantages of an automatic monitoring system are given by Secord (1995). Some of the advantages (cf. Secord, 1995) include the following:

- Reduction in manpower is achieved with regard to data acquisition and analysis.
- Data can be collected more frequently than in the case of manual data collection.
- Fewer errors are incurred in data recording and transmission over long distances.

Some of the disadvantages of automatic monitoring system include the following (cf. Secord, 1995) :

- There is a problem of having to manage a large volume of data generated by the system over time.
- There are usually high initial cost of setting up the system and a high cost of maintaining the system over time.
- There is usually high temptation to use the collected data directly in analysis without considering the possible errors that may have been due to lack of human intervention over time.
- There is a need for specialized personnel to perform regular field checks and carry out system maintenance.

The success of any mining and monitoring approach depends on how reliable, accurate, and timely the information provided by the monitoring system is. For an example, in Highland Valley Copper mine in British Columbia, mining occurs by open-pit methods. Stability of the benches, which form the massive walls encompassing the pit, is monitored using automated monitoring systems. The RTS theodolites (such as TCA1800S) are the primary measurement sensors in the automated monitoring system with approximately 500 slope monitoring target prisms located strategically throughout the pit; target prisms are placed at 100 m increments away from a specific RTS. The pit wall areas are continuously monitored 24 h a day. Monitoring prisms are mounted to the pit walls, and distances, horizontal directions, and zenith angles are measured to them in order to uniquely determine 3D positions of the prisms. The RTSs can be programmed for sequential self-pointing to a set of target prisms at predetermined time intervals and the measurements can be transmitted to remote stations via a telemetry link. The approach used in minimizing the effects of both refraction and random pointing errors based on the use of robotic total station measurements is given as follows (Bond, 2004):

- Maintain short distances from the RTSs to the target prisms.
- Take observations in several sets and spread the observations over long periods to randomize the effects of refraction. In monitoring system, meteorological sensors are interfaced with a computer to create a fully automated stand-alone monitoring system that corrects measurements for meteorological influences such as refraction.
- Keep lines of sight away from strong sources of heat radiation.

There are concerns that if RTSs are unstable, the displacements determined for the wall points will be biased. If the RTSs are suspected to be unstable, the RTS can be used together with the GPS antennas in either of the following arrangements:

1. Collocate the GPS antenna with the main RTS sheltered in the unstable mining region, then do the following:
   - Choose two other GPS antennas collocated with 360° prisms and locate one of such combinations on a stable point *S* probably outside the mining region and the other within the unstable region at point *U*.
   - The stable point *S* must be within 200–1000 m for best results with ATR in the RTS.
   - Use the GPS antenna (with prism) at stable point *S* to provide orientation for the main RTS in the shelter; the GPS collocated with the RTS will update the position of the RTS; and the main RTS will use the orientation and the updated position to make "correct" measurement to the targets on the unstable object points.
2. Collocate the GPS antennas with 360° prisms and position them on at least three stable points (forming reference points), probably outside the mining region

(but within 200–1000 m of the main RTS for best ATR results), then do the following:

- The sheltered main RTS (in an unstable region) will measure to three or more reference points with mounted 360° prisms in free station computation to determine its position and orientation before making measurements to the targets on the unstable object points.
- The GPS antennas collocated with the prisms (the reference points) are to be positioned so as to form a strong geometry in order to ensure that free station calculates with high accuracy.

In either of the above arrangements, the corrections to the RTSs are derived in fully automated mode. The typical measurements made by the RTSs to the target prisms are the direction, zenith angle, and distance measurements. The main challenges in using GPS for open-pit monitoring were given (Bond, 2007, 2004) as follows:

1. The steep walls of an open pit limit the effectiveness of satellite positioning technologies by masking some satellite signals and thereby diluting the geometric strength of solutions.
2. Large height differences between master and rover stations can lead to significant height biases in baseline solutions. Residual tropospheric delay biases due to a large height difference between master and rover stations can contaminate the vertical component of GPS baseline solutions. Height difference between the RTS/GPS at the bottom of the mine and the reference GPS station may be in the order of several hundred meters (e.g., 700 m) and slope distance can be up to 2.5 km.
3. There is a need to develop a fully automated GPS processor for continuous updates in real time and to be able to provide communication links to transfer data between GPS receiver and a central processing computer.
4. The problem of multipath, resulting in inaccurate GPS measurements; sources of interference for wireless networking and radio repeaters will be of concern in automated monitoring systems.
5. Supplying power to GPS units when a large number of targets must be monitored. This can be difficult in inaccessible target areas. Noisy and fluctuating power supply may also be of concern.
6. Hostile and variable weather conditions may require that the instruments be sheltered in the case of automated monitoring systems.

The purpose of RTS/GPS hybrid system is to obtain corrections to the RTS position components (northings, eastings, and heights) with a standard deviation of less than or equal to about 2.5 mm in a fully automated mode of operation. The following are some of the total station equipment-related concerns when used in open-pit mines:

1. Refraction and pointing errors limit the accuracy of direction and distance measurements made by total station instruments and laser scanners where

the pit diameter exceeds 1 km as is usually the case in large-scale projects. Degradation in precision of geodetic technologies will be so large that the minimum detectable displacement can exceed the mine's requirements for displacement detection.

2. Complex behavior of the pit as it responds to changes in its environment (e.g., excavation, increase in water saturation, tectonic movement). There is a need to delineate zone of deformation in order to identify stable regions as reference points. Suitable locations for target points should also be determined and the expected displacement rates must be predicted. This creates a problem of connecting to stable reference points; each reference control point must be stable. The instrument locations may also become unstable; this must be monitored in the automated monitoring system.

The general conclusion is that in most cases, it is more practical and economical to use GPS to monitor the stability of other sensors (e.g., total stations, laser scanners) that can provide spatial resolution at a lower cost in localized areas. GPS units, however, are expensive; they should not be left in areas that are likely to fail, causing the units to be damaged or lost.

9.4.8 Deformation Monitoring of Tunnels

Deformations of usual interest in tunneling are movement of the tunnel walls (inward movement, settlement, heave, and often three-dimensional displacement), deformation in the ground around the tunnel and ahead of the tunnel excavation face, and deformation (such as settlement, tilt, lateral displacement, and usually three dimensional displacement) at, or near, ground surface (on structures and utilities). Deformation monitoring of tunnels is to ensure that structures at the ground surface are not harmed by the tunneling operations. The process is to provide early warning against the collapse of the tunnels. The possible causes of tunnel deformations include the following:

- Combination of adverse ground and groundwater regimes
- Large overburden pressures, such as the existence of sensitive and/or utilities within the zone of influence of the tunnel, especially in the case of shallow urban tunnels
- Intense tectonic activities.

Ground deformation monitoring has different objectives in mountain and urban tunnels. In mountain tunnels, the main objective of deformation measurements during construction is usually to ensure that there exists enough margin of safety against roof collapse, bottom heave, failure of the excavation face, yielding of the support system, and so on. In shallow urban tunnels, the main objective of ground deformation monitoring is to limit ground displacements to values that are sufficiently low so as to prevent significant damages to structures and utilities on the surface.

Deformation monitoring in tunneling projects is carried out with instruments that are installed or operated either from the ground surface or from within the tunnel. Typically, the majority of ground deformation takes place ahead and close to the tunnel face. This requires that the monitoring system be installed as early as possible. However, there is usually a limit to how close the monitoring system could be to the tunnel face to avoid interfering with the construction of the temporary support system for the tunnel (such as sprayed concrete, steel sets). The geodetic and geotechnical monitoring approaches can be used to complement each other. The geodetic measurements will provide absolute coordinates of the target locations in time, while the geotechnical measurements will usually provide relative displacements of the target locations in one direction only, with respect to an initial condition of the target. The geotechnical measurements can also provide absolute coordinates of target locations if the initial coordinates of the targets are already obtained using geodetic methods. Typical geotechnical measurements in the tunnel are based on the use of extensometers, which allow measurements only along a specific line; the equipment is easy to use and maintain, but construction process is usually obstructed during the reading of the equipment.

The geodetic approach in measuring the tunnel wall usually involves using total stations with optical reflector targets (up to 5–7 reflectors per section), which are installed at sections along the tunnel (e.g., 15–20 m). The walls are measured three dimensionally with respect to stable reference positions, which are located outside the tunnel. Measurement of the targets inside the tunnel is obtained by placing the total station on predefined brackets (typically bolted on the tunnel wall). As the brackets used for positioning the instrument often move, following long-term displacement of the tunnel wall, corrections to the positions of brackets are necessary; positions of brackets may be unstable due to creep deformation of the tunnel walls.

9.5 VERTICAL DEFORMATION MONITORING AND ANALYSIS

Precise geodetic leveling procedures such as special-order and first-order differential leveling procedures can be employed in deformation monitoring of an object in order to determine the following:

- Tilts based on height differences measured over extended bases of virtually limitless lengths between pairs of benchmarks located in or on the monitored object
- Vertical expansion (settlement, uplift, or subsidence)
- Absolute height changes with respect to stable points.

With regard to ground surface subsidence monitoring of active areas, such as mining areas, it is common to employ special-order or first-order differential leveling procedure. Leveling with parallel glass plate micrometer with invar graduated rods or first-order digital levels with bar-code invar rods can be used for this purpose. Since level reference in geodetic leveling is created by an optical line of sight through a telescope, vertical atmospheric refraction will become the major source of systematic errors. Vertical refraction errors, however, may accumulate up to a few millimeters along moderately inclined long routes if there are unequal heights of the forward and backward horizontal lines above the terrain. Other source of errors in geodetic leveling approach is the usual increase in random errors due to rod scale error and settlement of instrument and rods when a large number of setups are involved. Trigonometric leveling method may also be used for economic reasons since the method is more economical than the conventional differential leveling procedure, especially when used in regions with rapidly changing elevations. If the trigonometric method is used, measurements must be made reciprocally in order to minimize the effects of refraction, which is usually the main source of errors with the method.

Space-borne GPS survey techniques can also be employed in vertical deformation monitoring. However, in order to optimally compromise between geodetic leveling and GPS determination of vertical components of displacements, the use of a three-baseline method has been found (Chrzanowski and Szostak-Chrzanowski, 2010; Chrzanowski and Bazanowski, 2011) to have produced reasonable results. The method economizes GPS surveys since one can carry out the field operation unassisted. This method is illustrated in Figure 9.14 and in the following steps. In Figure 9.14, let network points C_1, C_2, and C_3 represent the control points with continuously operating GPS receivers, while point P is the point whose position is to be determined.

The steps for positioning point P using three-baseline approach can be given as follows:

- Set up three GPS antennas on the three control points C_1, C_2, and C_3 to operate continuously during the whole survey campaign (of several days).
- Set up another GPS antenna on the monitoring point P and determine its position by measuring the three GPS baselines (B_1, B_2, and B_3) between the three control points and the monitoring point. Discrepancies between the three baseline results serve as a verification of the actual positioning accuracy. Additional monitoring points can be added and their positions determined independently based on the same three control points.
- In order to achieve subcentimeter accuracy in vertical positioning at 95% confidence level, the duration of the observation sessions has to be up

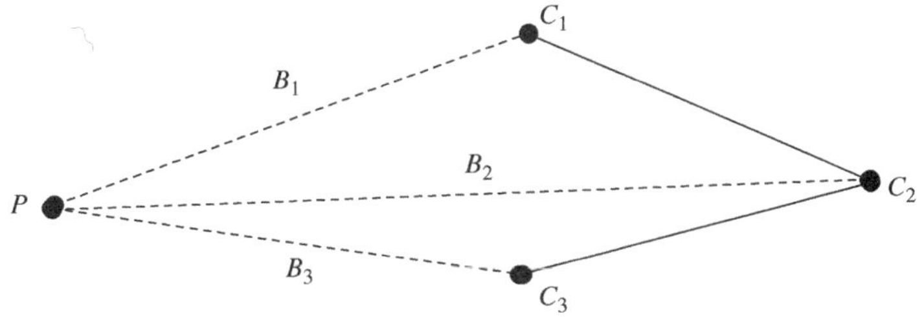

Figure 9.14 Three-baseline GPS survey method.

to 12 h (Chrzanowski and Szostak-Chrzanowski, 2010; Chrzanowski and Bazanowski, 2011).

9.5.1 Tilt, Strain, and Curvature Determination from Geodetic Leveling

Tilt is defined as the deviation of a surface relative to a horizontal reference surface, as shown in Figure 9.15(a). The terms *tilt* and *inclination* are often used interchangeably to mean the same thing. *Inclination*, however, is the deviation of a monitored surface from a reference vertical plane (i.e., the plane containing the direction of gravity) as shown in Figure 9.15(b). In Figure 9.15(a), if the tilt angle is α for a surface length of s_1, the tilt displacement can be expressed as $\ell_1 = s_1\alpha$ (where α is in radians); similarly, for inclined angle β with surface length of s_2, the corresponding inclination displacement can be given as $\ell_2 = s_2\beta$ (where β is in radians).

Tilts of a monitored surface can be determined using geodetic leveling approach. If this approach is used, tilt over an extended base of virtually limitless length can be determined. This is an advantage over the geotechnical approach (e.g., using tilt-meters), which is only able to determine tilts over a very short base length. For example, a typical geodetic-order digital level is capable of achieving 0.3 mm standard deviation over 1 km double leveling run with invar rod; this error translates to less than 0.1″ in angular tilt. The geodetic leveling approach for determining tilts of structural components can be illustrated using Figure 9.15(a). In the figure, let the elevations of points P_1 and P_2 determined in epoch t_1 be h_1 and h_2, respectively. The measured or determined height difference between the two points at epoch t_1 can be given as $\Delta h_{12t_1} = h_2 - h_1$. Similarly, the measured height difference in epoch t_2 can be given as Δh_{12t_2}. The tilt angle (α) in radians between the two points from epoch t_1 to t_2 can be calculated as

$$\alpha = \frac{\delta \Delta h}{s_1} \tag{9.69}$$

where s_1 is the horizontal separation between P_1 and P_2, and $\delta\Delta h = \Delta h_{12t_2} - \Delta h_{12t_1}$. If point P_1 remains stable between the two epochs, then $\delta\Delta h$ will be equal to ℓ_1 in

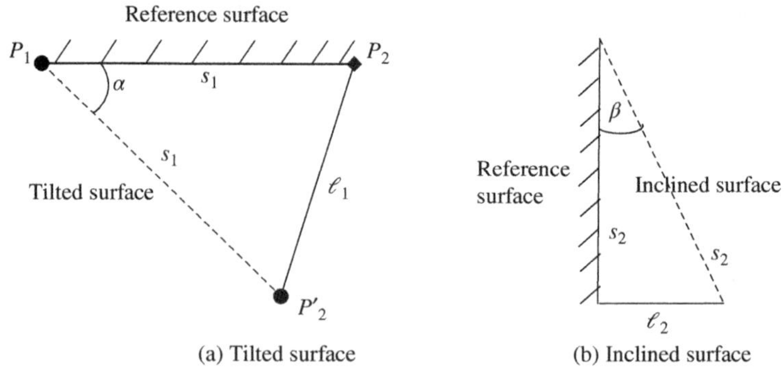

Figure 9.15 Tilted and inclined surfaces.

Figure 9.15(a). According to Chrzanowski and Secord (2000), the height difference and the distance between the two points need not be measured directly between the two points. The leveling can be done along any convenient route and the distance can be obtained in a variety of ways, for example, inversing from the coordinates of the two points.

Apart from being able to derive vertical displacements and tilts of terrains from ground subsidence monitoring data, other deformation parameters associated with ground subsidence can also be determined, such as extensions between ground points (or changes in horizontal strain) and curvature of subsidence bowl. Deformation tolerances for assessing the impact of ground subsidence on infrastructure are usually specified based on some of the deformation parameters or criteria, such as tilt (or vertical displacements), horizontal strain (or horizontal displacements), and curvature of the subsidence trough. The typical specifications with regard to deformation criteria are that the acceptable tilt and horizontal strain should not exceed deformation tolerance of 2.5 and 1.5 mm/m, respectively; and the radius of ground curvature should be larger than 20 km (Chrzanowski and Szostak-Chrzanowski, 2010). If these specifications are exceeded with regard to surface structure, then it could be concluded that the subsidence has significantly impacted the structure. Figure 9.16 can be used to illustrate how tilt, strain, and curvature of a subsidence bowl can be determined. In the figure, let the vertical displacements (determined from two epochs of survey) at two points P_1 and P_2 in the bowl be dh_1 and dh_2, respectively, and S is the length separating the two points. The tilt of one point relative to the other can be given as (Chrzanowski and Szostak-Chrzanowski, 2010)

$$T = \frac{(dh_2 - dh_1)}{S} \tag{9.70}$$

Similarly, if the horizontal displacements (resulting from the vertical displacements) in the direction of *x-axis* at the two points are dx_1 and dx_2, the homogeneous strain component in the *x-axis* direction between the two points can be given as

$$\varepsilon = \frac{(dx_2 - dx_1)}{S} \tag{9.71}$$

The curvature (K) of the subsidence bowl (an impact of mining on surface structures) can be given as (Chrzanowski and Szostak-Chrzanowski, 2010):

$$K = \frac{(T_2 - T_1)}{L} \tag{9.72}$$

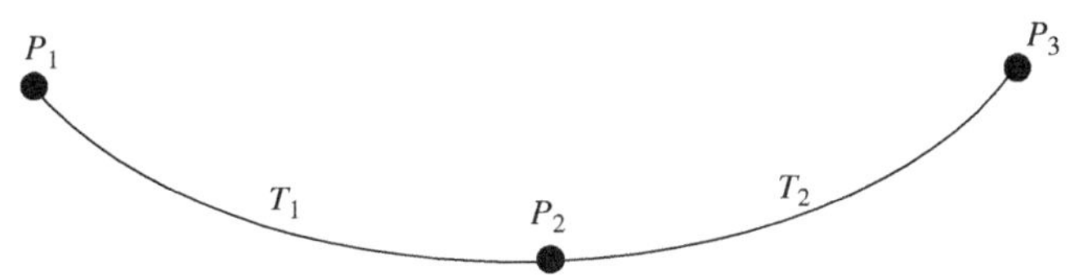

Figure 9.16 Subsidence bowl.

where T_1 is the tilt between P_1 and P_2, T_2 is the tilt between P_2 and P_3, and L is the distance between the midpoints of P_1-P_2 and P_2-P_3 (representing the locations of T_1 and T_2) as illustrated in Figure 9.16.

If geodetic leveling surveys are conducted inside the galleries of dam structures and the surveys are connected to a deeply anchored rod of a borehole extensometer, the resulting leveling measurements can be used to determine the absolute height changes as well as tilts of components of the dam structures. This is illustrated in Figure 9.17.

In Figure 9.17, for example, the vertical control station BM at Floor 1 and the precise level instruments at Floor 1 and Floor 2 can be used to establish the elevations of P_i and the borehole extensometer collar location P_j by making measurement to a suspended long invar scale line. After the initial establishment of the elevations of the points, the invar scale line can be replaced with a suspended plumbline with two scales attached, as shown in Figure 9.17. This is to inexpensively allow subsequent monitoring of the established points (P_i and P_j) for relative vertical movements. If the borehole is stable, changes in height between points P_i and P_j in two epochs can

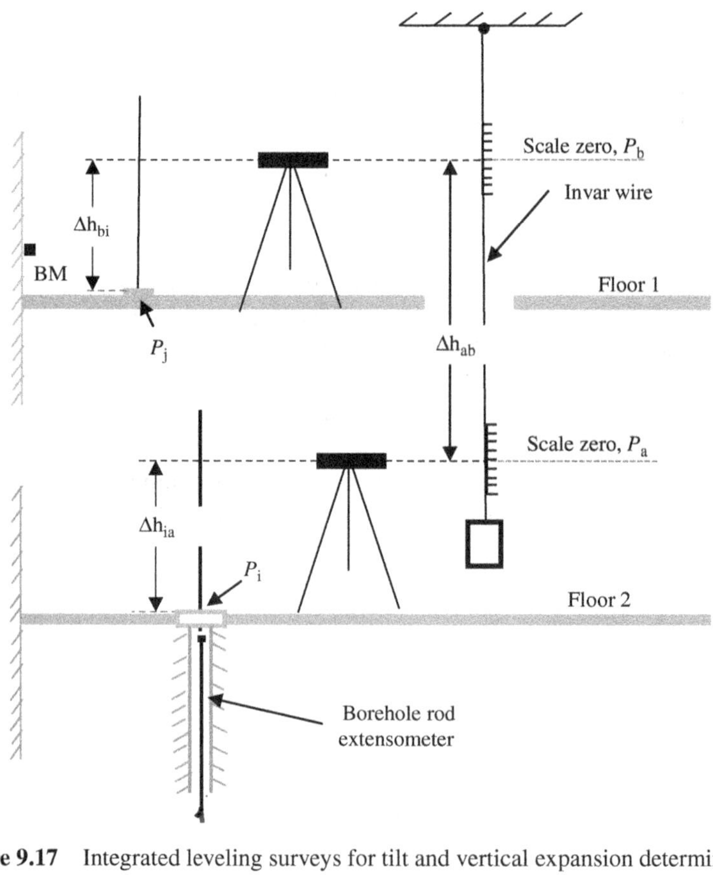

Figure 9.17 Integrated leveling surveys for tilt and vertical expansion determination.

be determined as follows. For each campaign k, let the adjusted elevations of points P_i and P_j be E_i^k and E_j^k, respectively; the adjusted height difference (Δh_{ij}^k) is derived from the adjusted elevations by

$$\Delta h_{ij}^k = E_j^k - E_i^k \tag{9.73}$$

A change in height difference ($\delta(\Delta h_{ij}^k)$) between campaign k and the initial campaign $k=1$ can be created for the kth campaign as

$$\delta(\Delta h_{ij}^k) = \Delta h_{ij}^k - \Delta h_{ij}^1 \tag{9.74}$$

The subsequent campaigns will give changes in the height difference with respect to the value at the initial campaign $k=1$. Taking y_k as $\delta(\Delta h_{ij}^k)$ and x_k as a given point in time of the campaign, y_k can be plotted against x_k (for $k=1, 2, \ldots, n$). By fitting a sinusoid to the data series, the rate of change of height difference (mm/year) is transformed into *tilt rate* (mm/m/year) by dividing the rate by the horizontal separation between the two points P_i and P_j (using their coordinates). The vertical displacement rate (mm/year) is transformed into an *extension or strain rate* (mm/m/year) by dividing the rate by the vertical separation between the two points P_i and P_j (using their elevations). If P_j is moving upward with respect to P_i, then the height difference between them is increasing, consequently both the displacement rate and the tilt rate and strain rate will be positive. The leveling results, however, are usually listed in the form of vertical displacement rates with respect to the reference horizontal surface of the borehole extensometer (the deepest anchored invar rod) accepted as stable.

Although in Figure 9.17 the actual vertical distance between a pair of scale zeroes along a plumbline may be unknown, the relative change between them can be considered as constant. The scale zeroes would not likely remain at the same elevation because of possible movement of the suspension point of the plumbline, but the readings with respect to the scale zeroes can be used to determine the changes in the height difference as time changes. From Figure 9.17, if the elevations of points P_i and P_j are E_i and E_j, respectively, the change in height can be given as

$$\Delta h_{ij} = E_j - E_i \tag{9.75}$$

From Figure 9.17, it can be expressed that

$$E_j = E_i + \Delta h_{ia} + \Delta h_{ab} - \Delta h_{bj} \tag{9.76}$$

Substituting Equation (9.76) into Equation (9.75) gives

$$\Delta h_{ij} = \Delta h_{ia} + \Delta h_{ab} - \Delta h_{bj} \tag{9.77}$$

If Equation (9.77) is considered at the kth campaign, height difference Δh_{ij}^k will be obtained, and the change in the height difference at the kth campaign (compared to the first campaign) can be given as

$$\delta\left(\Delta h_{ij}^{k}\right) = \left[\Delta h_{ia} + \Delta h_{ab} - \Delta h_{aj}\right]^{k} - \left[\Delta h_{ia} + \Delta h_{ab} - \Delta h_{bj}\right]^{1} \tag{9.78}$$

or

$$\delta\left(\Delta h_{ij}^{k}\right) = \left[\Delta h_{ia} - \Delta h_{bj}\right]^{k} - \left[\Delta h_{ia} - \Delta h_{bj}\right]^{1} \tag{9.79}$$

where Δh_{ab} is considered as constant. A data series is then created similarly as in the case of Equation (9.74).

10

DEFORMATION MONITORING AND ANALYSIS: HIGH-DEFINITION SURVEY AND REMOTE SENSING TECHNIQUES

OBJECTIVES

At the end of this chapter, you should be able to

1. Describe the operation principles of laser scanning systems
2. Classify terrestrial laser scanners according to data gathering techniques, angular coverage, and range coverage
3. Discuss the properties, advantages, and limitations of various classes of terrestrial laser scanners
4. Explain the sources of error in terrestrial laser scanners
5. Discuss the application of terrestrial laser scanners in deformation monitoring, including the limitations
6. Discuss the concepts of synthetic aperture radar and the application in interferometry
7. Discuss the basic principles of satellite-based interferometric synthetic aperture radar (InSAR), including the data processing techniques
8. Explain the approach for creating InSAR interferogram
9. Interpret InSAR interferogram with regard to deformation monitoring
10. Explain the different techniques for improving InSAR results with regard to the choice of scatterers

Precision Surveying: The Principles and Geomatics Practice, First Edition. John Olusegun Ogundare.

11. Discuss the applications, advantages, and limitations of InSAR and ground-based InSAR (GB-InSAR)
12. Compare LiDAR and InSAR systems

10.1 INTRODUCTION

High-definition survey (HDS) and remote sensing techniques for deformation monitoring and analysis discussed in this chapter are based on some aspects of laser scanning and radar systems that are capable of accurately detecting deformations in the order of centimeters to millimeter levels. The HDS aspect refers to mapping methods that produce a dense set of three-dimensional data points on large objects. It is based on the concepts of close-range photogrammetric technologies, which include terrestrial laser scanner (TLS) and ground-based interferometric synthetic aperture radar (GB-InSAR) systems. The remote sensing aspect is based on satellite-based interferometric synthetic aperture radar (InSAR) technologies.

10.2 LASER SYSTEMS

10.2.1 Properties of Laser

LASER is an acronym for light amplification by stimulated emission of radiation. This is another type of light source, just as a candle is a light source when it is burning. Laser makes use of processes that increase or amplify light signals after those signals have been generated by other means. Laser source consists of an amplifying medium (where stimulated emission occurs) and a set of mirrors to feed the light back into the amplifier for continued growth of the developing beam.

Comparing laser with a burning candle, the burning candle radiates light in all directions. This means that it illuminates various objects equally if they are equidistant from the candle. A laser takes light that would normally be emitted in all directions, such as from a candle, and concentrates that light into a single direction, for example, into a single beam of the diameter of probably a few millimeters. If you were standing a distance of 1 m from the candle, then the light intensity would be several thousand times as bright as the light that you normally see radiating from the candle. Note, however, that a candle is not the kind of medium that produces amplification and thus there are no candle lasers. Lasers span the wavelength ranging from the far infrared part of the spectrum ($\lambda = 1000\,\mu\text{m}$) to the soft-X-ray region ($\lambda = 3$ nm), such as far infrared, middle infrared, near infrared, visible (blue, green, red), ultraviolet, soft X-rays. Lasers are considered to be more hazardous than ordinary light because of their *monochromatic*, *directional*, *coherency,* and *intensity* properties (i.e., it can squeeze a massive amount of energy into its narrow beam). The high output power of some lasers is useful in cutting through steel and even ceramics, whereas narrow and straight beam properties are useful in surveying.

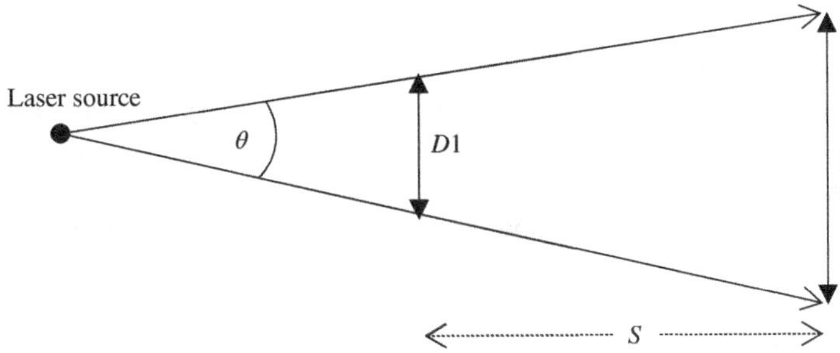

Figure 10.1 Propagation of laser beam.

10.2.1.1 Monochromatic Property of Laser Laser light is said to be *monochromatic* (or spectrally pure) because it consists of one wavelength or one color hue. Generally, it can be said that it has a spectral width (or narrow frequency distribution) much less than that of other light sources, that is, it has a very narrow bandwidth. Dispersion of light in the atmosphere is less for a monochromatic source. In contrast, ordinary white light is a combination of many different wavelengths or color hues.

10.2.1.2 Directional Property of Laser Laser is *directional* (or has low angular divergence) since it is emitted as a relatively narrow beam in a specific direction unlike ordinary light, which is always emitted in many directions away from the source. The beam divergence (θ, usually in milliradians or microradians) is illustrated in Figure 10.1 and by Equation (10.1):

$$\text{Beam divergence } (\theta) = \frac{D_2 - D_1}{S} \tag{10.1}$$

where D_1 is the diameter of the objective lens of telescope for focusing the laser beam and D_2 is the diameter of the laser beam at distance S m away from the objective lens. For example, if the laser has a 1-mm-diameter output (i.e., the diameter of the objective lens) and a 120 μrad divergence (or angular accuracy of ±60 μrad or ±12.4″), at $S = 100$ m, the beam will have a diameter of 13 mm.

Lasers, particularly HeNe gas lasers, propagate as Gaussian beams and have a very well-defined symmetry across their propagated wavefronts. The center of the laser beam can be detected accurately. Usually, the diameter of the laser spot at a target is described as the diameter, D_2, of the Airy disk only, which can be calculated from (Chrzanowski and Ahmed, 1971; Chrzanowski and Egberongbe, 1971):

$$D_2 = \frac{2.44\lambda S}{D_1} \tag{10.2}$$

where

S = distance of the target from the laser source;

D_1 = diameter of the objective lens of telescope for focusing the beam;

λ = wavelength of the laser (0.63 μm for He–Ne laser; 0.55 μm for green light used in some scanners).

Practical experiments (Chrzanowski and Ahmed, 1971; Chrzanowski and Egberongbe, 1971) show that in reasonable atmospheric conditions, the laser spot diameters are about 25% larger than the calculated values from Equation (10.2). In strong thermal turbulence of air, the diameter may be as much as 80% larger.

10.2.1.3 Coherency Property of Laser Laser is *coherent* because it consists of waves having a high degree of similarity of phase, direction, and amplitude. The wavelengths of the laser light are in phase in space and time. There is no mixture of many frequencies in different directions, and, spatially, all parts of a laser wavefront in a plane perpendicular to the direction of propagation are in phase. This means that lasers are stable (move with almost the same frequency) with less interference among themselves in space. For example, a 10 W laser will burn one at several meters away while one has to touch a 100 W light bulb before one gets burnt.

10.2.1.4 Output Intensity Property of Laser Laser has high output intensity (amplitude or brightness). This means that lasers have more energy or are stronger than any other visible light. Output power is expressed in watts and the output energy is quoted in Joules. Output energy is the output power multiplied by the pulse duration.

10.2.1.5 Degradation of Laser Properties When transmitted through the atmosphere, a certain degradation of laser properties occurs. Some of the factors causing the degradations are as follows:

1. Amount of water vapor or particles present in the air. Range will be decreased in a foggy or dusty environment. Attenuation is generally lower for infrared diode and solid-state lasers than for HeNe lasers.
2. Influence of the atmospheric refraction. The atmospheric refraction affects laser radiation the same way it affects any other light or infrared source.
3. Influence of air turbulence in the laser path. Air turbulence can affect quite seriously the beam emitted by fixed and rotating laser instruments; it affects coherence of beams and causes the laser spot to be broken into separate parts that appear to spark. Air turbulence is basically caused by random temperature fluctuations, which range from a few tenths to several degree Celsius in the atmospheric temperature close to the ground. The fluctuations result in density changes of air, and these in turn result in fluctuations in the refractive index. At distances greater than 150 m, it is often quite difficult to establish the center of laser beam.

Note also that intensity of laser depends on the distance to the target, angle of incidence with the target, and the surface properties (e.g., color, roughness) of the target. The intensity of laser beam actually reduces with distance so that phase shift cannot be reliably detected at some distances.

10.2.1.6 Application of Laser One of the common applications of laser today is in producing reflectorless total station equipment and three-dimensional laser-scanning instruments, which are useful for various geodetic and engineering applications. Reflectorless total station instruments are now widely used in geomatics and will not be discussed any further in this chapter. With regard to laser-scanning instruments, two types can be identified as *airborne laser scanners* and *TLSs*. Since only TLSs have demonstrated sufficient accuracy to justify their use in deformation monitoring and analysis, TLSs will only be discussed further in this chapter; the curious readers can refer to Section 8.5.1 for a brief discussion provided on some aspects of airborne laser scanners.

10.2.2 Terrestrial Laser Scanners

TLSs belong to the family of active sensors, such as the well-known radar. The scanners are considered active because they illuminate their targets themselves. Their carrier wave is laser (which may contain modulated signals), and they cannot survey specific points, but instead, provide nearly continuous scanning of the target object around the scanners.

TLSs are known by various names, such as *ground-based laser scanners*, *static TLSs, or terrestrial Light Detection And Ranging* (*LiDAR*) *systems*. They are close-range surveying systems, which are typically operated from fixed locations on the ground.

10.2.2.1 Measuring Techniques of Terrestrial Laser Scanners The TLS system is a combined hardware and software package. The hardware component consists of a tripod-mounted laser distance measuring system, the horizontal angle and vertical (zenith) angle measuring device, and a mechanical scanner for measuring light reflectance or intensity reflected from the targets. The hardware is coupled with polygonal mirrors to facilitate beam deflection in the horizontal and vertical directions, while the angle encoders record the orientation of the mirror. In most of the scanners, however, three-dimensional (x, y, z) coordinates are provided as output instead of the measured quantities (distances, horizontal and vertical angles). For the measurement of distances, different types of scanners adopt different distance measuring techniques, such as *time-of-flight* (*pulse*), *phase-shift, and laser triangulation techniques*. The description of each of these techniques is given as follows.

Time-of-Flight or Pulse Technique In time of flight (or pulse) technique, a scanner measures the time elapsed between emission and detection of laser pulse to produce the distance between the scanner and the target. This type of instrument fires rapid laser pulses of about 10×10^{-9} s long at a surface and then records the amount of time it takes each of the laser pulses to travel the distance to the surface and back. The time measurement is then converted into a distance measurement. Current laser systems operate at pulse rates from a few hundred pulses per second to 10,000 pulses per second. A higher pulse rate allows for a wider coverage swath or a closer spacing of elevation points and increases the laser power requirements. Each pulse covers a finite area determined by the instantaneous field of view (IFOV) of the scanner. A single pulse may have multiple returns, as in the case when it is partially reflected from tree canopies, undergrowth, or the ground. Some scanner systems record these multiple reflections to aid in removing vegetation reflections from the terrain model. The time-of-flight method, however, is limited by the precision of the timing device (clock) in the system. More details on this technique can be found in Section 5.3.1.

Phase-Shift Technique In phase-shift technique, a scanner compares the phase shift in the laser light reflected from the scanned object to standard phase, which is also captured for comparison. The phase-shift scanner employs an amplitude-modulated continuous waveform laser so that when the laser beam interacts with the target, the phase is reset, and the returned phase-shifted signal is processed to derive the distance. The phase shift can be resolved from 1/4000 and up to 1/8000 of the wavelength. More than one frequency is used to resolve distance ambiguity; the lowest frequency defines the maximum distance and higher frequencies are to improve the coarse distance provided by the lowest frequency within the desired precision. Some scanners use two frequencies for ambiguity resolution. More details on this technique can be found in Section 5.3.2.

Laser Triangulation Technique In laser triangulation technique, the scanner determines the distance between instrument and the target by using the triangle formed by the laser source, the target, and the instrument's recording unit (Boehler and Marbs, 2002). In this process, the base formed by the distance between the laser source and the recording unit and the two angles subtended with this base by the propagated laser beams are used in calculating the distance. This technique is used in the industrial field with the measuring distance up to 5 m.

10.2.2.2 Georeferencing Principles of Scanner Data The distance measurements, the vertical angle, and horizontal direction measurements are used in determining the (x, y, z) coordinates based on the scanner's internally defined coordinate system as discussed in Chapter 8. The set of (x, y, z) coordinates for several points constitute what is known as *point cloud or scan*. This term refers to a large collection of densely spaced and regularly measured points, appearing as a rendering of the project scene. The points are often colored according to the intensity of the laser return signal, and as the resultant image on the computer screen appears as many unconnected but closely spaced dots, it is often referred to as a point cloud.

The point cloud must be transformed from the scanner coordinate system to the ground coordinate system through georeferencing process as discussed in Chapter 8. The two methods of georeferencing the scan data are *direct method* and *indirect method* (Gordon and Lichti, 2004). The direct method was already discussed in Chapter 8 and will not be repeated in this chapter; instead, the indirect georeferencing method will only be discussed.

The indirect georeferencing method involves the use of scanners that usually do not have hardware facilities for positioning or orienting the scanners as required in the direct method scanners. It can be divided into two approaches: *two-step approach* and *one-step approach* (Reshetyuk, 2009). The two-step approach of georeferencing requires performing registration of pairs of point clouds (scans) to form a registered point cloud of the whole object as a first step and then following that with georeferencing of the registered point cloud of the whole object as a second step. It should be mentioned that in scanning large objects, it is common that many scans are made from different setup stations in order to completely capture the objects. The usual problem with this is that there is a need to "stitch" the different scans together to form one large scan, based on one common scanner setup coordinate (x, y, z) system. The process of doing this is known as r*egistration*.

The point clouds are registered based on clearly identifiable points that are sample in both point clouds. One of the point clouds is transformed in such a way that the distance between the tie points is minimized, usually, through least squares adjustment of Helmert transformation. The seven unknown parameters (three translations, three rotations, and one scale change) of the registration are solved by using at least three tie (or control target) points in the region of overlap. The high point density of the TLS acquisition of the target is used to determine the coordinates of a target point. The targets should be located in such a way that they form a good geometry, and when they are rotated or tilted, their centers remain in the same locations; they could be planar or spherical in shape. Using the provided registration software, the centers of the targets are determined.

These tie or control points are needed for each pair of scans in order to determine the seven unknown parameters needed to transform one scan into another and then tie all the scans into one coordinate system of a chosen reference scan. After the registration of all the point clouds, the registered point cloud of the whole object is transformed into a chosen external (ground) coordinate system through the process of georeferencing to complete the two-step approach. Helmert transformation may also be used in this step to do the transformations (Lichti et al., 2002). In the one-step approach of indirect georeferencing, point clouds are transformed into external ground coordinate system using just the control points; at least three control points are required in each scan for this purpose. This approach georeferences each scan independently and does not require any overlap between scans.

Both the indirect and direct approaches of georeferencing have their own advantages and limitations. The indirect georeferencing have the advantages of being more accurate than the direct georeferencing approach; centering, leveling, and measuring heights of instruments and targets are not required in the approach; and it is more flexible about instrument location. Some of the disadvantages of the approach are that it requires overlap areas and it is not always possible to achieve the needed good geometry in the overlap areas (Reshetyuk, 2009). In the case of direct georeferencing, overlap areas between pairs of scans are not needed; and since the approach is well known to surveyors, it is widely acceptable to surveyors, who are able to integrate the approach with the traditional survey practice.

10.2.2.3 Classification of Terrestrial Laser Scanners The 3D laser scanners can be classified into three categories according to the techniques adopted in measuring ranges (distances) to the targets (Schulz and Ingensand, 2004; Goor, 2011; Catalina and Andreea-Florina, 2013), such as *laser triangulation based scanners*, *pulse-based* or *time-of-flight scanners*, and *phase-based scanners*. The maximum ranges for time-of-flight systems are usually longer than phase-shift systems, but phase-shift systems usually have a higher measurement rate and higher accuracy. In general, phase-shift systems are well suited for high precision and detailed measurement of nearby scenes, such as industrial objects, heritage sites and crime scenes, while pulse systems are well suited for 3D reconstruction of scenes farther away from the scanner, for example, creating 3D models of plants, entire cities, and so on.

The TLSs can also be classified into three groups according to the angular coverage of the scanners as follows (Petrie and Toth, 2009):

(i) *Panoramic-type scanners* give a full 360° angular coverage within the horizontal plane passing through the instrument's center and typically have a minimum 180° coverage in the vertical plane at right angles to the horizontal plane.
(ii) *Hybrid-type scanners* have 50–60° vertical angular coverage and unrestricted horizontal scanning movement.
(iii) *Camera-type scanners* have much more limited angular range.

The other classifications of TLSs can be given based on the ranges of the scanners as follows (Petrie and Toth, 2009):

(a) Short-range scanners with maximum range of 50–150 m.
(b) Medium-range scanners with maximum range of 150–350 m.
(c) Long-range scanners with maximum range greater than 350 m.

The examples of short-range, medium-range, and long-range scanners are given in Tables 10.1–10.3, respectively, according to specifications quoted from Lemmens (2009), Petrie and Toth (2009), POB (2006), Leica (2013a, 2013b), and Z+F Imager (2013).

10.2.2.4 Procedures for Terrestrial Laser Scanning Project The field procedure for laser scanning project will require the following:

1. Placing highly visible targets around the project site and coordinating them by using the conventional traversing methods or by the GPS survey method.
2. Setting up the scanner system and scanning the object to be measured; if direct georeferencing method is used, the backsight target will be used to orient the scanner.

TABLE 10.1 Short-Range Laser Scanners.

Manufacturer	Trimble	Callidus	FARO Scanner Production GmbH	Leica Geosystems
System	Trimble FX	CPW 8000	Photon 120	Leica ScanStation P20
Range measurement principle	Phase shift	Pulse/phase	Phase	Pulse
Minimum/ maximum range	46 m	80 m	0.6 m/120 m (at 90% target reflectivity)	0.4 m/120 m (at 18% albedo)
Standard deviation of range	2.4 mm at 15 m (90% target reflectivity)	2 mm at 30 m	Range error at 25 m: 2 mm	±1.5 mm at 100% target reflectivity up to 100 m
Vertical angular field of view	270° Std: ±30″	300°	320°	270° Std: ±8″
Horizontal angular field of view	360° Std: ±30″	360°	360°	360° Std: ±8″
Measurement rate	190 kHz average	50 kHz	976 kHz	1 MHz
Applications	Civil, as-built surveys, archaeology	Plant, civil, archaeology	Industry products, residential; not for topography	Roadways, buildings, human-made objects
Tilt compensator	No	No	No	Yes

3. Processing the scan data, which will include the following:
    - Using the precise coordinates of the target points to perform point cloud registration and georeferencing;
    - Determining the coordinates of the backsight target center based on the point cloud and using them to improve the registration precision;
    - Performing data resampling to ensure that the point cloud data are evenly distributed;
    - Editing and tiling up the scan data for further analysis, such as deformation analysis, digital elevation model (DEM) generation, and so on.
4. Producing the deliverables (final products), which may include the contours of the measured area or deformation map of the area.

The highly visible targets must be placed around the project site in such a way that they can be included in multiple scans for use in stitching the overlapping scans

TABLE 10.2 Medium-Range Laser Scanners.

Manufacturer	Leica Geosystems	Z+F GmbH	Trimble	Maptek
System	Leica ScanStation C10	Imager 5010C	Trimble VX	I-Site 4400CR
Range measurement principle	Phase-shift measurement	Phase shift	Time of flight	Pulsed laser
Minimum/ maximum range	300 m (with 90% target reflectivity)	0.3/187 m	1/250 m	2/350 m
Standard deviation of range	±4 mm over 1–50 m	1.6 mm at 100 m (with 80% target reflectivity)	3 mm at <150 m	20 mm
Standard deviations of vertical and horizontal angles	±12″	±25″	1″	N/A
Vertical angular field of view	270°	320°	270°	80°
Horizontal angular field of view	360°	360°	360°	360°
Measurement rate	50 kHz	1 MHz	Up to 0.015 kHz	4.4 kHz
Applications	As-built, topographic, incident scene, monitoring surveys	Property survey, industry, forensics, archaeology	Conventional survey and scanning; land survey, civil	Underground, tunnel survey, infrastructural mapping, topography
Tilt compensator	Yes (accuracy: 1.5″)	Yes (accuracy: 25″)	No	Yes

together later. Suggested targets should include the objects that are visible on all scans for aligning different scans. Such targets could be different geometric objects, GPS antenna, and Styrofoam spheres.

Careful selection of scan station is very important in topographic surveying. At each location, the operator must identify the areas that will be obscured and decide whether there is a previous or future station that will "see" that area. The difference from a theodolite survey is the lack of prism rod to extend above obstructions. While data gathering is simplified, some expertise for this process is still required, especially when selecting scan stations and coverage range.

TABLE 10.3 Long-Range Laser Scanners.

Manufacturer	Optech Incorporated	RIEGL Laser Measurement Systems GmbH	Leica Geosystems	Maptek
System	ILRIS-HD	RIEGL LMS-Z620	Leica HDS8810	I-Site 8810
Range measurement principle	Pulsed, time of flight	Time of flight	Pulsed laser	Pulsed
Minimum/ maximum range	3/1800 m (80% reflectivity)	2/2000 m (80% target reflectivity)	Max. 2000 m (500 m on coal with 10% reflectivity)	2.5 m/2000 m (up to 1400 m with 80% target reflectivity)
Standard deviation of range	7 mm (4 mm averaged)	10 mm	8 mm at 200 m, 20 mm at 1000 m (under laboratory conditions)	±8 mm
Vertical angular field of view	40° (with 360° option) Std. of angle: ±17″	80°	80° Std. of angle: ±36″	80°
Horizontal angular field of view	360°	360°	360°	360°
Measurement rate (kHz)	10	11	8.8	40
Applications	Geological, civil, forensics, mining	Topography and mining, monitoring, civil, archaeology	Mine and topographic surveying	Mining and topographic surveys; monitoring
Tilt compensator	No	Yes	Yes	Yes

Scan data processing is primarily aimed at reducing the data set to a manageable size; as a topographic mapping tool, the scanner exceeds most users point density needs. The large data sets created by laser scanners demand special treatment in processing, such as the following:

1. Powerful computer with large hard drive is required for the manipulation of the large data sets.
2. Viewing the combined images on screen in the scanners' own software offers a rich collection of options – rotating the image, applying colors, and so on.

For those who may want to import scanned data into CAD software, the special problem is that many software packages are unable to process the multimillion point data sets created. One way of overcoming this is to apply *thinning filter* to the data before export. This requires specifying a minimum separation between points, resulting in deletion of extra points. At its highest density setting, the scanner can generate points at a spacing of 0.12 m over a distance of 100 m. In this instance, thinning to 0.25 m will eliminate approximately 50% of the data at this range. Even at 0.25 m separations, topographic data is far more densely packed than any surveyor would think of providing by traditional means.

3. Data from the project may be of a low quality if long grass covers most of the site at the time of survey. The laser detects the top surface, and so maps the grass cover. In this regard, a prism and theodolite survey would be superior.

10.2.2.5 Sources of Error in Terrestrial Laser Scanners Each point cloud measured with a laser scanner will likely contain a number of points that are affected by errors. The sources of those errors can be given as follows (Nguyen and Liu, n.d.; Lichti and Gordon, 2004; Soudarissanane et al., 2008; Cosarca et al., 2009):

1. Instrumental errors, which are difficult to determine and applied to angle and distance measurements. They reduce the precision of angle and distance measurements. Some of the instrumental errors are as follows:
   a. *Laser beam divergence error*, which will affect the angle measurements and also the point cloud resolution and positional uncertainty. Apparent location of the observation is along the centerline of the emitted beam; the uncertainty of the centerline could be as much as a one-quarter of the beam diameter.
   b. *Zero error*, which will affect the distance accuracy (based on EDM approach). This error will occur as a result of imprecisely known phase center of laser unit. The zero error also varies depending on the reflectivity of the scanned surface; a universal correction for zero error cannot be determined (a generally acceptable calibration and certification of laser scanner is not possible).
   c. *Positioning errors of rotating angle-measuring devices*, which will impact the angular accuracy. Angular errors will generate errors in coordinates of points when used for computing them.
   d. *Axial errors* (double-centering is not possible in this case) due to the possibility of the scanner axes not being perfectly aligned. Any axial error will result in angular errors. The typical axial errors are as follows:
      - *Vertical axis* (rotation axis) *error* occurs if the rotation axis of the scanner (about which the instrument horizontally rotates the laser beam) does not correspond with the vertical axis of instrument, causing eccentricities.
      - *Collimation axis error* occurs due to scan center not being the same as the vertical axis, which may cause the center of scanning mirror and the center of laser spot not being the same.

- *Horizontal axis* (rotating axis of deflecting mirrors) error occurs if the axis of the deflecting mirrors does not correspond with the horizontal axis of scanner.
- *Wobble of rotation axes* of the scanner during scanning operation will further compound the effects of the axial errors.

2. Errors related to the form and nature of scanned object, such as *boundaries effect,* will cause the returned signal to be a weighted average of both the reflection from the edge and the main surface of the scanned object. Boundaries effect depends on the reflective ability of the surfaces involved with the white surfaces providing the strongest reflections and less noise. This effect will lead to systematic errors in distance measurements with multipath effect contributing additional constant errors to the distance measurements.
3. Environmental errors due to the ways the atmospheric parameters modify the characteristics of the laser beam as it travels through the atmosphere. Some of the atmospheric parameters are as follows (Cosarca et al., 2009):
   - *Temperature of instrument*, surrounding atmosphere and scanned surface. Some errors are introduced to measurements if the instrument is operated in an unfavorable weather condition, or if it is not allowed to acclimatize before use; signal received from a scanned surface may vary depending on the temperature of the surface.
   - *Temperature gradient, pressure, and humidity* affect measurements as they do in EDM measurements; the effects are usually small on small distance measurements.
   - *Refraction and turbulence of beam.* Turbulence causes beam to land at different spots on the scanned surface while still maintaining the laser spot size.
   - *Effect of atmospheric attenuation* is not as much as in total station since short distances are involved.
   - *Interference from external radiation such as sunlight and lamp.* Scanning at night is recommended to minimize this effect.
   - *Effects of particles in the atmosphere.* Laser beams penetrate much worse through dense fog than through heavy rain (fog is a more severe problem for laser ranging than rain).
4. Methodological (or scanning geometry) errors, which are related to the method used to acquire and register multiple point clouds, affect point clouds. In scanning a surface, the incidence angle has the most influence on the data quality by affecting the signal-to-noise ratio; the received signal level of the measurements (which influences the precision of distance measurements) decreases with increasing incidence angles. Ideal setup for scanning a surface of an object (so as to increase the accuracy of distance measurements) is to position the laser scanner in a way that the laser beam is near perpendicular to the surface. The following are related to the scanning geometry: grid density (or resolution), which cannot be higher than the laser point accuracy; and distortions of measurements due to motions or vibrations during scanning, which can cause errors in angle measurements.

5. Errors due to sensor georeferencing at different epochs usually worsen the quality of results when used in deformation measurements. However, TLSs can be used to evaluate seasonal deformations of structures within points featuring a few centimeter displacements but not for the continuous monitoring as in GB-InSAR.
6. In laser scanners, scanning areas with elevated objects will result in "shadows," or missing data, behind the objects. The terrain in the shadow must be interpolated from adjacent points. To minimize shadowing, the total scan angle is kept narrow; however, this will reduce the area covered, increasing the number of scans required.

10.2.2.6 Advantages and Limitations of Terrestrial Laser Scanners Some of the advantages of using TLSs include the following:

1. *Accessibility problem is solved.* Laser scanner has an ability to gather data in a noncontact manner. This property can be important when an area is culturally sensitive or hazardous. It can be used for measuring distances where accessibility is a problem, for monitoring subsidence, or shifting, of wreckage piles (especially when the pile is inaccessible).
2. *Large density of data points collected for an area.* Compared with conventional methods, laser scanning technology helps in removing the need to make decisions regarding specific detail points to measure. With the scanner, one is concerned with coverage of regions rather than choosing break lines, significant feature edges or points. The scanner has an ability to collect very dense, precise three-dimensional data over a large area. The traditional methods based on the use of total station equipment and GPS record a limited amount of measurement points compared to laser-based technologies. In many cases with traditional methods, final products or deliverables lack key information, vital points, and coordinates required to accurately complete the task at hand. In deformation monitoring, the scanners are able to measure enough object points to represent the object being monitored.
3. *Speed of data capture is increased.* Due to the rapid nature of data capture, the scanners offer a cost-effective option for larger sites by allowing time in the field to be reduced significantly, as the area covered in a single scan can be in the order of several square meters. Only the time to place measurement targets and take images is usually required; it allows nearly real-time data collection and coordinates generation. Apart from providing minimum impact on industrial processes in an industrial area, the speed it provides can translate into increased safety of the personnel in a hazardous environment.
4. *Can be used anytime.* Laser mapping offers an advantage over conventional aerial photography because laser is not affected by light and shadow, and it can operate both during the day and at night.
5. *Easy to set up and use, making the technique less labor intensive.* Scanner does not need to be set on level ground, as long as one has three defined points

in the scene, one can get coordinates from every point, and the software can coordinate the data.

6. *Permanent record is provided.* Laser scanners provide a detailed record of the object at the time of inspection, allowing later viewing or measurement of other aspects of the object.
7. RBG data collected with integrated camera in a scanner can be used together with 3D scanning data in deformation monitoring survey to confirm movement identified in an analysis any single-point movement, for example, a piece of riprap toppling in a slope stability survey.

Some of the possible limitations of the TLS systems can be summarized as follows:

(i) With TLSs, it is possible to record the same object several times from different instrument setup points; however, it is *impossible to record the very same points in these repeated surveys*, making it nearly impossible to compare data collected with laser scanners.

(ii) Since laser scanners *do not support direct determination of coordinates of discrete points*, single points of scans cannot be analyzed and compared. The most appropriate object for laser scanning in terms of deriving coordinates of a discrete point is a sphere, whose center and diameter are defined. A minimum number of four points will be required to determine the center and the diameter of a sphere; if redundant scanning data are collected, the least squares adjustment method may be applied for the determination of the center and the diameter of the sphere.

(iii) Each laser scanner flight scan covers a relatively narrow swath, necessitating merging data from multiple scans to map large areas. Resolution is the ability to detect small objects in the point cloud due to the effects of the smallest possible increment of the angle between two successive points and the size of the laser spot itself on the object. An object representation in the point cloud is such that the point is recorded at the angular position of the center of the ray even if the object is hit only with the edge of the ray. *Wrong points are inevitable since the laser "spot" cannot be focused to a point size.*

(iv) There are issues with the *effect of poor geometry* of scanning and how scanner positions will be marked for the future use. In reality, without any identifiable physical points between two survey campaigns, one will be able to determine only a change in shape of the observed surface, which will give us deviations between the observed surfaces with respect to the reference mesh. It means only displacements in the direction perpendicular to the surface will be detected. This will be achieved by comparing point cloud area with the baseline mesh. For this purpose, spherical targets can be used at control points (stable points) and tie points for registration of point clouds; in correlating two epochs, the control points will be matched together; only changes in the tie points will be investigated in 3D; changes in other unmarked points can be viewed if RBG images are used and the intensity values from the scanner can be converted into RBG images. If all points in the point clouds are matched, it will still be impossible to determine the 3D displacements; determination of the perpendicular change in the shape of the two point clouds will only be possible.

(v) *Issue with calibration of TLSs*. Calibration is to provide values for the additive constant, vertical index error, horizontal collimation error; tilting axis error. It is not possible to supply a calibration or certification for laser scanners since the parameters and procedures influencing the result of a measurement are too many.

(vi) Issue with *how to screen scanning data for blunders* since redundant measurements are impossible except at control points (as in modeling a sphere). Blunders and noise of the point clouds that are not completely eliminated prior to registration will directly influence registration algorithm. There is also a need to understand how redundant measurements to the same point can be made.

(vii) Issue with how to *design optimal TLS locations and tie points*.

(viii) Issues with how to compare areas within a tile include the following:

- How to create matching tiles, identifying possible movement of points between epochs; it seems that it is currently feasible to check only the height components.
- How to determine the accuracy of creating mesh surface in the first epoch of laser scanning with the aim of being able to compare it with another corresponding mesh in the second epoch.
- How to quantify point accuracy relating to centering of TLSs, measurements (edge effect, effects of varying reflective surfaces on distances, and divergence) and geometry.

(ix) Processing the TLS data is challenging and time-consuming, requiring trained data processors (Goor, 2011). Laser scanning system produces large volumes of data, which requires several processing steps to produce the final products, such as DEM and the data acquisition/data postprocessing time ratio is usually set at 1:10. After acquiring the TLS data, several processing steps are required before deformation analysis can be done.

(x) The cost of three-dimensional terrestrial scanner systems may be very high, ranging from less than \$30,000–\$400,000 with the usual operational ranges being between a few centimeters to over 3 km.

(xi) Shadowing can be a problem in areas with large terrain relief or in urban areas with tall buildings.

10.2.2.7 Application of Terrestrial Laser Scanners in Deformation Monitoring

The TLS deformation monitoring technique seems to have an advantage over the other traditional survey techniques since it yields observations in three dimensions unlike those techniques and it provides a large redundancy in observations that potentially allow one to detect deformations well below the nominal individual point quality (Vezočnik et al., 2009). Until recently, the use of TLS in monitoring measurements was precluded because of its perceived poor precision, but the new generation of scanners as well as the emerging technologies in this field is helping to change this perception. It is, however, still considered that TLS be used as a complementary method, providing useful additional information, but not to completely replace the traditional point-wise techniques. Two main problems, however, exist if dense point clouds from laser scanner measurements are used for deformation analysis (Hesse and Kutterer, 2006):

- Handling of huge data volume
- Lack of fully automated deformation analysis methods.

Handling and especially reduction of high volume of scan data without losing relevant information is a challenging task. For detecting deformations of an entire object, it is recommended (Vezočnik et al., 2009) that the surface of the object be properly modeled by exploiting the high data redundancy since individual sampled points will only provide lower precision results.

The typical workflow and data processing steps involved in the monitoring and deformation analysis based on TLS system are as follows (Vezočnik et al., 2009; Goor, 2011):

- Design of measurement scheme
- Collection of point clouds
- Registration of point clouds
- Segmentation of the registered point cloud
- Deformation information extraction.

Design of Measurement Scheme At the design stage of deformation monitoring with terrestrial scanners, location and number of monitoring points should be arranged to ensure that the scanning distance is less than the effective measurement range of the scanner and to reduce the number of stations so as to improve the field data acquisition speed, as well as to reduce data registration error between different stations. The typical observables of laser scanners are three-dimensional coordinate information, intensity information, color information, and echo waveform. Since scanners use targets coordinated by other means, such as total station survey or using GNSS method, efforts must be made to provide sufficient control points for use in scanner resection as well as in georeferencing the point clouds. For deformation analysis purpose, some of the target locations must be stable between epochs; those points will be used to define the datum for the point clouds in two epochs.

Stability of the datum is very important. It is needed in order to separate the displacements from the noise produced by errors within the georeferencing process and to prevent unstable datum from biasing the computed deformation parameters (translation, rotations, and other structural distortions) when the 3D surface models from TLS data are compared. The geodetic datum will be realized by the geodetic points located on geologically stable ground. These points are to serve as *ground reference system*, and some are to serve as *ground control points* for georeferencing and as independent *check points*, when comparing scans captured at different times. The points can be aluminum disks with a central reflecting circular shape to be fixed to the structure being monitored or to some stable rocks in the nearby laser scanner stations; some target-tape can be fixed temporarily on some areas of the structure during each campaign for use with the scanners.

The TLS, precise total station, and GNSS positioning techniques can be used in a complementary way in the measurement setup to solve the problem of datum correctly. The use of total station and GNS positioning techniques are to design and control the stability of the datum for the evaluation of point cloud displacements acquired with TLS. If a stable set of ground control points has been established, the TLS technique can be used without continuity since the points will allow the repositioning of the scanner into the same datum each time it is used. It should be noted that laser scanners are mainly useful for periodic monitoring unlike geotechnical sensors, which can be used for continuous monitoring.

Coordinates of points in a scan are uniquely determined in TLS. It is important to have some redundancy in order to check for blunders and to improve precision of scanning. It is claimed (Gordon et al., 2003) that averaging repeat scan clouds will give better precision than single-point precision. In this case, multiple scans acquired sequentially and averaged have been found to create a cloud that may be two to three times more precise than an individual cloud, according to the root of the number of repeat scans (Gordon et al., 2003). The measurement scheme must consider the possibility of making multiple scans.

In general, the design should establish a network of reference points with geodetic network designed near the object being monitored. In order to control the quality and stability of the reference frame, GNSS observations should be used to provide absolute positions in a well-defined geocentric system. For high-precision projects, GNSS observations must be planned and processed according to recommendations for high-precision coordinate estimation. The purpose of GNSS observations is for the realization of a stable reference frame for further terrestrial observations in all measurement campaigns. Another possibility of controlling the reference frame is to use precise total station measurements, requiring that there are enough reliable orientation points in the line of sight (LoS).

The reference frame is linked with the TLS measurements (i.e., acquired point clouds) on the basis of the reference points forming the geodetic network. The network includes the reference points, scanner target positions, and control points; the control points can be used for comparison with the TLS results or may be used to determine the representative points of the object. The total station measurements include several sets of horizontal and vertical angles and slope distances to be used in estimating high-precision three-dimensional coordinates of network points; height differences are determined trigonometrically. All necessary corrections are applied to the measurements before use in coordinate estimation. If identical (at 95% confidence level) coordinates of pillars are obtained from GNSS between epochs, then the pillars are considered stable and can be used as stable datum. For example, if the downstream of a dam is to be monitored with TLS, a set of signalized targets should be distributed on the whole front of the structure, so that their 3D positions could be measured by multiple intersections using a total station to serve as controls and tie points.

Collection of Point Clouds TLS should be performed where good coverage of object is possible. Measurement approach must allow a complete and effective control over the individual segments involved as well as the error propagation process. It is important to have sufficient object coverage, that is, point cloud density, which depends on angular resolution and scanning geometry (i.e., incidence angle and distance to the object) of the scanner, and the chosen locations of the scanner. The quality of relative orientation of scans is closely related to the proper configuration of scanner targets in the geodetic network. The targets must be measured in the scans, with some of them serving as independent check points and some as ground control points. To exclude the possible errors due to variations in the network configuration, the scanner targets could be placed in the same locations in both epochs. The scanner targets can be placed on survey pillars as well as on tripods. The scanning could be performed from different (three) viewing angles with stations regularly arranged so that adjacent point clouds are to have some overlap. There must be dense point sampling of the object being monitored.

In each measurement campaign, GNSS equipment could be installed onto the reference pillars shortly after total station and TLS measurements have been performed and the GNSS observations could continue nonstop for additional few days after. Forced-centering procedure must be adopted to minimize centering error on measurements. Each measurement technology must be processed separately; the coordinates of the reference control points determined by GNSS are to be used in total station adjustment procedure, which will provide scanner target positions for the registration of the point clouds as well as observation points. In GNSS network adjustment, one of the reference points is considered stable.

Georeferencing of Point Clouds In georeferencing process, all the acquired point clouds are transformed to one common ground coordinate system. The accuracy and stability of georeferencing are very important in making comparisons between different multitemporal scans in order to detect deformations. Georeferencing of scans in two epochs may require many ground control points positioned correctly on the monitored object so that the scans are georeferenced into a stable reference frame. The accuracy of georeferencing depends on the geometric distribution of ground control points as well as on the accuracy of their measurement in ground reference system.

Georeferencing is affected by errors and these errors propagate into the georeferenced point clouds and influence the ability to detect deformations. There is, therefore, a need to perform georeferencing very accurately to the order of a few millimeters so as to be able to check which part of the detected displacement is really due to a structure's movement.

Segmentation of Registered Point Clouds Segmentation of registered point clouds is a process of grouping points of the point clouds on the basis of their homogeneity property. This process reduces the object model to single, representative points by dividing the given surface into segments, which can be planes, spheres, cylinders, or more complex surfaces. The determination of identical representative points in all measurement campaigns is very important in treating their displacements correctly. If a surface is changed significantly between acquisitions, the segmentation results will be different; the segmentation is considered the same, if surfaces have the same orientation, position, and size, when taken from the same setup point. If the object's shape has deformed, the representative points must be determined on the surface itself. The representative points could be obtained on the basis of modeling the shape of the object using appropriate surfaces, including discontinuities. The model, however, should resemble the actual shape to a required degree.

According to Remondino (2004), polygons are usually the ideal way to accurately represent the results of measurements and are able to provide an optimal surface description. For analyzing the deformation measurements of a structure acquired using TLS, the point cloud data must be interpolated and reconstructed as a three-dimensional surface model. For example, surface modeling of a dam that has a complex geometry with possible defects, damages, and deteriorations will be complex.

Deformation Information Extraction The computation of deformation based on the acquired point clouds is not an easy task. The traditional approach of obtaining displacements cannot be used in laser scanning approach since it is impossible to scan the same point in different measurement sessions, because of the imperfect repositioning of the instrument and the uncertainty associated with laser beam width (Lichti and Gordon, 2004). The individual laser pulses of repeated scans would not hit exactly the same locations. This requires that deformations be analyzed based on different approaches.

Deformation of an object between two epochs based on scanning data collected in the two epochs can be analyzed based on the following methods (Goor, 2011):

- Point cloud to point cloud method
- Point cloud to surface model method
- Surface model to surface model method.

POINT CLOUD TO POINT CLOUD METHOD In the point cloud to point cloud method, the high point density provided by TLS is not fully utilized; and it is difficult or impossible to fully do direct point-to-point comparison since one is not sure that the exact same point is sampled at two different epochs. As a compromise, the method usually uses the local neighborhood of points to estimate points for comparison. In doing this, the scans are first transformed to the same setup point, and the distance is calculated by subtracting the range image of that pixel from the other range image. The range difference is then used to quantify the deformation (Little, 2006). The other approach is to select corresponding spherical targets from point clouds in two epochs and contrast their deformations based on their fitted centers. This approach, however, is labor intensive, the analysis results are restricted, and the results are affected by noise.

POINT CLOUD TO SURFACE MODEL METHOD In the point cloud to surface model method, surface reconstruction techniques convert the irregular discrete point of the reference point cloud into three-dimensional surface model, and the surface deformation is detected by calculating the distance between a point in the second point cloud and the surface model. Deformation is then calculated for every point in the second point cloud. The surface reconstruction techniques are to find a surface model that represents the surface with the sampled points assumed lying on the surface. Since the number of sampled points is usually limited, it is most likely that the surface model generated will not exactly represent the original surface. In this case, the second point cloud can be broken into segments of neighborhood so that for every point in a point cloud, the distance to a local surface representation of the local neighborhood in other point cloud is computed.

The point cloud to surface model method, which requires a large amount of computer memory, is very time-consuming and is only suitable for simple objects. The method was proposed by Van Gosliga et al. (2006) in a deformation analysis of a tunnel, where a cylinder was used to model the tunnel. Another typical model that can be created from a point cloud is a digital terrain model (DTM). In this case, after generating the DTM with the first epoch measurements, the following phases of the point cloud data are segmented into small grids, and by contrasting the elevation of corresponding grid point with the DTM of the first epoch, changes are determined.

SURFACE MODEL TO SURFACE MODEL METHOD The surface model to surface model method requires a surface representation, according to Lindenbergh and Pfeifer (2005), for the point clouds in the two epochs of deformation analysis. The difference between the surfaces is detected after resolving the cell division and point cloud density issues. The deformation is not calculated for the original points in the point clouds, but for points at a fixed interval (cell division). A segment of the surface may be divided into grid cells, and for each cell, a plane can be fitted to all the points contained in that cell. The plane parameters and their covariance are used for the deformation analysis, resulting in a surface model to surface model method. Advantage of a planar surface model, in this case, is the simplicity of the model, so that a 3D plane is defined by only four parameters of the plane.

As an example, if a survey pillar is to be monitored for inclination and horizontal movement, after registration of the point clouds in the two epochs, the data not belonging to the pillar surface can be manually removed from the point clouds; and the remaining points for the pillar can be used to model the shape of the pillar in the two epochs. Cylindrical model can be used for the pillar in the least squares adjustment process to determine the cylinder parameters and to provide the best-fit surface for the pillar. In this case, the patterns of the pillar in the two epochs will be consistent in both epochs, affecting the cylinder parameters in the same way. The TLS data can then be used to analyze the trends in the pillar inclination in order to get a better understanding of how the terrain movement affects the pillar. From the computed inclination, it is possible to derive if the displacements of the observation points on the top of the pillar reflect the actual movement of the ground on which the pillar is situated.

10.2.2.8 Propagated Error for Computed Deformations The estimated deformation of a surface is the sum of registration errors, deformations, measurement errors, plane fitting errors, and unmodeled errors. On this basis, the propagated error on the calculated deformation will be due to error contributions from registration errors, measurement errors, plane fitting errors, and unmodeled errors, such as errors introduced by the laser scanning system. For example, the error of fitting a plane to a surface is directly related to the sum of the residuals in fitting the plane to that surface.

On the basis of current development in TLS, it is possible with TLS to determine deformations to the same order of magnitude as the ones measured with total stations and probably geotechnical instrumentations. However, there are still processing problems with TLS technique as a result of computer memory and limitations of software in segmenting point clouds, modeling the surfaces, and so on. Deformation analysis usually requires the nearest neighborhood analysis as well, which is a computationally heavy task when segments contain many points. Moreover, deformation analysis is only sensitive to deformations perpendicular to the local surface. For deformations in order directions, corresponding points in both point clouds have to be identified.

Good distribution of setup points is essential for usable point cloud. It guarantees a good point density for the areas of interest, good scanning geometry (range and incidence angle), and a minimum of occlusions. The scanning geometry for the targets must be in such a way that there are enough targets well distributed in the overlap areas of the scans and that those targets will have high point density on them; these are needed in order to reduce the registration error for the targets. More discussions on propagated variance–covariance matrix of directly georeferenced coordinates of points in registered point clouds are provided in Chapter 8.

10.3 INTERFEROMETRIC SYNTHETIC APERTURE RADAR TECHNOLOGIES

InSAR, which is based on the concepts of synthetic aperture, will be discussed under two main technologies, such as satellite-based InSAR and GB-InSAR. Before the details of these two technologies are given later in this chapter, the foundation for understanding them are first laid by reviewing the concepts of synthetic aperture radar (SAR) and the basic principle of interferometry.

10.3.1 Concepts of Synthetic Aperture Radar

The concepts of SAR are based on the concepts of *radar*, which is an acronym for radio detection and ranging. Being an active illumination system, radar transmits and receives microwave radiation, which is a part of the electromagnetic spectrum (consisting of both electric and magnetic fields whose intensities follow a sinusoidal pattern) in the frequency range of 10^8–10^{11} Hz with corresponding wavelengths of order 1–1000 mm. The different radar frequency bands are given in Table 10.4.

As an active illumination system, radar is capable of illuminating the ground feature (with the illuminated area known as *antenna footprint*) thereby functioning both

TABLE 10.4 Different Radar Frequency Bands.

Frequency Band	Frequency (GHz)	Wavelength (cm)
Ka	40 to 26.5	0.8–1.1
K	26.5 to 18	1.1–1.7
Ku	18 to 12.5	1.7–2.4
X	12.5 to 8	2.4–3.8
C	8 to 4	3.8–7.5
S	4 to 2	7.5–15
L	2 to 1	15–30
P	1 to 0.3	30–100

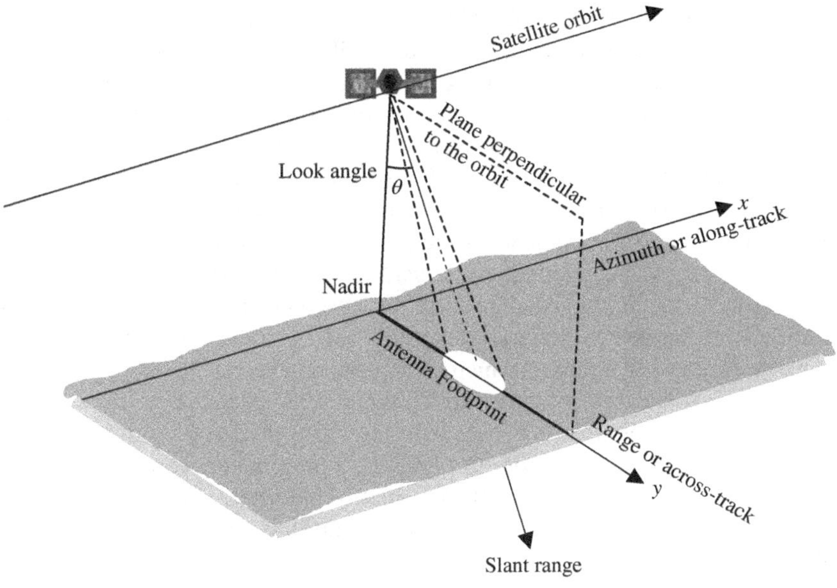

Figure 10.2 Radar system operating from a satellite.

day and night; and since its radiation is microwave, it is capable of penetrating clouds and precipitation. The radar illumination direction is side-looking with respect to the direction of motion of the aircraft or spacecraft carrying the radar. The main elements of a typical radar system are illustrated in Figure 10.2. In the figure, two coordinates most often used to describe a radar image of the ground are shown as x and y, where x-coordinate is the direction of platform motion known as the *azimuth* direction or *along-track* coordinate and y-coordinate is the direction of radar illumination known as radar *range* or *across-track* coordinate. The direction along the "line of sight" (LoS) from the radar to the target is known as *slant-range* direction, and the *range resolution* is based on the arrival time of the radar signal (echo) and the timing precision of the radar.

The range resolution is also dependent on the transmitted radar pulse width with a narrow pulse producing a fine range resolution. The azimuth dimension is perpendicular to the range dimension and its resolution is dependent on the position of the platform carrying the transmitting antenna and the beam width of the radar. As the antenna beam fans out, the azimuth resolution deteriorates. Different rows of points (pixels) of radar image are associated with different azimuth locations, while different columns of pixels of the image indicate different slant range locations. For a radar system to image separately two ground features that are close together in the range direction, it is necessary for all parts of reflected signals of the two features to be received separately at different times by the radar antenna. If the time interval between the receptions of the two signals is too short, the images of the two features will become blurred together. Ground features in the range direction is resolved by precisely timing the returns of radar energy, while the features in the azimuth are resolved by tracking changes caused by the Doppler effects.

Airborne or satellite-based radar system collects data with its single physical antenna element at different positions at different times while moving in the azimuth direction; these data are stored as functions of locations (while ignoring the time variable) and processed later as if they have been collected by one physically long real antenna element. The distance moved along-track by the antenna for the processed data is known as *synthetic aperture* and the radar equivalent to this traveled distance is an extremely large electronically simulated antenna aperture called SAR. In this case, the data processing technique is considered to have effectively lengthened the antenna along-track direction and the term SAR is coined from this signal processing technique. On this basis, the term "aperture" refers to the forward motion of the antenna over many radar pulses, which are combined to create the image of a ground scatterer. The synthesized antenna is much larger than its real aperture, which helps in improving the resolution of the radar in the azimuth direction. In an ideal case, the achievable azimuth resolution of a SAR is approximately equal to one-half the length of the actual antenna if the effect of platform altitude neglected (ESA, 2007). For a typical civilian satellite, SAR's range resolution is about 20 m and its azimuth resolution is about 5 m (Pritchard, 2006).

SAR sensors are able to transmit more than a thousand pulses per second, illuminate millions of pixels in the radar beam at each pulse time, and require thousands of processor operations per pixel in order to resolve an image. Each *pixel* in an SAR image gives a complex number that carries *amplitude* corresponding to the *intensity* of the returned radar energy and *phase* information representing a fraction of a complete wavelength. The amplitude and phase measurements are the properties of the microwave radiation backscattered toward the radar by all the scatterers (rocks, vegetation, buildings, etc.) within the corresponding pixel. Amplitude mostly depends on the roughness than on the chemical composition of the scatterers on the terrain; for example, exposed rocks and urban areas usually show strong amplitudes, whereas smooth flat surfaces such as quiet water basins show low amplitudes. The amplitude images show recognizable features of the ground (similar to optical images) while the phase images look like random noise. A pixel in a SAR image will change its phase due to a number of factors, such as the antenna-scatterer relative position, possible temporal changes of the target (reflectivity of scatterer), and the atmospheric variations (Ferreti et al., 2001). A typical radar image displays only amplitude (or brightness) data, but a SAR system is able to retain both amplitude and phase information in the radar echo during data acquisition and subsequent processing. The amplitude measurements will have "noisy" aspect since individual reflections contributing to one pixel can add together and make the overall reflection stronger or they can cancel one another out. This noise-like characteristic in the reflection of coherent radiation is called *speckle*. The general characteristics of SAR images can be given as follows:

1. Smooth surfaces, such as calm surfaces of water bodies, will appear black in SAR images since the incident radar reflects away from the spacecraft.
2. Surface variations of the size close to the radar's wavelength can cause strong backscattering.
3. A rough surface will backscatter more brightly when it is wet.
4. Due to the reflectivity and angular structures of buildings, bridges, and other human-made objects, these targets tend to behave as corner reflectors and will show up as bright spots in a SAR image.

10.3.2 Basic Principles of Interferometric Synthetic Aperture Radar

InSAR is a SAR imaging system, which has interferometric configuration. *Interferometry*, with regard to SAR, is a group of techniques in which phase shifts of reflected microwave signals are combined and the patterns formed through the combination process are investigated in order to extract useful information associated with the signals. Two SAR interferometry methods can be identified as follows (Keydel, 2005):

1. *Single-pass interferometry method* in which two antennas (one a master and the other a slave) are placed on the same platform and are simultaneously acquiring images of the same scene from two different angles. The relative phase differences from the two images are used to construct DEM.
2. *Repeat-pass interferometry method* in which a pair of images from the same sensor is taken at different times. In this method, the scenes are acquired at different times with likely different viewing geometry. The two passes, however, must have rather similar geometry in order to allow the extraction of the relative phase differences, requiring that the satellite be on an exact repeat orbit. The term InSAR is most commonly associated with repeat-pass interferometry.

From the two methods of SAR interferometry, it can be deduced that not all SAR platforms are capable of producing images suitable for interferometric use. Some representative InSAR platforms are given in Table 10.5. These InSAR systems work in microwave C-band, L-band, or X-band.

The basic measurement made by an InSAR system is the single-look complex (SLC) image consisting of both the amplitude and phase of the return signal from

TABLE 10.5 Approximate Parameters of Some Representative InSAR Platforms.

Sensor	Nominal Altitude (km)	Wavelength (cm)	Repeat (days)
Canadian RADARSAT-1	798	5.66	24
Canadian RADARSAT-2	798	5.55	24
European Union EnviSat/ASAR	790	5.63	35
Italian COSMO/SkyMed	619	3.125	16
German Aerospace Center/EADS Astrium TerraSAR-X/TanDEM-X	514	3.125	11

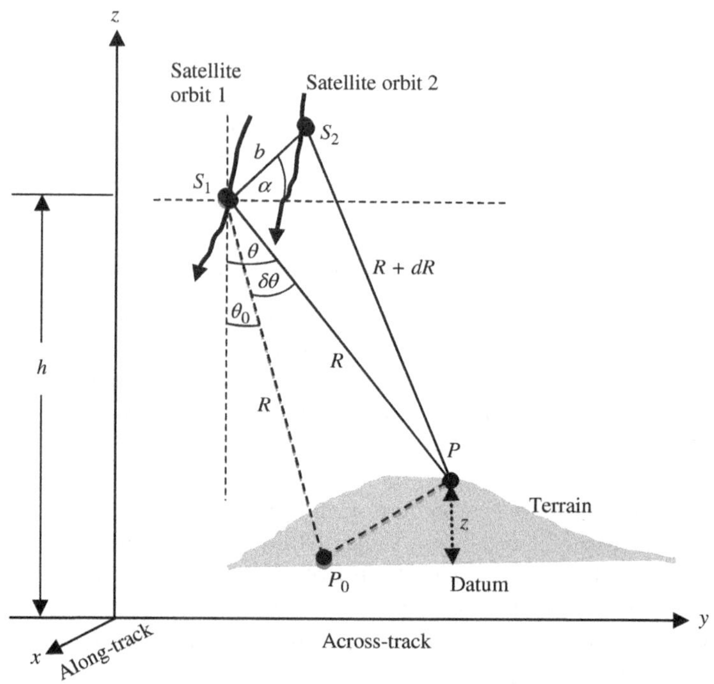

Figure 10.3 Basic geometry of SAR interferometry for topographic height determination.

the investigated surface. The measured phase values, however, can only take values between 0 and 2π since the integer number of 2π inherent in phase measurements (i.e., the number of whole wavelengths) to the satellite is usually unknown and only phase shift (some fractions of wavelengths) can be precisely measured. The basic imaging principle of interferometry is explained in Figure 10.3. In the two figures, it can be seen that (y, z) location of every surface point is reduced to range R and the radar look angle θ in the SAR image. Considering Figure 10.3, two radar antennas S_1 and S_2 are simultaneously viewing the same scene with the interferometric baseline as b, the attitude angle as α (measured between the baseline and the horizon), and the antenna S_1 located at height h above the datum. The figure can also represent a single antenna viewing the same scene on two separate passes. In the case where the two antennas are viewing the same scene simultaneously, one antenna (considered as the master) will both transmit and receive radar signal, while the second one will only receive signal with no capability to transmit signal. From Figure 10.3, the elevation (z) of point P above the datum can be given as

$$z = h - R\cos\theta \tag{10.3}$$

From cosine law,

$$(R + dR)^2 = R^2 + b^2 - 2Rb\cos(90 - \theta + \alpha) \tag{10.4}$$

From Equation (10.2), it can be deduced that

$$R = \frac{b^2 - dR^2}{2(b\sin(\theta - \alpha) + dR)} \tag{10.5}$$

According to Madsen and Zebker (1994) and Keydel (2005), the interferometric phase (ϕ) from the corresponding pixels in two SAR images with their measured phases ϕ_1 and ϕ_2 and corresponding ranges R_1 and R_2, can be given as

$$\phi = \phi_2 - \phi_1 = \frac{2\pi}{\lambda}m(R_2 - R_1) \tag{10.6}$$

where $m = 1$ when the antennas share the same transmitter, and $m = 2$ if each antenna act as transmitter and receiver. With regard to Figure 10.3, if the antennas S_1 and S_2 share the same transmitter, the range difference $dR = R_2 - R_1$ can be given from Equation (10.6) as

$$dR = \frac{\lambda\phi}{2\pi} \tag{10.7}$$

Substituting Equations (10.5) and (10.7) into Equation (10.3) gives

$$z = h - \frac{b^2 - \left(\frac{\lambda\phi}{2\pi}\right)^2}{2\left(b\sin(\theta - \alpha) + \frac{\lambda\phi}{2\pi}\right)}\cos\theta \tag{10.8}$$

From Equation (10.4), if it is assumed that $b \ll R$ (for very small baseline), then the approximate range difference can be given as

$$dR \approx -b\sin(\theta - \alpha) \tag{10.9}$$

From Equations (10.7) and (10.9), the interferometric phase can be given as

$$\phi = \frac{-2\pi b\sin(\theta - \alpha)}{\lambda} \tag{10.10}$$

From Figure 10.3, assume θ_0 represents the radar look angle to a point P_0 on a datum (flat earth surface), the following can be expressed from the figure:

$$\sin(\theta - \alpha) = \sin(\theta_0 + \delta\theta - \alpha) \tag{10.11}$$

or

$$\sin(\theta - \alpha) \approx \sin(\theta_0 - \alpha) + \cos(\theta_0 - \alpha)\,\delta\theta \tag{10.12}$$

The interferometric phase in Equation (10.10) can be rewritten as

$$\phi = \frac{-2\pi b \sin(\theta_0 - \alpha)}{\lambda} - \frac{-2\pi b \cos(\theta_0 - \alpha)}{\lambda}\delta\theta \tag{10.13}$$

where the first term in Equation (10.13) represents the flat earth (or topography-free) phase difference. If the flat earth phase difference is removed from the measured interferometric phase, what is left is known as flattened interferogram, which is expressed as

$$\phi_{\text{flattened}} = \frac{-2\pi b \cos(\theta_0 - \alpha)}{\lambda}\delta\theta \tag{10.14}$$

The flattened interferogram relates to the height variation of the scene relative to the flat earth. In conventional InSAR terrain mapping, this is used to transform interferometric phase to change in relative height from one pixel to the next. A height map is formed by choosing a reference point in the image, assigning a height value to the point, and then using the change in relative heights derived from Equation (10.14) to determine the heights of other points based on the value of the reference point. Equation (10.8) is a case when the SAR interferometry is used for determining the elevations of terrain points as in DEM. For repeat-pass approach, where single-antenna SAR system (acting both as transmitter and receiver) revisits the same position and images the same scene after some time (assuming no significant change in the scene between acquisition of the two images), dR in Equations (10.5), (10.7), and (10.9) will be replaced by $dR = \lambda\phi/4\pi$.

Considering a repeat-pass approach in which ground deformation (due to an earthquake or volcano swelling) has displaced many of the resolution elements in the second pass as shown in Figure 10.4; if an object is imaged from the same location at two different times (the same orbit in two passes) and phases of the backscattered signals differ, it can be inferred that the object has moved about Δh, which can be given from Figure 10.4, as

$$\Delta h = \frac{dR}{\cos\theta} \tag{10.15}$$

where R_1 and R_2 are the ranges at two locations S_1 and S_2, respectively; dR is the range change; Δh is the movement in the direction of the satellite (change in ground height); and θ is the look angle.

From Equation (10.6) and taking $m = 1$, the interferometric phase can be given as

$$\phi = \frac{4\pi(dR)}{\lambda} \tag{10.16}$$

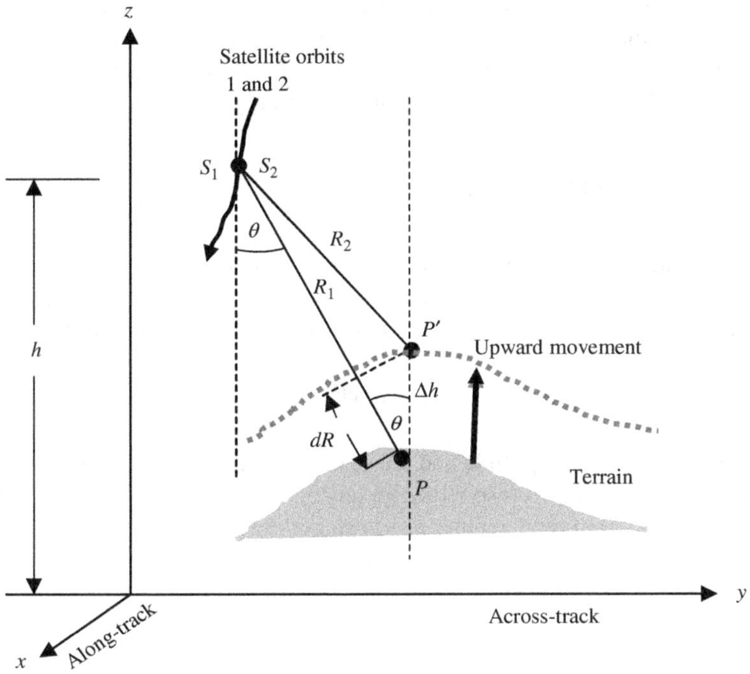

Figure 10.4 Basic geometry of SAR interferometry for displacement determination.

where $\phi = \phi_2 - \phi_1$ and $dR = R_2 - R_1$. Substituting dR from Equation (10.15) into (10.16) and rearranging, the small height change Δh, which occurs between the times of acquiring the images (P moving to P'), can be expressed as follows:

$$\Delta h = \frac{\lambda \phi}{4\pi \cos \theta} \tag{10.17}$$

In Figure 10.4, the range difference $dR = R_2 - R_1$ produces the ground uplift Δh. Radar, however, measures only changes (dR) in the LoS direction; in order to estimate deformation in any other direction (vertically or in three dimensions), the ascending, descending, and adjacent satellite orbits, together with certain assumptions depending on the case, must be used. In practice, to determine the displacement due only to the uplift in the vertical direction, the flat phase contribution must first be subtracted from the interferometric phase difference by a process known as *flattening the earth*; the effect of topography is then subtracted. Generally, the interferometric phase difference may be more complex and may consist of up to seven layers of contributions (which are usually not perfectly known) such as flat earth (or orbital effect), topographic contribution, effects of atmospheric inhomogeneity, ground displacement, orbital uncertainties, random noise, and decorrelation (Stevens et al., 2001). In order to get rid of any unwanted layers of information within each pixel, the most basic step of SAR interferometry consists of subtracting corresponding phase-shift values from the original interferometric phase of two successive radar images of the same area. It should be mentioned that the orbital parameters of the spacecraft and the spacecraft orientation during SAR image acquisition is considered well known if the antenna baseline length b and the orientation angle α are known.

10.3.3 InSAR Data Processing Overview

Referring to Figures 10.3 and 10.4 again, it can be seen that two single-look complex (SLC) images are required in SAR interferometry in order to determine the nature of the investigated surface. For a given complex image 1 (the master), the signal from a pixel (u_1) of the image can be represented as a complex function (Bamler and Hartl, 1998; Stilla et al., n.d.):

$$u_1 = a_1\, e^{j\phi_1} \tag{10.18}$$

where a_1 and ϕ_1 are the amplitude and phase of the signal from the pixel in image 1, respectively. For any two corresponding pixels of any two SLC images of the same scene, the interferogram representation (u) can be given as the product of the master signal u_1 with the complex conjugated slave signal u_2^* of u_2, which can be given as

$$u = u_1 \cdot u_2^* = a_1 \cdot a_2 e^{j\phi_1} \cdot e^{-j\phi_2} \tag{10.19}$$

or

$$u = ae^{j\phi} \tag{10.20}$$

where $a = a_1 \cdot a_2$ and $\phi = \phi_1 - \phi_2$ are the amplitude of the interferometric signal and the interferometric phase, respectively, for the pixel of the interferogram. The overall interferogram is determined by pixel-by-pixel complex multiplication of the master signal with the complex conjugated signal; the amplitude of the interferogram is the product of the amplitudes of the two initial SLC images; and its phase is equal to the phase difference of the images. It should be mentioned that the interferometric phase in Equations (10.7), (10.16), and (10.20) is the predicted unwrapped phase, which cannot be measured directly; what is measured in practice is the wrapped phase (ϕ_w) from which the unwrapped phase (ϕ) is determined. The wrapped phase can be expressed as

$$\phi_w = \phi - 2\pi(n) \tag{10.21}$$

where n is the unknown integer number, which has to be determined independently through the process known as *unwrapping*. Due to the cyclic nature of interferometric phase-shift values, interferometric phase-shift values are recorded as repeating fringes with values ranging between zero and a full 2π cycle. But according to Lu (2007), the phase value of a single pixel in an interferogram cannot be practically used in determining the range difference dR, but the phase difference ($\delta\phi$) between two adjacent pixels of the interferogram is what is used in estimating the range difference to a subwavelength precision.

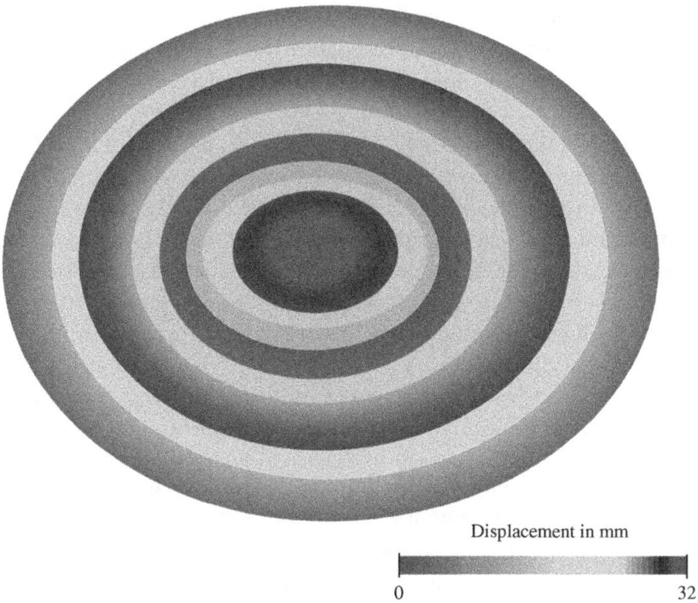

Figure 10.5 Possible interferogram showing three fringes of modeled uplift.

Considering the SAR interferometry for displacement determination as illustrated in Figure 10.4 and Equation (10.17), if for example two RADARSAT-2 images recorded before and after the ground uplift (with $\lambda = 5.55$ cm, nominal average look angle as 30°, and the calculated phase difference $\phi = 2\pi$) are analyzed, the corresponding uplift value (Δh) will be calculated from Equation (10.17) as 3.2 cm. If the total ground uplift is 10 cm (or 100 mm) as illustrated in Figure 10.4, the two RADARSAT-2 images, when combined, will generate an interferogram with a colorful pattern of three fringes as shown in Figure 10.5. In the figure, each of the three fringes (from red to blue) represents a change in the direction of gravity (z-axis) of about 3.2 cm (32 mm) for a full phase cycle 2π. If the displacement along the LoS is being considered, the look angle in Equation (10.11) will be set to zero so that each fringe of interferometric phase will correspond to a change in the satellite-to-ground distance of half the radar wavelength (for RADARSAT-2, it will be 2.8 cm).

The interferometric phase, however, contains components due to topography (or baseline) and the ground displacements, assuming some of the point scatterers on the ground slightly change their relative position in the time interval between the two SAR observations, for example, in the event of subsidence, landslide, earthquake, and so on. If a DEM of the region is available, the topographic contribution can be subtracted from the interferometric phase, thus generating the so-called *differential interferogram*, in which the remaining ground displacements can be measured. The DEM can be generated by using the interferograms formed from two SAR images of the scene before the movement. In this case, three radar images as two pairs of interferograms are combined to separate topographic component so that the displacement field can be obtained to millimeter level (Zebker et al., 1994; Prati et al., 1992).

Differential interferometry synthetic aperture radar (or D-InSAR) is the common term for the production of interferograms from which the topographic contribution has been removed. Interferometric phases, however, are only resolvable relative to other points in the interferogram. In this case, absolute deformation can be inferred by assuming one area in the interferogram (e.g., a point far away from expected deformation sources) experienced no deformation, or by using a ground control based on conventional surveying techniques (e.g., GPS or total station positioning) to establish the absolute movement of a point.

Some of the corrections usually applied to D-InSAR interferogram in the process of producing surface deformation (displacement field) can be summarized as follows:

1. Baseline corrections. These corrections are to account for the slightly different locations of the satellite antenna during the two consecutive coverage of the given region. For interferometry to work, the satellites must be as close as possible to the same spatial position when the images are acquired. This means that images from two satellite platforms with different orbits cannot be compared and that satellite data from the same orbital track are ideal and most desired. In practice, the perpendicular distance between the two orbits (known as *baseline*) is often known to within a few centimeters. This slight difference causes a regular difference in phase that changes smoothly across the interferogram and can be modeled and removed. The slight difference in satellite position also alters the distortion caused by topography, meaning an extra phase difference is introduced by a stereoscopic effect. The longer the baseline, the smaller the topographic height needed to produce a fringe of phase change, known as the *altitude of ambiguity*. The altitude of ambiguity is the amount of topographic error required to generate one interferometric fringe in a topography-free interferogram (Massonnet and Feigl, 1998). This effect can be exploited to calculate the topographic height and used to produce a DEM.
2. Corrections for topography. If the height of the topography is already known, the topographic phase contribution can be calculated and removed.
3. Corrections for spatial and temporal variations in the atmospheric condition, for example, due to air temperature, atmospheric pressure, and water vapor content variation, between observations. These variations cause delay in phase propagation through the troposphere, so that the atmosphere contributes the atmospheric phase contribution to the interferogram. The effects of such contribution impact on both altitude (especially in the case of small baselines) and terrain deformation measurements.
4. Corrections for other phase noise sources. The other phase noises are related to the presence and types of many scatterers per pixel and their changes in time. Four main contributions to the phase noise can be given as follows (ESA, 2007):

- Phase noise due to temporal change of scatterers. Water basins or densely vegetated areas as scatterers may change after a few milliseconds whereas exposed rocks or urban areas remain stable over several years.
- Phase noise due to different look angle. There is a critical baseline over which the interferometric phase is pure noise. The critical baseline depends on the dimension of the ground range resolution cell, which is a function of terrain slope, radar frequency, and sensor-target distance.
- Phase noise due to volume scattering. The critical baseline reduces in the case of volume scattering when elementary scatterers are not disposed on a plane but occupy a volume (e.g., the branches of a tree). The speckle change will then also depend on the depth of the volume occupied by the elementary scatterers.
- Nature of interaction with the ground, such as changes in the refractive index of the medium, transition at an interface, uniform changes of the electrical conductivity within the surface covered by the radar pixels. The reflected signal back from 1 pixel is the summed contribution to the phase from many smaller targets in that ground area, each with different dielectric properties and distances from the satellite.

Given two SLC images of the same area (labeled as "master" and "slave") that are focused and with preserved phase, the interferogram processing steps are given as follows (Keydel, 2005; Dixon, 1995) and illustrated in Figures 10.6 and 10.7:

1. Filter the complex images to optimize coherence and interferometry phase purity and to minimize baseline-induced decorrelation, and so on.
2. Coregister two images to 1/8th to 1/20th pixel accuracy (Ouchi, 2013) using a correlation procedure to find the offset and difference in geometry between their two amplitude images. In this case, the slave SAR image is resampled to match the geometry of the master image so that each pixel now represents the

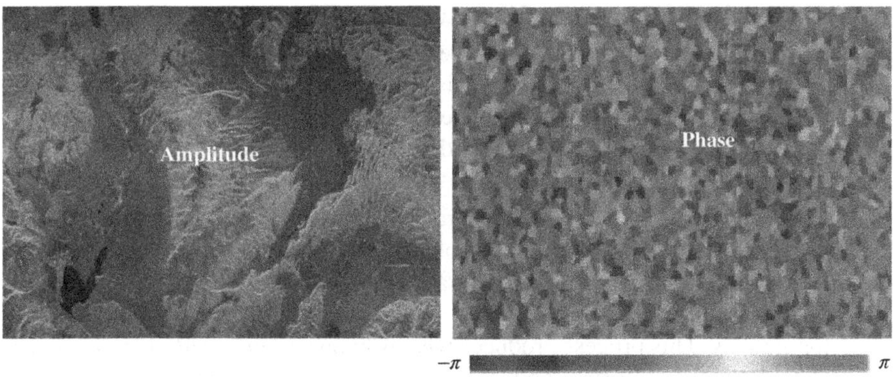

Figure 10.6 Typical InSAR complex image of a scene. *Source*: Amplitude image is due to the Courtesy of NASA/JPL-Caltech.

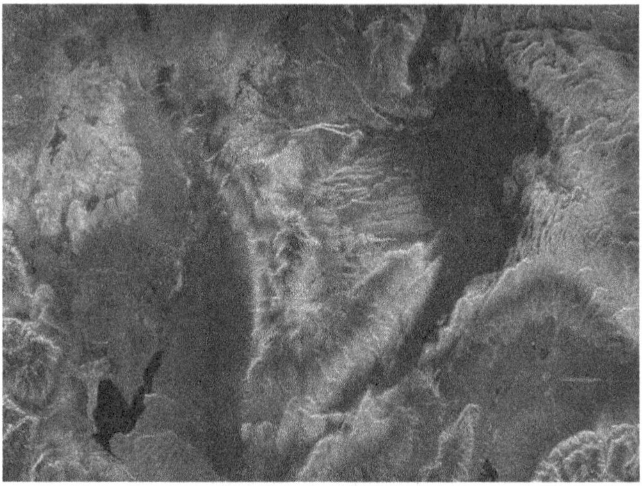

Figure 10.7 Typical InSAR interferogram of a scene. *Source*: Courtesy of NASA/JPL-Caltech.

same ground area in both images. This step is to ensure that each ground target contributes to the same (range, azimuth) pixel in both the master and the slave images, thereby increasing interferometric coherence.

3. Compute raw (or complex) interferogram (Figure 10.7) by cross-multiplying pixel by pixel the complex master image by the coregistered complex conjugate of the slave image (Equation (10.20)). The amplitude of the complex interferogram becomes the amplitude of the master image times that of the slave image and is used to produce *coherence image*, and its phase, called the *interferometric phase*. This interferometric phase is the phase difference between two images, which is used for contour generation (called interferometric fringes or the interferogram). If the coherence is low, the contrast of the interferometric fringes becomes low, or no fringes are produced at all.
4. Perform flat-earth or orbital phase removal to produce *flattened interferogram*. The flat phase depends on the baseline separation of the successive SAR images, which translates to change in ground-range distance, assuming the earth surface is flat; the corresponding interferometric fringes are called the *orbital fringes*.
5. Perform topographic phase removal by simulating the contribution of the topography to the interferometric phase and removing the topographic effect from the interferogram. If an accurate DEM, sampled at the SAR resolution, for the area imaged is available, it can be used to estimate and compensate for topography. This process produces what is known as *differential interferogram* suitable for monitoring and detecting ground displacements. In order to compute the topographic phase, the flat phase must have been removed from the interferogram (from step 4).

6. Perform coherence image estimation by determining the correlation between the master and the coregistered slave images. The cross-correlation operation is done over a small local area surrounding each pixel in the interferogram.
7. Reduce the phase noise from the interferogram so as to facilitate phase unwrapping. This involves filtering the residual interferogram by using an averaging window with a size of several resolution elements in both range and azimuth directions. The coherence image can be evaluated to check if there is loss of coherence due to temporal decorrelations in the complex images used; if there are decorrelations, it would become difficult to produce a proper analysis of deformation of the surface of interest. Sometimes, wrapped phases may not be available everywhere in the images since there may be pixels without significant radar return. This will require that bilinear interpolation of phases be performed to fill in the gaps.
8. Perform phase unwrapping of the consecutive fringes present in the interferogram by adding the correct integer multiple of 2π to the interferometric fringes. This process is to determine the absolute phase relationship between all pixels in an interferogram. In an interferogram, the 2π phase discontinuities are usually clearly visible as black/white transitions, which can be eliminated by adding or subtracting an integer multiple of 2π to each pixel of the original interferometric phase image. One of the most difficult problems in interferometry, however, is how to extract absolute phases from the available ambiguous (wrapped) values.
9. Geocode the image to produce the interferogram in a desired geographic projection. Usually, the scale bar on the interferogram-based deformation map represents one fringe, that is, one cycle of phase variation from 0 to 2π rad.

In Figure 10.6, it can be seen that the phase of pixels seems to be randomly distributed, which can be attributed to the large number of scatterers usually contained in the SAR resolution cell. The phase, however, can only be exploited (forming an interferogram) if two images of the same scene, in which the scatterers remain unchanged in the resolution cell, are combined. Figure 10.7 shows the interferogram of the same scene shown in Figure 10.6, displaying interferometric phase shift at each pixel (color-coded) as a function of position, represented as color fringes. The fringe pattern in the figure can be interpreted as a contour map of the LoS component of displacement (as opposed to displacement component along the direction of gravity (z) given in Figure 10.4) of the ground surface point in relation to the spacecraft over certain period, with contour interval of half the wavelength of the imaging radar. In the figure, the "contour interval" represented by the complete color sequence is equivalent to 360° or 2π phase. In principle, there is a need to correct for surface topography before surface displacement can be determined. The effect of surface topography can be ignored if the surface being considered is nearly flat (Goldstein et al., 1993).

An important condition for successfully creating an interferogram is that the scattering properties of the ground surface must be relatively constant between observations so as to maintain high coherence of the radar returns. Radar returns are coherent when they are in phase, that is, they vibrate in unison. In an *interferogram*, coherence is a measure of correlations, which range from 0 (where there is no useful information in the interferogram) to 1.0 (where there is no noise in the interferogram and the phase information is reliable). High coherence results in attractive and less-noisy interferograms, while low coherence results in unattractive and noisy interferograms; the areas of no fringes usually correspond to areas of little coherence. The degree of coherence can be used as a quality measure since it significantly influences the accuracy of phase differences and the quantities derived from them. Some of the factors that may affect the interferometric coherence (the degree of correlation) include the following (Ouchi, 2013):

- *Local slope* of the surface, with steep slopes producing low coherence; a flat surface tilted toward the imaging radar will produce decreased coherence, while the surface tilted away from the radar increases the coherence.
- *Temporal decorrelation* due to the interaction of incident microwave with the scattering objects. This includes the decorrelation by the temporal changes of scattering objects as well as multiple scattering associated with different incidence angles. For example, surface of a body of water will change between passes and produce near-zero coherence; surface of solid ground that does not change between passes will produce high coherence; and vegetation and forests will have moderate to low coherence. Properties of the surface being mapped, such as due to construction, erosion, and ground movement, will also result in low coherence.
- *Time lag* between passes, with long lags leading to low coherence.
- *Baseline separation*, with long baselines resulting in low coherence and the shorter one giving higher coherence, but at the expense of interferometric resolution. This means that small baseline or zero baseline will result in reduced baseline decorrelation.
- *Additive system noise* will lead to loss of coherence if the signal-to-noise ratio (SNR) is small.
- *Coregistration and resampling techniques*, with poor coregistration or resampling resulting in low coherence.

10.3.4 Persistent or Permanent Scatterer InSAR Technique

Persistent or permanent scatterer (PS) points are sparsely distributed phase-stable point targets that provide consistent and stable radar reflections back to the satellite. The PS points are preexisting reflectors such as buildings, radio masts, or prominent features. A PS point, usually of the size of a pixel or a subpixel, is expected to remain coherent over the entire observation interval and to be present in every image in a stack of InSAR images for the observation interval; pixels showing a stable sequence of amplitude will be considered as permanent scatterers, provided the same targets appear in the same pixels for all the campaigns. The techniques of studying such images and interferograms for pixels that display stable amplitude and coherent phase throughout every image of the data set are usually referred to as *persistent or permanent scatterer InSAR (or PS-InSAR) techniques*. The pixels with PS points are to be used to overcome the shortcomings due to temporal and geometrical decorrelation (Noferini et al., 2005) or to evaluate the atmospheric disturbance with the aim of removing it. Commonly, the PS-InSAR techniques are most useful in urban areas where there are a lot of permanent structures. A millimeter-level accuracy has been quoted (Noferini et al., 2005) for displacement determination using arrays of PS points.

10.3.5 Artificial Scatterer or Corner Reflector InSAR Technique

If permanent scatterers are not available for an InSAR technique, artificial ones known as corner reflectors (CR-InSAR) can be used. An artificial corner reflector, shown in Figure 10.8, is a simple structure, which is stable with respect to amplitude and phase. According to Hanssen (2011), these types of scatterers have some disadvantages, which are mainly due to their big and heavy sizes; difficulties in deploying and maintaining them; their susceptibility to disturbance by weather, animals, vandalism, or theft during long-term measurements; their likelihood of being affected by snow, rain, and debris; their likelihood of undergoing local movement if not properly anchored to the ground; and their likelihood of poorly reflecting back to the satellite if they are not properly oriented toward the satellite. The CR-InSAR techniques, however, are capable of monitoring the movement of specific structures and locations with millimeter-level LoS accuracy possible (Chrzanowski, 2009).

Figure 10.8 Typical artificial corner reflector. *Source:* Courtesy of NASA/JPL-Caltech.

10.3.6 Limitations of InSAR Techniques

Some of the limitations of InSAR are essentially due to the following (Zebker and Villasenor, 1992; Dixon, 1995; Zebker et al., 1997; Chen et al., 2000; Ferreti et al., 2001):

1. *InSAR only detects deformation in the LoS direction of the radar beam.* This implies that only one component of deformation (movement toward or away from satellite) can be measured in an individual interferogram, which is the anticipated limitation of interferometric technology.
2. *Not all SAR images are suitable for interferometric use.* The suitability of SAR images depends on the view angle, geometrical decorrelation and baseline decorrelation, time of data acquisition, coherence, and atmospheric inhomogeneity. Geometrical decorrelation and the baseline decorrelation due to differences in antenna's viewing positions between two observations (also known as spatial decorrelation) will limit the number of image pairs suitable for interferometric applications. Accuracy of centimeter level required in baseline estimation is not possible with the current orbit parameters. With regard to coherence, temporal decorrelation, which is due to lack of temporal coherence, is a major problem since the electromagnetic profiles or the positions (or both) of the scatterers usually change with time within the resolution cell. For successful comparison of two SAR images, Dixon (1995) specified that the standard deviation of position of the surface scatterers within a pixel must remain constant within a fraction, of around 10–20%, of the radar wavelength. The atmospheric inhomogeneity (such as change in humidity or variations of atmospheric water vapor), which creates atmospheric phase delays on each SAR image, can seriously compromise accurate deformation monitoring with InSAR techniques.
3. *Not all SAR-image-producing satellites are suitable.* Only a few operating satellites (Table 10.4) are currently able to provide stable data for interferometric use. Other important considerations in choosing suitable SAR platforms are the revisit times of the platform and the availability of suitable DEM for two-pass InSAR method.
4. *A mix of several different layers of geometrical information in a given signal may be difficult to separate.* Measurement accuracy in InSAR methods is driven by the terrain stability and the ability to separate the various components in the SAR signal. The main limiter of the basic accuracy of the measurement is change in the geometric and physical properties of the ground during the time intervals between the observations, for example, if the moisture content of the soil changes, or there is a local motion. There is also a difficulty of seeing through vegetation to the ground beneath and a problem of distinguishing deformation signals from orbital uncertainties.
5. *Coregistration problem.* In the process of matching two pixels of the same point target from two different SAR images, there may be serious errors or coherence loss if the matching is not done to subpixel level (Fornaro and

Fraceschetti, 1995). In this case, the mismatched pixels will represent slightly different scattering targets and different interference patterns, producing incoherent pixel phase (the phase becomes essentially random and noisy from pixel to pixel rather than varying smoothly). Anything that will change the contributions to the phase within each pixel (such as changes to the ground targets in each pixel due to vegetation growth, landslides, agriculture or snow cover, variation in atmospheric condition) will essentially destroy coherence. Theoretically, accuracy of coregistering two images to a subpixel level to ensure that the same ground targets are contributing to that pixel can be reached, but it may be impossible in practice.

6. *Phase unwrapping algorithm used may be unreliable.* Unreliable phase unwrapping algorithm may introduce considerable errors to the processed data.
7. *InSAR can only be used periodically (not continuously in time domain) and difficult for remote control.*
8. *Achievable accuracy in deformation monitoring with InSAR techniques is still at the medium level.* For example, each orbital path of SAR platforms usually deviates slightly from the previous one to form a spatial baseline between the imaging centers; this will introduce some errors into deformation determination.
9. Large movement exceeding certain interval (Dixon, 1995) may be difficult to detect. This may be the case when there are glacier flows or large deformation by earthquakes, volcanic eruptions, landslides, and so on, where phase may change more than one cycle within the slant range resolution cell, making the determination of surface movement more challenging.
10. With regard to deformation analysis, a posteriori assessment of unstable reference points can only be done with InSAR techniques based on information derived from the SAR data itself. In the case of geodetic techniques, a priori assessment of unstable points is possible, which can be based on physical inspection of the reference points for their stability or using other practical techniques.

10.3.7 Applications of InSAR Techniques

InSAR is an emerging technology that is capable of measuring a variety of observables. Some of the important applications and advantages of InSAR include the following:

1. Conventional InSAR technique is now being applied in DEM generation. Interferometric phase comparison of SAR images gathered at different times and with different baselines is capable of providing DEMs with meter accuracy (Prati et al., 1992).
2. Differential interferometry (D-InSAR) technique is also applied in ground displacements monitoring. The earth's surface displacements from glaciers,

earthquakes, and volcanoes to subcentimeter levels are measured by comparing phase information from radar images taken at different times. The D-InSAR technique gives scientists a large-area image of the deformation field, not just deformation at a series of points on a map. It has been shown (ESA, 2007) that radar interferometry can be used for problems related to legal issues and for monitoring damages to the environment. Subsidence caused by natural gas storage, oil extraction, irrigation water pumping, or mining has also been monitored based on D-InSAR technique; and the displacement of a dam has been monitored using D-InSAR methods over a long period with an accuracy in the order of a fraction of millimeter claimed (Tarchi et al., 1999). The technique has been reported (Prati et al., 1992) to be capable of providing terrain deformations with millimeter accuracy. However, landslides (always located on slopes) have been found difficult to monitor with the technique considering the angle of incidence of the imaging radar.

3. Data obtained from InSAR techniques have also been used (Pritchard, 2006) in many major discoveries, such as ground moisture changes, groundwater movements beneath major cities, magma movement, ocean currents, and so on.
4. InSAR techniques can be used without endangering human beings or expensive instruments and can be used quickly to survey extremely remote and otherwise unmonitored areas in the order of thousands of square kilometers and achieve a spatial resolution of a few meters. The techniques are used without subjecting field crews to hazardous conditions on the ground.
5. Unlike other techniques such as geotechnical instrumentation and many geodetic techniques that rely on measurements at a few points at very high costs, InSAR techniques produce a spatially complete map of ground deformation with centimeter-level accuracy at low costs. It can provide deformation information continuously in space domain and it can show spatial patterns of deformation in remarkable detail. Since the area usually covered by the two radar images is typically a square of 50 or 100 km on a side, this method allows scientists to look at deformation over large areas, including monitoring long bridges. With InSAR techniques, it is possible to detect deformations at locations where deformations is not anticipated, unlike in the case of geodetic techniques where (because of costs) deformation measurements are only made at locations suspected of possible deformations.

10.3.8 Ground-Based InSAR (GB-InSAR) Techniques

GB-InSAR is a remote sensing radar technique that uses a microwave transmitter and receiver that travel back and forth on a mechanical rail (usually 2–3 m long) to map ground movement. The interferometric concept used in the GB-InSAR techniques is essentially the same as that of space-borne InSAR techniques. In a similar way, the GB-InSAR antenna emits microwave signal and measures the complex image consisting of amplitudes and phases of the ground pixels from the returned signals. The difference of two phase images of the object observed at two different times is used

TABLE 10.6 Summary of the Differences Between GB-InSAR and Space-Borne InSAR.

	Space-Borne InSAR	GB-InSAR
1. Rate of image acquisition	Several days or weeks	Few minutes (as often as 5 or 10 min)
2. Working range/altitude	Several hundred kilometers away (about 800 km in altitude)	Few kilometers away (up to 4 km) in line-of-sight distance of the area being mapped
3. How synthetic aperture of radar is obtained	Obtained by the antenna moving round an orbit	Obtained by an antenna traveling back and forth on a mechanical rail of about 2–3 m long
4. Ground horizontal spatial resolution size	Depends on satellite, radar instrument, and look angle of radar; ranges from 3 to 30 m (TerraSAR-X has a variable resolution, typically 3 m by 3 m)	Few decimeters to several meters depending on the equipment and the monitoring distance (a typical commercial equipment has a resolution of about 0.5 by 4 m at 1 km)

to determine the displacements in the line of sight (LoS) directions (from the sensor head to the surface to be monitored) for each resolution cell of the interferogram formed. In this case, only one-dimensional variations in LoS range are evaluated; the variations, however, can be decomposed along other lines if the local geometry is known. Usually, negative displacement values indicate movement toward the sensor (shortening along the LoS), and positive displacement values indicate movement away from the sensor (lengthening along the LoS). Since only relative phase differences are formed in this process, with the number of full phase cycles unknown, phase unwrapping is also done to the phase differences in order to determine the phase ambiguities. The unwrapped phase difference of the interferogram is used to determine the LoS range changes with 2π (or one cycle) phase difference corresponding to half the radar wavelength. GB-InSAR technique allows two-dimensional color radar image of the investigated area to be achieved with a high-range resolution along the instrument LoS and cross-range resolution along the scan direction. It also allows the displacement time series of each pixel to be plotted. Some of the differences between GB-InSAR and space-borne InSAR are summarized in Table 10.6.

There are currently two generic types of GB-InSAR systems that are commercially available:

1. SAR type where a small radar antenna slides along a rail collecting data, which is processed to form multiple fan-shaped beams. This type of system scans only in the azimuth direction with a fan beam that simultaneously covers all elevation angles. With a fine grain DTM, range measurements can be mapped

into elevations. The examples of this type of system are IBIS-L and IBIS-FS from the Italian company Ingegneria dei sistemi (IDS).

2. Real-beam aperture radar type, which uses a conventional dish antenna to mechanically scan a pencil beam in raster fashion over the region of interest. This type of system allows three-dimensional representation of the rock face with the displacement characteristics superimposed in color. The examples of this type of system are Slope Stability Radar (SSR) developed by GroundProbe Pty in Australia and Movement and Surveying Radar (MSR 300) by Reutech Mining in South Africa.

The SAR type does not form a narrow transmitter beam like the real-beam aperture radar; it transmits and receives with a wide pattern as the antenna traverses a horizontal rail placed side-on to the scene. After completing the transition, all the beams are synthesized by processing the data set. This technique forms a set of fan beams, which have a narrow pattern in azimuth and a broad pattern in elevation. An important advantage of this technique is that the horizontal rail can span a much wider horizontal aperture than is practically possible with a dish antenna. One of the disadvantages of this technique is that the vertical fan beam gives no information about the elevation angle of any returned signals and the radar has to rely on range resolution to help separate returns from different elevations. The fact that vertical resolution is variable and dependent on the slope will affect the accuracy of measured displacement, making the system less precise compared with the real-beam aperture radar system. The summary of the differences between the two types of GB-InSAR systems is given in Table 10.7. Generally, it is the technology behind the image formation process,

TABLE 10.7 Summary of Differences Between Synthetic Aperture Radar and Real-Beam Aperture Radar.

Constraints	Synthetic Aperture Radar (IDS IBIS)	Real-Beam Aperture Radar (GroundProbe SSR and Reutech MSR)
1. Mapping displacement	Less robust and not accurate in 3D	More robust and more accurate in 3D
2. Coverage of slope	Typically 70% of the scanned area is covered with measurement errors and less reliability	100% of scanned area is covered
3. Maximum sector scanned	Up to ±30° in azimuth direction only	Typically ±120° in front, left side and right side directions with potential for 360° coverage
4. Monitoring shallow slopes at long range	Suitable for all ranges	Suitable but not for extremely long ranges (many kilometers)

which differentiates these two radar systems and their suitability for particular applications. Both of them, however, are not suitable for monitoring steep slopes relative to LoS.

The accuracy of GB-InSAR systems is affected by a number of factors, which include the following:

(a) Propagation anomalies such as shimmer seen on a hot day
(b) Movement of vegetation cover
(c) Interfering returns from reflectors at other angles.

The first two factors affect both types of radar in a similar way, but interfering reflections will impact them differently. Generally, the theoretical accuracy of GB-InSAR equipment is in the order of tenths of millimeters to a few millimeters depending on the monitoring distance and the atmospheric conditions (Mazzanti, 2012).

The best applications of the ground-based SAR interferometry include continuous displacement monitoring of unstable slopes and dams, bridges, localized subsidence, rock scarps, volcanoes, landslides, infrastructures, and so on. In the case where non-continuous measurements are carried out at different times after dismounting and repositioning the GB-InSAR system, it is necessary to perform image coregistration. If the system is dismounted and repositioned at each observation campaign, the GB-InSAR will not be able to unambiguously detect displacements corresponding to more than one π (or half of a cycle) in terms of phase (Crosetto et al., 2011). In this case, properties of GB-InSAR will be very similar to those of the space-borne type, requiring the use of artificial stable (coherent) reflector targets such as passive corner reflectors (PCRs) or the use of natural targets. With the use of PCRs, GB-InSAR system can be mechanically repositioned with centimeter accuracy without compromising the precision obtained from the coregistration in the final displacement maps by measuring displacements at the PCRs. The integration of TLS and GB-InSAR data is expected to open up further interesting applications in the future, where TLS is able to detect lower frequency deformations with a higher point density and GB-InSAR is able to monitor higher frequency deformations at a lower spatial resolution.

10.3.8.1 Examples of SAR Systems: IBIS-L and IBIS-FS The SAR systems, such as IBIS-L and IBIS-FS systems consist of the following units:

- *Radar unit,* a portable unit for generating, transmitting, and receiving microwave electromagnetic signals, which are processed in order to determine the displacement of the investigated object.
- *Linear scanner* consisting of a 2.5-m-long aluminum track with support system, along which the radar unit (the sensor) is moved under the control of a step-by-step motor.

TABLE 10.8 IBIS-L Main Features.

Feature	Value
Operating frequency/wavelength	17 GHz (Ku band)/17.6 mm
Best spatial resolution (range × azimuth) at 1 km range	0.5 × 4.4 m
Displacement sensitivity (accuracy)	Up to 0.1 mm
Minimum data collection rate	5 min
Maximum operation distance	4000 m

- *Control PC,* running the system management software for configuring the acquisition parameters, managing and storing measurements and for displaying first results just after ground data acquisition.
- *Power supply unit* for providing power to the system through a pack of two 12 V batteries or through the connection to an external energy supplier.

Considering IBIS-L system as an example; it can be bolted on concrete block for stability when in use; and when it is being used to monitor landslide, the system, which is able to continuously survey the landslide without human intervention, is usually located on the crest of the landslide headscarp with downslope view of nearly the entire landslide. The main features of IBIS-L system are given in Table 10.8. The other important features of the system include the ability to process radar data with automatic atmospheric corrections in real time; ability to provide fully georeferenced outputs in the form of displacement and velocity maps; ability to generate alarms based on velocity data and user-defined levels and also allow multiple alarm criteria for user-defined spatial zones; and ability to provide all its software outputs in the formats that can be exported to common GIS, CAD, or mine planning software (Crosetto et al., 2011). The major technical advances of the system are its interferometric processing techniques based on advanced algorithms that use statistical analyses to select a grid of high-quality pixels (persistent scatterers, PS) for removing atmospheric artifacts from the interferometric signal. Since IBIS-L has two-dimensional measuring feature, it is suitable for measuring surface displacement such as in slope collapses, erosion of volcanic edifices, landslides, and large-scale structures such as dams. The 4 km operating range of the system, which allows for complete coverage of a range of landslide sizes and remote installation in stable locations, makes the system ideal for most landslide applications. Conventional approaches used in other radar systems and optical systems such as robotic total stations generally do not work well over long ranges or with highly variable atmospheric conditions.

The main advantages of IBIS-L over the space-borne SAR can be summarized as follows (Rödelsperger, 2011):

- The position of the linear scanner can be determined and be monitored accurately so that IBIS-L interferograms are free of orbit errors.

- The IBIS-L system uses zero baselines, so that it does not require the use of DEM to retrieve displacements.
- The sampling rate of 5–10 min of IBIS-L is high compared to the space-borne SAR systems that have revisit time of several days. This high sampling rate simplifies phase unwrapping considerably.
- It is better in monitoring steep slopes than the space-borne SAR systems do.

Some of the disadvantages of IBIS-L over the space-borne SAR can be summarized as follows (Rödelsperger, 2011):

- IBIS-L system is limited to monitoring displacements at local scale, while space-borne SAR systems can monitor large area at any place on the earth.
- IBIS-L is not as good in monitoring subsidence as space-borne SAR systems.

10.3.8.2 Examples of Real-Beam Aperture Radar Systems: SSR and MSR 300
The SSR developed by GroundProbe Pty in Australia in 2001 for remotely tracking movement of slopes in open-pit mines, is an example of real-beam aperture radar system. The SSR system uses real aperture radar (RAR) to scan the investigated object with a mechanically rotating dish antenna up to 270° in horizontal direction and up to 100° vertically over a range of 2800 m. It is also capable of giving alarm warnings if the movement of the slope being monitored is accelerating toward failure. The system provides two operating ranges (GroundProbe, n.d.1): SSR–T (0.9 m dish) with an operational range of 30–1400 m and SSR–XT (1.8 m dish) with an operational range of 30–3500 m. According to GroundProbe (n.d.2), SSR technology is now generally accepted as a tool for high-risk slope management with its deployment and application in many mines in Australia, Indonesia, Africa, Chille, Canada, and the United States. It should also be mentioned that the largest practical sized dish antenna limits the effective operating range of RARs to several kilometers, beyond which failures can no longer be resolved. The beam cone angle is set by the size of the antenna aperture measured in wavelengths, and for a 1-m-diameter dish at 3 cm wavelength (10 GHz) this is 2°. So a requirement for a 10–15 m resolving power limits the maximum range of radar with these parameters to about 300–450 m. The maximum range will increase proportionately as the size of the dish or the operating frequency of the radar is increased.

The other type of real-beam aperture radar system is MSR 300 by Reutech Mining in South Africa. This system, which is similar in design to the SSR system, is an all-weather system capable of operating in harsh mining environments and also capable of real-time detection of submillimeter slope movements (Reutech Mining, 2014; Little, 2006). The manufacturer-claimed operating range is 2500 m.

10.3.8.3 Example: Fast Ground-Based Synthetic Aperture Radar (FastGBSAR)
MetaSensing, a remote sensing company in the Netherlands, offers very-high-resolution airborne and ground-based sensors for real-time deformation monitoring of structures, slopes, dikes, and bridges. The ground-based system, known as fast

ground-based synthetic aperture radar (FastGBSAR) system, is a fully portable product consisting of a compact radar sensor, an environmentally resistant 2-m-long rail, and the processing and power units. It is an active system, capable of producing its own illumination over the monitored scene day and night. The FastGBSAR instrument can be used in two different operational modes:

1. SAR mode when the FastGBSAR is mounted on 2-m-long linear drive (a ground-based rail). This mode is used when large unstable areas need to be monitored, as in the case of landslides, open-pit mines, dikes, and dams.
2. RAR mode in which the instrument is placed on a tripod or any other fixed installation and can take measurements at a rate of 4000 profiles per second (4 kHz).

The FastGBSAR produces displacement maps, which are given in the LoS direction joining the FastGBSAR sensor and the observed point. Negative displacements indicate the movement of pixels toward the radar, while positive displacements indicate that the pixels are getting further away with a possibility of the monitored region collapsing. The other specifications for this system are given (Rodelsperger et al., 2013; MetaSensing, 2013) as follows:

- It is a frequency-modulated continuous wave radar.
- It operates in Ku frequency band (17.2 GHz).
- Spatial resolution pixel: 0.75 m × 4.5 mrad (or 4.5 m at 1 km distance).
- Accuracy: ±0.1 mm (for SAR) and ±0.01 mm (for RAR) up to 4 km distance from the scene even on grass-covered slopes (with this level of accuracy not yet achievable with other ground-based SAR systems).
- Operating temperature: −20 to 60 °C.
- Data acquisition duration: only 4 s.

10.3.8.4 Advantages and Disadvantages of GB-InSAR Techniques The advantages of GB-InSAR system over the conventional monitoring techniques, such as GPS and total station equipment and TLSs, can be summarized as follows (Rödelsperger, 2011; Hanssen, 2011; Chen, 2011):

1. It has an ability to monitor displacements from a remote position of up to 4 km away without the need to install targets or sensors on the monitored ground or structure. In this case, an access to monitored object, which is required in total station survey, is not needed with regard to GB-InSAR, so that hazardous areas or inaccessible parts of a structure (large towers, dams or landslide areas) up to 4 km away can be monitored remotely.
2. It is capable of high-precision detection of relative displacement to submillimeter level with all targets simultaneously monitored.
3. Its operations are automatic and the system is capable of real-time monitoring of large areas of several square kilometers.

4. As an active system, it is able to carry out measurement under any lighting and weather conditions, including rainfalls, clouds and fog.
5. It has a high data-sampling rate (in the order of few minutes).
6. Custom software enables the user to pinpoint movements with the help of a high-resolution visual image, and to set alarm threshold to warn of any unstable conditions.

The main disadvantages and limitations of GB-InSAR with respect to the conventional monitoring techniques, such as GPS, total station and laser scanner systems are as follows (Rödelsperger, 2011; Hanssen, 2011; Chen, 2011):

1. It requires complex management, processing, and interpretation of data. The qualities of all major processing steps (e.g., image processing, image registration, interferogram filtering, phase unwrapping) are time-consuming.
2. The size of the equipment, which may be up to 3 m long or more makes it difficult to deploy and maintain.
3. Limited cone of view (in the order of some tenths of degrees in both the horizontal and vertical planes).
4. Measurement of displacements is along the instrument one-dimensional LoS, so that deformations that are basically three dimensional are projected onto one dimension. Some assumptions are usually required in order to resolve one-dimensional displacements into three-dimensional or two-dimensional displacements.
5. The effect of signal phase ambiguity; the measured displacements are usually ambiguous since phase shifts, and not absolute phases, are measured. Moreover, displacement higher than certain amount between two images may not be easily detected.
6. In geodesy, three important quality aspects are usually considered as *precision*, *accuracy,* and *reliability* estimates. The results of most D-InSAR applications are derived using a single interferometric pair; there is no redundancy involved so that one can say that the deformation estimates are not reliable. Moreover, the errors associated with the D-InSAR observations have different origins, such as unwrapping-related errors, residual topographic component due to DEM errors, and the effects of the atmosphere.
7. Reflectivity of the monitored surface will impact the monitoring results, for example, in the areas of high-ground vegetation, the radar waves may not penetrate the vegetation so that only the vegetation surface is observed, leading to loss of coherence.
8. Difficulty in localizing point displacement from volume displacements provided by the system. The exact locations of the measuring points are unknown (where artificial scatterers are not used) in InSAR techniques since network is usually randomly formed based on the accidental presence of coherent scatterers. This is unlike in the case of geodetic techniques in which network points are carefully chosen at the design stage.

TABLE 10.9 Comparison of LiDAR Systems with InSAR Systems.

Property	InSAR Systems	LiDAR Systems
Signal	Returned microwave signal	Reflected infrared laser pulse
Measuring technique	Phase difference	Time of flight
Wavelength	Centimeter level	Micrometer level
Illumination	Side-looking	Nadir or side-looking
Range measurement	Weather independent	Attenuation (by rain or fog) in atmosphere limits the range
Elevation accuracy	Variable (more sensitive to noise)	Decimeters (higher than InSAR)
Pixel resolution	Decimeters to meters	Decimeters to meters
Similarity	• Active system, illuminating the scene • Reflectance depends mainly on surface properties	• Active system, illuminating the scene • Reflectance depends mainly on surface properties
Coverage	Large area coverage in a short time and from a long distance	Less coverage over a short distance

10.4 COMPARISON OF LASER (LiDAR) AND RADAR (InSAR) TECHNOLOGIES

LiDAR systems and InSAR systems are compared in Table 10.9 (Stilla et al., n.d.).

11

DEFORMATION MONITORING AND ANALYSIS: GEOTECHNICAL AND STRUCTURAL TECHNIQUES

OBJECTIVES

At the end of this chapter, you should be able to

1. Describe the observables and the operation principles of various geotechnical instrumentations for deformation monitoring
2. Discuss the advantages and disadvantages (or limitations) of various geotechnical instrumentations for deformation monitoring
3. Discuss the various applications of geotechnical monitoring techniques, using extensometers, four-pin gauges, joint meters, plumblines, inclinometers, tiltmeters, fiber-optic sensors (FOS), and micro-electro-mechanical system (MEMS) sensors
4. Design geotechnical deformation monitoring schemes
5. Identify the differences between geotechnical and geodetic deformation monitoring schemes
6. Perform basic analysis of geotechnical (extensometer, plumbline, joint meter, tiltmeter) deformation measurements
7. Appreciate the accuracy specifications for various geotechnical instrumentations with regard to deformation monitoring
8. Explain how geotechnical monitoring techniques complement geodetic monitoring techniques

Precision Surveying: The Principles and Geomatics Practice, First Edition. John Olusegun Ogundare.

11.1 INTRODUCTION

Three types of measuring techniques are used in monitoring a dam: traditional geodetic techniques (discussed in Chapter 9); high-definition surveying and remote sensing techniques (discussed in Chapter 10); and geotechnical and structural techniques, which are the subject of this chapter. The knowledge of surveyors in data acquisition (in relation to traditional geodetic and remote sensing and photogrammetric techniques) and the intimate knowledge of the behavior of structures, soils, and rock by other specialists, such as geotechnical, structural, and rock mechanics engineers, are currently being advocated to complement each other in successfully analyzing and interpreting deformation monitoring data. According to Avella (1993), it is usually concluded with regard to dam deformation monitoring and analysis that surveying engineers are not promoting or educating themselves adequately in the field of deformation monitoring with regard to other techniques outside their discipline. With regard to this conclusion, it has become important for surveying engineers to become enlightened on the available methodologies of the design and implementation of deformation monitoring and analysis of deformable objects, including the monitoring of hydroelectric dams. The geotechnical and structural techniques of deformation monitoring discussed in this chapter mainly focus on monitoring and analysis of hydroelectric dams.

Geotechnical and structural techniques for deformation monitoring have the following characteristics:

1. They provide more localized information (usually in one dimension) at discrete locations with no physical correlation with other instrument locations. The typical information they provide includes deformation, load, stress and ground water pressure, about deformable structure. With this, the techniques alone cannot provide overall behavior of the structure being monitored. When an observation is isolated, it is important to ensure its acceptable precision and consistency. Recent technologies of nanometrology and MEMS sensors allow miniaturization of sensors, which can be used to monitor movement in any particular direction with a possibility of three-dimensional results.
2. They require frequent calibration of the instruments used for the effects of environmental temperature, drift of the readout unit, and conversion constant of the readout unit. Calibration of equipment is usually very critical to the long-term reliability of the equipment, especially for equipment that will be left on-site for automatic data collection (e.g., the in situ equipment); such equipment usually has the equipment testing as part of the measuring procedure. Because of the isolation and localization of geotechnical instruments, testing and calibration of the instruments are very important. Geotechnical instruments, however, are usually poorly calibrated and their readout units are susceptible to drift.
3. They do not provide redundant measurements so that their measurements are considered less reliable; they only provide repeated measurements of the same observable, which may be subject to deformation. As a result of this, locally

disturbed information is obtained without any check unless compared with some other independent measurements. In the data collection procedure, a campaign consists of a single measurement and could be observed relatively frequently; any geotechnical data series can be depicted as a simple series plot of the data.

4. Those that can output electrical signals are easily adapted for automatic and continuous monitoring in time domain than conventional geodetic survey techniques. The techniques also allow automatic and remote control of operation and data transmission with the instrumentation and are capable of providing monitoring information more frequently than the geodetic survey techniques.
5. They can provide relative displacements within limited ranges; the relative displacements could also be considered as absolute displacements if the measurements are made relative to a stable reference such as the bedrock below the deformation zone.
6. They are expensive to install since they are very sensitive to local instability of their installation components, such as the anchor point in solid bedrock. Most geotechnical instruments need careful, complicated, and expensive installations by experienced personnel; if instrument malfunctions, it may be difficult to get it repaired or replaced.
7. Once installed, geotechnical instruments require only infrequent checks on their performance so that skillful observers are usually not required.
8. The techniques can measure points that the operator has no access, such as the internal part of the structure, for example, boreholes and foundations of structures.
9. Some of the geotechnical instruments, such as borehole extensometers, are affected by local disturbances, such as thermal influence and instability of anchor points and vibrations; these require that adequate care be taken in order not to interpret the local disturbances as deformations.
10. They are able to provide highly precise readings that are taken to within a few one hundredth of a millimeter (0.01 mm); this precision is higher compared to that of geodetic techniques. Because of the conditions under which measurements are made in geotechnical instrumentation, the observational accuracy of instruments, which is often expressed as a tolerance or resolution, does not always correspond to the claims by the manufacturer.
11. Short bases of geotechnical instruments, such as tiltmeters, make them sensitive to local tilt, which may not be representative of the tilt of the structure being monitored. They are, therefore, not recommended over a large area, although they can be used as supplementary tool in monitoring mining effects on large infrastructure, such as building, shafts, boreholes, pipelines.

In the total effort of deformation monitoring, the quality of the analysis of the behavior of the object being monitored depends on the location, frequency, type, and reliability of the data gathered. The current trend in dam deformation monitoring is to integrate various geotechnical/structural and geodetic survey techniques into integrated monitoring scheme (Chrzanowski and Secord, 1987).

11.2 OVERVIEW OF GEOTECHNICAL AND STRUCTURAL INSTRUMENTATION

During the construction of large dams, geotechnical instrumentation is installed to monitor loads, stresses, and deformations to confirm design assumptions and to determine if there would be need for changes or remedial measures (Hunt, 2005). After the completion of dam construction, stresses and deformations are monitored to provide early warning against possible failure of the dam structures. The instrumentation is to monitor movements of rock and soil slopes, groundwater conditions, crustal activity, deflections of structural components, and so on. The selection of geotechnical/structural instrumentation type for a particular phenomenon depends on the condition of the dam that has to be monitored, that is, surface movements, subsurface deformation, and in situ pressure and stresses. Different types of geotechnical instruments are needed for varied locations of the structures being monitored. Some of the geotechnical/structural instrumentation and their applications with achievable accuracies at 95% are given (Hunt, 2005; Chrzanowski, 2009) in Table 11.1.

The main cores of geotechnical instruments are transducers, which are of two types: *mechanical transducers* and *electrical resistance transducers*. The mechanical transducers are of two types: *dial indicator* and *micrometer* types. The ranges of dial indicators are about 50 mm up to 300 mm, and their reading resolutions are generally within ±0.0025 mm to ±0.025 mm. The micrometers of transducers measure displacements by measuring rotations of finely threaded plungers when they travel in or out of their housing. They measure fractional revolutions with vernier. Depth micrometers (which are considered more robust than dial indicators) have accuracies that are limited to about ±0.025 mm, and their ranges can be extended to 5 m (for long length inside micrometers). In the case of electrical resistance transducers, measurements are made using the basic property of electrical conductor that the resistance of conductor is directly proportional to the length of conductor.

11.2.1 Extensometers

Extensometers are designed to measure extension (displacement) that takes place with time between pairs of points of a structure. They measure relative movements between points and they can be used to measure movements across a crack or inside or on the surface of a structure.

Two mechanical devices most frequently used for measuring movements in geotechnical instrumentation are *dial indicators* and *depth micrometers* shown in Figure 11.1. The ranges of dial indicators are about 50 mm up to 300 mm with the reading resolutions generally within ±0.0025 mm to ±0.025 mm. A micrometer, which is considered to be more robust than a dial indicator, measures displacements

TABLE 11.1 Some of the Geotechnical Structural Instrumentation and Their Applications with Achievable Accuracies at 95% Confidence Level.

Method/Instrument	Applications	Achievable Accuracy at 95% Level
Extensometers or strain meters (rods, wires, tapes)	Installed singly or in series (MPBX) in boreholes (in any orientation) or anchored at foundations of dam structures, to monitor deflections or settlement of structures or and to detect subsurface shear zones	±0.05 mm/10 m
Plumblines (suspended and inverted types)	Monitor structures (dam, column/beams) for surface horizontal movements or tilts	±0.05 mm/10 m
Vibrating strain meters	Measure linear strains downslope or across faults or joints	10^{-5}
Shuttle probes (biaxial or uniaxial inclinometer)	Measure subsurface deformation such as lateral deflections or shear zones; locate failure surface in a slope and monitor slope movements	±0.5 mm $\sqrt{L}$ m (L is measurement interval or longitudinal distance between inclinometer wheels)
Tiltmeters	Measure rotational component of deflection of structures, for example, dams	±0.2"
Fiber-optic strain meters	Measure linear strains	Conduit up to 10 km is possible
MEMS-inclinometer string	Measures three-dimensional displacements and strain components	Accuracy degrades with square root of length (over hundreds of meters range)

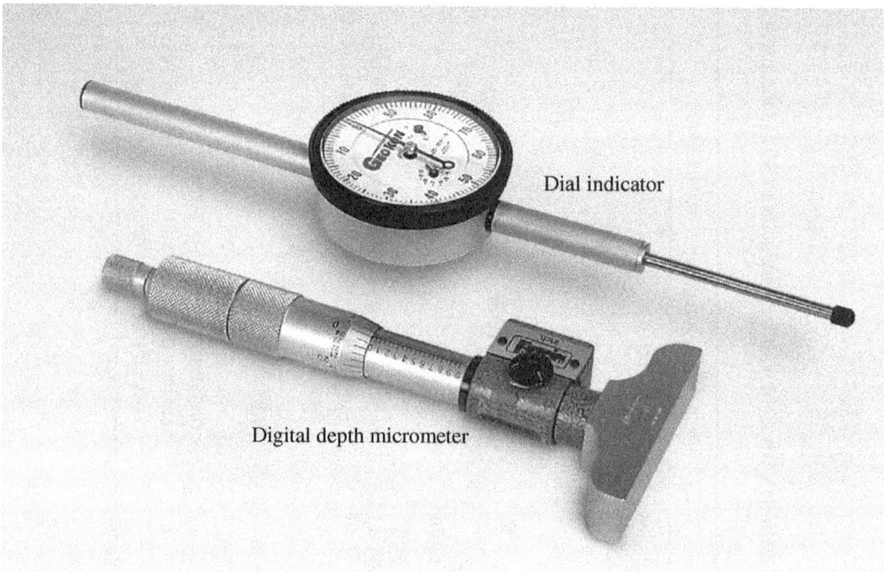

Figure 11.1 Two mechanical devices for reading rod extensometers. *Source:* Used with permission from Geokon Inc.

by measuring the rotation of a finely threaded plunger when it travels in or out of its housing. Fractional revolutions are measured with a vernier. The range of a micrometer can be extended to 150 mm with accuracies limited to about ±0.025 mm.

Apart from providing high-resolution measurements, electrical sensors will allow unattended monitoring by the data logger. The electronic readings are based on linear-displacement transducers, such as linear potentiometer or linear variable differential transformer (LVDT) also known as linear variable displacement transducer (LVDT), which can be connected to the extensometer rods and the readout box or data logger. This LVDT is a type of electrical transformer used for measuring linear displacement with high precision and repeatability. It is capable of operating in harsh environment and under high vibration and shock levels. It is also able to retain its measurements even after the power is switched off so that on restarting it, it will still show the same measurement without losing any information. The movement ranges with LVDT sensors are generally about 3 cm to several centimeters and their sensitivity is about 10 μstrains. LVDT sensors are considered to be much less sensitive to moisture and less affected by temperature than are linear potentiometers.

The LVDT system also provides a means of connecting all the components of instrumentation system into one central network application, so that one can use the Ethernet-to-serial controllers and have complete control of the instrumentation system from a local computer. For example, the LVDT system can be used to measure x and y displacement components of plumb-line installations and the measurements transmitted remotely to the readout box; and it is also able to provide one readout box that can display measurements from plumb line, extensometer, and temperature sensors installed at different remote points of the structure being monitored. The accuracy of measurements of borehole extensometers with LVDT is about ±0.1 mm (J. Fletcher, personal communication). Most of the extensometers currently in use today have a digital readout and can store data digitally and are capable of being linked to an alarm system. Limitation to its use, however, may be imposed by the narrow range of the measuring equipment, which is typically between 25 and 100 mm with a total range of about 25 cm possible if both the depth micrometer and the electric sensors will be reset (which may introduce some additional error) to obtain that range.

One of the advantages of using extensometers is that it can be automated (allowing continuous automated monitoring of extension) with a capability of being interfaced with temperature sensors to correct for wire expansion or contraction. The automated system may also transmit alarms to the remote station if certain conditions occur, such as broken wire, movement exceeding a particular specified amount. The remote station collects and processes the data and generates screen displays and reports of rock movements. This type of monitoring system does not require periodic site visits by technicians to undertake measurements, unlike the probe-type monitoring systems; and it is easy to use.

Various types of extensometers are available, such as rod, tape, wire, video, laser extensometers. Two common types of extensometers are *fixed borehole rod extensometers* and *portable wire line extensometer*. Borehole rod extensometers are the most reliable when compared with the wire extensometer measurements; invar rod may be used in rod extensometers to reduce the effects of thermal expansion with achievable accuracies of ±0.1 mm or better for up to 100 m range.

11.2.1.1 Rod Extensometers The fixed borehole rod extensometer is used for measuring extension parallel to the borehole axis between a fixed number of anchor points and a reference point on the same measurement axis. The basic components of a fixed borehole extensometer are an anchor, a linkage, and a reference head. The reference head is installed at the borehole collar; the linkage system, which may be composed of fiberglass, alloy, invar, or of stainless steel, spans the distance between the reference head and the anchor. The rod is installed with one end anchored in natural ground in the borehole. The position of the outer end of the rod can be monitored relative to a fixed collar on the borehole face. By comparing the current reading to the previous reading, the operator can calculate the change between the anchor point and the fixed collar point. Readings on the rod extensometers can be taken manually using depth micrometer or using electrical resistance transducers (sensors), which store the readings in an electronic data logger to be transferred to a computer later.

Two types of rod extensometers can be identified as *single-point rod extensometer* and *multipoint extensometer*. A single-point rod extensometer consists of an anchor, a rod, and a reference head as shown in Figure 11.2. A multipoint extensometer consists of many rods of various lengths anchored at different points with the rods monitored at one reference head as shown in Figure 11.3. The reference head is installed at the borehole collar. The rod spans the distance from the anchor to the reference head.

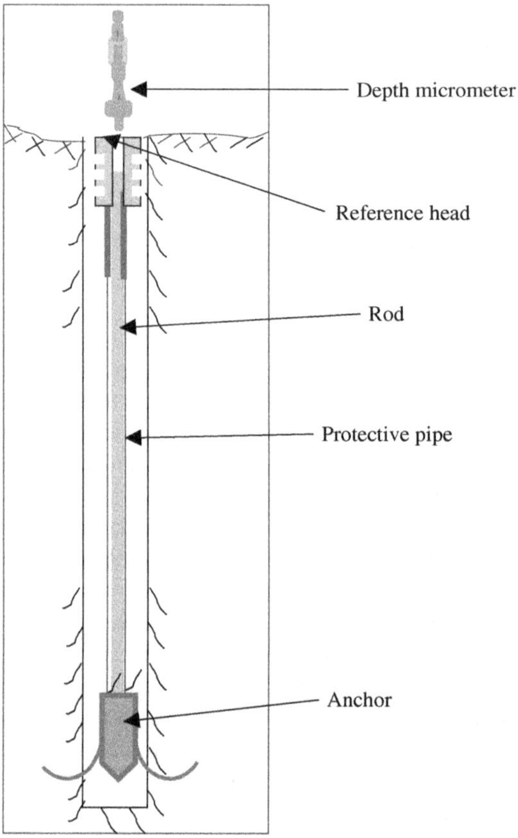

Figure 11.2 Sketch of a single-point rod extensometer.

Readings, which can be taken mechanically or electronically, are obtained at the reference head by measuring the distance between the tip of the rod and a reference head. A change in the distances between two epochs of measurements indicates that movement has occurred.

Figure 11.3(b) illustrates how readings can be taken mechanically (using depth micrometer) at a reference head of a multipoint extensometer assembly. For electronic readings of extensometers, LVDT system is commonly used; a typical system consisting of borehole extensometers, LVDT, and a readout box is shown in Figure 11.4.

Multipoint measurements of vertical deformation using multipoint extensometer can reveal the distribution of movement along the axis of the borehole in addition to providing total movement as it is usually done with a single-point extensometer. Referring to Figure 11.3, for example, if R_{10} and R_{11} are the initial and current readings at the reference head for Rod 1, which is anchored at depth d_1, and R_{20} and R_{22} are the initial and current readings at the same reference head for another Rod 2 anchored at depth d_2 (with d_2 at a greater depth than d_1), the change in length between

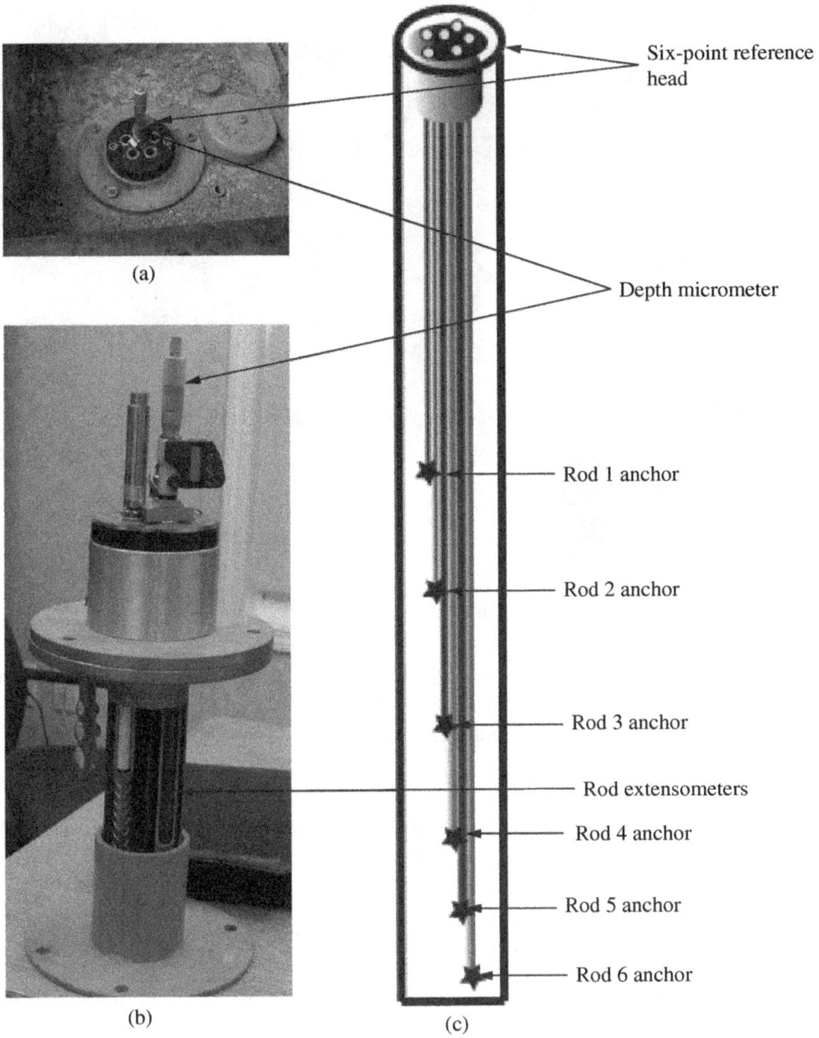

Figure 11.3 (a) Reference head for a six-point rod extensometer installation with depth micrometer in one of the reference points. (b) A six-point rod extensometer assembly with depth micrometer in one of the reference points for illustration. (c) A sketch of six-point rod extensometer installation.

anchors 1 and 2 will be $(R_{22} - R_{20}) - (R_{11} - R_{10})$ over a distance of $d_2 - d_1$. It is usually assumed that the deepest anchor is in a stable ground so that the rod associated with it will record no movement.

Depending on the orientation of the extensometer rod in the borehole, some borehole extensometer measurements will provide mainly the information on the vertical or horizontal expansion of the monitored structure, soil or rock. In Figures 11.2–11.4,

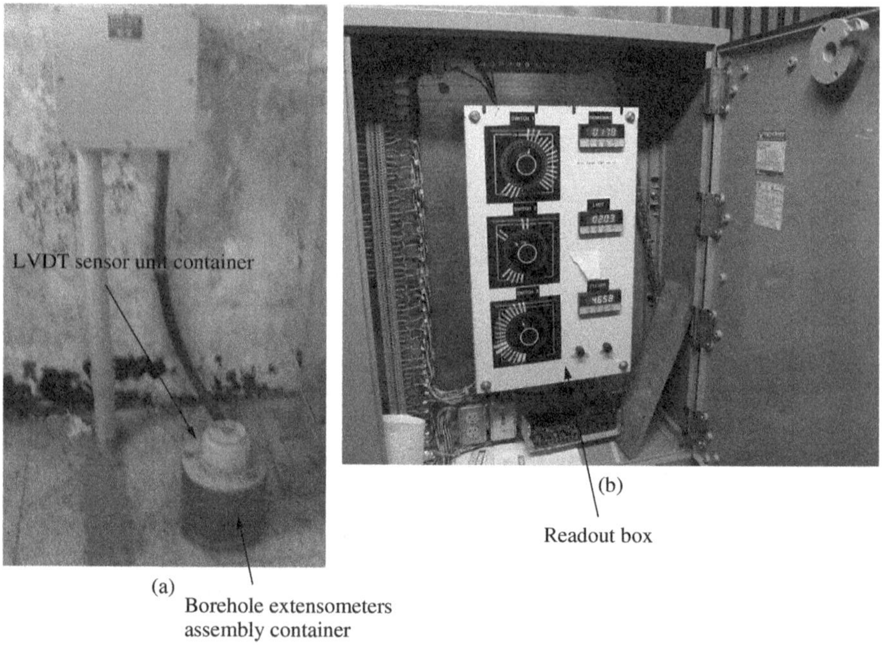

Figure 11.4 (a) Borehole rod extensometer equipped with LVDT sensors for automatic monitoring of rod extensometers. (b) Centralized LVDT readout system for automatic measurements of LVDT installations at different locations.

the rod extensometers are placed vertically to monitor vertical deformations, which may be required in order to verify the following:

- If soil consolidations due to structures being monitored are consistent with the soil predictions.
- If engineered foundations of monitored structures are performing according to expectations.
- If the impact of monitored settlements on infrastructures is at the acceptable level.

If a rod extensometer and a weighted plumbline can be arranged as shown in Figure 11.5, it is possible to use them to establish a relationship between the vertical movement of a section of a structure (measured with the rod extensometer) and the horizontal movement of that section (measured with the weighted plumbline). In this arrangement, the vertical displacement of a reference surface (grouted to the structure) relative to the rod extensometer anchor point is measured with a micrometer as illustrated in Figure 11.5(b); the plumbline is then measured as usual to determine the horizontal movement of the structure relative to the plumbline anchor point, which is related to the extensometer anchor point.

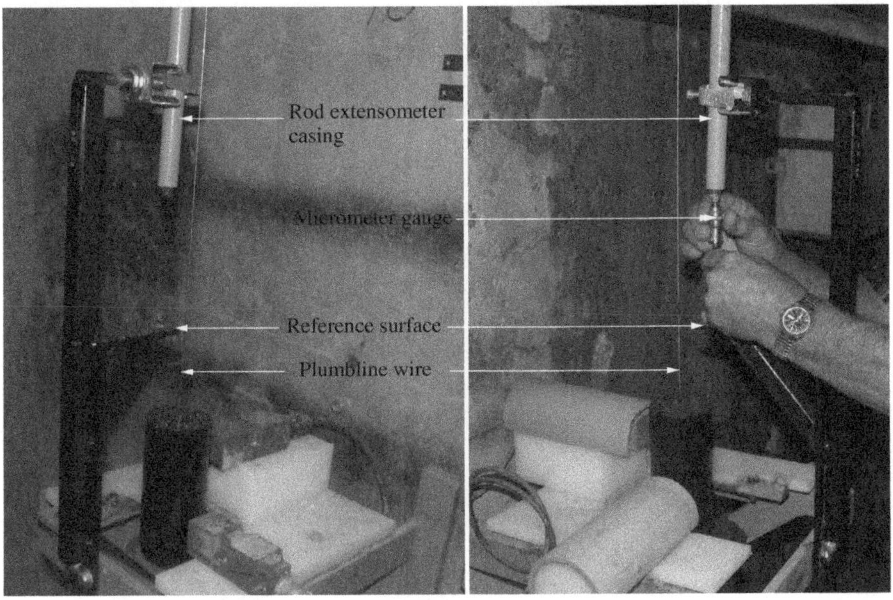

(a) (b)

Figure 11.5 (a) Arrangement of suspended pendulum and invar rod extensometer. (b) Micrometer measurement of relative vertical displacement between the extensometer anchor point and the bracket grouted to the wall in the Intake structure.

A nonvertical installation of rod extensometers is shown in Figure 11.6, in which two invar rod extensometers are installed horizontally from two anchor points on the walls of a gallery of a hydroelectric generating station and connected to two measuring heads; the relative movements of the measuring heads of the two extensometers are then measured with micrometer gauge. The measuring heads can also be equipped with LVDT sensors for automatic reading of possible movements. If the rods are connecting the downstream and upstream walls of a Powerhouse, then the downstream/upstream movement of the walls will be monitored. It should also be mentioned that the operational principle of invar rod extensometers is the same as that of borehole extensometers except that invar rods, which are mounted on the surface in this case, are used instead in order to reduce the effects of thermal expansion of the rods on deformation measurements. The deformation measurements, however, may be erroneous if the anchor points are loosened and if free movement of rods are disallowed due to bending of tubing.

A nonvertical application of rod extensometer is shown in Figure 11.7, where the horizontal changes in distance (across a joint in a structure) between two measuring heads are measured using long invar rod micrometer gauge.

11.2.1.2 Tape Extensometers A tape extensometer, consisting of a measuring invar tape, tape-tensioning device coupled to a sliding scale, and a dial gauge,

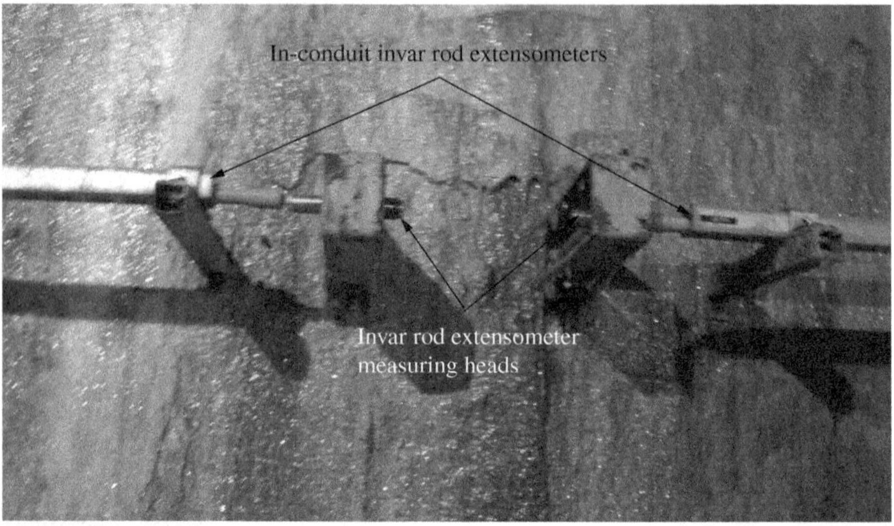

Figure 11.6 Invar rod extensometer installation with the measuring heads (with micrometer measurements usually taken between the two heads).

is used for measuring distances between pairs of points. The measuring tape has equally spaced punched holes of about 25 cm apart with a location pin at the tip of the instrument. This location pin is inserted into a hole on the tape and secured in place with a retaining clip and the tape tension adjusted to a desired level before the extensometer is read. In reading the extensometer, the position of the pin along the tape is read first and then the dial gauge, which is precise to ±0.01 mm, is read after. Many tape extensometer and invar rod extensometer measurements are often made at the generator and turbine floors; and many horizontal invar rod extensometers are often installed along the upstream and downstream walls of the Powerhouse structure and also along the walls of the Intake structure. A typical tape extensometer measurement procedure in the Intake and Powerhouse structures of a generating station are shown in Figure 11.8(a) and (b).

The use of tape extensometer in deformation measurements, however, has proved to be most troublesome with the following possible problems (Chrzanowski and Secord, 1987):

- Tape may be improperly calibrated over its entire length, before and after each survey campaign.
- Tape may be broken and repaired resulting in length change, or tape holes may become deformed over time due to frequent insertion of steel pin in the invar tape holes and tensioning of the tape.
- Tensioning device for the tape may have weakened over time or may have changed due to repair or exchange so that achieving required tension for wire extensometers may become difficult. Also, as the tensioning springs of the

Figure 11.7 Measuring the change in the joint on an Intake structure of a hydroelectric generating station using invar rod micrometer gauge.

instrument ages, the instruments may indicate false expansion results unless they are carefully calibrated on a suitable baseline before and after each measuring campaign.

- Inconsistent alignment of the tensioning marks during readings, which may affect the sag of the tape and the accuracy of measurements.
- The effects of inaccurate temperature reduction and wind on the tape measurements.
- Tape or wire extensometers will hang in the shape of a catenary if not fully supported, requiring sag correction.

In general, tape or wire measurement results may lack continuity, resulting in gaps or slips, which have to be included as nuisance unknown parameters in fitting cyclic function to the measurements. Without the tape extensometer measurements, however, deformation analysis of a Powerhouse structure could become incomplete. Some of the examples of wire or tape extensometers include Kern distometer and CERN distinvar, which are accurate to 0.05 mm or better, over 1–100 m base if properly calibrated (Chrzanowski, 1986).

11.2.2 Four-Pin Gauges

Four-pin gauge is a monitoring device for determining three-dimensional displacements of one side of a crack or opening with respect to an opposite side. In this

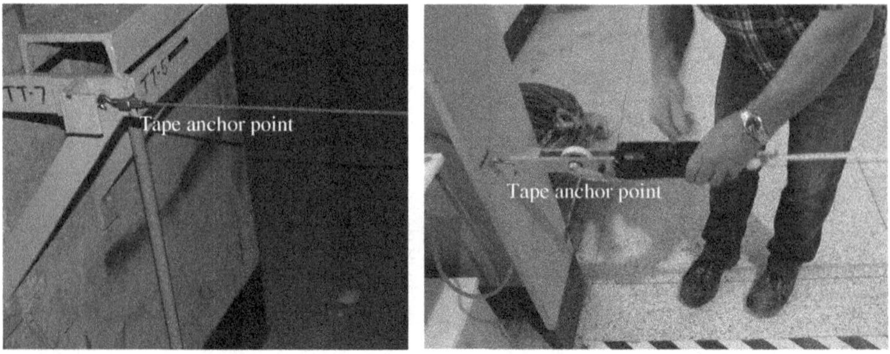

(a) (b)

Figure 11.8 (a) Tape extensometer measurement between two wall anchor points. (b) Tape extensometer measurement between the upstream and downstream columns (anchor point on end side of one column is shown) in a Powerhouse.

technique, which is illustrated in Figure 11.9(a), three pins (P_1, P_2, P_3) in a triangular pattern on a steel plate is anchored on a surface on one side of a joint or crack; the fourth pin (P_4) is installed on another plate that is anchored to the other side of the joint or crack. The vertical displacement between the base plate that is put flush with the three pins on one side and the top of the single pin on the other side is measured using depth micrometer with a base plate as shown in Figure 11.9(b). The horizontal components of displacement are determined using a micrometer caliper by measuring distances P_2-P_3, P_2-P_4, and P_3-P_4 with the distance P_2-P_3 being a constant; the measurement of distance P_4-P_3 is shown in Figure 11.9(c). The accuracy of displacements determined through this process is quoted (J. Fletcher, personal communication) as ±0.1 to 0.2 mm.

11.2.3 Joint Meters

A joint meter is a device for measuring displacements across a crack or joint. In this technique, two steel brackets are placed on each side of a crack or joint with the shape of the bracket designed to measure displacement in one, two, or three directions (X, Y, Z) as shown in Figure 11.10. The X, Y, and Z coordinates are measured with the X, Y axes based on the local coordinate system and Z based on mean sea level elevation. Micrometer gauge or linear variable displacement transducer (LVDT) can be used to measure these displacement components to an accuracy of better than ±0.1 mm.

11.2.4 Plumb Lines

A plumbline (or pendulum) is a wire with a weight or plumb bob hanging on it, which is used to provide a vertical reference line. It can be used to measure tilts of structures and relative horizontal movements of reference points with respect to

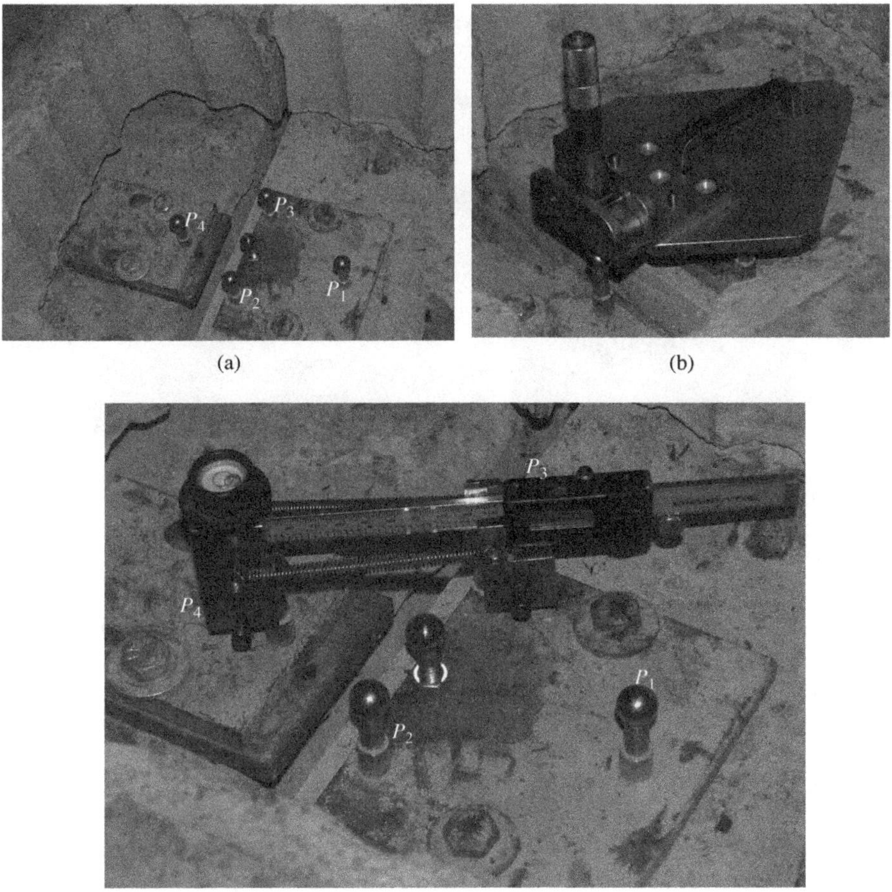

(a) (b)

(c)

Figure 11.9 Four-pin gauge for displacement measurement. (a) Four-pin monitoring points. (b) Four-pin vertical movement measurement. (c) Four-pin joint measurement across points P_4 and P_3.

a vertical reference line (or wire). There are two types of plumblines: *suspended* (or *weighted*) and *inverted* (or *reversed*). Two types of wire commonly used in the plumbline systems are the cheaper 1.05-mm-diameter stainless steel spring wire and the very expensive 1.68-mm-diameter invar wire. The invar wire type is used where dimensional stability is important as in the case of height transfer. The length of a plumbline is usually limited to 60 m in order to minimize vibration of the plumb wire due to air current and wind from the surroundings. To monitor, for example, the deflection of a complete profile of a high concrete dam whose height is much greater than 60 m, several plumblines may have to be installed in vertical alignment one above the other starting with an inverted plumbline anchored in the borehole located in the dam foundation. The anchors for inverted plumblines are usually several meters

(a) (b)

Figure 11.10 (a) Joint meter mounted over a joint with vertical reading taken with a micrometer gauge. (b) Joint meter mounted over a joint with the horizontal reading taken with a micrometer gauge.

deep (up to 50 m or even more) below the dam foundation in order to obtain absolute displacements of the surface points.

A plumbline system consists of two types of readout devices: manual and automatic. In the manual device types, readings are taken with a steel measuring tape to an accuracy of ± 0.5 mm or to an accuracy of ± 0.03 mm using traveling vernier microscopes (Dunnicliff, 1988). In the case of automatic readout types, positions of plumb wires are continuously sensed, with remote readings and recordings possible. These position measurements are made with no physical contact with the pendulum wire. Most automatic readout types are based on *induction with frequency output* or *optical* principles. Those based on induction principles are designed with a target inserted along the plumb-line wire that passes through the inductive table. The readout type based on optical principles usually have a light source incorporated on one side of the wire such that the shadow of the wire is projected onto a linear photodiode array attached to the table frame on the opposite side. Accuracies of 0.01 mm to 0.05 mm are quoted (ASCE, 2000) for the readout types based on the optical principles; the induction readout types are likely to have accuracy better than 0.1 mm.

The common sources of error in using plumblines to monitor deformation are the influence of air currents and the spiral shape of wires. The amount of horizontal displacement caused by air current can be given (Chrzanowski and Robinson, 1981) as

$$e = \frac{30(h)(H)(d)v^2}{Q} \text{ mm} \tag{11.1}$$

where H (m) is the distance between the anchor point of the plumbline and the center of the underground excavation exposed to the air current; h (m) is the length of the wire, approximately equal to the height of the underground opening exposed to the air current; d (m) is the diameter of the wire; v (m/s) is the air current velocity at the cross section h; Q (kg) is the weight of the plumb bob. Spiral shape will affect all wires that are not specially straightened. The horizontal distance (s) between the actual position of the spiral wire and the expected plumbline (direction of gravity) is given (Chrzanowski and Robinson, 1981) as

$$s = \frac{\pi d^4 E}{64RQ}\,\text{mm} \tag{11.2}$$

where E is Young's modulus of elasticity (about 2×10^{11} Pa for steel); R (cm) is the radius of the unloaded spiral wire; d (cm) is the diameter of the wire; Q (kg) is the weight of the plumb bob.

The use of plumblines in monitoring deformations has some advantages: they are simple to use and they are able to operate over several years with little or no maintenance. Because of these advantages, they are usually permanently installed in Powerhouses, Intake structures, and Diversion sluiceway structures of hydroelectric generating stations. The suspended (or weighted) and inverted plumbline methods of monitoring deformations are discussed in the following sections.

11.2.4.1 Suspended (or Weighted) Plumb Lines A suspended pendulum or plumbline (shown in Figure 11.11) consists of an anchor point, a free-hanging wire, and a weight of about 10–20 kg immersed in a tank filled with liquid. Immersing the plumbline weight in liquid is to help reduce vibrations and swinging of the plumbline, which may be due to the motion of turbines in the Powerhouse structure or due to some other sources. The suspended plumbline is used for measuring the relative horizontal displacement of suspension point of the plumbline (or the pendulum wire) with respect to reference marks (reference frame) anchored to a structure at various levels. Figure 11.11 is an example of a suspended plumbline system installed to monitor the inclination of the top (anchor point P) of a column of a Powerhouse structure of a hydroelectric generating station, with respect to the bottom (the reading table position Q). It is common to have many columns (in both upstream and downstream sides) and other components of a Powerhouse structure of a generating station equipped with weighted plumblines.

Some of the weighted plumbline installations (the stairwell and the hoist well plumblines) in a Powerhouse of a generating station are shown in Figure 11.12; the plumblines are to monitor inclinations of certain walls of the Powerhouse.

The operation of a suspended or a weighted plumbline is similar to that of an inverted plumbline except that the process is reversed as shown in the schematic diagram of a weighted plumbline installation in Figure 11.13, in which the anchor point is at point P and the measuring table is bolted to the structure at level Q. The readings at level Q will reflect the relative horizontal movement between the anchor point P and the measuring table Q. The movement is measured by the measuring unit at Q as x- and y-displacements in a three-dimensional x, y, z coordinate system with its origin at Q; the z-axis corresponds with the direction of the plumb wire, while the x- and y-axes are on a plane which is orthogonal to z-axis at Q. Similar to the case of

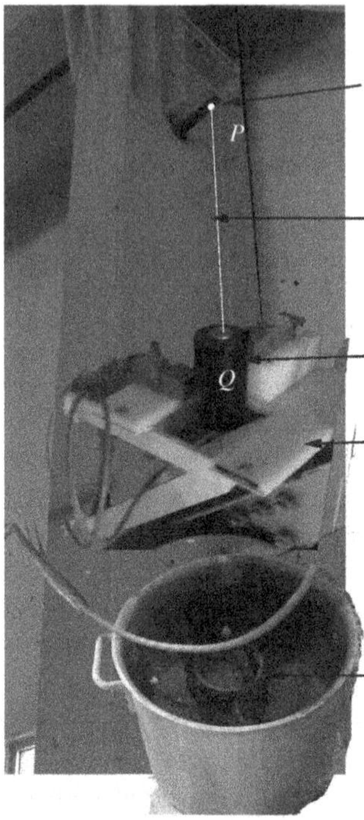

Figure 11.11 A weighted plumbline system to measure the inclination of a column.

inverted plumbline, the tilt of the structure is calculated as the difference between the "top" and the "bottom" measurements and presented as a nonnegative value.

To illustrate how a weighted plumbline operates, consider the schematic view of the weighted plumbline installation, shown in Figure 11.13(a), and let (b) be the state of the plumbline system at the second epoch of measurement. In Figure 11.13(b), the anchor point P and the plumb wire PQ moved to point P' and line P'-P'', respectively; and the reference point Q moved to Q'. The displacement reading at the reading table Q will be dr_Q, which is the relative movement between points P and Q (or length $Q'P''$). The displacement will always be measured from the instantaneous position of Q (which is Q' since it moves with the measuring unit) to the instantaneous position of plumb wire P'. This displacement can be measured either electronically (using sensors) or mechanically (using micrometers).

The accuracy of measuring displacement in plumbline systems depends on the reading system and the location of the instrument. In an area where airflow is a problem, it is possible to achieve an accuracy of ±0.2 mm to ±0.4 mm and where a table system is used in a moderate airflow condition; the accuracy of ±0.1 mm is possible. A typical device for electronically measuring displacements between reference marks and a wire (black hollow cylindrical object) is shown in Figure 11.14, in which the

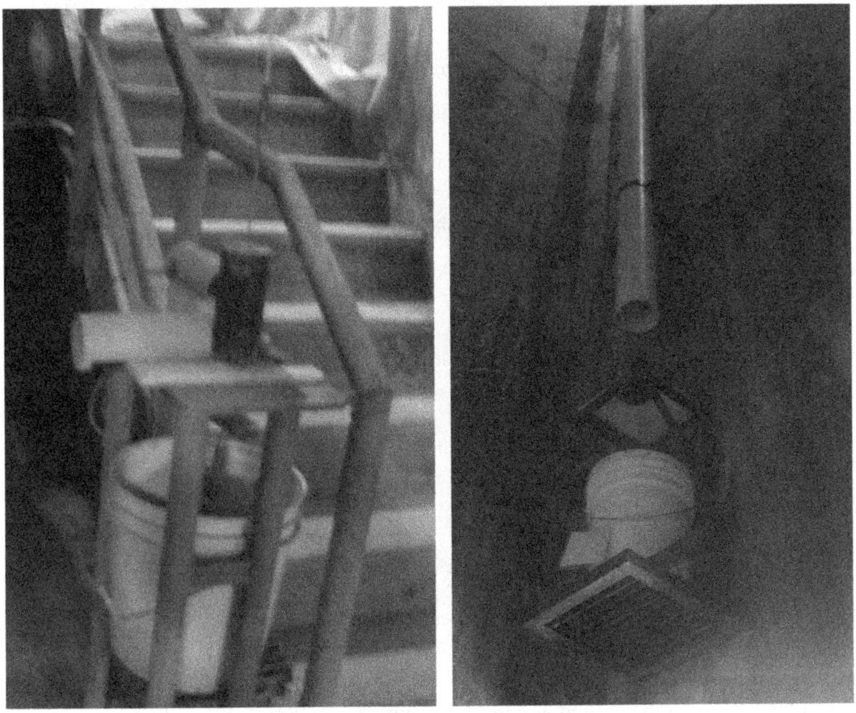

(a) (b)

Figure 11.12 (a) Typical measurement location of stairwell plumbline in a Powerhouse. (b) Typical measurement location of hoist well plumbline in a Powerhouse.

x-displacement measurement is currently being displayed in the digital readout unit; the y-displacement is measured separately by disconnecting the digital readout unit from the x-displacement unit and hooking it to the y-displacement unit.

The relative horizontal movement between two levels (P and Q) as shown in Figure 11.13 can be converted into tilt by dividing the horizontal movement by the distance between the two levels. Since the displacement dr_Q is in relation to the vertical distance (dh) between points P and Q, the tilt (in radians) can be given as dr_Q/dh. The displacement dr_Q can also be resolved into two (x, y) perpendicular directions and read as x and y components so that the tilt in each direction can be determined. It is typical to have x-direction in the downstream direction and the y-direction perpendicular to it.

The primary disadvantages of using weighted plumblines are as follows: they require trained personnel to operate the measuring units and to acquire displacement readings; the metal components of the systems are prone to corrosion, which may impact the accuracy of measurements; and they cannot be used at the construction stage of the structure being monitored (the structure being monitored must be completed before the plumblines can be installed and used).

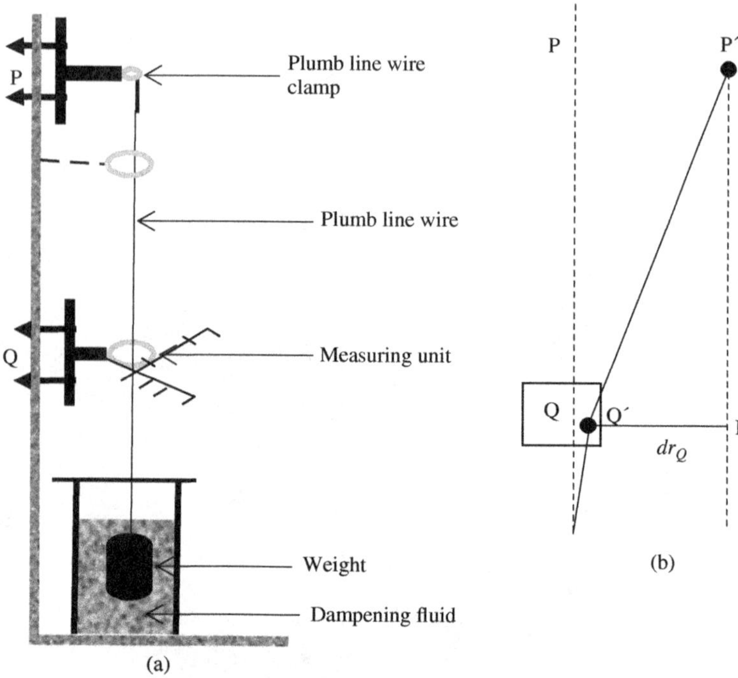

Figure 11.13 (a) A schematic diagram of a weighted plumbline installation. (b) Horizontal displacement of point P with respect to point Q.

11.2.4.2 Inverted Plumblines Inverted plumbline (pendulum) is used to measure the relative horizontal displacement between a reference datum (the wire or the anchor point) and the reading table on the metal frame bolted to the structure (the floor) whose movement is being monitored, as illustrated in Figure 11.15. The plumbline consists of a stainless steel spring wire anchored in the structure foundation with a float tank fixed at its upper end. The float (shown in Figure 11.16), which is free to move in the tank, is to provide tension on the wire and keep it vertical; this pendulum wire will retain the vertical position as long as the movement of the float is unrestricted.

In the typical installation of an inverted plumbline shown in Figure 11.15, the pendulum wire is guided through a vertical pipe of 15 cm or more in diameter from the gallery of the dam to the desired depth up to the stable anchor point with the wire free to swing within the pipe. A right amount of tension is maintained in the plumbline wire with the liquid in the tank acting as damping medium to prevent any to-and-fro oscillation of the plumbline. A tilt or relative horizontal movement between the anchor point and the metal frame bolted to the structure will bring about a shift in the float. This shift is measured on the measuring unit on the reading table by reading the scales on the unit in two (x and y) orthogonal directions. The tilt of the metal frame is calculated as the difference between the "anchor point" and the

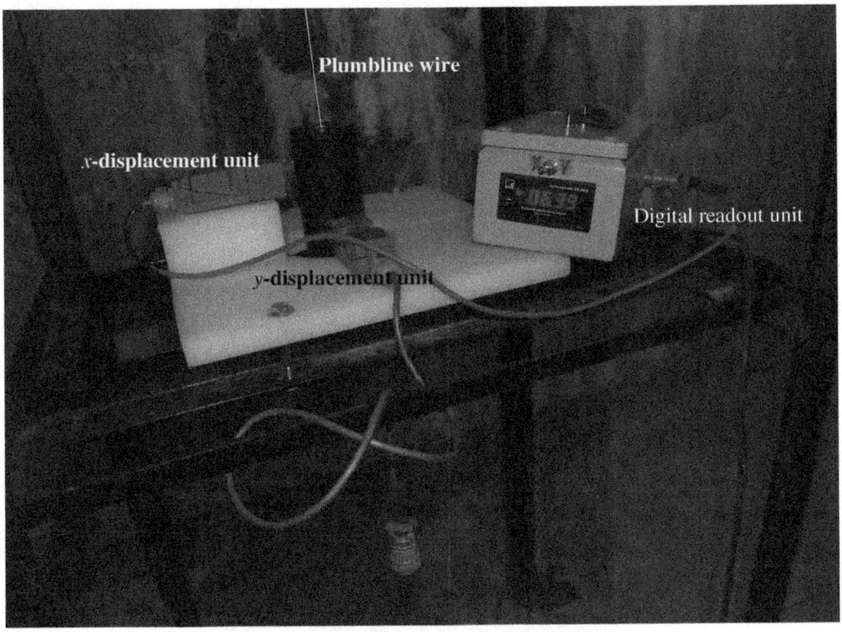

Figure 11.14 Reading the x- and y-displacement of a weighted plumbline.

“reading table” measurements, presented as a nonnegative value. If the anchor point is stable, the inverted pendulum will provide information on the absolute horizontal displacements as well as vertical displacements of the reading table at the given level with respect to the stable point. By connecting the inverted plumbline readouts with the readouts of suspended plumblines at other locations in the structure, the absolute horizontal movements of the reading tables at those other locations can also be determined.

The schematic diagram of a typical installation of an inverted plumbline is shown in Figure 11.17(a), in which the displacement of the table at level Q relative to the anchor point P is to be determined. In Figure 11.17(b), the displacement reading of point Q with respect to stable point P can be given as $dr_Q = r_{Q2} - r_{Q1}$ (where r_{Q1} and r_{Q2} are the measurements at epochs 1 and 2, respectively). This displacement, being over the vertical distance (dh) between point P and Q, will translate to a tilt, dr_Q/dh (in radians). The displacement can also be resolved into two (x, y) perpendicular directions and read as x and y components so that the tilt in each direction can be determined. It is typical to have x-direction in the downstream direction and the y-direction perpendicular to it.

Two inverted plumbline installations in one of the galleries of an Intake structure of a generating station are shown in Figure 11.18. The installations are for measuring the tilt of the wall at this level of the gallery relative to the lower level of the structure. An example of a device for monitoring relative horizontal displacements between a suspended or inverted pendulum wire and a structure is the Roctest RxTx telependulum.

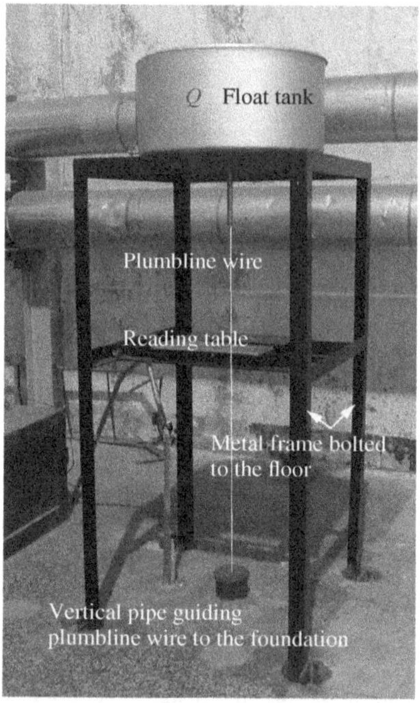

Figure 11.15 An inverted plumbline installation in a Powerhouse of a dam.

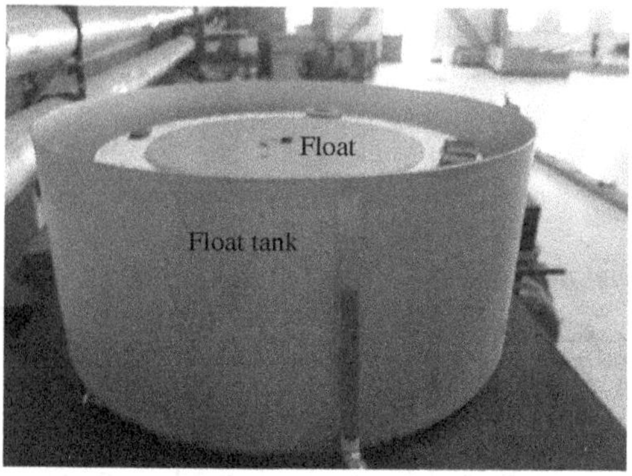

Figure 11.16 A plumbline tank containing a float and liquid.

This device, which has two communication ports for data transmission to computer or remote station via a modem, is capable of optically measuring the relative position of a pendulum wire in the *X*, *Y*, and *Z* axes with a precision of ±0.05 mm (Roctest, 2007). The accuracy of the device, however, is usually considered to be ±0.25 mm to ±0.30 mm. The setup of the RxTx telependulum device in an Intake structure of a generating station to measure the position of an inverted pendulum wire is shown in Figure 11.19.

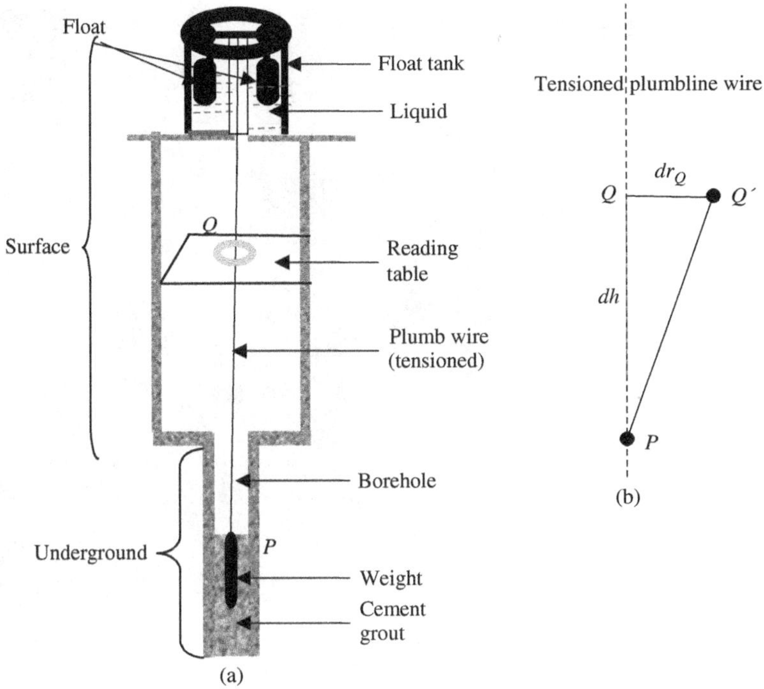

Figure 11.17 (a) A schematic diagram of inverted plumbline installation. (b) Displacement of point Q with respect to point P.

The main advantage of inverted plumbline approach over weighted plumbline approach is that it is possible to determine the absolute horizontal displacement of structures with respect to stable bedrock or foundation with the use of inverted plumbline, which is impossible with weighted plumbline. The main disadvantage over weighted plumbline approach is that the installation of inverted plumblines requires drilling vertical holes of between 6 and 50 m deep (depending on the anchor points) with the diameter of the holes varying from 70 to 170 mm to allow a vertical pendulum cylinder of about 50 mm in diameter to be free within the holes; it is usually difficult to drill straight vertical ducts or boreholes into bedrocks of foundations so as to allow plumb wires to swing freely in the drilled boreholes. The other point to note about inverted plumblines (and also the suspended plumblines) is that it will fail to recognize displacement when the entire object being monitored including the anchor point translates as an entity; in this case, there will be no change in tilt and no displacement will be detected.

11.2.5 Inclinometers

Inclinometers are used to monitor differential subsurface deformations by means of a probe that measures changes in inclination along the length of a borehole. Inclinometers are the main geotechnical instruments for monitoring horizontal

Figure 11.18 Inverted plumbline installations in one of the galleries of the Intake structure of a generating station (with brackets bolted to concrete wall).

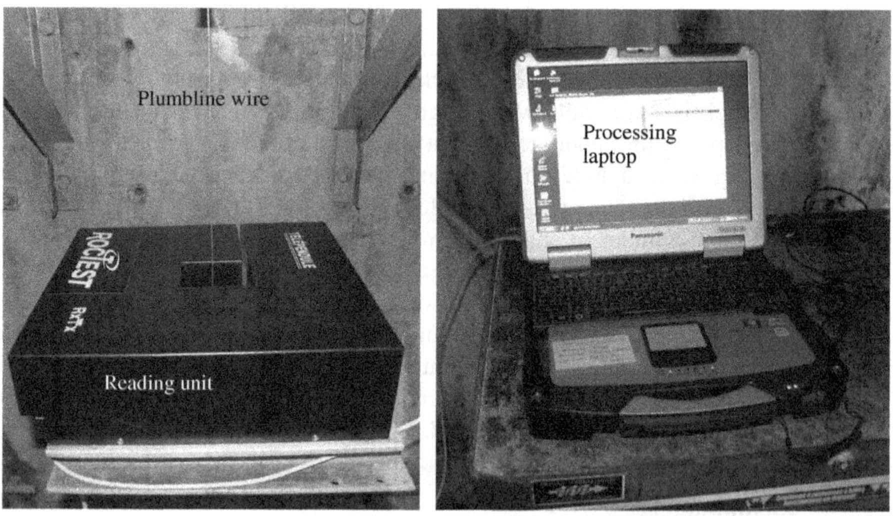

Figure 11.19 Roctest RxTx telependulum device interfaced with a computer for reading relative position of an inverted pendulum wire.

subsurface movements of soil and rocks and for profile measurements. They consist of servo-accelerators as sensors and require the use of inclinometer casing. The casing of a traditional inclinometer is a special-purpose, grooved pipe that provides access for the inclinometer probe so that the probe can obtain subsurface measurements. It can be installed in a borehole that passes through suspected zones of movement or embedded in fill, cast into concrete, or attached to structures being monitored. The casing, which is designed to deform with movement of the adjacent ground or structure, is usually made of ABS plastic material so that it can retain its shape and flexibility over a wide range of temperature. The useful life of the casing, however, ends when continued movement of the ground shears the casing so that the passage of the inclinometer probe is hindered. The grooves inside the casing are for controlling the orientation of the probe and to provide a surface from which repeatable measurements can be obtained; and they should be placed in such a way that one set of opposite grooves are aligned with the direction of expected movement. The quality of the grooves inside the casing directly influences the measurement accuracy of the inclinometer, and the inclinometer sensors work best in casing that is installed within 3° of vertical.

The traditional types of inclinometers may be classified according to the number of servo-accelerator it consists, such as uniaxial inclinometers consisting of one servo-accelerator, for measuring inclination in one direction; and biaxial inclinometers consisting of two servo-accelerometers, for measuring inclination in two perpendicular planes. Two main types of inclinometers are *portable, traversing probe system* (*or the shuttle probe*) and *dedicated, in situ sensor system*. The *traversing probe system* (*or shuttle probe*) consists of sensors (usually of servo-accelerometer types), a portable wheeled probe, graduated control cable, and a portable readout as shown in Figures 11.20–11.22. The shuttle probe helps to centralize the plumbline wire at each measuring point in the borehole being monitored with the electronic sensors used for reading the positions of the wire.

Examples of inclinometer probes are Geokon Model 6100D digital inclinometer probe and Digitilt inclinometer probe systems, such as the type shown in Figure 11.22. Geokon Model 6100D digital inclinometer probe is quoted (Geokon, 2013) to have a resolution of ±0.025 mm/500 mm casing; and Digitilt inclinometer probe systems are quoted (DGSI, 2013) to have a resolution of ±0.02 mm/500 mm casing and a system accuracy of about ±6 mm/25 m of casing. Some of the advantages of the traversing inclinometer probe system are given as follows:

- Ability to provide a detailed survey of the entire length of the inclinometer casing, which will allow multiple shear zones to be identified.
- Possibility of reusing the same probe to monitor other installations.

Some of the limitations of the traversing probe system are as follows:

- Inconvenience of having to carry bulky and heavy probe cable and readout unit from one installation to another.

Figure 11.20 A shuttle probe being lowered into a borehole guiding tube.

- Need for direct access to borehole installations for data acquisition since it is impossible to read the system remotely.
- Inability to allow automatic data acquisition as in the case of *in situ* system.
- Cost of borehole for inclinometer casing and ongoing cost of sending out a technician to read the installations.

The *in situ inclinometer system* consists of one or more dedicated sensors connected to a data logger. It is installed in a structure when continuous monitoring of the structure is required. Inclinometers that are placed permanently at important locations to log data continuously are operating similar to tiltmeters and can be referred to as tiltmeters. The sensors in the system are positioned to span the zones where deformation is likely to occur (a traversing probe system may be used initially to detect such zones). This type of inclinometer is capable of good angular resolution but is very expensive and inflexible. Because of high cost, it is impracticable to

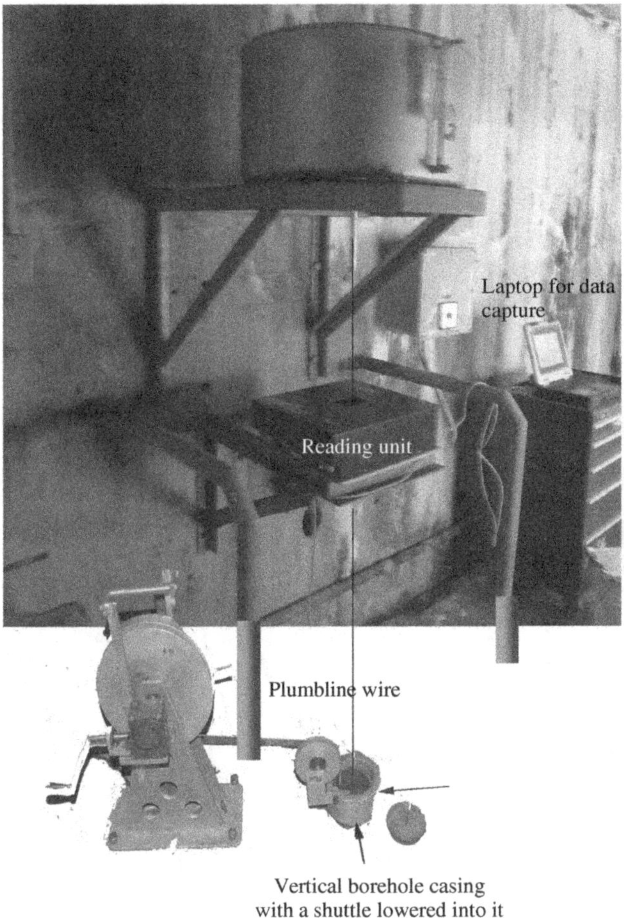

Figure 11.21 Inverted pendulum measuring head and laptop computer for data capture while a shuttle is lowered into a casing.

install multiple of such sensors on a single line. Some of the main limitations of *in situ* inclinometer system include the following:

- Need to constantly protect long horizontal runs of cable from possible electrical interference.
- Cost of borehole for inclinometer casing with a minimal ongoing cost of sending out a technician to read the installations.

According to Ding and Qin (n.d.), borehole of up to 200 m in depth can be measured using inclinometers or shuttle probes. The shuttle probe measurements are made in a given near vertical borehole by lowering the probe (placed at a certain orientation) through the casing to the base of the borehole and then pulling it up

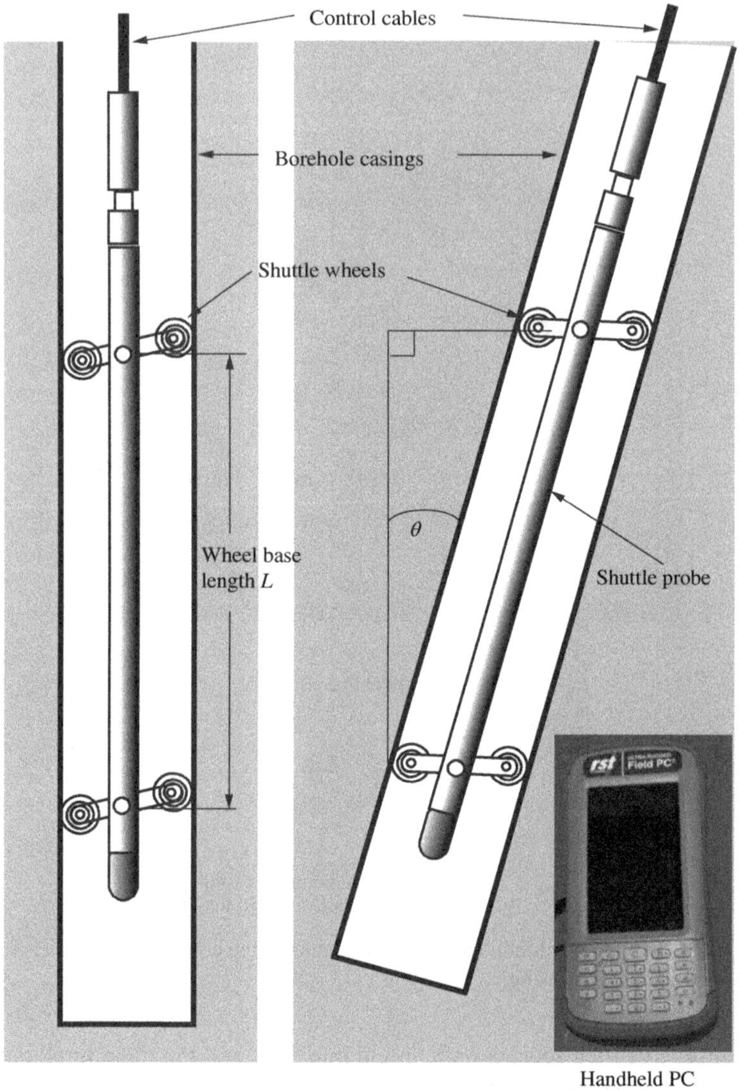

Figure 11.22 Typical shuttle probes in borehole casings.

while the inclination information of the probe in two orthogonal planes is registered at certain intervals (about 0.5 m interval, which is the typical length between the two wheels of the probe). The orientation of the probe is then changed by rotating the probe 180° and a second survey is done down the borehole. The resulting average data at each probe point provides a detailed profile of the casing. If ground movement occurs, subsequent surveys will reveal changes of the profile. These changes can be plotted to determine the magnitude, depth, direction, and rate of ground movement.

The software component of the inclinometer system is used for data reduction and graphing of data, showing the location of sensors, readings, alarm status, and trend plots.

11.2.6 Tiltmeters

Tiltmeters are for measuring angular tilt at specific locations on a structure or over relatively small base length. They consist of gravity-sensing transducer within a housing case, a beam, and bubble level (similar to a theodolite's level) with leveling adjustment device at the end of the beam. The beam will be in a horizontal plane when the bubble is leveled. Before tiltmeter readings of a monitored structure are taken, tiltmeter must be properly oriented on the part of the structure that is representative of the whole structure and it must be placed in an exactly reproducible position on a reference plate (securely bolted to the monitored surface). By comparing current tiltmeter readings with the previous readings, changes in tilt of the structure monitored are determined.

There are two main types of tiltmeters: *portable* and *in situ* types. The portable tiltmeter types are mounted on brackets and attached securely on the structures they monitor, in order to measure the tilts of the structures. In order to properly measure the tilts, the tiltmeter must be attached (using alignment bars) to the bracket in a particular position that can be repeated in the next epoch of measurements. Usually, a portable tiltmeter is carried from one bracket to another to obtain readings. In the case of *in situ* tiltmeter types, they are more expensive and their use is limited to only the most critical applications while the portable types can be used anywhere. One of the disadvantages of the portable tiltmeter is that it is slow and requires an on-site operator unlike in the case of *in situ* type. Depending on the type, tiltmeters are capable of providing digital or analog outputs.

Tiltmeters are also available in *uniaxial* and *biaxial* types. Uniaxial tiltmeters are capable of measuring tilts in one direction, while biaxial types are capable of measuring tilts in two perpendicular directions. An example of uniaxial tiltmeter is Model 801 Tuff Tilt by Applied Geomechanics Inc. (2005) with a quoted repeatability of $0.72''$ to $1'12''$ and resolutions of $0.36''$ to $36''$, depending on the version. Examples of biaxial types are Leica Nivel210 precision biaxial inclination sensor (a form of tiltmeter) capable of simultaneous measurement of inclination displacement and direction of inclination of GPS reference station installation, and In-place MEMS Tilt Meters by RST Instruments Ltd (2010). The quoted resolution of Leica Nivel210 sensor is 0.001 mrad with achievable accuracy of 0.47 mm/100 m range (or 0.0047 mrad) and instrument range of $\pm$1.51 mrad (Leica, n.d.). The In-place MEMS tiltmeter (shown in Figure 11.23), which can be uniaxial or biaxial, has a variable range of up to $\pm 15°$, a resolution of 0.01–0.025 mm/m and a repeatability of 0.06–0.03 mm/m (RST Instruments Ltd, 2010). Typical precision of some tiltmeters is between $\pm$0.013 mm for 200-mm-long beam and $\pm$0.13 mm for 900-mm-long beam (Dunnicliff, 1988).

Some of the applications of tiltmeters include slope stability studies and dam deformation monitoring. The deformation profiles of a dam structure, for example,

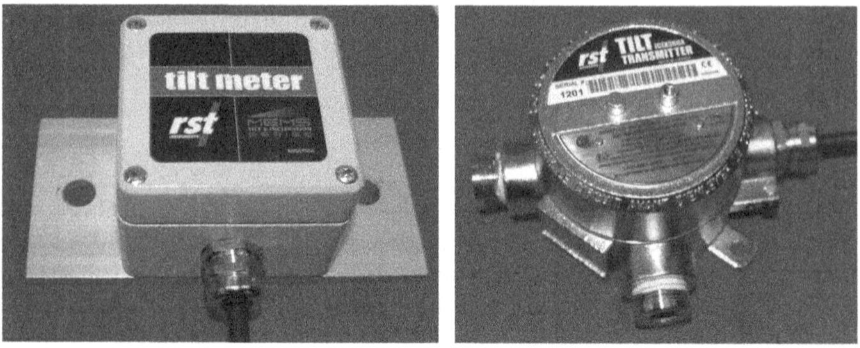

Figure 11.23 Typical MEMS Tilt Meters by RST Instruments Ltd.

may be determined by placing a series of tiltmeters at different levels of the structure. Some of the factors affecting the accuracy of tilt-sensing devices when they are applied include (Chrzanowski, 1986; Chrzanowski and Secord, 2000) the following:

- Temperature changes, which can change the dimensions of the mechanical components or the viscosity of the liquid in the case of liquid tiltmeters.
- Drifts in the tilt indicator and fluctuations of the readout, which could be due to many reasons, including aging of the tilt-sensing device; thorough testing and calibration of the instrument will be needed to establish the effects.

Thorough testing and calibration of tiltmeters or inclinometers at the start and end of measurements are suggested (Chrzanowski and Secord, 2000) for minimizing the effects of the aforementioned sources of errors.

Some of the important advantages of using tiltmeters include the following:

- Their use is not labor intensive, and no intervisibility between survey stations are required as in geodetic surveys.
- They can be left in place at the observing station with a telemetry monitoring system allowing for communication to the remote processing station.
- They are capable of higher accuracy than are geodetic surveys.

11.2.7 Fiber-Optic Sensors

Fiber-optic sensors (FOS) or optical fiber sensors are fiber-based devices for sensing temperature, mechanical strain, displacements, and so on. Optical fibers are long, thin, and flexible threads (much less than a millimeter in diameter) made of glass or plastic that allow electro-optical waves to be propagated through them; they are optical waveguides, which are usually packaged in larger cables, much like copper conductors are packaged in electrical cables. Optical fibers are usually made of glass material and can be fairly flexible and can be several hundreds of kilometers long.

To launch light into glass fiber, collimated laser beam is focused into the fiber core. The light then propagates along the core (with the intensity distribution of the light possibly extending beyond the core) and comes out of the fiber at the other end as divergent beam. The propagation of light through the fiber generates low propagation losses so that the optical intensity can be maintained over the whole length of the fiber that may be several kilometers long.

The application of FOS is due to their advantages over more conventional electrically based sensors. Some of the advantages are (Paschotta, n.d.) as follows: FOS are safe to use in hazardous environments; they have excellent resistance to chemicals and can be used in highly corrosive environment; they are free from problems associated with lightning strike, electromagnetic interference; they may be used in high-temperature areas where electronic systems would not survive; there are usually no electronics or power required at remote-sensing point; optical fibers are small, light, and relatively cheap; and they are able to sense, communicate, and multiplex signals within single optical network.

11.2.7.1 Basic Principle The basic principle of fiber-optic sensing is feeding light (usually from laser with close to single frequency) into an optical fiber. The light fed into the fiber is modulated through its interaction with what is being sensed (such as pressure, strain, temperature, Bragg gratings), and the modulated light is transmitted back to a detector arrangement, which will detect and demodulate the light and measure the perceived changes. It is then believed that there will be one-to-one correlation between the phenomenon being sensed and the demodulated signal.

FOS can be divided into two large categories according to the types of modulation implemented for the sensors: *Intensity-modulated FOS* and *phase-modulated FOS*. Very popular type of intensity-modulated FOS is based on *fiber Bragg gratings* (FBGs) and the phase-modulated types are those based on *interferometric principles*. FOS can also be classified into four types according to their operation techniques as follows (Inaudi and Glisic, 2007a):

- *Point sensors* with single sensing points located at the end of the fiber line. This type uses the Fabry–Pérot interferometric technique, which is based on monitoring the gap change in spacing (of about 10 mm) between two fibers attached to a capillary tube near its two extremities. The monitored gap change is then translated to average strain variation.
- *Partially distributed sensors* with multiple sensing points along a single fiber line are based on FBGs techniques.
- *Long-base sensors,* producing single measurement integrated over a long measurement base. They are based on Michelson interferometric technique.
- *Fully distributed sensors,* capable of multiple sensing along a single fiber line. They are based on Brillouin scattering and Raman scattering techniques.

11.2.7.2 Partially Distributed Fiber-Optic Sensors Partially distributed FOS are also referred to as FBG sensors since they are based on FBG technology. They are the most important type of multiplexed sensors, allowing measurements to be made at multiple sensing points along a single fiber line (Inaudi and Glisic, 2007a). A FBG is a pattern of disturbances in the index of refraction fabricated or written in the core of a short segment (a length of few millimeters or centimeters) of special type of optical fiber; the majority of commercial gratings are fabricated using intense ultraviolet (UV) source, such as UV laser. By exposing the core of the fiber to UV laser, the refractive index of the core is changed in the process. The pattern (grating) fabricated consists of multiple fringes with specific spacing between them and varying refractive index, and it serves as partial reflector, reflecting certain wavelengths of light (blocking them off) and transmitting all others. In this case, each fringe acts as a partially reflective mirror reflecting at least a small amount of light.

All reflected light waves from grating fringes combine coherently to one large reflection (with a maximum amplitude) at a particular wavelength when the spacing between grating fringes is approximately half (or a quarter of) the input light's wavelength, making the elements act as a high-quality reflector. This property is referred to as *Bragg condition*, and the wavelength at which this reflection occurs is called the *Bragg wavelength*. If the light signal travels at wavelengths other than the Bragg wavelength, the light will essentially propagate through the grating with negligible attenuation or signal variation. Only those wavelengths that satisfy the Bragg condition are affected and strongly back reflected. The range of wavelengths that are reflected by the grating elements is also called the *photonic stopband* and any light within this range of wavelengths will not be transmitted through the grating. The ability to accurately preset and maintain the grating wavelength is a fundamental feature and an advantage of FBGs.

The basic principle behind the operation of a FBG is Fresnel reflection, where light traveling between media of different refractive indices may both reflect and refract at the interface. Consider Figure 11.24, in which an optical fiber consists of three refractive indices: n_1 for the outer part of the fiber, n_2 for the core, and n_3 is the effective refractive index of the grating in the fiber core. In the figure, n_0 is the refractive index for the surrounding air; Λ is the grating spacing, which can be of the order of hundreds of nanometers (1×10^{-9} m), or much longer for long fiber gratings; and λ_B is the reflected Bragg wavelength. The reflected Bragg wavelength can be given by the standard Bragg equation for light at normal incidence as (Paschotta, n.d.; Lee, 2003):

$$\lambda_B = 2n_3\Lambda \qquad (11.3)$$

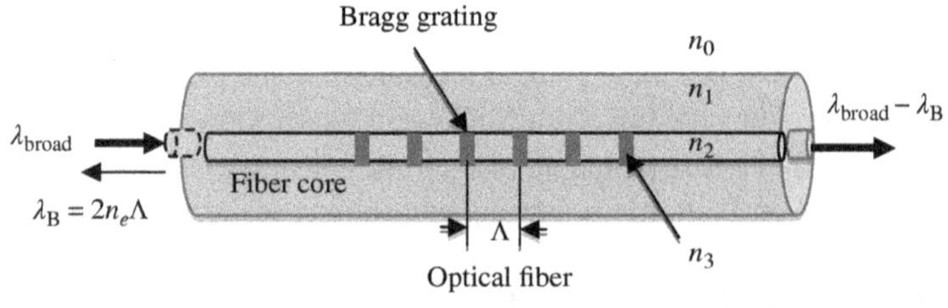

Figure 11.24 Operational principle of fiber Bragg grating (FBG).

It can be seen in Equation (11.3) that Bragg wavelength is changed with a change in the grating spacing (period) or the effective refractive index. The change in spacing relates to strain while the change in effective refractive index relates to temperature variation. Since the grating reflects a spectral peak based on the grating spacing, any change in the length of the fiber due to tension or compression will change the grating spacing and the reflected Bragg wavelength at which one obtains maximum reflectance of light. In this case, by measuring the wavelength of the maximum reflectance (reflected Bragg wavelength), it is possible to quantitatively determine strain, making FBG strain sensor. When used as a strain sensor, however, it must be compensated for the temperature influence since the parameters n_3, Λ, and the reflected Bragg wavelength are dependent on temperature and strain of the fiber. This dependency is well known and allows the determination of temperature or strain from the reflected FBG wavelength. In this case, every physical parameter that can be converted into strain can be measured by FBG.

The main interest in using Bragg gratings is due to their *multiplexing* potential. Multiplexing is a method of sending multiple signals or streams of information on a carrier at the same time in the form of a single, complex signal and then recovering the separate signals at the receiving end. In some optical fiber networks, multiple signals are carried together as separate wavelengths of light in a multiplexed signal. FBG sensors can be multiplexed in one fiber using wavelength division multiplexing (WDM) or time division multiplexing (Chen, 2011). Many gratings can be written in the same fiber at different locations and tuned so that each sensor reflects a specific wavelength. This allows the measurement of strain at different places along a fiber using a single cable with gratings sharing the spectrum of the source used to illuminate them. Typically, 4–16 gratings can be measured on a single fiber line (Inaudi and Glisic, 2007a). Since FBGs have short base lengths, they can also be used as conventional strain gages, which can be installed by gluing them on metals and other smooth surfaces.

Networks of FBGs written into a single fiber length have been extensively used in multiple-point mode as arrays of strain and temperature sensors for load and condition monitoring. But the cost of fabricating FBG sensors can be high, being in the order of a thousand US dollars apiece (Chen, 2011), and they may be unreliable especially in environments where temperature is high.

11.2.7.3 Long-Base Fiber-Optic Sensors Long-base FOS use FBGs as mere reflectors in Michelson interferometric application. As long-base sensors, they integrate measurements over a long measurement base that can be up to several meters long. They are phase-modulated FOS types or fiber-optic interferometers, which are commonly used when extreme sensitivity is required. In this type of FOS, a change in length or refractive index or both of the fiber, under temperature influence, or fiber strain will cause a phase change in the fiber. The phase change can be large even if the change in fiber length is small (in the order of a wavelength) or if the change in refractive index at long sections of the fiber is very small. Since optical phase changes cannot be directly detected, the principles of interferometry are used in order to determine them.

The operation of long-base sensor system is based on the principle of Michelson interferometry in which the amplitude of a light wave is split into two components, which propagate along different paths and later recombined to create interference that can be observed with a detector (Measures, 2001). The system measures the difference of travel times within fibers or the phase difference. In this case, the displacement or strain information is derived from the coherence properties of the light and not from its intensity. Based on the interferometric principle, the long-base sensor system uses two fiber-optic beam splitters. The first splitter splits the optical wave into two and directs them into two separate fibers. One of the fibers known as the measurement fiber is pretensioned and mechanically coupled to the structure at two anchor points so as to be able to measure both the elongation and shortening of the structure. The other fiber, the reference fiber, is free in a tube (isolated from the surrounding) and no displacement of the structure should strain it. Since the measurement fiber will also change its length and its refractive index due to temperature changes, the reference fiber is to help compensate for this effect since both fibers will be affected by the same amount, being in the same tube. The reference fiber is also fixed to the anchor points but due to its extra length it will not experience strain if the structure changes. If the measurement fiber is undisturbed, then both fibers will have exactly the same length and the optical waves in the second splitter will be in phase and coherently add to give a maximum intensity output. If the measurement fiber, however, experiences some kind of strain, the optical length of the measurement fiber increases and the optical path difference changes and the intensity output decreases due to destructive interference. The system readout unit measures the optical path difference changes by compensating it with a matching length difference in its internal interferometer. Each measurement gives a new compensated position reflecting the elongation or shortening of the structure relatively to the previous measurement points.

Fiber-optic monitoring system based on long-base (long-gauge) FBG sensors can detect submicrometer level deformations in both triggered-dynamic and continuous measurements. They can be installed as fully embedded borehole sensors or as surface extensometers. The single readout unit of the sensors can be used with high precision and high stability to monitor several fiber pairs in multiple structures. In this case, the readout unit can be disconnected and used again to monitor other fiber sensors and other structures; and in some cases, it is possible to connect a number of sensors to the same readout unit.

An example of long-base sensor type is the SOFO (a French acronym for Surveillance d'Ouvrages par Fibres Optiques or Structural Monitoring using Optical Fibers) system by SMARTEC, IMAC-EPFL in Switzerland. It is a white-light Michelson interferometric fiber-optic sensor acting as a precise extensometer over gauge lengths from a few centimeters up to a few tens of meters with long-term stability (over years of measurements) and a precision mechanical readout in microns. The system has been used in bridges, tunnels, dams, piles, anchors, historical monuments, nuclear power plants, and so on (Inaudi et al., 1994, 1999; Inaudi and Vurpillot, 1999). The interferometric sensor is used in multipoint and continuous modes. Some of the important components and properties of the SOFO system are as follows (Inaudi et al., 1999):

- Stand-alone FOS, which are most suitable for harsh environment with lots of mud, dust, and so on
- Gauge length of 20 cm–10 m for standard sensors up to 50 m with special-type long sensors
- Cable network with a cable length of up to 10 km
- Reliable reading unit with a resolution of 2 μm (2/1000 mm), which is independently from the gauge length (for static measurements)
- Measurement speed of less than 10 s/measurement
- Data acquisition and analysis software
- Long-term stability with drift not observable over at least 4 years.

11.2.7.4 Fully Distributed Fiber-Optic Sensors Fully distributed FOS provide the ability to measure from a single readout unit strains and temperatures at several measuring points over several tens of kilometers long single fiber line. The typical spacing between measuring points is 1 m, which is also referred to as *spatial resolution* of the sensor; and it is possible to have a fiber with a length of up to 30 km, which is termed the *range* of the sensor (Inaudi and Glisic, 2007a). This type of sensor is considered capable of being used for deformation monitoring landslides, dams, dikes, levees, pipelines, tunnels, and so on.

The fully distributed FOS use the optical fibers themselves as sensing media. They use the interaction between the intense light propagating through the media and the glass material of which the fibers are made. Most of the sensors are then based on the principles of Rayleigh scattering, Raman scattering or Brillouin scattering, and the principle that if an intense light at a given wavelength is propagated through a fiber, a very small amount of the light will be scattered back from every location along the fiber itself (Inaudi and Glisic, 2007b). The back-scattered light are said to contain information about the strain and temperature that were present at the location where the scattering occurred. The original propagated light is called the Rayleigh component, while the two components contained in the back-scattered light are known as Raman and Brillouin components. The scattered components are known to have wavelengths that are higher and lower than the original Rayleigh component (Inaudi and Glisic, 2007b). The sensors based on Brillouin scattering are considered better in strain measurements than those based on Raman scattering (Inaudi and Glisic, 2007b). The Brillouin scattering technique is used in long sensing distance in the order of tens of kilometers, but usually accompanied by low spatial resolution with its measurement sensitivity far worse than that based on FBG sensor array.

The general operation principle of the fully distributed FOS is similar to that used in radar techniques; the light pulses are sent to interrogate the fiber so as to be able to discriminate different points along the sensing fiber using the different time-of-flight of the scattered light. By combining the radar technique and the spectral analysis of the returned light, the complete profile of strain or temperature along the fiber is obtained. Distributed fiber sensors, however, are still limited to temperature and strain measurements; with the sensors, it is difficult to achieve high spatial resolution in measurements, producing poor results when used as point sensors. The following specifications are quoted (Inaudi and Glisic, 2007b) for typical distributed sensor systems:

- Systems based on Raman scattering are able to achieve a temperature accuracy of ±0.1 °C and a spatial resolution of 1 m over a measurement range of up to 8 km.
- Systems based on Brillouin scattering are able to achieve a temperature accuracy of ±0.1 °C, a strain accuracy of ±20 μstrains, and a spatial resolution of 1 m over a measurement range of 30 km.

11.2.8 Micro-Electro-Mechanical System (MEMS) Sensors

Micro-Electro-Mechanical Systems (MEMS) is a term used in North America to mean a manufacturing technology for creating tiny integrated devices or systems that combine mechanical and electrical components (Loughborough University, 2002). The technology is known as *Microsystems Technology* (*n*) in Europe or *Micromachines* in Japan.

The MEMS devices or systems, which are usually fabricated using integrated circuit (IC) batch processing techniques, can range in sizes from a few micrometers to millimeters with their components usually of much smaller sizes. A typical MEMS device will consist of mechanical microstructures, a central unit that processes data (the microprocessor or the microelectronics) and several components that are able to interact with the surroundings such as transducers, all integrated onto the same chip. A device, however, is only considered MEMS device based on how it is made; its electronics must be fabricated using IC technology and the micromechanical components must be fabricated by sophisticated manipulations of the appropriate wafer, using micromachining processes. In this case, MEMS technologies are designed to take full advantage of the electrical and mechanical properties of the wafer.

According to Bryzek (2005), MEMS does not refer to a specific product, but to the technology that includes many processes needed for three-dimensional shaping of wafers or stacks of wafers. While most of the MEMS applications use silicon wafers, many other materials have been used, including glass and quartz wafers. Technically, some of the factors making MEMS attractive as manufacturing technology are listed (Bryzek, 2005; Loughborough University, 2002) as follows:

- Interdisciplinary nature of MEMS technology and its micromachining techniques, which includes designing, engineering, manufacturing, integrated circuit fabrication technology, material science, and so on.
- Provision of basis for manufacturing products that cannot be made by other means.
- Potential for integrating devices with IC circuitry to create integrated systems on a chip.
- Use of silicon with excellent mechanical properties, which are comparable or superior to steel.

- Low-cost and high-volume production of devices made possible through batch wafer processing technology.
- Potential for producing devices with reduced physical sizes.
- Availability of cutting-edge IC processing equipment and high-volume IC packaging technologies.
- Availability of needed knowledge, skills, and expertise in MEMS technology.

The current MEMS devices include accelerometers for airbag sensors, inkjet printer heads, computer disk drive read/write heads, projection display chips, blood pressure sensors, optical switches, micro-valves, biosensors, and many other products that are all manufactured and shipped in high commercial volumes (PFP, 2002). Many manufacturers are now offering MEMS-based inclinometer probes, such as the digital horizontal MEMS inclinometer system by RST Instruments Ltd (RST Instruments Ltd, n.d.). Digital MEMS inclinometer probe is considered to be more accurate with a quoted accuracy of ±2 mm/25 m and with a higher thermal stability and rugged durability compared with the older technologies based on servo-accelerators (RST Instruments Ltd, n.d.). Another MEMS-based device is known as ShapeAccelArray (SAA) by Measurand Inc. (Measurand, 2013). As a typical example of MEMS sensor, more details on SAA are provided in the following sections.

11.2.8.1 Example of MEMS Sensor: ShapeAccelArray (SAA) Sensor SAA is a real-time monitoring system that can be installed in a borehole or embedded in a structure to be monitored (Danisch et al., 2008). The major elements of the SAA monitoring system are 500- or 305-mm-long segments and joints, communication cable, PEX tubing, eyelet, *X*-marks, on-reel markings, and a label, as illustrated in Figure 11.25. The cable is used to provide power to the instrument and to allow for communications between SAA and a data logger or computer; PEX tubing is to protect the communication cable from damage, to provide a secure way of retrieving the SAA, and also to set azimuth of the SAA; and the software-calibrated *X* axis for each sensor is aligned with the *X*-marks. The *X*-marks are to provide a sense of direction of deformation, or alternatively, the user can choose to include magnetometers in the SAA construction for that purpose.

A typical package of SAA for shipment to customers is shown in Figure 11.26. The package, which includes a wooden reel on which the device is wound, is designed to ease installation and storage of SAA. The weights of SAA reels vary from 43 kg for 32 m (an array of 104 segments) SAA to 113 kg for the 100 m (an array of 328 segments) SAA. An important recommendation by the manufacturer is that SAA should always remain on its reel, after use, for storage with the joints bent the same way as at the factory and should only be off the reel when it is being used to measure the shape. The recommendation is to help preserve the mechanical integrity of the joints of SAA.

SAAs have been used in many different applications, which include in situ monitoring of unstable slopes, monitoring of civil engineering structures, monitoring of mines and excavations, and measuring drill-hole shape (Measurand, 2013).

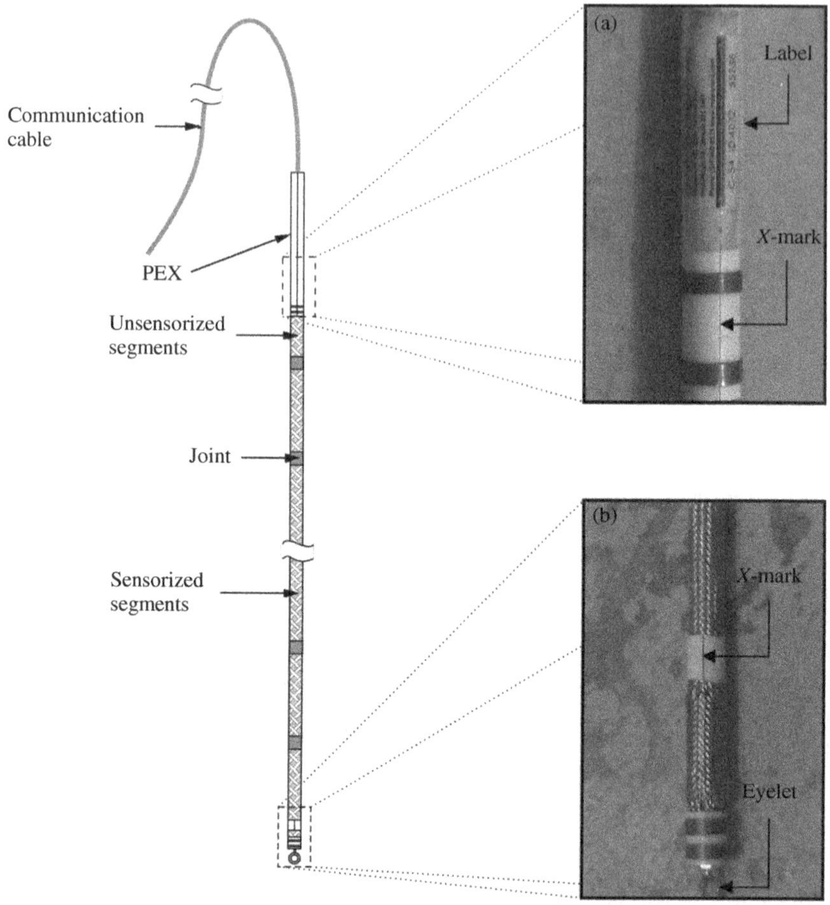

Figure 11.25 Anatomy of an SAA, showing the placement of X-mark, label, and eyelet on the SAA tubing. *Source:* Reproduced by permission of Measurand Inc.

Some of the important properties of SAA are as follows (Danisch et al., 2008; Barendse, 2012; Measurand, 2013):

1. SAA is composed of an array of rigid hollow segments connected end to end by nontwisting flexible joints with each rigid segment (defining the spatial resolution) being 0.305 or 0.500 m long, measured from one joint center to another joint center, as shown in Figure 11.25.
2. Each segment of SAA contains triaxial MEMS accelerometers with their axes set to form orthogonal *X*-, *Y*-, and *Z*-axes of a local coordinate system in which the *X*, *Y*, *Z* components of the joint centers are defined and known. The *X*-, *Y*-, and *Z*-axes are aligned in the software according to calibration files created at the manufacture of the system. The accelerometers are for determining the tilt of

Figure 11.26 SAA placed on a reel for storage. *Source:* Reproduced by permission of Measurand Inc.

individual segments of SAA with respect to gravity: the *X* and *Y* accelerometers are for calculating the tilt of a segment in a near-vertical orientation; and the *Z*-accelerometer is needed for better accuracy for a case when the segment lies in a near-horizontal orientation.

3. Every eighth segment contains a microprocessor, analog-to-digital converter (for collecting data from the groups of accelerometers), and digital temperature sensor with a resolution of 0.0625 °C for accelerometer temperature-dependence correction.
4. The SAA measurements can be recorded using data loggers or using computers running specialized Measurand software. If connected to an uplink device, SAA measurements can be collected in real time and uploaded wirelessly to the Internet, with a capability to send alarm warning via cell phone if a certain threshold movement is exceeded. The raw measurements collected by SAA; the tilt calibration parameters; and the factory characterization data for each individual sensor for offset, gain, and temperature dependence, are used on the dedicated computer in building a 3D or 2D polyline that represents the shape of the array.
5. In building the polyline that represents the shape of SAA in a borehole, the tilt vector of each SAA segment calculated relative to gravity and the azimuth of each resulting tilt vector known relative to the first segment because of the twist constraint are used. Using the calculated angles, the known segment lengths,

and by calculating and summing the displacements of segments from the bottom up, the shape of the entire array (SAA) is determined. This shape will resemble the deformed shape of the borehole axis, appearing in the form of a polyline. Figure 11.27 illustrates how tunnel deformations are processed and displayed by SAA. In this figure, SAA is used to simulate deformations of tunnel with real-time computer display of the deformations (in white outline) in three-dimensional (X, Y, Z) coordinate system.

6. The manufacturer of SAA considers the installation of SAA as one of the most important aspects of using the monitoring system. Typical SAA installations have the SAA placed inside 27 ± 1 mm inside-diameter polyvinylchloride (PVC) electrical nongrooved vertical conduit held in soil or structure with grout. When the SAA is being installed in the vertical conduit, the SAA joints will be in extension (with smaller diameter of about 23 mm) and the joints will be smaller than the inside diameter of the conduit. When the SAA is resting on the bottom cap of the conduit, it will be in compression with its diameter increasing sufficiently so that it is in contact with the inner wall of the PVC. A schematic representation of a typical SAA string installation is shown in Figure 11.28.
7. The SAA system is guideless in the sense that it does not utilize any special grooved casing or wheel assemblies as in the traditional inclinometer casing with a manual servo-accelerometer–based inclinometer probe.
8. The technique (known as "Averaging In Array") in which SAA averages multiple samples of readings together into a single reading is claimed by the manufacturer to improve the accuracy of SAA measurements to the order of square root of the number of samples averaged together to produce the single reading (Measurand, 2013).

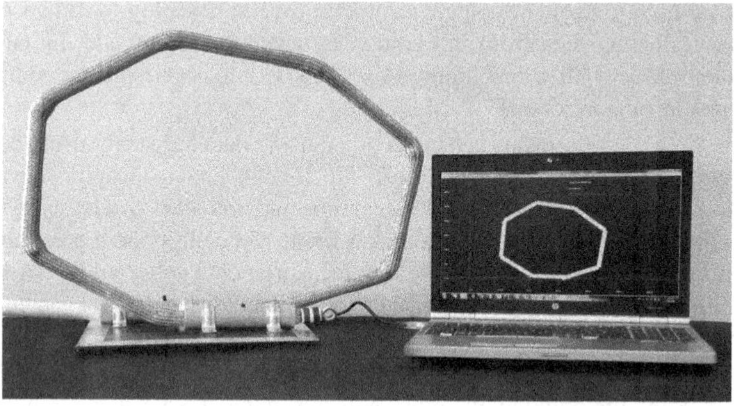

Figure 11.27 Simulation of tunnel deformations with an SAA, and the corresponding real-time display of the deformations (in white outline) on a laptop computer. *Source:* Reproduced by permission of Measurand Inc.

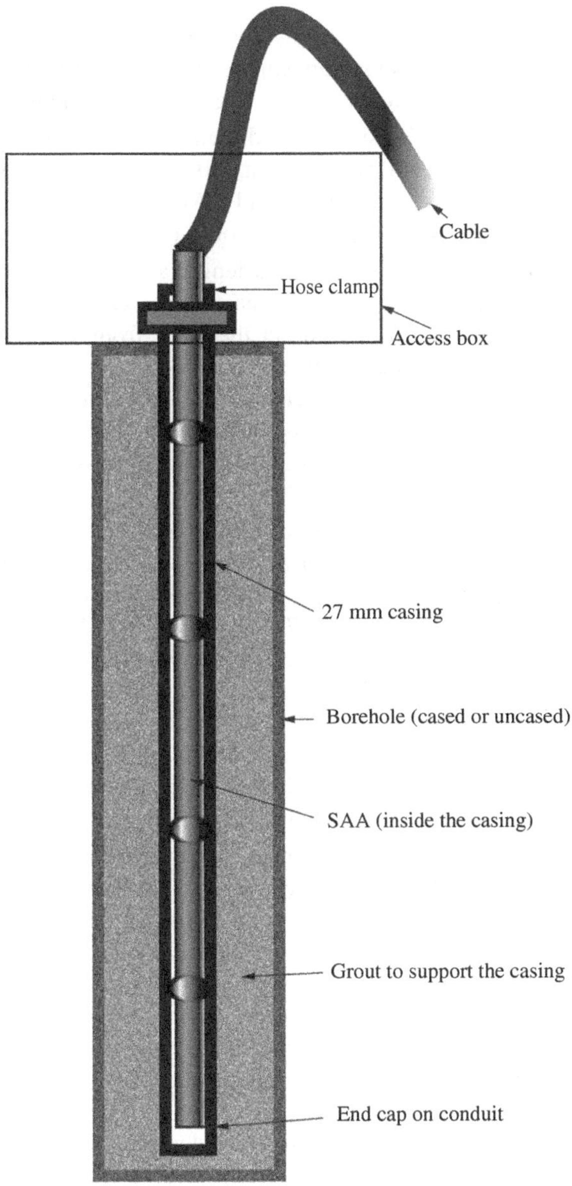

Figure 11.28 Schematic representation of a typical SAA string installation. *Source:* Adapted from Measurand Inc.; http://measurandgeotechnical.com/Installation_Guide_2011.pdf.

Some of the important advantages of SAA are given (Danisch et al., 2007, 2008; Barendse, 2012) as follows:

- It is a long MEMS-inclinometer string (about 32 m) that does not utilize grooved casing or guide wheels. The system can be adapted to fit into smaller diameter drill holes or cased holes not originally intended for inclinometers.
- It is able to measure in situ three-dimensional ground deformation and two-dimensional lateral soil acceleration at 0.5–1 m intervals. Since the sensor spacing used in SAA is short (0.305 and 0.5 m), it is able to achieve detailed deformation profiling to detect multiple zones of ground deformation.
- By virtue of the SAA's shorter segment length and smaller diameter of casing (about 25 mm), the system is able to measure a larger bending deformation in the borehole and it is easier to extract the array from significantly deformed casings and reuse it on other applications. This is an advantage over the conventional grooved casing inclinometer probes, which have a limitation where multiple shear zones exist. In this case, an upper deformation zone could cause the guide casing to bend excessively and obstruct the probe from being lowered to measure deeper shear zones.
- MEMS technology leads to low cost per sensor. For example, MEMS accelerometers, which are used in SAA system, are miniaturized in such a way that there are many of them to a wafer, leading to low cost per sensor. According to Danisch et al. (2007, 2008), an MEMS package measuring 1.5 mm by 4 mm by 4 mm will include two or three orthogonal accelerometers along with circuitry, to produce analog or digital outputs corresponding to some acceleration. The SAA system is therefore said to take advantage of small sizes of MEMS accelerometers (with the largest dimension being less than 0.5 mm) and the properties of single-crystal silicon on which many of the accelerometers are manufactured, which are superior to those of steel.
- SAA has the potential of providing more accurate results compared with conventional inclinometers. Since MEMS sensors used in SAA do not drift and are usually held in perfect registry with the structure being monitored, the main error source remaining in SAA is noise. The technique of averaging many samples into a single reading effectively reduces this noise. With regard to random noise, the manufacturer claims that the error will grow along the length of SAA according to the square root of the number of segments contained in that length or the length of SAA. Based on this understanding, Measurand (2013) predicted the accuracy of deformation value over a 32-m-long SAA as ± 1.5 mm with this value increasing to 2.1 mm over 64 m length. In practice and based on some studies (Rollins et al., 2009; Birch et al., 2011), the accuracy of SAA for deformation measurements, in comparison with results from conventional inclinometers, has been shown to be in the range of 1.5–2 mm.

The guideless design property of SAA has been studied and found (Barendse, 2012) to create some disadvantages as follows:

- After installing the SAA device, sand (instead of grout) is sometimes used to backfill the casing so as to put the device in intimate contact with the soil; this approach, however, is not recommended since this will be a source of variability in lateral support of the SAA. In this case, an incomplete backfilling or backfill settlement may cause spurious casing movements. Measurand now recommends grout instead of sand to alleviate this problem (Measurand, 2014, personal communication).
- The possibility of the axial rotational alignment of individual SAA sensors deviating from their factory-calibrated condition may constitute a source of error since the correct orientation of the sensor system is very important for correct results. Maintaining the rotational alignment of the many segments of SAA system is crucial in the same respect that avoiding or identifying a spiraled inclinometer casing is. After SAA is assembled, each segment is calibrated at the factory to a "zero azimuth," which is then marked near the top of the instrument. Any twists between segments would lead to incorrect directional readings and incorrect summation of the displacements (as with an uncorrected spiraled inclinometer casing). Field checks on calibration status are usually recommended on a retrieved instrument prior to each subsequent installation to verify the sensor alignment and to ensure that all the components are functioning as originally intended. The SAA manufacturer now affirms that software packages are available for performing field calibration of rotational alignment of SAA, and triaxial magnetometers can now be installed with SAA to facilitate identification of twists in situ.

11.3 DESIGN OF GEOTECHNICAL AND STRUCTURAL MONITORING SCHEMES

In Geomatics, the design of monitoring schemes is usually done based on the criteria such as precision, reliability, and overall cost of measurements. These criteria, however, are usually considered differently in the design of geotechnical monitoring schemes. For example, in geotechnical instrumentation procedures, reliability is considered as a consequence of human, instrument, and environment factors. The human factor is based on the quality of performance of personnel during the instrument installation; the precautions taken during data collection, processing, and interpretation; and the maintenance and calibration procedure adopted in ascertaining the correctness of the instrument. The instrument factor is dependent on the durability, simplicity, and self-checking ability of the instrument in the installed environment. In this case, the instrumentation will be considered durable if the cables, tubes, or pipes connecting the sensor to its readout unit are able to survive imposed pressure changes, deformation, water, sunlight, corrosion, and so on. With regard to simplicity of an instrument, Dunnicliff (1988) claims that optical sensors are simpler than mechanical sensors and mechanical sensors are simpler than electrical ones; on this basis, the author claims that mechanical sensors are generally more reliable than electrical sensors.

Geotechnical instruments are usually installed to solve a specific problem. The instruments may be installed at the construction stage of the structure to initially monitor the structure for purposes that may include the following:

- Evaluating and applying appropriate modifications to the uncertainties in the design of the structure as the construction is being carried out.
- Checking the safety of adjacent infrastructures with regard to the ongoing construction.
- Demonstrating to the public (for gaining their trust) that the impact the construction will have on their infrastructures are being closely watched and that their interests are being protected.
- Providing data that may be used as legal protection for the engineer in case of any litigation.

The monitoring process during the operation of the structure will serve similar purposes as in the case of monitoring process at the construction stage of the structure, except that safety of life and the infrastructures are paramount during the operation of the structure. Some of the steps involved in the design of geotechnical deformation monitoring schemes include the following (Dunnicliff, 1988):

1. Prediction of what will constitute the primary mechanisms that are likely to determine the behavior of the monitored objects, such as stress deformation, possible shear strength.
2. Preliminary evaluation of the construction site conditions with regard to stability of the site in time and space; any physical evidence of deformation such as cracks; possible locations of instruments; possible lengths of connecting tubes and cables for instrument installations; magnitude and distribution of loads, and so on.
3. Defining what purpose the instrumentation will serve to provide initial data for design purpose on temporary basis, to provide data during the construction stage on temporary basis, or to provide data on the performance of the object after construction on short-term or long-term basis. This is done to determine the life expectancy of the instruments to be used so that appropriate protective measures may be designed for the instruments.
4. Identification of types of parameters to be monitored, such as one-dimensional deformations, two-dimensional deformations, three-dimensional deformations, tilt, strain, stress, pore water pressure, surface movements, subsurface movements.
5. Prediction of the magnitudes of changes expected. This is to be used in predicting the range, sensitivity, accuracy, and alarm-warning tolerance for the monitored object.
6. Selection of instrumentation type depending on the following:
   a. Availability of resources and the skills required by the personnel in order to be able to use the instruments.

b. Availability of adequate support facilities for maintaining and calibrating the instruments after installation.
c. Data acquisition techniques whether automatic or manual data acquisition will be appropriate. There will be a need for automatic data acquisition in situations where readings are required very frequently, real-time monitoring and automatic alarms are needed, if easy accessibility to sensor locations are limited or impossible, if installed sensors are too many for timely manual readings or if qualified technicians for manual reading of sensors are not available.
d. Cost of instrument, which must be comparable to other costs associated with the installation of the instrument. For example, the quality of borehole extensometer instrument must be high enough so that its cost is comparable to the cost of drilling and backfilling a borehole, which can be up to 10–20 times greater than the cost of the borehole extensometer that goes into the borehole.
e. Instrument performance on the basis of reliability, simplicity, durability, good past performance record, sensitivity, range (the lowest and highest readings possible with the instrument), resolution (the smallest change that can be displayed on a readout device), and precision or repeatability of instrument, which is usually very important when monitoring changes. In general, in assigning values to the parameters being monitored, the instrument used should not unduly change the values of the parameters.

7. Selection of instrument locations based on the identified zones of primary concern with an In-place system that will allow cross-checking of readings at those locations.
8. Establishment of a methodology to ensure that instrument readings are checked for errors and are backed up regularly.
9. Designing a plan for instrument installation and procurement of necessary materials and tools.
10. Designing a plan for regular calibration and maintenance of installations, including readout units and embedded components.
11. Designing a plan for data collection in the form of data collection schedule, provision of appropriate field data sheets, and how often the data will be collected.
12. Designing a plan for data processing with a consideration for automatic data processing, types of data format, and the staff training requirement.
13. Designing a plan for data presentation, interpretation, and reporting. These should include the nature of conclusions to be made and the reporting requirements, contents, and frequency.
14. Designing a plan for implementation of reports.

There are no unified monitoring standards and specifications in the world for conducting deformation monitoring works apart from those designed by individual countries, some organizations, and some dam owners and operators (Avella, 1993). Since the design of a monitoring system is entirely dependent on the expected behavior of the dam, it is practically impossible to prepare unified standards and specifications that will be applicable to all dams. The sample specifications for geotechnical monitoring of concrete dams were recommended by the Swiss National Committee on Large Dams (SNCOLD) (Biedermann et al., 1988; Avella, 1993). In the recommendations, for example, variations in lengths and deflections along boreholes should be measured with rod or wire extensometers to an accuracy of ±0.5 mm and movement of cracks should be measured with micrometer to an accuracy of ±0.05 mm.

In order to achieve the objective of being able to detect any sign of abnormality in the behavior of a structure reasonably early is to design the monitoring program such that the number (frequency) of measurements is sufficient and not overly abundant as to become uneconomical. The frequencies of measurements generally vary, depending on the type of parameter to be monitored. In the case where a fully automatic data acquisition system is used, the frequency of measurements does not impose any problems since the frequencies can be preprogrammed for any desired time interval. There is no generally accepted range of frequencies of monitoring dams (Avella, 1993). Summaries of some commonly implemented ranges of frequencies of monitoring concrete and embankment dams are given by Avella (1993). From the summaries, ranges of frequencies of monitoring concrete dams depend on the type of instrument being used and the stages of the dam being monitored, which can be construction, initial filling, and the normal operation stages of the dam. The range of frequencies of monitoring could vary from once per day to once per year.

11.4 ANALYSIS OF GEOTECHNICAL MEASUREMENTS

The geotechnical measurements are usually considered as contaminated with the following effects (Chrzanowski and Secord, 1987):

- Observation errors.
- Seasonal (thermal) cyclic expansions of the measured objects.
- Changeable thermal expansion of the mechanical components of the geotechnical; instrumentation (particularly, tape and borehole extensometers).
- Uncertainties associated with installing instruments and observing deformation parallel to the borehole axis; the use of borehole extensometers may be limited by the orientation, depth, and size, of deformations in the borehole.
- Friction between rods and the protective pipes in the case of nonvertical installations of rod extensometers; borehole rod extensometers usually have protective plastic pipe to prevent it from bonding with the grout backfill; most in-wall extensometers extend to about 30–40 m, and boreholes are typically of 60–100 mm in diameters.
- Other systematic errors arising from lack of proper calibration of the instruments.

The actual deformation trend and its accuracy are then analyzed and evaluated to remove the aforementioned effects. Geotechnical measurements may have combined effect of the cyclic expansions of the structures and thermal expansion of the instruments reaching or exceeding 2 mm in amplitude. The following were generally concluded (Chrzanowski and Secord, 1987; Chrzanowski et al., 1989) with regard to results and interpretations of geotechnical measurements:

- There are short-term irregularities of deformations. Any extrapolation of the smoothed results of the long-term analysis for detailed prediction purposes is not reliable.
- Any interpretation of the short-term deformation trends seems to be meaningless unless rigorous thermal expansion corrections could be applied to both the instrumentation and structural material. This will be based on direct temperature measurements at each survey location, including anchor points of borehole extensometers (high accuracy of temperature measurements is not usually needed, approximately ±0.5 °C accuracy is sufficient). These corrections can only be derived empirically from long-term observations at each location separately.
- Cyclicity of the temperature influence has a period of about 1 year, requiring that at least 2 years of observations be collected in order to derive reliable temperature corrections.
- The phase of cyclic changes depends on delays in the heat transfer inside the structural material, which is not the same at each location.
- Amplitude of seasonal changes usually varies from one location of the structure to another and may vary from year to year depending on the annual average temperature of the environment, such as air and water.

Attaining appropriate accuracy for the monitoring involves (Chrzanowski et al., 1989) the following:

- Extensive calibration procedures; the design of tape extensometer that does not allow for a direct calibration of the tensioning instrument against a known tensioning force is improper for the high accuracy measurements. Tape calibration is expected to be done each day during the survey campaigns.
- Documentation for all measuring sensors, especially to maintain continuity in case of instrument failure, repair, exchange, or replacement.

One of the commonly used instrumentation data management software packages for long-term performance monitoring of dams is known as DamSmart by URS Systems Engineering Company in the United States. DamSmart handles a variety of tasks from data collection and reduction to data storage/archiving, data analysis, reporting, and plotting. Field data are usually collected electronically and stored in Excel spreadsheets in the field computer and later, in the office, uploaded to a desktop PC that has DamSmart software installed (J. Fletcher, personal communication). Some of the important capabilities of the software include the following (URS, 2012):

- Automating instrumentation data collection, reduction, plotting, and reporting for multiple projects.
- Storing raw and historical data with the corresponding instrumentation and all associated formulas.
- Allowing also manual data entry into the system.
- Allowing in-house routine to be integrated for instantaneous viewing of selected instruments with sinusoidal regression plot.
- Allowing in-house routine to be integrated for discontinuous regression lines when the slope shows a sudden change due to a cut or some other activity.
- Allowing GIS integration and publishing of data to the web.
- Allowing analytical plots and reports to be prepared; and the communication of results of analyses to be facilitated.

Generally, graphs of measurements as shown in the following sections can be used by project engineers to design appropriate remedial measures; and by extrapolating the graphs, they can predict the future behavior of the monitored objects.

11.4.1 Analysis of Extensometer Measurements

According to Chrzanowski (1986), if an extensometer is installed in a structure having a homogeneous strain field, the measured change $\Delta\ell$ of the distance ℓ in two epochs gives directly the strain component in the direction of the measurement. The strain component (ε) for the homogeneous material between the two anchor points can be given as

$$\varepsilon = \frac{\Delta\ell}{\ell} \tag{11.4}$$

where $\Delta\ell$ is the change in lengths of two extensometer rods anchored at two different points and ℓ is the distance between the two anchor points. For the homogeneous structure, let the multipoint extensometer readings at the two anchor points (taken at the collar of the structure) at epoch 1 be r_{11} and r_{12}; for epoch 2, let the readings be r_{21} and r_{22} (where for a given r_{ij} reading, i represents epochs 1 and 2, and j represents the extensometer anchor points 1 and 2). The relative movement of the anchor points can be given as $\Delta\ell = [(r_{22} - r_{21}) - (r_{12} - r_{11})]$. This relative movement can be considered as distance change ds_{12} between the anchor points 1 and 2, which can be represented as a function of the three-dimensional coordinate changes (dx_1, dy_1, dz_1) and (dx_2, dy_2, dz_2) of the anchor points, assuming the coordinates of the anchor points 1 and 2 are (x_1, y_1, z_1) and (x_2, y_2, z_2), respectively. Let the distance (s_{12}) between the two anchor points be given as

$$s_{12}^2 = (x_2 - x_1)^2 + (y_2 - y_1)^2 + (z_2 - z_1)^2 \tag{11.5}$$

The partial derivative of Equation (11.5) can be given as follows:

$$\frac{\partial(s_{12}^2)}{\partial s_{12}}ds_{12} = \frac{\partial(s_{12}^2)}{\partial x_1}dx_1 + \frac{\partial(s_{12}^2)}{\partial y_1}dy_1 + \frac{\partial(s_{12}^2)}{\partial z_1}dz_1 + \frac{\partial(s_{12}^2)}{\partial x_2}dx_2 + \frac{\partial(s_{12}^2)}{\partial y_2}dy_2 + \frac{\partial(s_{12}^2)}{\partial z_2}dz_2 \quad (11.6)$$

Equation (11.6) simplifies to the following equation relating change in distance (ds_{12}) to the coordinate changes (dx_1, dy_1, dz_1) and (dx_2, dy_2, dz_2) as follows:

$$ds_{12} = \frac{(x_2 - x_1)}{s_{12}}(dx_2 - dx_1) + \frac{(y_2 - y_1)}{s_{12}}(dy_2 - dy_1) + \frac{(z_2 - z_1)}{s_{12}}(dz_2 - dz_1) \quad (11.7)$$

Similarly, if the azimuth of the line 1-2 is given as α_{12}, the vertical angle from 1 to 2 is v_{12}, and the horizontal distance is HD $= s_{12} \cos v_{12}$, the following can be formulated:

$$\sin \alpha_{12} \cos v_{12} = \frac{(x_2 - x_1)}{s_{12}}; \quad \cos \alpha_{12} \cos v_{12} = \frac{(y_2 - y_1)}{s_{12}}; \quad \sin v_{12} = \frac{(z_2 - z_1)}{s_{12}} \quad (11.8)$$

Substituting Equation (11.8) into Equation (11.7) gives extensometer observation equation as follows:

$$ds_{12} = \sin \alpha_{12} \cos v_{12}(dx_2 - dx_1) + \cos \alpha_{12} \cos v_{12}(dy_2 - dy_1) + \sin v_{12}(dz_2 - dz_1) \quad (11.9)$$

Equation (11.9) as the extensometer observation equation can easily be integrated with the geodetic observation equations in least squares adjustment for the solution of the displacement vector or vector of coordinate changes ($dx_1, dy_1, dz_1, dx_2, dy_2, dz_2$).

The coordinates of the anchor points 1 (x_1, y_1, z_1) and 2 (x_2, y_2, z_2) can be determined by first of all coordinating the collar of the extensometer borehole and determining the azimuth and dip of the borehole. The steps for determining the dip and the azimuth of the hole are illustrated in Figure 11.29, where *CT* is a straight pipe placed in the borehole, *C* is in the collar of the borehole, and *T* is the tail of the pipe.

In order to determine the dip and the azimuth at the collar of the borehole in Figure 11.29, the following steps can be taken:

1. Establish control point *A* (x_A, y_A, z_A) where points *C* and *T* and the backsight control point *B* can be clearly seen. The three-dimensional coordinates of the control points must have been determined from the previous survey.
2. Set the total station instrument over point *A* and make the following measurements:

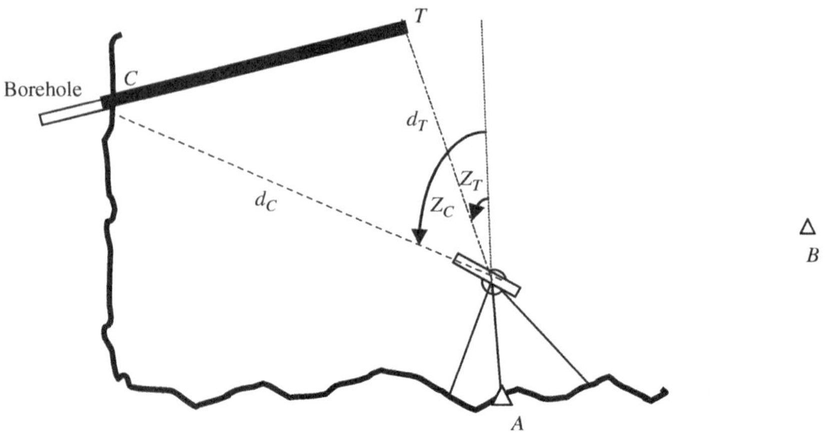

Figure 11.29 Determination of azimuth and dip at the collar of a borehole.

- Height of instrument (HI)
- Zenith angles Z_T and Z_C to points T and C, respectively
- Horizontal angles θ_T and θ_C from control point B to points T and C, respectively
- Slope distances d_T and d_C to points T and C, respectively.

3. Calculate the elevations (z_T and z_C) of points T and C, and the elevation difference Δh between them as follows:

$$z_T = z_A + \text{HI} + d_T \cos Z_T \tag{11.10}$$

$$z_C = z_A + \text{HI} + d_C \cos Z_C \tag{11.11}$$

$$\Delta h = z_C - z_T \tag{11.12}$$

4. Compute the horizontal distances s_T and s_C to points C and T, respectively:

$$s_T = d_T \sin Z_T \tag{11.13}$$

$$s_C = d_C \sin Z_C \tag{11.14}$$

5. Compute the horizontal coordinates for points T (x_T, y_T) and C (x_C, y_C) as follows:

$$\begin{aligned} x_T &= x_A + s_T \sin \alpha_{A-T} \\ y_T &= y_A + s_T \cos \alpha_{A-T} \\ x_C &= x_A + s_C \sin \alpha_{A-C} \\ y_C &= y_A + s_C \cos \alpha_{A-C} \end{aligned} \tag{11.15}$$

where α_{A-T} and α_{A-C} are the azimuths from point A to points T and C, respectively. Note that these azimuths are determined using the measured angles (θ_T and θ_C) and the calculated back bearing from control point A to control point B.

6. Compute the azimuth (α_{T-C}) and the horizontal length (HD_{T-C}) of the line TC as follows:

$$\alpha_{T-C} = \tan^{-1}\left(\frac{x_C - x_T}{y_C - y_T}\right) \tag{11.16}$$

(with quadrant analysis applied)

$$\text{HD}_{T-C} = \sqrt{(x_C - x_T)^2 + (y_C - y_T)^2} \tag{11.17}$$

7. Compute the dip angle (Dip) or grade (in percent) as follows:

$$\text{Dip} = \tan^{-1}\left(\frac{\Delta h}{\text{HD}_{T-C}}\right) \tag{11.18}$$

8. Use the computed azimuth of the borehole (Equation (11.16)), the computed dip (Equation (11.18)), and the distances from the collar to anchor points 1 and 2 to determine the coordinates of the points. For example, given the length of the extensometer from the collar to anchor point 1 as ℓ_{C-1}, the coordinates of anchor point 1 can be computed as follows:

$$x_1 = x_C + \text{HD}_{C-1} \sin \alpha_{T-C} \tag{11.19}$$

$$y_1 = y_C + \text{HD}_{C-1} \cos \alpha_{T-C} \tag{11.20}$$

$$z_1 = z_C + \ell_{C-1} \sin(\text{Dip}) \tag{11.21}$$

where HD_{C-1} is the horizontal distance from the collar point C to the extensometer anchor 1, expressed as

$$\text{HD}_{C-1} = \ell_{C-1} \cos(\text{Dip}) \tag{11.22}$$

The above approach can also be used in the case of an open-pit mine where many boreholes are drilled with the aim of locating minerals underground. Usually, the drill holes may not be straight down with some of them deviating in azimuth or dip/climb in inclination. This means that the drill holes, which are likely to be evenly spaced across the surface, may have different arrangement underground. In this case, the holes must be surveyed (and their three-dimensional coordinates determined as earlier) in order to determine where the holes are underground. This is to confirm the spacing of each drill hole underground in order to improve and economize blasting and the associated operations. The coordinates of borehole locations underground and on the surface are also used to provide the three-dimensional view of the path of the borehole.

11.4.1.1 Calibration Aspects of Rod and Tape Extensometers In order to successfully use invar rod micrometer gauge for displacement measurements, micrometer must be calibrated regularly on a dedicated calibration beam in the environment where the measurements will take place. LVDTs also need to be calibrated on the calibration benches before they are installed. Typical invar rod and tape extensometer calibration beams (or benches) in a Powerhouse of a hydroelectric generating station are shown in Figure 11.30. The beams consist of a system of invar rods with known values and anchor points, for calibrating rod and tape extensometers over their working ranges.

Tape extensometer measurements are calibrated using reference invar rod distances on dedicated calibration table shown in Figure 11.30(b). The calibration table is to allow the invar tape to be compared to common and stable reference, thereby eliminating problems due to tensioning and creeping of the tape extensometer. A

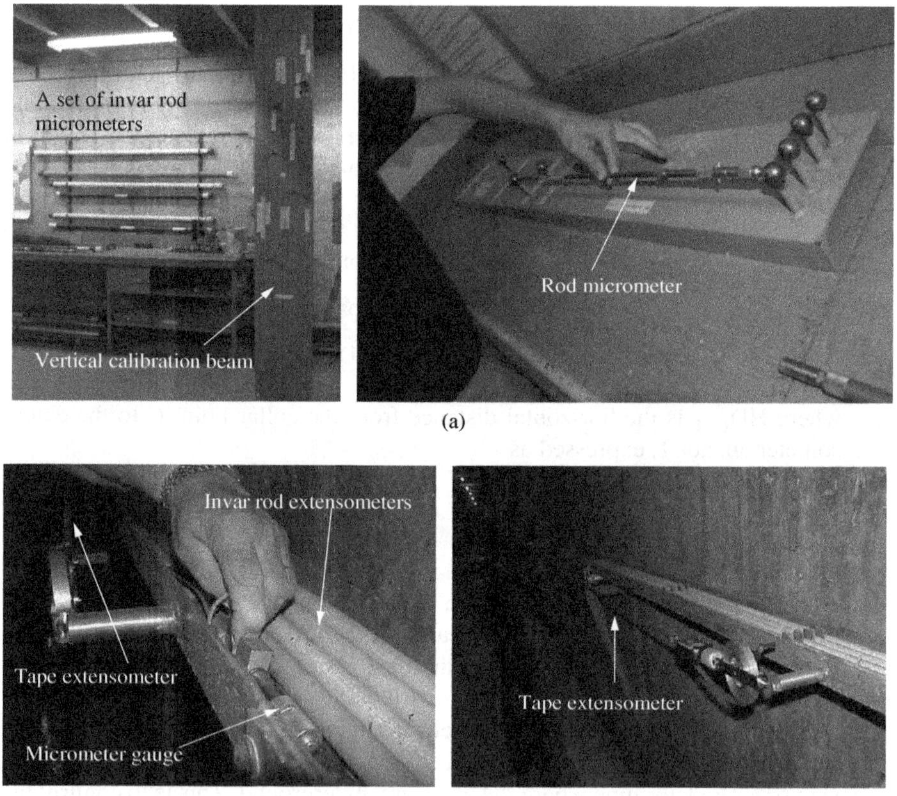

Figure 11.30 (a) Invar rod micrometers and the typical vertical and horizontal calibration benches installed in a Powerhouse of a hydroelectric generating station. (b) Horizontal calibration bench for tape extensometer calibration.

typical calibration table (usually made of invar) consists of five 0.25″ diameter invar rods (5, 10, 15, 20, and 25 m long) placed on a flat leveled steel beam as shown in Figure 11.30(b). The rods are anchored to the beam at one end and are inserted into a stainless steel reading head. The unrestricted beam may expand or contract according to temperature changes without affecting the length of the invar rods; the invar rods and the table react equally to an ambient temperature since both are made of the same invar material. If the tape extensometer is allowed to acclimatize with the calibration table before readings are taken, there will be no need to include temperature effects in the measurements. The calibration is done by comparing the distance between the anchor and reading head obtained from micrometer measurements to the same distance measured by the tape extensometer. It is assumed that changes in length of the tape are linear over 5-m intervals; thus, obtained corrections are applied proportionally over the same 5-m intervals. Every other factor that may affect the calibration is supposed to be taken care of with the micrometer measurement that is taken every time the calibration is done.

The calibration procedure of tape extensometers can be illustrated as follows. Given the invar rod micrometer readings as m_0 (initial micrometer reading), m_t (final micrometer reading) and the corresponding extensometer readings as e_0 (initial extensometer reading) and e_t (final extensometer reading), the correction (C_t) to extensometer reading can be determined as follows (Chrzanowski et al., 1989):

$$m_t - m_0 = (e_t + C_t) - e_0 \tag{11.23}$$

or

$$C_t = (m_t - m_0) - (e_t + e_0) \tag{11.24}$$

In order to calibrate tape extensometer, the tape extensometer anchors are monitored using a series of invar rods at intervals of 5, 10, 15, 20 and 25 m. The calibration process requires that 1 μm reading be taken on each calibration line and several extensometer readings be taken on each line. If there are calibration irregularities, it is more likely that they are due to changeable tension of the extensometer instrument. The calibration bar measurements are then used to derive corrections to be applied to the measurements. For distance measurements other than the multiples of 5 m taken during the calibration of the extensometer, the corrections to be applied are commonly interpolated.

11.4.1.2 Borehole Rod Extensometer Measurements A typical borehole rod extensometer measurement (y) is analyzed using the following cyclic functional model (Chrzanowski and Secord, 1987):

$$y = a_1 \cos(wt) + a_2 \sin(wt) + a_3 + a_4 t \tag{11.25}$$

where a_1 and a_2 are components of the amplitude, w is the phase angle, a_4 is the rate of the length changes of the extensometer rod, and a_3 is a constant at the start of epoch. The value of a_4 and its standard deviation are the most important parameters

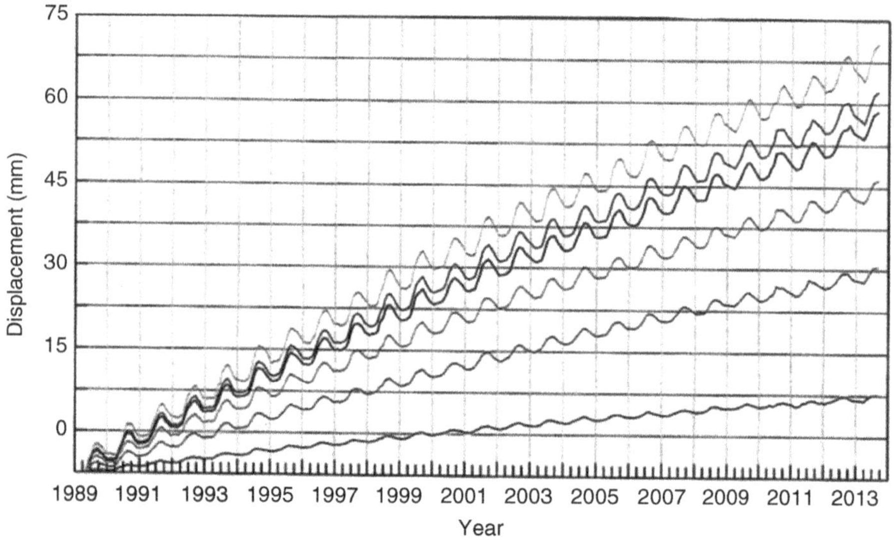

Figure 11.31 Sample display of 1989–2013 displacements from six-point borehole extensometer installed in a single borehole.

in this type of deformation analysis. The usual problem of fitting the model to the measurements may be due to errors in measurements and also nonuniform expansion of the extensometer rods. There is a need to incorporate a correction term for the effect of temperature changes on the aluminum rods of the borehole extensometers; this correction term is represented by the first two terms in Equation (11.25).

The sample measurements of vertical movements at six-point invar rod borehole extensometers (oriented vertically along the direction of gravity) in one borehole are plotted over the intervals from 1989 to 2013 (inclusive) as shown in Figure 11.31. The figure displays the sample data from six extensometers in one borehole with the fluctuations attributed to changes caused by temperature variations. The slopes of the plots represent the rate of expansion of the object being monitored. The lengths of the six extensometer rods from their respective anchor points to the same collar point (reading head) are 6.0, 14.0, 20.5, 27.4, 29.5, and 42.0 m. The typical monitoring interval with the extensometer is 4 weeks.

11.4.1.3 Tape Extensometer Measurements Tape extensometer measurements (y) are analyzed by fitting a cyclic function to the measurements. A typical function that can be used is given as follows (Chrzanowski and Secord, 1987):

$$y = a_1 \cos(wt) + a_2 \sin(wt) + a_3 t + a_4 + a_5 + a_6 + \cdots + a_n \tag{11.26}$$

where a_1 and a_2 are components of the amplitude; w is the phase angle; a_3 is the rate of the length changes of the extensometer tape; a_4 is a constant at the start of epoch; and $a_5, a_6, \ldots, a_n$ are the unknown slips, which could be due to periodic breakage of

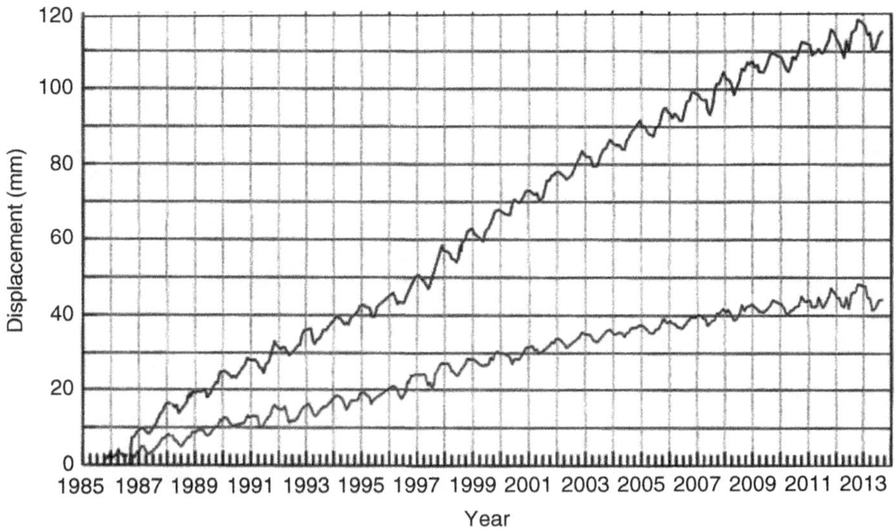

Figure 11.32 Sample display of 1985–2013 tape extensometer measurements between two pairs of columns in a Powerhouse.

the tape or other factors that may influence the length change of the tape. Just as in the case of borehole rod extensometers, the value of a_3 and its standard deviation are the most important parameters in this type of deformation analysis. Most of the factors that may affect the accuracy of the tape extensometer measurements, however, can be taken care of by proper calibration of the instrument.

The typical example of measured horizontal movements (in the Y-axis direction of the local coordinate system or the downstream direction of a Powerhouse) using tape extensometer between two pairs of anchor points (bolted to columns in a Powerhouse) located along the X-axis direction of the local coordinate system is plotted over the intervals from 1985 to 2013 (inclusive) as shown in Figure 11.32; the distance between the first pair of columns is 24.660 m (corrected for calibration) and the distance between the second pair of columns is 23.050 m (corrected for calibration) with the typical monitoring interval of 4 weeks.

The display in Figure 11.32 is similar to that in Figure 11.31 and can be interpreted similarly. In the figure, it can be seen that the plots seem to be decelerating toward 2013 (as the slope seems to be deviating from being constant).

11.4.2 Analysis of Joint Meter Measurements

The measured horizontal movements in a Powerhouse of a hydroelectric generating station (in the Y-axis direction of the local coordinate system) at three points along a joint extending in the X-axis direction of the local coordinate system are plotted over the intervals from 1984 to 2014 (inclusive) as shown in Figure 11.33; the typical monitoring interval is 3 weeks.

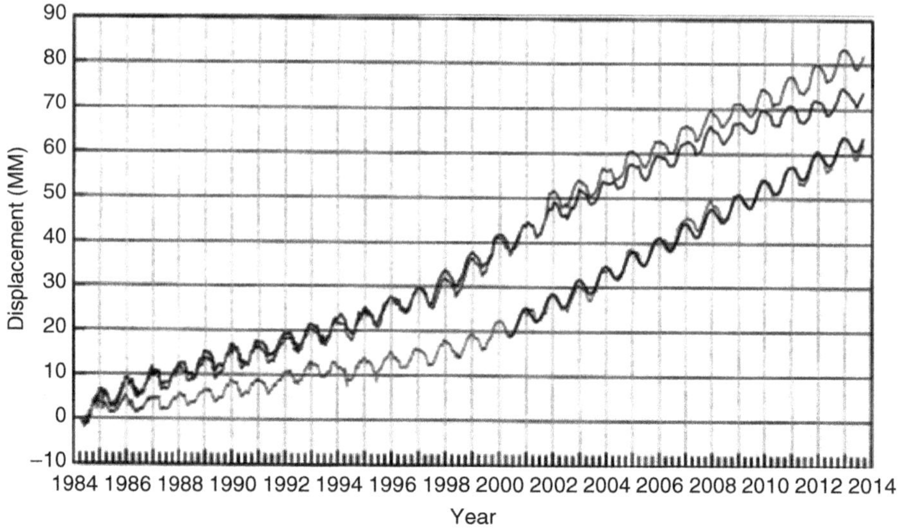

Figure 11.33 Sample display of 1984–2014 Joint meter measurements for three units of a Powerhouse.

As can be seen in Figure 11.33, the display of Joint meter measurements is similar to those of extensometer measurements in Figures 11.31 and 11.32. The interpretation and analysis of Joint meter measurements, therefore, can be done in the same way as in the case of extensometer measurements. In Figure 11.33, however, the slope of the plots changed around 1999, showing some form of acceleration at that time.

11.4.3 Analysis of Plumbline Measurements

The measured horizontal movements (in the *X*- and *Y*-axes directions of the local coordinate system) at various measuring positions of plumblines with respect to certain anchor point (at elevation −22.83 ft.) are plotted over the intervals from July 2011 to July 2013 (inclusive) as shown in Figures 11.34 and 11.35; the typical monitoring interval using shuttle probe is 6 weeks. The inverted pendulum data is collected using a laptop computer with the software that monitors the *X* and *Y* movements of the plumb wire after the shuttle is moved to each successive measuring position. The software usually allows measurements to be recorded at the measuring position after the plumbline wire has come to rest or has completely stopped swinging to ensure that readings are not taken prematurely. The software also allows data to be collected only when the readings are within ±0.1 mm for 10 successive readings. With an inverted pendulum viewer that accompanies the software, any set of inverted pendulum readings can be used as a baseline with the other sets of readings reduced relative to the baseline.

In Figure 11.34, using the plot of July 2011 measurements as reference, the structure shows some movement toward right (along the positive *X*-axis) between 2011 and 2013 around February and March with the greater movement (+9.0 mm)

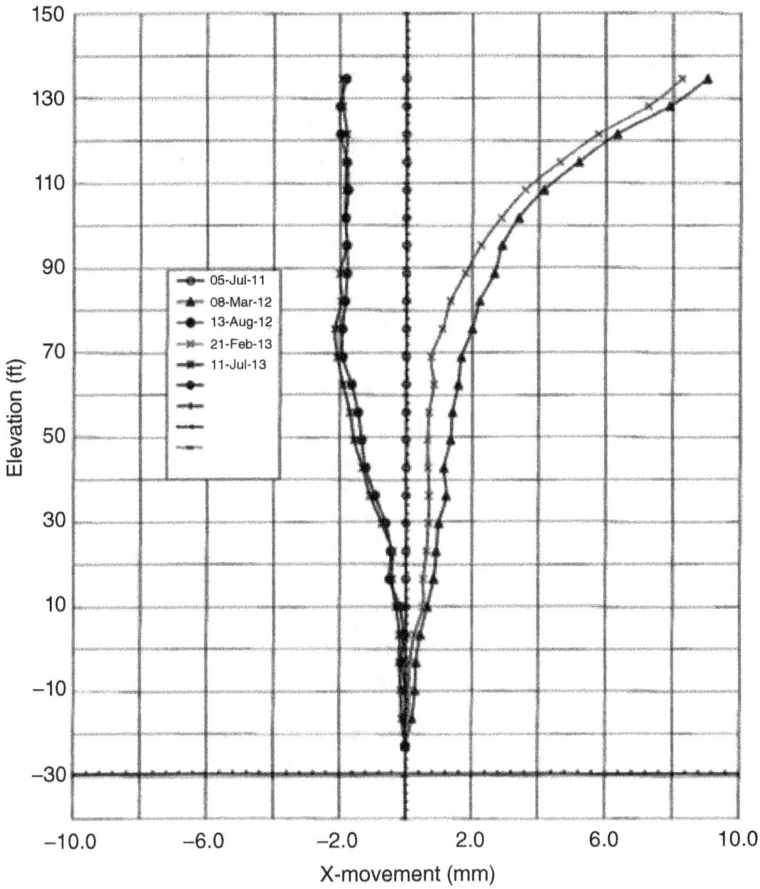

Figure 11.34 Sample display of inverted pendulum *X*-movements profiles from 2011 to 2013 based on shuttle probe measurements with July 2011 measurements as baseline.

in March. The structure then shows some movement toward left (along the negative *X*-axis) between 2011 and 2013 around July and August with the greater movement (−2.0 mm) in July; the movements in July and August, however, are very close.

Similarly, in Figure 11.35, using the plot of July 2011 measurements as reference, the structure shows some movement upstream (along the positive *Y*-axis) between 2011 and 2013 around February and March with the greater movement (+0.5 mm) in February with the movements in March close to zero and most of the time downstream. The structure shows movements downstream (along the negative *Y*-axis) in July and August with the greater movement (−3.0 mm) occurring in July.

11.4.4 Analysis of Tiltmeter Measurements

According to Chrzanowski et al. (1980), ground subsidence along a terrain profile of a mining area can be monitored using a series of tiltmeters arranged along the

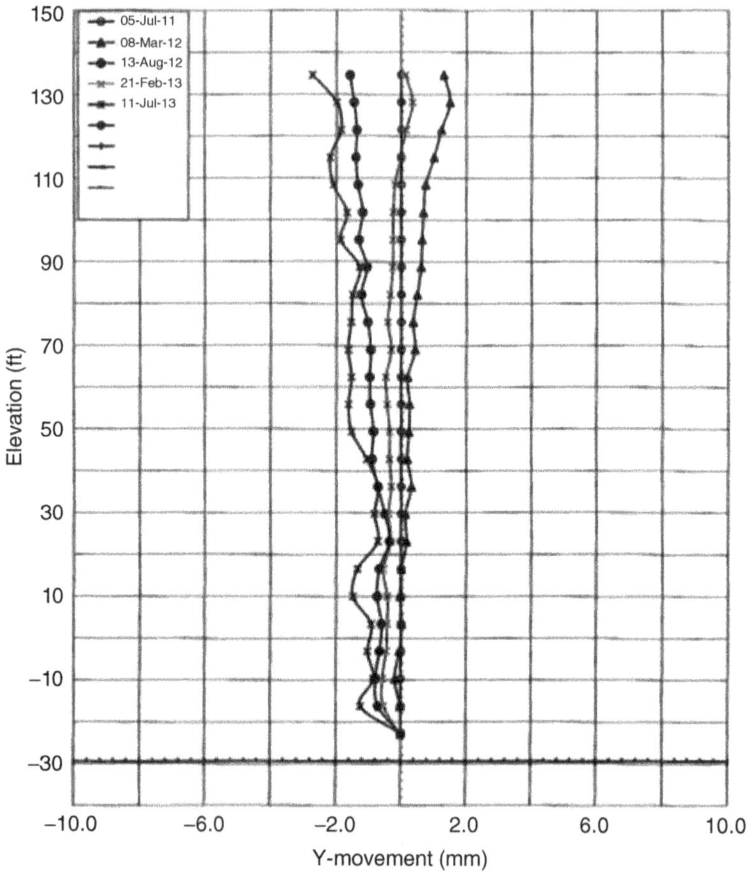

Figure 11.35 Sample display of inverted pendulum Y-movements profiles from 2011 to 2013 based on shuttle probe measurements with July 2011 measurements as baseline.

profile. In such an arrangement, if, for example, points 1, 2, 3, and 4 were originally on the same level ground surface 1 -*P*, the subsidence (dz_4) (i.e., the variation from the level ground surface) at point 4 with respect to point 1 can be given (Chrzanowski, 1986) as:

$$dz_4 = \frac{S_1(\alpha_1 + \alpha_2)}{2} + \frac{S_2(\alpha_2 + \alpha_3)}{2} + \frac{S_3(\alpha_3 + \alpha_4)}{2} \tag{11.27}$$

where α_1, α_2, α_3, and α_4 are the changes in tilt observed by the tiltmeters at points 1, 2, 3, and 4, respectively; and S_1, S_2, and S_3 are the corresponding distances between pairs of tiltmeter positions. The accuracy of this method, however, will depend on the density of tilt measurements along the profile and the continuity of the profile (a constant change in slope of the terrain between measurement points is assumed).

Tilt observations (similar to extensometer observations discussed in Section 11.4.1) can also be integrated with geodetic observation equations in least squares adjustment of the displacement vector. In this case, a tilt observation may be considered as a special case of a vertical angle observation when points 1 and 2 are close enough. If the tilt measurement α_{12} (in radians) from point 1 to point 2 between two epochs of survey is very small (as usually expected), it can be expressed as

$$\alpha_{12} = \frac{dz_2 - dz_1}{S_{12}} \tag{11.28}$$

where dz_1 and dz_2 are the vertical displacements at the two points 1 and 2, respectively; and S_{12} is the distance separating the two points.

11.4.5 Numerical Examples

Example 11.1 A deformable structure is expected to expand uniformly and linearly by 0.001 m at 95% confidence level (in 1 year) along its length of 20 m. Answer the following:

a. What is the strain component per year (at 95% confidence level) for this structure?

Solution (a): From Equation (11.4), the strain component at 95% confidence level per year can be given as

$$\varepsilon = \frac{\Delta \ell}{\ell} \quad \text{or} \quad \varepsilon = \frac{0.001\,\text{m}}{20\,\text{m}} = 5 \times 10^{-5} \quad \text{per year}$$

b. The traditional assumption in the design of deformation surveys is that the survey should be able to detect one-third of the expected deformation, meaning that the maximum error (error at 95% confidence level) of detecting the deformation should be one-third of the deformation at 95% confidence level. Based on this assumption, what is the standard deviation of detecting the relative movement between any two points on this structure?

Solution (b): Based on the assumption, the precision of the survey can be calculated as follows:

The strain should be detected to $(5 \times 10^{-5})/3$ per year (or 1.67E−5/year) at 95% confidence level;

From the statistical testing procedure in Chapter 2, from Equation (2.15), it can be seen that $z_{0.975} \times \sigma_\varepsilon = 1.67\text{E}{-5}$ (where $z_{0.975} = 1.96$ is the normal statistical distribution value at 95% confidence level and σ_ε is the standard deviation of strain determination per year). Based on this, the standard deviation for the strain determination per year will be

$$\sigma_\varepsilon = \frac{1.67\text{E}{-5}}{1.96} \quad \text{per year} \quad (\text{or } 8.50\text{E}{-6} \text{ strain per year})$$

Consider the relative movement as the change in lengths ($\Delta\ell$) of the two extensometer rods anchored at two different points with the standard deviation of $\sigma_{\Delta\ell}$. From error propagation of Equation (11.4), the following is obtained:

$$\sigma_\varepsilon^2 = \frac{1}{\ell^2}\sigma_{\Delta\ell}^2 + \left(\frac{\Delta\ell}{\ell^2}\right)^2 \sigma_\ell^2 \tag{11.29}$$

For change in length, $\sigma_\ell = (\sigma_{\Delta\ell}/\sqrt{2})$; substitute this and $\varepsilon = (\Delta\ell/\ell)$ into the above Equation (11.29) to obtain the following:

$$\sigma_\varepsilon^2 = \left(1 + \frac{\varepsilon^2}{2}\right)\left(\frac{\sigma_{\Delta\ell}}{\ell}\right)^2 \tag{11.30}$$

Since $(\varepsilon^2/2) \approx 0$ compared with 1, Equation (11.30) can be reduced to

$$\sigma_{\Delta\ell} = \pm\ell \times \sigma_\varepsilon \tag{11.31}$$

Substituting $\ell = 20\,\text{m}$ and $\sigma_\varepsilon = 8.50\text{E} - 6$ strains into Equation (11.31) gives the standard deviation of relative movement of 1.70E−4 m per year.

c. If the survey will be performed twice a year (with the same precision each time), what are the expected standard deviations of detecting relative movement and the strain component between any two points on this structure?

Solution (c): Assuming the precisions of detecting relative movement for the two monitoring sessions in the year are $\sigma_{\Delta\ell_1}$ and $\sigma_{\Delta\ell_1}$ (being the same), following the variance–covariance propagation:

$\sigma_{\Delta\ell}^2 = 2\sigma_{\Delta\ell_1}^2$; from Solution (b), $\sigma_{\Delta\ell} = 1.70\text{E}-4\,\text{m}$ per year

$\sigma_{\Delta\ell_1} = \frac{\sigma_{\Delta\ell}}{\sqrt{2}} \rightarrow \sigma_{\Delta\ell_1} = 1.20\text{E}-4\,\text{m}$ per every half a year.

Similarly for the strain component: $\sigma_\varepsilon = 8.50\text{E}-6$ strain per year, so that $\sigma_{\varepsilon_1} = (\sigma_\varepsilon/\sqrt{2}) \rightarrow \sigma_{\varepsilon_1} = 6.01\text{E}-6$ strain every half a year.

d. Express answers in Question (c) in terms of tilt angles at 95% confidence level, giving your answers in arcseconds.

Solution (d): The standard deviation for the strain per half a year can be considered as tilt if the error is considered in the vertical direction, perpendicular to the surface of the structure. In this case, the tilt value (at standard level) can be determined by using the standard deviation $\sigma_{\Delta\ell_1} = 1.20\text{E}-4\,\text{m}$ per every half a year and the given length ($\ell = 20\,\text{m}$) of the structure:

$$\text{Tilt} = \frac{\sigma_{\Delta\ell_1}}{\ell} \times 206{,}265''(\text{or } 1.24'')$$

Tilt at 95% confidence level is obtained by multiplying this by 1.96, giving 2.43″. Similarly, the tilt value can be obtained using the error calculated for the strain detection, so that the tilt value is given as $6.01\text{E}-6 \times 206,265''$(or 1.24″). The tilt value at 95% confidence level is obtained by multiplying this by 1.96, giving 2.43″. The values calculated in both cases should be the same as can be seen above.

e. Assuming the available instrumentation for the tilt measurement in Question (d) has a maximum error of 2″ (at 95% confidence level) and taking into consideration the amount of tilt that can be detected in 1 year, will it be justified to repeat the tilt measurement three times a year (e.g., at regular 4-month intervals)?

Solution (e): Following the steps in Question (c), the expected precision of detecting relative movement for three monitoring sessions in the year can be given as $\sigma_{\Delta \ell_1}$ (being the same for the three sessions), following the variance–covariance propagation:

$$\sigma^2_{\Delta \ell} = 3\sigma^2_{\Delta \ell_1};\ \text{from Solution (b)},\ \sigma_{\Delta \ell} = 1.70\text{E}-4\,\text{m per year}$$
$$\sigma_{\Delta \ell_1} = \frac{\sigma_{\Delta \ell}}{\sqrt{3}} \rightarrow \sigma_{\Delta \ell_1} = 9.81\text{E}-5\,\text{m per every 4 months.}$$

The tilt value for every 4 months will be given as

$$\text{Tilt} = \frac{9.81\text{E}-5}{20} \times 206,265'' = 1.0''\ \text{(at standard level)}$$

Multiply this value by 1.96 to obtain the value at 95% confidence level as 1.98″. Since the precision (2″) of instrumentation is not as good as what is to be detected (1.98″), it will not be justified to measure the tilt value three times a year.

11.5 INTEGRATED DEFORMATION MONITORING SYSTEM

Integrated deformation monitoring system is a highly flexible monitoring system that combines geodetic, geotechnical, and meteorological sensors to match the needs of a monitoring challenge. Some of the basic problems (or limiting factors) of deformation monitoring that the system is likely to address include the following (cf. Chrzanowski, 1993):

(a) Inadequate instrumentation. Sparse instrumentation (or the monitoring system not measuring key features) will not provide expected quantitative information to identify or narrow down the mechanisms that triggered the deformation.

(b) Poorly designed monitoring schemes such as the monitoring schemes not including stations at the points where maximum deformations have been predicted and/or the measurements not being accurate enough due to instrument precision. In absolute monitoring over decades, some points in the poorly designed network may lack intervisibility or some markers may be destroyed, so that the geometry of the network becomes very weak. New constructions might block the intervisibility of some markers over time or some markers tampered with over time. In this case, the network has become more unreliable for monitoring to continue successfully. It is expected that the original deformation monitoring network will be strong enough to provide large redundant measurements at the initial stage so that over decades the network will still remain fairly strong.

(c) Effect of atmospheric temperature, refraction, and so on. In absolute monitoring case over several decades, the effect of secular changes in atmospheric temperature may create permanent deformation, which is apart from the usual process of deformation of the structure. This is usually a concern that can be approached by modeling the effects of atmospheric change in temperature. This will involve observing the trend of the deformation over the decades, producing a time series, which is then modeled appropriately; the result of time series analysis is used to reduce the measurements for the effect of secular changes. The terms (considered as due to secular changes) in time series function that are time dependent are important and must be corrected for in the deformation measurements and not interpreted as deformation.

(d) Environmental influences such as thermal effects on the mechanical, electronic, and optical components of the instruments. A temperature change will produce dimensional changes of the mechanical and other components of tiltmeters, causing drifts of tilt indications and fluctuations of the readout.

(e) Lack of or improper calibration of the instruments (or lack of adequate knowledge of calibrating geotechnical instruments and lack of sufficient calibration facilities)– aging of the instruments may result in a drift of the instrument readout. The permanently installed instruments are very often left in situ for several years without checking the quality of their performance. In long-term measurements, instrument precision may be affected by aging of the electronic and mechanical components, resulting in a drift of the instrument readout. It is also possible that different instruments need to be used due to new technology or the need to upgrade the instrument due to changing accuracy (or precision) of the instrument as a result of aging. It will be important that instruments be calibrated regularly and the calibrated precision values used in the subsequent least squares adjustment and deformation analysis.

(f) Local instability of the observation stations (due to improper monumentation of survey stations and improper installation of the in situ instrumentation). As a rule, the reference network should consist of at least six points so that they can be used in identifying possible instability in the network. According to Chrzanowski (1993), the identification of the unstable points may be difficult or even impossible if the reference network consists of less than six points. When considering absolute deformation monitoring over several decades, stability of reference points becomes a problem. This problem may be approached by considering the following:

- Ensure there are sufficient relocation points (at the design stage) for monitoring the reference points.
- Use free network constraints adjustment without fixing any of the network points; perform weighted similarity transformation to bring coordinates in both epochs to the same datum. This will provide good results if there are enough redundant points and measurements for identifying unstable points.
- The reference marks must be well protected from persistent heat from the sun.

From the advantages and limitations of geodetic and structural/geotechnical instrumentations, it seems that integrating the two techniques will enhance the ability to determine the status of a monitored object. The integration of the measurements from both techniques at the processing level usually complements each other in achieving better accuracy of deformation monitoring (Chrzanowski, 1993). Some of the examples of how the integration may complement each other are as follows:

1. For the detection of an expected tilt in a structure, leveling surveys may be supplemented by tiltmeter and plumbline measurements. Geodetic leveling, with an achievable accuracy of better than ±0.1 mm over distances of 20 m (or equivalent of ±1″) may provide better accuracy for the tilt determination than local measurements with electronic tiltmeters.
2. In the case of an expected expansion between two points of a structure, extensometer measurements may be supplemented by geodetic surveys using electromagnetic distance measurement (EDM) device to determine the relative displacement of the two points. Among the most precise wire extensometers are Kern distometer and CERN distinvar; if properly calibrated and used, can give accuracies of 0.05 mm or better in measurements of change in distance over lengths in the range of 1–30 m. Digital tape extensometers may provide relative movement over short distances to precisions as high as 0.1 mm with less accumulated random errors, unlike geodetic approach. Precision electro-optical geodetic instruments such as Kern ME5000 with accuracies of ±0.3 mm over short distances may serve as extensometers in relative deformation surveys. However, geodetic surveys with optical and electro-optical instruments are always contaminated by atmospheric refraction, which limits their positioning accuracy to about ±2 ppm (at one sigma level) of the distance. With the average distance between the object and reference points of about 500 m, the absolute displacements of the object points cannot be determined with an accuracy better than about ±3 mm at the 95% probability level.
3. In the case of when information on absolute displacements of a structure is needed, geodetic positioning surveys may be supplemented by measurements with an *inverted plumbline* or *borehole rod* extensometer, anchored deep in the bedrock. Inverted plumblines and borehole extensometers, if anchored deeply enough in bedrock outside the deformation zone, may serve better than geodetic surveys for determining the absolute displacements of objects. For power dams,

the depth of the anchors must be 30 m or even more below the foundations. The basic concern with using an inverted plumbline is ensuring verticality of the boreholes so that the wire has freedom of motion, and the influence of air currents and the spiral shape of wires.

4. Geotechnical instrumentation is more likely to be used to monitor the relative movement, especially where it is practically impossible to use geodetic approach. For example, in order to determine the relative movement of points inside a gallery of a dam (e.g., if the local stability of foundation of a dam needs to be monitored), intervisibility will be impossible for geodetic approach to be used; extensometer may have to be used in the foundation of the structures, for example, dams, Powerhouses, walls, and so on, which are inaccessible for geodetic measurements.
5. There will be sufficient redundant measurements by using different measuring techniques; geometry of the scheme will also be self-checking. Geodetic observables are usually interrelated to form a network, while geotechnical observables are located in isolation from other observables, and the only check on an observable is usually an assessment of the immediate repetition of an observation.
6. Geodetic surveys may be inadequate and uneconomical when a high frequency of repeated observations is needed. The movements may be desired more frequently so that it will be too expensive to carry out repeated measurements at short intervals using geodetic approach. Geotechnical methods can easily be adapted for continuous monitoring of relative movements and can easily be used to transmit data remotely by radio to off the site offices. It is also possible to use geodetic approach to transmit data remotely, but it is more expensive involving more expensive equipment.

12

MINING SURVEYING

OBJECTIVES

At the end of this chapter, you should be able to

1. Describe survey standards and procedures for mine surveys
2. Define some mining terminology
3. Discuss various techniques (including instrumentation) for transferring position and orientation underground
4. Describe the advantages, disadvantages, and limitations of various mining orientation techniques
5. Solve problems related to orientation transfer in mining (including tunneling) surveys
6. Discuss the sources of errors in various mining orientation techniques and explain the methods of minimizing their effects
7. Discuss the operations of gyrotheodolite/gyro station in orientation transfer, including various gyro orientation methods, sources of errors, and how their effects are minimized and various reductions applicable to gyro measurements
8. Determine volumes of materials moved in mining activities

Precision Surveying: The Principles and Geomatics Practice, First Edition. John Olusegun Ogundare.

12.1 INTRODUCTION

The main focus of this chapter is on underground mining surveys rather than surface mining surveys that are basically the same as the well-known conventional surveys. Usually, mine baselines, which are permanently marked survey lines on the surface, are established using conventional survey methods or GPS survey techniques. An established baseline is then extended underground through subsidiary control surveys to define the direction and position of the workings of a mine.

A mine is a pit or an excavation made in the earth from which mineral ores are extracted. An ore is a mineral deposit that has enough worth to be mined at a profit. Generally, a mineral is a nonrenewable resource such as petroleum, natural gas, water, and mining has to do with extracting the minerals. Mining techniques can be divided into two common excavation types: *surface mining* and *underground mining*. Surface mining is done by removing (stripping) surface vegetation, dirt, and if necessary, layers of bedrock in order to reach buried ore deposits. Techniques of surface mining include *open-pit mining,* which consists of recovery of materials from an open pit in the ground, and *strip mining*, a form of open-pit mining, which consists of stripping surface layers off to reveal ore/seams (flat-lying ore bodies) underneath. Underground mining consists of digging tunnels or shafts into the earth to reach buried ore deposits. Note that production of liquids and gases, as in the petroleum industry, is not generally considered mining, and *mining claim* is a portion of mining land, usually 40 acres in size.

According to the International Society for Mine Surveying (ISM, n.d.), "Mine surveying is a branch of mining science and technology which includes all measurements, calculations and mapping which serve the purpose of ascertaining and documenting information at all stages from prospecting to exploitation and utilizing mineral deposits both by surface and underground working." The ISM (n.d.) gives the following as the list of the main activities expected of a mine surveyor:

- *The interpretation of the geology of mineral deposits in relation to the economic exploitation thereof.*
- *The investigation and negotiation of mineral mining rights.*
- *Making and recording and calculations of mine surveying measurements.*
- *Mining cartography.*
- *Investigation and prediction of effects of mine working on the surface and underground strata.*
- *Mine planning in the context of local environment and subsequent rehabilitation.*

From this list, it can be deduced that mining surveying includes surface surveying associated with underground and open-pit mining and underground surveying for mining purpose. This classifies mining surveying as an integral part of Surveying Engineering discipline involving all measuring activities connected with mining operations on or below the surface, representation (producing mine surveying plans for open-pit and underground workings) and management of data associated with a mining operation. Mine surveying, however, is different from the tunneling surveying in that the workings of a mine are far more irregular since the excavations must follow underground deposits of ore, coal, or minerals.

Typically, a mine surveyor must have geodetic and topographic skills in order to be able to carry out prospecting surveys; cadastral skills in dealing with mineral rights and mining lease boundaries; engineering survey skills for day-to-day operations of a mine; and cartographic skills for preparing surface and underground plans. Mining surveyors deal with many activities above and below ground surface using advanced techniques such as GPS surveying, classical surveying, aerial photogrammetry, terrestrial scanning, gyrotheodolite traversing. Mining surveyors are often in charge of producing and updating the database of the GIS for the mine; usually, mapping and volume determination is a daily job of mining surveyors. All surveys are plotted on a master mine map, which is updated daily. The map shows underground workings and the buildings and other facilities on the surface, as well as boundary and lease lines. Features that must be avoided such as wells are shown on the maps. Emergency escape routes are clearly marked, both underground and on the maps. In the event of a mine emergency, management and rescue personnel can use the map to save valuable time and perhaps lives.

Some of the specific and peculiar circumstances in underground mining surveys are as follows:

1. Survey networks follow narrow corridors and inclined drifts, requiring steep vertical sights with special equipment.
2. Control points are built up from short traverses (since lighting is sometimes very poor), with an unfavorable influence on the error propagation.
3. Three-dimensional coordinates may be required; the coordinate system (or the orientation of the underground surveying networks) must be correlated to that on the surface. Underground, the GPS, and astronomic methods cannot be used making control transfer and orientation difficult tasks.
4. Needs to detect rock movement; extensive rock mass deformation or surface movements can result in unnecessary expense or dangers of a far more serious nature, such as human injuries, cave-ins, and property damages. Detection of the existence, magnitude, and direction of these movements is another important task of the mining surveyor.
5. Rough working conditions. There may be high temperature, falling water, poor visibility, and heavy traffic. Surveyors underground are most of the time required to carry a lot of heavy items such as survey equipment (total station, tripod, plumb bobs, measuring tapes, etc.); and other items such as hammers, stakes, methane detector to avoid setting off explosion.
6. Survey control points of underground mines are generally located in the roof (backs) of mine workings. This is done by first drilling a small hole in the roof, into which a round wooden plug is driven. Locating points in the floor is not viable. The main disadvantages are that it is difficult to install and access

roof points (which may be up to heights of over 5 m) and it is challenging to the survey team to center the instrument (using optical or zenith plummet or plumb bob) under those survey points. This type of centering may contribute centering error of up to ±2 mm or more. Survey points are better located in the wall installation; this makes access much easier, safer, and faster and instrument centering under a point will not be required. This method, however, requires specially designed target prisms that retain central position through all rotations (Taylor Hobson spheres may be used).

12.1.1 Survey Standards and Procedures for Mine Surveys

All mining surveys and plans should be based on mine baselines, which are commonly established on the surface through control surveys. The baselines are usually about 25 m long with their end points permanently marked (usually in concrete) (Department of Mines and Petroleum, 2011). Surface and underground control surveys are subsequently carried out from these baselines to establish the position of mine workings. According to Davis et al. (1981), horizontal control network surveys in the mines are based on the following three orders, usually done in reversed order:

- Third-order open traverse with allowable accuracy of 1:1000
- Second-order open traverse with allowable accuracy of 1:5000
- First-order closed-loop traverse with allowable accuracy of 1:10,000 to 1:20,000.

It is typically required that the position of mine workings be established with second-order accuracy or better with respect to mine datum (Department of Mines and Petroleum, 2011). Lower order accuracy may be acceptable, however, in the case of inaccessible mine workings or where reflectorless total station equipment or laser-ranging equipment is being used for cavity measurements. Correlation between surface and underground surveys should be carried out with an accuracy that is better than third order.

The typical observables for mine traverse surveys are the direction (or angle) and distance measurements. In order to achieve the desired level of traverse accuracy, the direction or angle measurements (horizontal and vertical) should be done with a maximum standard error of $\pm 5''$; forced-centering instruments are not required for second-order and third-order works, but needed for first-order work. If the instrument cannot be set on a tripod on the floor, it can be hung from some special supporting bars. Centering an instrument beneath a survey marker that is on the back is done using a sharp string plumb bobs pointing on the specially marked point on top of the instrument telescope. The instrument telescope usually comes with an attachable (to the top of the instrument telescope) optical zenith plummets or interchangeable (same tribrach) with the instrument. Lighted targets are to be used for first-order surveys while miners light shone on a plumb bob string with light colored material behind the string is to serve as a target for other lower order surveys.

Distance measurement should be done with a maximum standard error of ±3 mm + 5 ppm. Depending on the order of job, measurements can be made with steel tape, total station that meets underground requirements (fire and damp proof in gaseous mines) and stadia measurements. Total station and taping are used for first-order and second-order works and stadia measurements for detail surveys. Surveyors can now use total station equipment to perform resection to two known survey points as an alternative to physically setting up on the points; in this case, no instrument and target heights need be measured. The resection process will involve measuring one horizontal angle, two vertical angles, and two slope distances to two known survey points that have been previously established in the mine. The resection method combined with the use of wall stations is also superior in transferring heights underground since no target or instrument heights are measured, thus eliminating most of the errors in height transfer. Height transfer typically should be done with a maximum standard error of leveling per kilometer of double run of ±4 mm. Where vertical measurement is required for height transfer underground, the maximum permissible error should not exceed 0.05 m (Department of Mines and Petroleum, 2011). Nowadays, terrestrial scanning laser systems are becoming common for detail (three-dimensional positioning) and some drift volume determinations.

12.1.1.1 Typical Survey Markers in the Mines Floor survey points in mines are rare because of traffic, mud, and water; they may be used in mines with concrete floors. Most of the survey points are located on the mine walls. The wall targets, creating reference points from which a theodolite can be set up on, over or under can be in different forms, such as forced-centered brackets installed in about 30-mm-diameter hole drilled in the wall; permanent structures bolted to the wall, similar to survey pillars on the surface They can protrude from the wall by as much as 30 cm and can easily be damaged. In the underground environment, however, it is not practical or economical to permanently mount electromagnetic distance measurement (EDM) prisms to the walls. Usually, a small hole is drilled into the wall of a stope with a piece of aluminum tube inside it; the aluminum tube would stay in the wall permanently. The prism and stem would be inserted during the survey and removed later. The wall station stem made of stainless steel is designed to fit into the aluminum sleeve at one end and to lock into the base of the prism at the other end. The prism should be such a way that if one rotates the target left or right, the back of the prism actually will not move off the line of sight. Wall station sleeves installed in underground mines could be a piece of aluminum tube installed into the wall or simply a hole drilled into the rock.

12.2 MINING TERMINOLOGY

Some of the mining terms that will be used in this chapter are illustrated in Figure 12.1 and discussed as follows. The term **adit** will be used to mean a horizontal or slightly inclined passage from the surface to a mine; it is similar to a **tunnel** except that the tunnel must be open to the atmosphere at both ends while the adit ends in the mine.

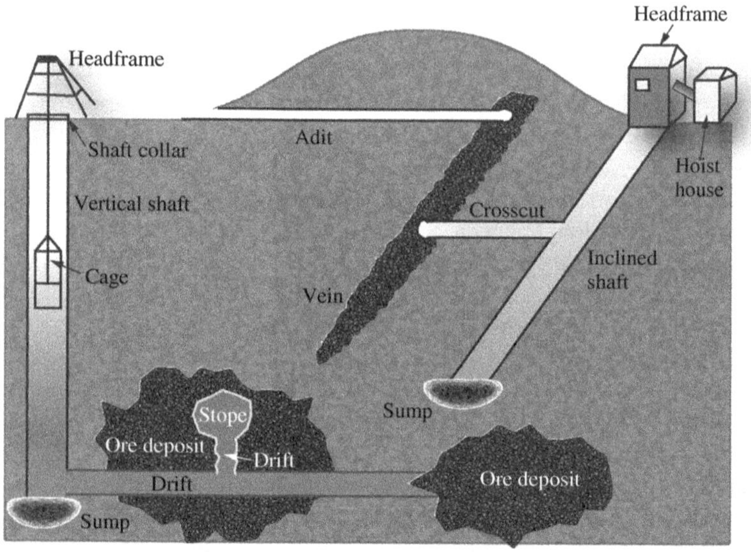

Figure 12.1 A cross section of a mine illustrating some mining terms.

In underground transportation system, tunnels are usually driven to connect inclined or vertical shafts whose relative locations are established by surface surveys.

Unlike a tunnel, which is usually horizontal or near-horizontal, a **shaft** is vertical or near-vertical; it is a primary vertical or inclined hole that starts from the surface and goes into the underground mine. The wood or concrete lining at the surface around the mouth of the shaft is known as **shaft collar**; and the process of excavating the earth vertically (or near-vertically) from the surface to the underground is known as **shaft sinking**. The structure erected over the shaft for supporting the machine, which raises and lowers the cage or other conveyance in a shaft, is known as **headframe**.

The other important mining terms are **sump**, which is an underground excavation used as a collecting point for drainage water; **crosscut**, a hole driven from a shaft, cutting across the substance to be mined; and **drift**, which is a horizontal underground hole driven along the rock formation of the substance being mined. For further information, the mining glossary of terms, such as KMI Glossary (2009), can be consulted.

12.3 HORIZONTAL MINE ORIENTATION SURVEYS

The subject of mine surveying also includes surface surveying for mining claims and surveying for patent; orientation surveys may also be needed for tunneling between two mines or between two shafts or for sinking of shafts. Orientation surveys are also important in the process of protecting surface and underground objects and structures from the adverse effects of underground mining. The usual purpose of mine orientation surveys is to give coordinates (X, Y, Z) of at least one point of the underground network with reference to the surface coordinate system and to establish the azimuth

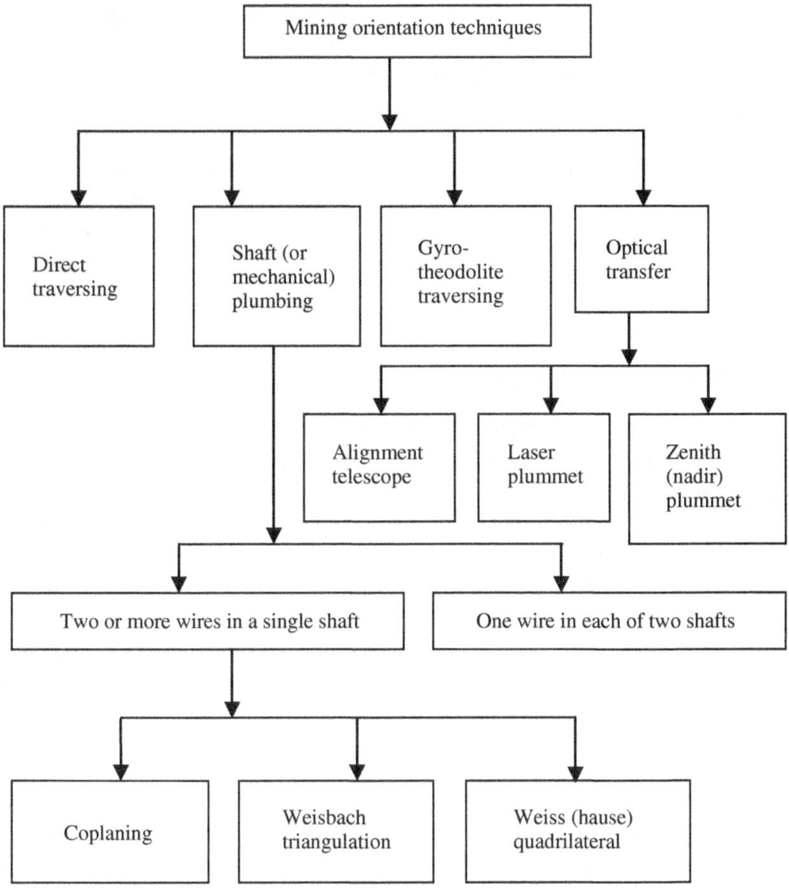

Figure 12.2 Different mining orientation techniques.

of one line of the underground network. In mining, this is usually referred to as *correlation of surface and underground surveys*. The chosen orientation survey technique depends on the method of gaining entrance into the mine, for example, by an adit, an inclined shaft, or a vertical shaft. The classification of main survey techniques is shown in Figure 12.2. The individual techniques are discussed in the following sections. Also note that the orientation methods discussed in this section are also applicable to tunneling surveys, where the tunneling starts from shafts, such as in the case of underground nuclear accelerator ring.

12.3.1 Direct Traversing Technique

Direct traversing method is done when nonvertical shafts lead to underground workings; in this case, open traverse procedure is carried out through an adit or a near-horizontal shaft. Three-dimensional traverse angles and distances measured are used to determine the coordinates of the underground control points, given the bearing of at least one of the traverse legs; elevation of each survey station may be determined by trigonometric method. Disadvantages of this method include the following:

(a) Steep sights might be involved – angle from horizon may be up to 80°; the effects of centering and leveling errors on angle measurements become serious problems; forced centering with increased number of angular observations at each station are required to reduce angular error per setup.
(b) Inclined sight may require increased number of setups.

12.3.2 Mechanical Technique

The mechanical correlation technique is also known as shaft plumbing method. The basic concept of this method is that wires hanging freely in a shaft will occupy the same position underground that they occupy at the surface and the bearing of the line connecting the wires will remain constant through the shaft. Shaft plumbing is therefore a process of transferring one or more points (or bearing) at the surface of a shaft to plumbline points at the bottom of the shaft to ensure that the shaft is sunk in the direction of gravity or to transfer bearing underground. In shaft plumbing approach, piano wires hanging in a vertical shaft are used. It is recommended that weights (usually made of lead) not greater than 50% of the breaking strength of the wire used be applied to the wire when it is hanging in a shaft. The following are true concerning weights to be hanged on the wires:

- Sufficient weight is needed to keep the wire from coiling and to help reduce swinging. Magnetic attraction (due to surrounding rock containing large amount of magnetic minerals) will influence steel bobs and steel wires. It may be necessary to use bronze wires to prevent attractive forces that could displace the wires slightly and warp the plane. Weights must be carefully made so that excessive swinging of the wire is minimal. In deep shafts (in excess of 900 m), gravitational attraction between bobs and nearby masses or voids may become very significant, causing more displacement of the wire.
- As shafts deepen, heavier weights must be applied to correspondingly thicker wires. One disadvantage of using thicker wires is that the thicker the wire the more difficult it is to make accurate pointings to it.
- Plumb bob is usually immersed in a drum containing water, viscous oils, or other suitable liquids in order to steady the plumb bob; water is the most commonly used since it is cheaper to get.
- Angles and distances to plumb wires cannot be measured very reliably and accurately in underground because of movements of the plumb wires.

Generally, equipment for shaft plumbing includes reels, wire-centering devices, plum bob, piano wire, and immersion liquids. Reels are for preserving, lowering, and winding up the wire. Usually, shaft sinking is the most expensive part of a mine development, requiring that sites of the shaft be carefully selected. Diameters of shafts range usually from 4 to 8 m depending on the planned transportation capacity.

The most dangerous factor causing errors in the mechanical correlation method is the deflection of wires by air current of the ventilation system; any mechanical ventilation system should be closed down while the wires are suspended. To reduce further oscillation of the wires due to the ventilation, plumb bobs (weights) are usually immersed in a drum containing water, light oil, or other liquid; water is naturally used since it is readily available at the bottom of the shaft. In spite of this, it may still take several hours for the swinging weight to come to a complete stop.

Apart from mining orientation, other applications of shaft plumbing include sinking a new shaft and also in a control survey to determine the deformations of a shaft and its equipment owing to rock mass movement (but plumbing wires may disturb work schedule in this case).

As can be seen in Figure 12.2, the two subtechniques involved in mechanical plumbing are

- orientation transfer using two or more wires in a single vertical shaft;
- orientation transfer using one wire in each of two vertical shafts.

Some of the advantages with mechanical plumbing method include the following:

(a) It gives much higher accuracy of mine orientation. The accuracy achievable depends on the method (Weisbach or quadrilateral) and the geometry of the measurement network (refer to Section 12.3.2.1).
(b) Mechanical method is simple compared with other methods.
(c) Automatic data transfer is possible when the method is used in shaft deformation monitoring; it can easily be adapted to continuous monitoring of deformations using inductive sensors of structural deformations over distances up to a few hundred meters (with achievable accuracies of 0.1 mm).

12.3.2.1 Orientation Transfer with Two Wires in a Single Vertical Shaft In this method of shaft plumbing, the transfer of orientation (i.e., azimuth and position) underground (e.g., to a tunnel or an adit) is done down a single shaft using a pair of plumblines, P_1 and P_2. Establishing these plumblines (with a typical distance of about 2–4 m between them) in the shaft, however, requires utmost care and experience. Also note that a small deflection of the underground position of one of the plumblines in the direction perpendicular to the vertical plane of the two points on the surface will cause a large rotation of the underground control network. For example, a deflection of one of the plumblines underground by only 1 mm may result in the rotation angle of more than 1 min of arc.

Three methods of performing orientation transfer with two wires in a single vertical shaft are as follows:

- Coplaning method
- Weisbach triangulation method
- Weiss quadrilateral method.

Coplaning Method Coplaning method is a mine orientation technique in which the vertical crosshair of a theodolite's telescope is placed exactly in the vertical plane formed by two wires suspended in a vertical shaft (Frush, 1973). In the method, two points representing the centerline of the tunnel is set out on the ground surface and plumbed onto the bottom of the vertical shaft and then extended into the tunnel. The orientation procedure is such that a theodolite, set up on the surface within 3–4 m of the nearer of the two wires, is brought into line with the two wires. This is done by first observing the far wire and then moving the theodolite on line. Angles may be measured to a number of control points on the surface in order to establish the azimuth between the two wires. Similarly, the theodolite is aligned with the two wires at each level in the mine usually at the same time as it is done on the surface. However, the deeper the shaft, the more difficult it becomes to align the theodolite with the two wires since the wires are constantly swinging due to air current. The main advantage of the method is its simplicity with little chance of blunders.

Weisbach Triangulation Method Weisbach triangulation method is an attempt at minimizing the problems associated with exactly aligning the theodolite with two wires as required in coplaning method. The technique is different from the coplaning method since it only requires that the theodolite be set up close to the plane of the wire and not exactly on the plane of the wire (Frush, 1973). In this method, two piano wires P_1 and P_2 are led down a single vertical shaft as shown in Figure 12.3. In the figure, a theodolite is located on the surface at point B and another one underground at point C. The locations of the two wires are P_1 and P_2 on the surface and P_1' and P_2' underground. The direction P_1-P_2 is determined on the surface and then transferred

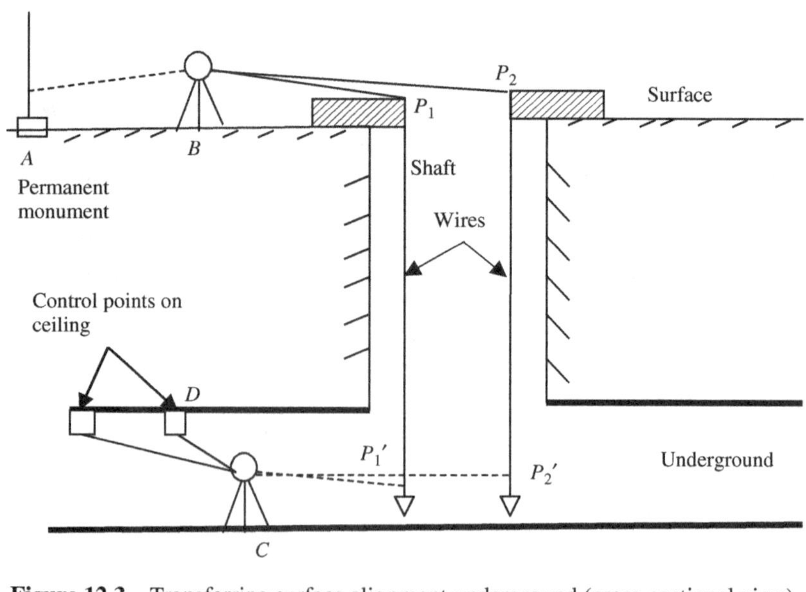

Figure 12.3 Transferring surface alignment underground (cross-sectional view).

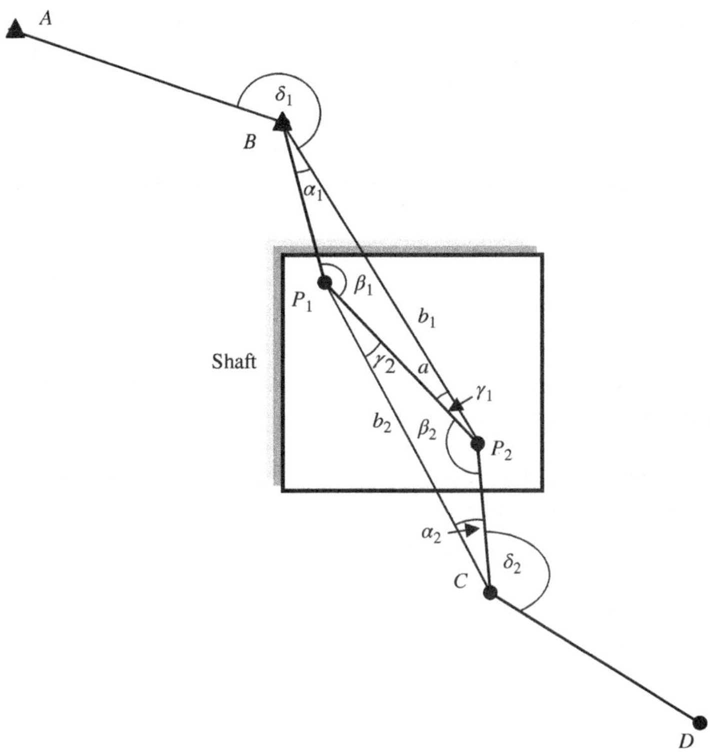

Figure 12.4 Weisbach triangle (plan view).

to P'_1-P'_2 underground, which is used to orient underground surveys. If vertical collimator is used, two points on the top of the shaft will be transferred on the floor through vertical lines of sight. These points are utilized in a precise double-centering operation to prolong the tunnel alignment. Specific methods of transferring alignment underground are discussed as follows.

The Weisbach method is illustrated further in Figure 12.4 (in plan view), where points A and B are surface stations, points C and D are underground stations, and P_1 and P_2 are the vertical shafts (or plumblines). In the Weisbach triangle in the figure, for surface surveys, the known surface data are the following:

- Coordinates of the surface stations A and B
- Measured surface angles δ_1 and α_1
- Measured surface distances B-P_1, B-P_2, and P_1-P_2.

For underground surveys, the known underground data are as follows:

- Measured angles α_2 and δ_2
- Measured underground distances C-P_1, C-P_2, and C-D.

The unknown underground parameters to be determined are the coordinates of points *C* and *D* and the azimuth of the line *C-D*. The required computation steps in determining the unknown underground parameters are as follows:

1. Solve for angles (γ_1 and β_1) in the surface triangle B-P_1-P_2 using Sine law.
2. Solve for angles (γ_2 and β_2) in the underground triangle C-P_1-P_2 using Sine law.
3. Solve for azimuths of surface and underground lines (B-P_1, B-P_2, P_2-P_1, P_1-P_2, P_2-C, P_1-C, C-D.
4. Traverse A-B-P_1-C-D or A-B-P_2-C-D.

Advantages, Disadvantages, Limitations Some of the important elements of the Weisbach method are as follows:

1. Theodolite (or Weisbach) stations B and C are as close as possible to the near wire almost in line with both wires. This may be seen as an advantage, since the shaft area is usually very small and cramped.
2. Angles β_1 and β_2 are equal to 180° and α_1 and α_2 are measured with high degree of precisions (the angle at the Weisbach station must be measured repeatedly). For maximum accuracy, the angles subtended by the wires should be measured in not less than three sets. Mean position of the wire is determined by placing a scale behind each wire, perpendicular to each line of sight. The observer notes a series of extreme positions for each wire, on both sides of the mean. The average is calculated, and this average scale reading used for subsequent pointings of the telescope for the angular observations. The need to measure the angles precisely is a disadvantage of this method, since more work and care are required in doing this.
3. Distance between the wires should be as long as possible. There is usually a limit to how long this distance can be, considering the usual diameter of the shaft of about 4–8 m.
4. Errors in distance as large as 10 mm may be neglected if the angles α_1 and α_2 are less than 30′ (Davis et al., 1981). This may be considered an advantage of using this method, since distances do not need to be measured very precisely. Usually, the distances between the wires and to the theodolite stations should be accurate to 0.3–3 mm.
5. The method is not applied when the angles subtended by the two wires (angles α_1 and α_2) must be greater than 10°, because the influence of the errors of the measured distances becomes critical. Typically, the angle subtended by the two wires should be less than 1°. This may be considered a disadvantage and a limitation since the choice of angles α_1 and α_2 is restricted when using the method.

Sources of Error in Weisbach Triangle The standard error of the transferred bearing ($\sigma_{\mathrm{Az}_{CD}}$) for line *CD* underground is made up from the following effects:

(i) Uncertainty in connecting the surface base to the wire base, σ_s, which is mainly due to the errors in measuring angles δ_1 and β_1 on the surface.

(ii) Uncertainty in connecting the wire base to the underground base, σ_u, which is mainly due to the errors in measuring angles δ_2 and β_2 underground.

(iii) Uncertainty in the verticality of the wire plane, σ_p, which is due mainly to random deflections of the two plumblines. The expected random error effect on azimuth (Az) determination using shaft plumbing method has been given (Davis et al., 1981) as

$$\sigma_p = \frac{206,265}{a}\sqrt{(e_1)^2 + (e_2)^2}\ (\text{arcsec}) \tag{12.1}$$

where e_1 and e_2 are small random deflections for the first and second plumblines (with respect to the plane defined at the surface) and a is the distance separation between the two plumblines. Assuming that the deflections of the two wires P_1 and P_2, at right angles to the line P_1P_2, are 1 mm each and that the wires are 2 m apart, then $\sigma_p = 146''$.

The combined error on the azimuth of line CD underground can be given as

$$\sigma_{\text{Az}_{CD}} = (\sigma_s^2 + \sigma_u^2 + \sigma_p^2)^{1/2} \tag{12.2}$$

or

$$\sigma_{\text{Az}_{CD}} = (\sigma_{\text{Az}_{BA}}^2 + \sigma_{\delta_1}^2 + \sigma_{\beta_1}^2 + \sigma_{\beta_2}^2 + \sigma_{\delta_2}^2 + \sigma_p^2)^{1/2} \tag{12.3}$$

If $\sigma_{\text{Az}_{CD}}$ in Equation (12.2) or (12.3) is required to be less than $120''$ and assuming $\sigma_s = \sigma_u$ and $\sigma_p = 100''$, the following can be calculated for σ_s or σ_u:

$$120 = (2\sigma_s^2 + 10{,}000)^{1/2} \quad \sigma_s = \sigma_u = 47'' \tag{12.4}$$

Referring to Figure 12.4, the angles β_1 and β_2 are not directly measured but are usually calculated using the Sine laws. The usual Sine law for calculating any of these two angles (β_i, $i = 1, 2$) can be given by a general formula

$$\sin\beta_i = \frac{b_i \sin\alpha_i}{a} \tag{12.5}$$

where a is the distance between the two plumblines, b_i is the length of the line facing the corresponding angle β_i, and α_i is the corresponding angle facing the line connecting the two plumblines. Following the rules of error propagation on Equation (12.5), the variance of any calculated angle β_i can be given as

$$\sigma_{\beta_i}^2 = \frac{\tan^2\beta_i}{b_i^2}\sigma_{b_i}^2 + \frac{\tan^2\beta_i}{a^2}\sigma_a^2 + \left(\frac{b_i^2}{a^2\cos^2\beta_i} - \tan^2\beta_i\right)\sigma_{\alpha_i}^2 \tag{12.6}$$

where σ_b^2, $\sigma_{a_i}^2$, and $\sigma_{\alpha_i}^2$ are the variances of distance b, distance a_i, and angle α_i (in radian), respectively. Of course, the angle β_i must have been determined directly from Equation (12.5) before the error propagation formula in Equation (12.6) can be used, since Equation (12.6) is derived directly from Equation (12.5).

Example 12.1 Table 12.1 shows the field notes taken in the process of orientation transfer down a single shaft by means of Weisbach triangle (referring to Figure 12.4). Let P_1 and P_2 represent the plumblines, B and C the respective surface and underground theodolite stations, and A and D the surface and underground reference points, respectively.

a. Determine the bearing of line CD assuming the bearing of line BA is 300°00′00″.

Solution: From Figure 12.5, representing the surface part of the Weisbach triangle, use Sine rule as follows:

$$\frac{\sin \alpha_1}{4.48} = \frac{\sin \gamma_1}{4.34} \tag{12.7}$$

or

$$\gamma_1 = \arcsin\left(\frac{4.34 \times \sin 0°0'4''}{4.48}\right) \tag{12.8}$$

$$\gamma_1 = 3.875''$$

Considering triangle BP_2P_1 in Figure 12.5:

$$\text{Angle } BP_1P_2 = 180° - (4'' + 3.875'') \rightarrow 179°59'52.125''$$

$$\text{Bearing } BA \text{ (given)} = 300°00'00''$$

$$\text{Bearing } BP_1 = \text{Bearing } BA\text{–Angle } \delta_1$$

$$= 300°00'00'' - 107°43'31'' \rightarrow 192°16'29''$$

$$\text{Bearing } P_1P_2 = \text{Bearing } P_1B + \text{Angle } BP_1P_2$$

$$= 192°16'29'' - 180° + 179°59'52.125'' \rightarrow 192°16'21.125''$$

TABLE 12.1 Field Notes for Orientation Transfer through a Single Shaft.

At	From	To	Distance (m)	Angle
B	P_2	A	B–P_2 = 8.83	107°43′35″
	P_1	A	B–P_1 = 4.34	107°43′31″
		P_1P_2	4.48	–
C	P_1	D	C–P_1 = 9.40	156°04′18″
	P_2	D	C–P_2 = 4.91	156°04′27″

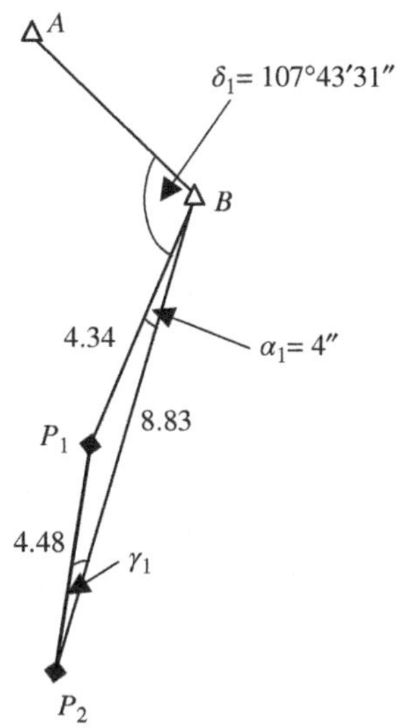

Figure 12.5 Plan view of Weisbach triangle (surface part).

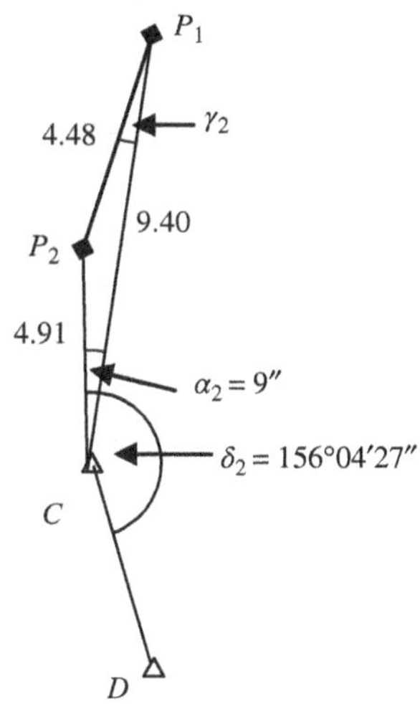

Figure 12.6 Plan view of Weisbach triangle (underground part).

The underground calculations are done similarly using Figure 12.6 as follows:
From Figure 12.6:

$$\frac{\sin \alpha_2}{4.48} = \frac{\sin \gamma_2}{4.91} \quad \text{so that } \gamma_2 = \arcsin\left(\frac{4.91 \times \sin 0°0'9''}{4.48}\right)$$

$$\gamma_2 = 9.864''$$

$$\text{Bearing } CD = \text{Bearing } P_1P_2 + \gamma_2 - 180^\circ + \delta_2 - \alpha_2$$
$$= 192^\circ 16'21.125'' + 9.864'' - 180^\circ + 156^\circ 04'27'' - 9''$$

The bearing CD is $168^\circ 20'49''$.

b. Using Figure 12.5 for the surface triangle, determine the angle at point P_1 (angle B-P_1-P_2) using the Sine law (Equation (12.5)) and its propagated standard deviation (Equation (12.6)). Take the standard deviations of the measured distances as 2 mm and the standard deviations for the measured angles as $1''$.

Solution: Remember that you have to use Sine law (Equation (12.5)) directly before you can use the error propagation formula (Equation (12.6))
For the surface part (Figure 12.5): $a = 4.48$ m (or 4480 mm); $b_1 = 8.83$ m (or 8830 mm) and $\alpha_1 = 4''$.
From Equation (12.5):

$$\sin\beta_1 = \frac{b_1 \sin\alpha_1}{a} \rightarrow \beta_1 = \sin^{-1}\left(\frac{8.83 \sin 4''}{4.48}\right)$$
$$\beta_1 = 0^\circ 00'7.9''$$

Since arcsine of a number cannot give values greater than 90°, the value for the angle will be $\beta_1 = 180^\circ - 0^\circ 00'7.9''$ or $\beta_1 = 179^\circ 59'52.1''$.
Using error propagation laws from Equation (12.6):

$$\sigma_{\beta_i}^2 = \frac{\tan^2\beta_i}{b_i^2}\sigma_{b_i}^2 + \frac{\tan^2\beta_i}{a^2}\sigma_a^2 + \left(\frac{b_i^2}{a^2\cos^2\beta_i} - \tan^2\beta_i\right)\sigma_{\alpha_i}^2 \tag{12.9}$$

where $a = 4.48$ m (or 4480 mm); $b_1 = 8.83$ m (or 8830 mm); $\alpha_1 = 4''$;

$$\sigma_{b_1} = 2\,\text{mm};\quad \sigma_a = 2\,\text{mm};\quad \sigma_{\alpha_1} = 1''(4.84813226\text{E–}6\,\text{rad})$$

$$\sigma_{\beta_1}^2 = 7.5256\text{E–}17\,\text{rad}^2 + 2.923531\text{E–}16\,\text{rad}^2$$
$$+ 3.88477061(4.84813226\text{E–}6)^2\,\text{rad}^2$$
$$\sigma_{\beta_1}^2 = 7.52562\text{E–}17\,\text{rad}^2 + 2.923531\text{E–}16\,\text{rad}^2 + 9.130915\text{E–}11\,\text{rad}^2$$
$$\sigma_{\beta_1}^2 = 9.13095\text{E–}11\,\text{rad}^2 \rightarrow \sigma_{\beta_1} = 9.55560\text{E–}6\,\text{rad or } \sigma_{\beta_1} = 2.0''$$

Approximate approach can be used to determine the standard deviation of the angle β_1 as follows. Assume $\beta_1 = 180^\circ$ in Equation (12.6), so that $\tan^2\beta_1 = 0$; $\cos^2\beta_1 = 1$; Equation (12.6) becomes reduced to

$$\sigma_{\beta_1}^2 = \frac{b_1^2}{a^2}\sigma_{\alpha_1}^2 \tag{12.10}$$

Substituting the corresponding values $a = 4.48$ m, $b_1 = 8.83$ m, and $\sigma_{\alpha_1} = 1''(4.84813226\text{E–}6\,\text{rad})$ into Equation (12.10) gives $\sigma_{\beta_1} = 9.55558\text{E–}6\,\text{rad}$ or $\sigma_{\beta_1} = 2''$, which is the same as the value obtained in Equation (12.9).

Weiss Quadrilateral Method In modern practice, both the coplaning method and Weisbach method have been replaced by quadrilateral method. In comparison with the Weisbach method, the surface connections to P_1 and P_2 (in Figure 12.3) are the same for the quadrilateral method but the underground connections are different. In the quadrilateral method, points P_1, P_2, C, and D are plumb points set out by plumblines (for P_1 and P_2) and the total station optical plummet or laser plummet (at C and D). Lines P_1 and P_2 can also be fixed precisely by laser plummet on both the ground surface and the shaft bottom. The plumb points are then tied with ground control points by total station, with distances being measured between them. Inside the tunnel, coordinates of the established control points are determined by measuring both angular and distance measurements to the four plumb points and computed by the least squares method. This method is further illustrated in Figure 12.7 (in plan view), where points A and B are surface stations, points C and D are underground stations, and P_1 and P_2 are plumblines in a vertical shaft.

Measured surface angles and distances:

Angles: α_1, α_2.
Distances: B-P_1, B-P_2, P_1-P_2.
Known surface data: Coordinates of surface stations A and B.

Measured underground angles and distances:

Angles: a_1, a_2, a_3, a_4.
Distances: only C-D needed, but can also measure C-P_1, C-P_2, D-P_2, D-P_1.

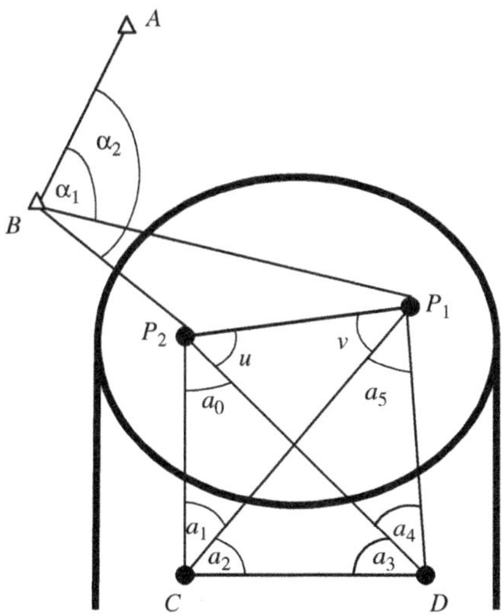

Figure 12.7 Quadrilateral method (plan view).

Unknown underground quantities:

- Coordinates of C and D.
- Azimuth of the line C-D.

The main problem in Figure 12.7 is determining angles u and v, which can be done in two ways: solving for the angles directly or using local coordinating approach to solve for the angles. The direct solution of the angles u and v can be given as

$$v = \cot^{-1}\left(\frac{x + \cos(y)}{\sin(y)}\right) \tag{12.11}$$

or

$$v = \tan^{-1}\left(\frac{\sin(y)}{x + \cos(y)}\right) \tag{12.12}$$

$$u = y - v \tag{12.13}$$

where

$$y = a_2 + a_3 \tag{12.14}$$

$$x = \frac{\sin(a_0)\sin(a_2)\sin(a_4)}{\sin(a_5)\sin(a_3)\sin(a_1)} \tag{12.15}$$

and a_0 and a_5 are determined by solving the triangles P_2-C-D and P_1-C-D, respectively.

In the local coordinating approach, the required computation steps for the determination of angles u and v are as follows:

1. Use angles α_1 and α_2 to compute bearings to P_1 and P_2 and their coordinates.
2. Solve for distance and azimuth of the line P_1-P_2 (by inverting their coordinates).
3. Use local coordinate system (with C as origin and line CD as x-axis with assumed bearing like 90°) and coordinate P_1 and P_2 by intersection; use the computed local coordinates of points C, D, P_1, and P_2 to determine the angles u and v.

After determining the angles u and v, the azimuth P_1 to C or azimuth P_2 to D can be determined; the traverse P_1-C-D or P_2-C-D is then run in order to determine the coordinates of C and D and the azimuth CD.

Advantages, Disadvantages, Limitations Some of the important elements of the quadrilateral or Weiss method are as follows:

- Error analysis indicates that the best shape for the quadrilateral is square; the error of orientation will increase if the ratio of the length CD to the wire base is increased. Achieving square shape may become a limitation/disadvantage since a square configuration may be impossible to achieve with limited space available in the shaft area. This approach is usually recommended when α_1 and α_2 are greater than 10°. This property may also be considered an advantage in that the method can be applied when the angles α_1 and α_2 are greater than 10°, when Weisbach method cannot be used.

- Calculation of distance *CD* is not critical in determining orientation angle at P_1 and P_2; and the errors of distances have no influence on the accuracy of the transferred azimuth; they do not need to be measured precisely. This is an advantage of using this method since the errors of distances will not affect the transferred azimuth.
- Centering of instruments and targets are very important. Since distance *CD* is just a few meters, the accuracy of centering the theodolite and the target is critical. Forced centering is recommended or else two theodolites should be used simultaneously at stations C and D, each pointing at the crosshairs of the other (telescopes focused to infinity). This is a disadvantage of this method since centering errors may bias the orientation result, if centering of instruments and targets are not properly done when the orientation data are being collected.
- Redundant measurements are possible. This will allow simultaneous least squares adjustment of measurements and statistical analysis of results. This property can be seen as an advantage as well as a disadvantage. It is an advantage since redundant measurements produce better reliability of result; it is a disadvantage since it involves more measurements, making the method more laborious than the Weisbach method.

Example 12.2 Employing the Weiss quadrilateral approach, using Figure 12.8 and the coordinates for the two surface points S_1 and S_2 provided in Table 12.2, and the distance and angle determinations given in Table 12.3, compute the north (*N*) and east (*E*) coordinates of the underground points 5 and 6 and the azimuth of the line 5-6. (Note: S_1 and S_2 are surface control points, W_A and W_B are the two wires in the shaft, and 3–6 are underground control points.)

1. Solve for the coordinates of W_A and W_B

$$\begin{aligned}
\mathrm{AZ}_{S_2S_1} &= \arctan\left(\frac{dE}{dN}\right) \rightarrow \arctan\left(\frac{11.135}{-9.50}\right) + 180^\circ = 130^\circ 28' 11'' \\
\mathrm{AZ}_{S_2W_A} &= 130^\circ 28' 11'' + 215^\circ 30' 40'' = 345^\circ 58' 51'' \\
\mathrm{AZ}_{S_2W_B} &= 130^\circ 28' 11'' + 269^\circ 01' 49'' = 39^\circ 30' 00'' \\
N_{W_A} &= 253,000.000 + 3.725\,\cos(345^\circ 58' 51'') = 253,003.614\,\mathrm{m} \\
E_{W_A} &= 54,010.000 + 3.725\,\sin(345^\circ 58' 51'') = 54,009.098\,\mathrm{m} \\
N_{W_B} &= 253,000.000 + 4.885\,\cos(39^\circ 30' 00'') = 253,003.769\,\mathrm{m} \\
E_{W_B} &= 54,010.000 + 4.885\,\sin(39^\circ 30' 00'') = 54,013.107\,\mathrm{m}
\end{aligned}$$

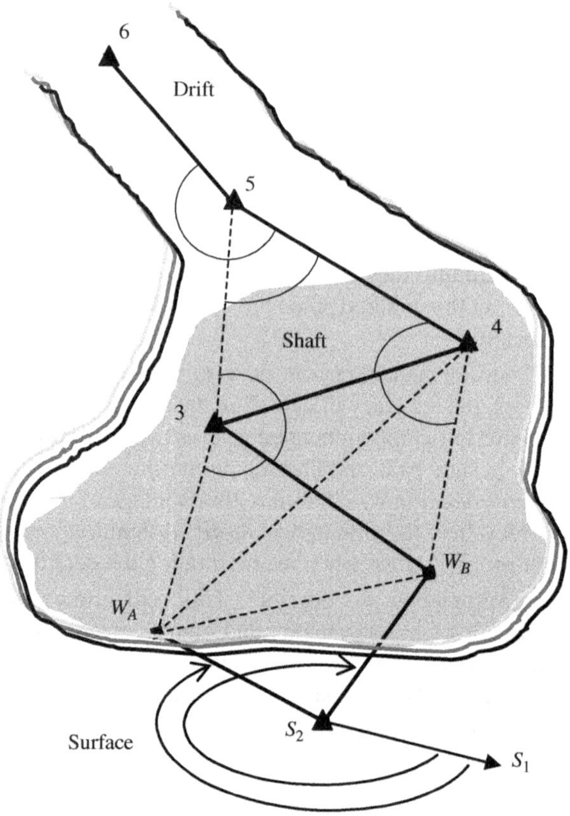

Figure 12.8 Example on quadrilateral method (plan view).

TABLE 12.2 Given Coordinates.

Point	Northing (m)	Easting (m)
S_1	252,990.500	54,021.135
S_2	253,000.000	54,010.000

2. Solve *for distance and azimuth of the line $W_A W_B$:*

$$d_{AB} = \sqrt{(4.009)^2 + (0.155)^2} = 4.012$$

$$\text{AZ}_{AB} = \arctan\left(\frac{4.009}{0.155}\right) = 87°47'09''$$

3. Using Equations (12.11)–(12.15) as follows:

TABLE 12.3 Field Measurements.

Setup (From-At-To)	Angle
S_1-S_2-W_A	215°30′40″
S_1-S_2-W_B	269°01′49″
W_B-3-W_A	60°50′24″
4-3-W_B	56°30′40″
W_B-4-W_A	56°59′40″
W_A–4–3	30°00′20″
5-3-4	58°45′10″
3-4-5	57°25′55″
4-5-3	63°48′55″
4-5-6	190°30′05″
Setup (At-To)	**Distance (m)**
S_2-W_A	3.725
S_2-W_B	4.885
W_B-3	4.598
3-4	2.728
4-5	2.599
5-6	3.495

From Figure 12.8 and Table 12.3:

$$a_0 = 3 - W_B - 4 = 36^\circ 29'20''(\text{solving triangle } W_B - 3 - 4)$$

$$a_1 = W_B - 4 - W_A = 56^\circ 59'40''$$

$$a_2 = W_A - 4 - 3 = 30^\circ 00'20''$$

$$a_3 = 4 - 3 - W_B = 56^\circ 30'40''$$

$$a_4 = W_B - 3 - W_A = 60^\circ 50'24''$$

$$a_5 = 4 - W_A - 3 = 32^\circ 38'36''(\text{solving triangle } W_A - 3 - 4)$$

From Equation (12.15), $x = 0.6883636$.
From Equation (12.14), $y = 86^\circ 31'00''$.
From Equation (12.11), $v = 53^\circ 06'41''$.
From Equation (12.13), $u = 33^\circ 24'19''$.

4. Calculate Azimuths W_B-3, 3-4, 4-5, and 5-6:

$$\text{AZ}_{W_B3} = 87^\circ 47'09'' + 180^\circ + 33^\circ 24'19'' = 301^\circ 11'28''$$

$$\text{AZ}_{34} = 301^\circ 11'28'' - 180^\circ - 56^\circ 30'40'' = 64^\circ 40'48''$$

TABLE 12.4 Traverse Computation.

From	Distance (m)	Bearing	Northing (m)	Easting (m)	To
			253,003.614	54,009.098	W_A
W_A	4.012	87°47′09″	253,003.769	54,013.107	W_B
W_B	4.598	301°11′28″	253,006.150	54,009.174	3
3	2.728	64°40′48″	253,007.317	54,011.640	4
4	2.599	302°06′43″	253,008.699	54,009.438	5
5	3.495	312°36′48″	253,011.065	54,006.866	6

$$AZ_{45} = 64°40'48'' + 180° + 57°25'55'' = 302°06'43''$$

$$AZ_{56} = 302°06'43'' - 180° + 190°30'05'' = 312°36'48''$$

5. Perform the traverse computation along W_B-3-4-5-6 as shown in Table 12.4.

12.3.2.2 Orientation Transfer with Two or More Vertical Shafts This is a method of shaft plumbing through two or more vertical shafts with one plumbline in each shaft. For example, for a case of two shafts, one wire P_1 will be in one shaft and wire P_2 will be in the other shaft. This method of orientation, which is also called *fitted traverse method*, determines coordinates of each wire on the surface by multiple intersections from as many surface control stations as possible. From the coordinates of the surface wire points, the bearing of the surface wire base is obtained. A fitted traverse is then run (using assumed bearing) from one wire to the other through an underground connecting tunnel. Since the angles at the underground wire points cannot be measured directly, the traverse is run as an open one based on assumed bearing of the first traverse leg. At the end of the underground traverse, the computed bearing between the underground wire points is compared with the bearing between the corresponding surface wire points; the underground traverse is then swung by the amount of the difference between the two bearings. Any possible linear error between the surface and underground traverses can be corrected by multiplying each underground traverse length by a scale factor that is equal to the ratio of the distance between the wires on the surface and the corresponding underground distance. The traverse is then recalculated based on the corrected bearings and distances in order to obtain new coordinates for the wires.

Orientation error due to the nonverticality of the wires is much smaller in this method than in the method discussed in Section 12.3.2.1, since the distance between shafts can be several hundred meters apart. This method, therefore, gives a higher accuracy of mine orientation than shaft plumbing through one vertical shaft; it may also give better accuracy than gyro orientation. Error in orienting the underground traverse using fitted traverse method consists of the following:

- Error in azimuth of line between two plumblines (from surface)

- Error in azimuth of line between two plumblines (from underground distance and angle measurements)
- Error due to deflections of plumblines.

Some of the typical problems with this method of mine orientation include the following:

(a) Not every mine has access to the surface through two or more vertical shafts from mining levels that require orientation. This method is useful only when the level is accessed by two vertical shafts or raises straight enough to allow one wire to be hung in each without contact with the sides of the shafts.
(b) It is time-consuming and requires the utmost care to fulfill the high-accuracy requirements (30–120″ in azimuth).
(c) Air currents need to be minimized.

12.3.3 Orientation Transfer Using Optical Method

The orientation transfer in this approach will have the optical lines of sight (based on theodolite, lasers, and zenith plummets) replacing the plumblines (in the case of orientation with two plumblines); sometimes, strong ventilation may make it difficult to set up under a station using a plumb bob, requiring that optical approach be used. This may require setting up a theodolite on the edge of the shaft or directly over the shaft with corresponding difficulties involved. The optical instrument can also be set at the bottom of the shaft to project line of sight vertically up to specially arranged targets at the surface and appropriate observations made directly to the targets. If the optical lines of sight are arranged in a form of well-configured Weisbach triangles, the orientation process will follow the Weisbach approach.

In some cases, the optical plummet may be set up underground and the target (usually in Taylor Hobson sphere) may be bracket-mounted at the top of the shaft; optical (or laser) plummets are used in conjunction with total stations and gyro station to accomplish survey control transfer to underground mining workings. The total station set up on the surface is used to locate the center of precise spherical target (target in Taylor Hobson sphere) whose position is transferred through the vertical shaft to underground point (at the nadir of the plummet) using zenith plummet set directly below the spherical target. The distance and azimuth to the located underground point are determined using total station and gyro station located underground away from the located point. Some of the disadvantages of optical sighting in a shaft are as follows:

- Limitation of visibility due to fog, causing increased pointing and focusing errors – accuracy of pointing improves with collimated lasers.
- Limitation due to the depth of the shaft (high-magnification telescope is needed; magnification of some optical plummet is 31.5×); there is usually a problem of correctly detecting the center of the laser beam in deep shafts or where there are variations in air density.

- Ensuring the verticality of laser beam; automatic compensator may be used for this purpose.
- Effect of refraction in the shaft.

Some of the advantages of optical sighting in a shaft are as follows:

1. Use of lasers allows automated alignment procedure for continuous data acquisition.
2. Use of optical plummet is diverse since it provides very precise line of sight when the depth involved is short. Apart from mining surveying, optical or laser plummets are used in the following:
   - Determination of verticality of tall building and tower construction
   - Tunneling (shaft sinking); underground highways and railways; water, sewer, and drainage systems, and scientific purposes such as the construction of super-conducting super collider rings
   - Deformation studies – dams and tall structures (buildings, towers, and chimneys)

The optical method of orientation transfer is divided into three submethods (Figure 12.4), such as

- Using alignment telescope
- Using laser plummet
- Using zenith (or nadir) plummet.

Alignment telescope cannot be used if visibility is poor in the shaft since the method requires a clear line of sight. In this method, pointing and focusing of telescope is limited by visibility in the shaft.

12.3.3.1 Using Laser Plummet In this method, the laser beam with its small angle of divergence and high intensity provides a good visible reference plumbline. In principle, if a laser beam projected straight down a shaft is reflected back (from a level surface underground) to the point of origin, it is clear that such a beam is truly vertical. For deeper shafts, collimated laser beam can be used as the plumbline. It is possible (Dazhi, 1988) to use laser guiding equipment for shaft plumbing up to 1200 m. The deviations of a laser beam will be caused mainly by the following factors (Dazhi, 1988):

- Nonverticality of the vertical axis of the laser equipment, which depends primarily on the sensitivity of the level of the instrument.
- Nonalignment of the laser beam axis with the direction of gravity, which can be controlled to a few arcseconds.
- Divergence or wavering of the laser beam in air due to refraction as a result of changing temperature gradient and humidity and the effects of moving air

currents. This makes it difficult to define sufficiently narrow beam of light to produce a point. This divergence, however, is small compared to that of the other light sources.

The total plumbing error (e_p) from using laser optical plummet can be expressed as

$$e_p = \sqrt{e_l^2 + e_t^2 + e_d^2}\ \text{arcsec} \tag{12.16}$$

where e_l is the leveling error (in arcsec) resulting in nonverticality of the vertical axis of the laser equipment (equivalent to how much the level bubble is off); e_t is the amount by which the standing axis of laser beam is off the direction of gravity (in arcsec), and e_d is the divergent angle of the laser beam (in arcsec). If the shaft is H m deep, the plumbing error (ε) in meters can be given as

$$\varepsilon = \frac{e_p \times H}{206,265}\ \text{m} \tag{12.17}$$

where the constant 206,265 is for converting the angle e_p from arcseconds into radians.

Laser equipment may be very useful in controlling shaft-sinking procedures and in transferring coordinates (shaft plumbing) when using the gyro or the two-shaft method of mine orientation. Remember that the plumblines defined optically will be affected by refraction since they may be close to the walls of the shaft.

12.3.3.2 Using Zenith Plummet Specially designed optical (zenith or nadir) plummet can be used in providing a vertical direction in a shaft. The use in shaft plumbing is limited, however, to a short range (about 100–200 m) only because of the poor visibility in the shaft atmosphere; moreover, optical measurement from the bottom up the shaft often causes problems because of water dropping down. A typical optical plummet is WILD/Leica ZL automatic Zenith plummet with a specified accuracy of 1:200,000 (Geodetic Supply & Repair, 2009). The main steps in using zenith plummet in orientation transfer are given as follows:

- Surface control point is established about 3 m from the shaft.
- Set up a theodolite over the control point, with the theodolite sleeve, tangent device, extension rods, and target.
- At the bottom of the shaft, a nail or bolt is placed in the decking covering the sump and the zenith plummet set over this point.
- The observer lines up the plummet so that the initial plane is approximately normal to the line of the extension rods on the surface and the horizontal angle observed.
- The observer then instruct via the shaft telephone to the instrument man on the surface to move the target away from or toward the transit until the target is

centered over the plummet crosshair – this is done by sliding the extension rod in or out through the theodolite sleeve and over the tangent device. Two persons are usually required in this method.

- Coordinates of target are calculated and taken as coordinates of the plummet and also the station over which plummet is set. The plummet has 90° mechanical stops instead of the horizontal circle allowing the measurements in the four positions to be carried out easier and much faster.
- From the underground baseline, the bearing of which has already been determined by a gyrotheodolite, the shaft station is observed and the distance measured.

12.3.3.3 Using Theodolite and Plummet Theodolite and zenith plummet can be used in transferring orientation underground in the case of shallow shafts (about 20–80 m deep). The procedure for transferring horizontal control underground in this case can be summarized as follows:

1. At the top of the shaft, set up Taylor Hobson spheres on survey brackets included in the shaft collar; the centers of the Taylor Hobson spheres will then be defined in three dimensions. Some of the important properties of Taylor Hobson spheres are
   - They can accommodate concentric ring targets or retro-reflectors so that direction, zenith angle, and distance measurements can be made directly to the centers of the spheres.
   - They can be set in any arbitrary orientation without introducing an eccentricity, for example, they can be set up to measure the distance vertically from the bottom of the shafts and can also be rotated in any other directions for surface measurements.
2. Locate two or more temporary tripod points within 40 m of the shaft collar as follows:
   - The tripods must be arranged forming strong geometry with the two Taylor Hobson spheres located at the collar of the shaft.
   - The tripod points must be visible from several control points already positioned as part of densification network on the surface.
3. Measure the directions, zenith angles, and distances according to the designed number of sets (with distances measured from both ends of each line), to connect the Taylor Hobson spheres, the temporary tripods, and the control points; these measurements are made in three dimensions using forced-centering system.
4. Set up two tripods at the bottom of the shaft with pairs of translation stages that will allow plumbing to be performed with good accuracy as follows:
   - Mount precision zenith plummet (such as Wild/Leica ZL plummet) on the translation stages for centering under the spherical targets on the surface.

5. After completing the plumbing operation underground, one of the plummets is removed and replaced by a total station; control is then extended from the plumb points to the permanent tunnel brackets using temporary forced-centered tripod points.

12.3.4 Orientation Transfer by Gyro Azimuth

In the method of orientation transfer by gyro azimuths, coordinates must still be transferred from the surface by shaft plumbing if more efficient methods are not available or using laser optical plummet. In this case, shaft plumbing is used only for the transfer of coordinates of one point and gyrotheodolites are used for transferring azimuth independently of shaft plumbing. The orientation transfer with gyro equipment is naturally more accurate than using plumblines since position determination and azimuth transfer are independently done.

12.3.4.1 Gyrotheodolite/Gyro Station Equipment Gyros are north-seeking devices that are mounted on theodolites. A gyro attachment consists of miniature gyro motor suspended on a thin tape with the driving current reaching the motor via thin leads. When spinning at high speed, the gyro is influenced by the horizontal component of the earth's rotation, making it to oscillate about the plumbline symmetrically to the meridian plane. The determination of true north entails finding the axis of symmetry of a sinusoidal oscillation of the gyro; this is a function of time and the angle between the spin axis and the meridian plane. To determine the meridian plane (the north direction), the time or the angle or both can be measured. Typical gyrotheodolites used are Wild/Leica GAK1 (manual type) with an accuracy of about $\pm 20''$, which is achievable in 20–30 min; SOKKIA GP3X Gyro station with an accuracy of $\pm 20''$, which is achievable in 20–30 min; and the precision gyrotheodolite Gyromat 3000 (fully automatic type) with an accuracy of about $\pm 3''$, which is achievable in 10–15 min. As an example, GP3X Gyro station by SOKKIA is illustrated in Figure 12.9 and discussed in the following subsections.

The GP3X Gyro station (shown in Figure 12.9) is used to locate true north and to determine the azimuth without any other aid. The gyro station consists of the GP-1 gyroscope unit mounted on the Set3X total station. The total station is to implement the gyro calculation program. The gyro station GP3X, for example, is made up of two main components: the GP-1 gyroscope unit and the total station Set3X. Some of the technical details of each of these components are summarized as follows.

Some of the specifications of the Sokkia GP-1 gyroscope unit are given (SOKKIA, 2004) as follows:

- The gyroscope unit allows the true north to be determined with $20''$ accuracy in 20–30 min. The GP-1 gyroscope unit, in principle, has a rotating rotor that maintains the direction of its original rotating axis in space with the earth's rotation making the axis appear to be changing in direction from 0° to 360° in 24 h. Due to the earth's gravity force acting to push down the forward tail end of the rotating axis, the gyroscope axis also rotates (or undergo precession) about

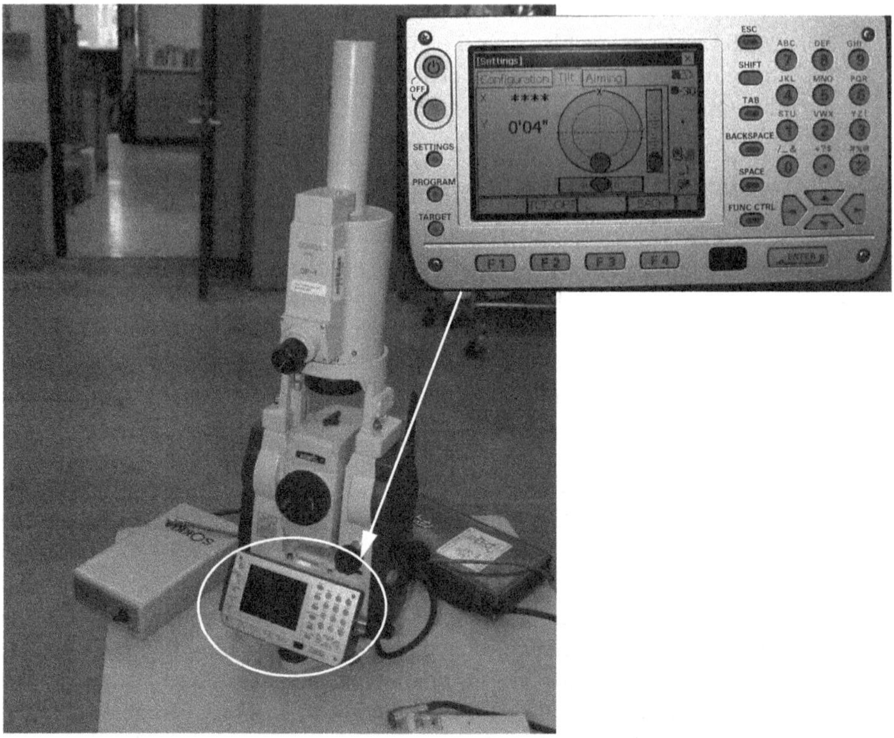

Figure 12.9 GP-1 gyro unit mounted on Set3X total station.

its local horizon about the north–south direction (the meridian). The precession of the gyroscope axis is used to locate the meridian plane at the given location.

- There is always some humming sound noticeable while the gyro rotor is spinning at 1200 rpm (the usual speed of the rotor).
- When the gyro axis is rotating, it projects the gyro mark, which is measured against a graduated scale. When the projected mark is exactly in the center of the scale, the spin axis of the gyro and the line of sight through the telescope of the theodolite are parallel.

Some of the specifications of the Sokkia Set3X total station part of the gyro station are given (SOKKIA TOPCON, 2009) as follows:

- Angular measurement accuracy (ISO17123-3) is 3″.
- Automatic dual-axis compensator has a working range of ±4′; and the sensitivity of the tubular level is 30′/2 mm; and the telescope magnification is 30×.
- Distance measurement accuracy with Prism in fine mode is $(2 + 2\,\text{ppm} \times D)$ mm (where D is the distance measurement).

- Refraction and earth-curvature correction can be applied automatically using coefficient of refraction of 0.14/0.20; or other values can be selected and used.

The main differences between the possible methods of determining the meridian plane using different gyro equipment depend on the following:

- Keeping the theodolite telescope permanently parallel to the gyro's spin axis (by following-up with the theodolite alidade), for example, reversal point methods (also known as turning point or **follow-up** methods).
- Keeping the alidade fixed in a direction close to north, for example, the **time** or amplitude and transit methods.

The procedure for azimuth determination using GP3X gyro station can be summarized as follows:

1. Preorient the telescope of the gyro station (or gyrotheodolite) approximately toward north using the methods discussed in Section 12.3.4.2.
2. According to Sokkia (2004), orient the telescope precisely (to ±20") in the direction of north using follow-up (with multiple turning points) method if the telescope is preoriented toward north to ±2° or transit method if the preorientation toward north is known to ±2′, as discussed in Section 12.3.4.3.
3. Transfer the true north azimuth to the total station horizontal angle-displaying device using the appropriate gyro station built-in facility.

12.3.4.2 Preorientation of Gyrotheodolite It is possible to orient the telescope of gyrotheodolite toward north to within ±30°, using the sun, maps, compass, or intuition. When the gyro is released with the telescope oriented 30–150° away from north, a very pronounced acceleration (toward the direction of north) of the gyro mark will already be seen after only a few seconds, from which it becomes obvious that the preorientation to north is completely wrong. In this case, the gyro must be clamped and the telescope swung through 30–45° in the direction of the north and the gyro released again. If after about 3 min of follow-up, the gyro still continues to accelerate and does not slow down at all, the gyro should be clamped and the alidade turned through a further 90° in the direction of the gyro mark's oscillation. When the telescope has been aligned reasonably well toward true north, two quick methods for the preorientation of the telescope are *quarter time method* and *two reversal point method* (also known as turning point method).

Quarter Time Method The quarter time method is dependent on latitude and requires that you know a priori the quarter oscillation period of the gyro to within ±1 s of time; the average value (for GAK1) in latitude 50° is 2 min 3 s. The quarter oscillation period ($T_U/4$), otherwise known as *swing time*, is the amount of time needed for a particular gyro to oscillate from a reversal point to a transit through the meridian. When the gyro is released in an approximate north direction, the oscillation of the gyro mark is followed up with the alidade so that the V-shaped index of the scale is always slightly ahead of the moving mark. When the slowing down of the mark before the reversal point is noticed, the alidade is clamped and the stop watch is started at the exact moment that the gyro mark passes through the middle of a scale (away and back); if this time interval read on the stop watch is t, then the gyro mark is followed up immediately and continuously, with the alidade unclamped, until the watch shows the time $(t/2 + T_U/4)$, when the alidade is reclamped. The telescope will now be pointing in an approximately north direction. The accuracy of this method depends on the amplitude of the oscillation and, therefore, mainly on the initial approximate orientation to north. For example, if the initial orientation to north is ±30°, the expected accuracy will be ±20′.

Two Reversal (Turning) Point Method After the gyro has been released, oscillation of the gyro mark is followed up smoothly by the alidade, with the horizontal clamp of the theodolite loosened, so that the moving gyro mark is always in the V-shaped index as shown in Figure 12.10. The mark slows down as it approaches the turning (reversal) point; shortly before the first turning point is reached, which is seen by a noticeable slowing down of the gyro in its movement, the alidade is clamped and the gyro mark followed up using the horizontal tangent screw of the theodolite until the turning point is reached. The horizontal circle is then read. The alidade clamp is loosened once more and the mark followed up until shortly before the other turning point, on the opposite side of the meridian. As soon as the oscillation is seen to slow down, the alidade is reclamped and the gyro mark is again followed up to the turning point, by using the tangent screw; the horizontal circle is read again. The mean of the two circle readings indicates the approximate true north direction. The gyro is then clamped, and the telescope is set in the direction of the mean of the two circle readings.

For two turning (or reversal) point measurements (a_1, and a_2), the corrected north direction $N = (N' + E)$ can be given as

$$N = \frac{a_1 + a_2}{2} + E \tag{12.18}$$

where E is the alignment constant for the gyroscope. The accuracy achieved in locating the north this way is ±2′ to 3′. If more than two turning points are measured, the Schuler Mean discussed in Section 12.3.4.3 can be used to determine the average north direction N.

12.3.4.3 ***Precise Methods of Gyro Orientation*** After the telescope has been approximately oriented in north direction using any of the quick methods, the precise orientation will be required in mine survey orientation. Two precise methods of observations are discussed: *multiple reversal point method* (multiple turning point or *follow-up* method) and *multiple transit* (or time) method.

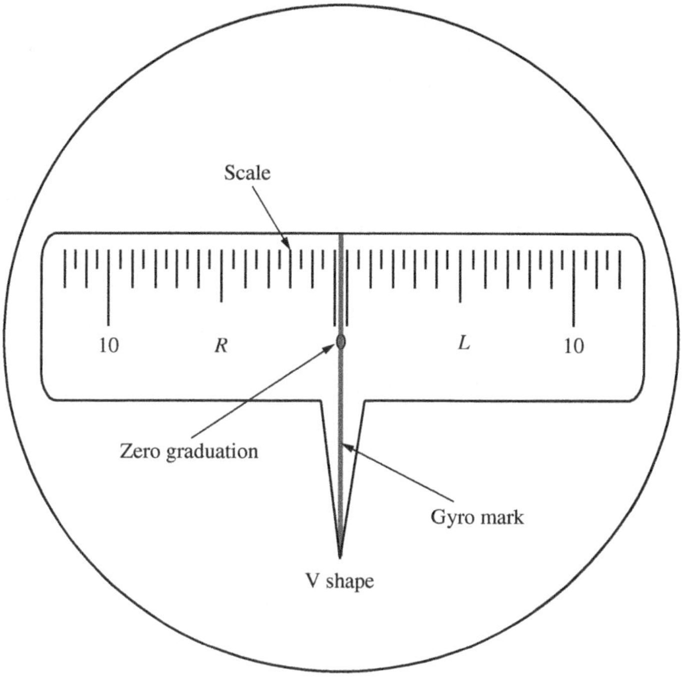

Figure 12.10 Gyro station eyepiece showing the gyro mark in the V shape.

Multiple Reversal Point Method (Multiple Turning Point Method or Follow-Up Method) In order to use this method for a precise north determination, the theodolite must already be oriented to within ±0.5° to 2°, depending on the range of the tangent screw of the theodolite or total station instrument (about 3° for T2 and about 10° for T16/T1A). This must have been done using any of the quick methods. In this method, after the gyro has been released, the oscillation of the gyro mark is followed up by the alidade, using the tangent screw with the moving gyro mark kept as sharply as possible in the V-shaped index in the middle of the scale. At the reversal point, where the gyro mark seems to be at complete standstill for a few seconds, the horizontal circle is read and the gyro mark is followed up again immediately in the opposite direction; the horizontal circle is read at each turning point, and from these values the mean oscillation position is calculated as Schuler Mean. Jerky movements of the gyro mark must be avoided by moving the tangent screw slowly and smoothly. The accuracy of this method is limited by the ability to maintain the coincidence between the gyro mark and the V-shaped scale index, which is possible with standard error of ±6″ to 10″. The general standard error expected for this method is ±15″ to 30″. The following example shows how the booking and calculation for multiple reversal point method are done. The four turning point values are y_0 to y_3; the mean north values are highlighted in columns 2 and 4; and the Schuler Mean (N') is given on the last row in column 5.

Example 12.3 Table 12.5 is the field sheet I consisting of four turning point measurements from GAK1 gyro equipment. The Schuler Mean is to be calculated for the measurements, assuming the gyro equipment is being used to determine the azimuth of line B1-RO (with the gyro equipment setup on station B1).

Solution: The solution to this problem is presented in Table 12.5. In the table, the Schuler Mean is the circle reading of the approximate north position based on the theodolite's line of sight, which can be given as N'. If the alignment constant (or the calibration correction) between the zero of the gyro (the V-shape) and the line of sight through the total station telescope is E, the corrected circle reading of the north point through the gyro axis (N) will be $N' + E$. If the mean angle measurement between the zero scale of the theodolite and the reference object (RO) is H, the corrected Gyro azimuth (A_G(RO)) to the RO can be given as

$$A_G(\text{RO}) = H\text{–}(N' + E) \tag{12.19}$$

or

$$A_G(\text{RO}) = (H\text{–}N')\text{–}E \tag{12.20}$$

where $(H - N')$ is the uncorrected gyro azimuth of the line of sight to RO. If the grid azimuth is desired from the setup point to RO, the convergence of meridian at the setup point must be determined with respect to the central meridian of the map projection. If the convergence of meridian at the setup point is calculated as γ, the desired grid azimuth, Br(RO), from a station to a reference object (RO) can be calculated as

$$\text{Br(RO)} = A_G(\text{RO}) \ - \gamma \tag{12.21}$$

or

$$\text{Br(RO)} = (H\text{–}N')\text{–}E\text{–}\gamma \tag{12.22}$$

Note that γ will have negative numerical value when the setup point is in the western side of the central meridian and will have positive numerical value when it is on the eastern side (depending on the type of map projection used).

TABLE 12.5 Gyrotheodolite Field Sheet I (Turning Point or Follow-Up Method).

Gyro: GAK1 Column 1	Observer Column 2 Turning Point (y) Left	Column 3	Date Column 4 Turning Point (y) Right	Column 5 Schuler Mean (Line Averages)
y_0	32° 02′ 31″			
Mean $(y_0 + y_2)$	32° 04′ 16.5″	y_1	39° 42′ 32″	35° 53′ 24.2″
y_2	32° 06′ 02″	Mean $(y_1 + y_3)$	39° 40′ 56.5″	35° 53′ 29.2″
		y_3	39° 39′ 21″	
Schuler Mean = Mean of column 5			N'	**35° 53′ 26.7″**

The calibration correction (E) can be determined on a baseline whose astronomic azimuth (A_{base}) is already known. If the gyro uncorrected azimuth of the baseline is $(H - N')$ according to Equation (12.20), the correction E can be determined as

$$E = (H - N') - A_{base} \tag{12.23}$$

Example 12.4 Continuing from Table 12.5, calculate the gyro azimuth (A_G) and the grid azimuth (Br) of line B1-RO. The other relevant field data are provided in the gyrotheodolite field sheet II in Table 12.6. Assume $E = -0°02'52''$ and $\gamma = 1°23'48''$ for the calculations.

Solution: The solution to this problem is presented in Table 12.6.

Also note that the method illustrated in Examples 12.5 and 12.6 are based on what is also referred to as "follow-up" method in the GP3X gyro station brochure (SOKKIA, 2004).

Transit (or Time) Method Transit method, which is also referred to as *Time* method in the GP3X gyro station brochure, is based on the time of transit of the approximate north N' (along which the gyro's zero graduation is currently aligned) as illustrated in Figure 12.11. In Figure 12.11, T_L (or T_R) is the length of time taken by the gyro mark to transit the V-shaped index to the L (or R) direction and back, DL (or DR) is the oscillation amplitude value of the gyro mark to the turning point in the L (or R) direction and θ is the offset of the true north (N) from the direction where the telescope is currently pointing (N').

In order to use the transit method for a precise north determination, the total station equipment must already be oriented to within ±20′ of the true north using any of the quick methods described in Section 12.3.4.2. The total station equipment and its horizontal circle are then kept clamped throughout the measurement session, while the time epoch is recorded each time the gyro mark transits through the middle of the

TABLE 12.6 Gyrotheodolite Field Sheet II (Azimuth Determination).

Name: A001			
Station: B1			
RO:			
RO (FL)	245°	28′	25″
RO (FR)	65°	28′	23″
Mean RO (H)	245°	28′	24″
Gyro north reading (N')	35°	53′	27″
Calibration correction or alignment constant (E)	0°	02′	52″
Gyro azimuth: $A_G(\text{RO}) = H - (N' + E)$	**209°**	**37′**	**49″**
Meridian convergence (γ)	1°	23′	48″
Grid azimuth: $\text{Br(RO)} = A_G(\text{RO}) - \gamma$	**208°**	**14′**	**01″**

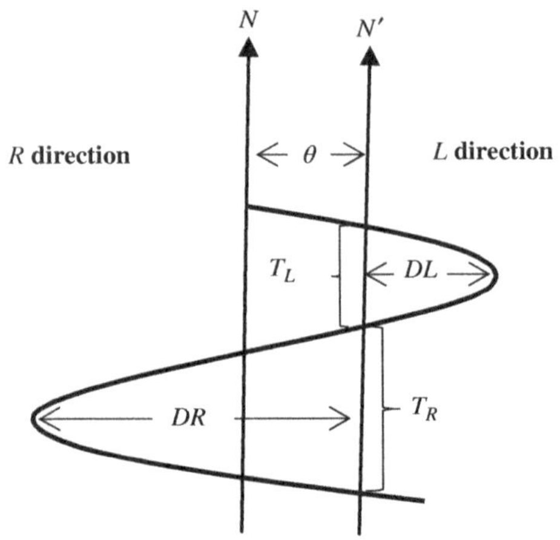

Figure 12.11 Time method of gyro azimuth determination.

V-shaped index of the gyro unit. The auxiliary scale, which can be seen through the gyro eyepiece, is used to estimate the values of the oscillation amplitude of the gyro mark to the L and R directions in the gyro unit. The correction (θ) to be applied to N' in order to obtain the corrected north $N = N' + \theta$ is given in Equation (12.24):

$$\theta = -(k \times D \times \Delta t + E) \tag{12.24}$$

where k (given for some GP-1 as 3.452) is the instrumental constant or instrument's proportionality factor (also known as torque ratio constant) for converting measured times into angular equivalent values; E (given for some GP-1 as $-10''$) is the alignment constant for transferring direction from the zero graduation of gyro to optical axis of the total station equipment;

$$D = \frac{(DR + DL)}{2}$$

Δt is the simple average of all the time differences ($\Delta t_i = (T_R - T_L)_i$) between successive zero graduation transits as shown in Figure 12.11.

The constant k in Equation (12.24) is dependent on the latitude and should be determined for the gyro instrument whenever the suspension tape of the gyro has been replaced or whenever the measuring location is significantly different in latitude from the original location of the instrument calibration. The determination of k and E is done automatically and simultaneously in GP-1 gyro station using the gyro station program. For the sake of explaining the underlying concepts involved in determining k, the procedures for determining k empirically for GP-1 gyro station are discussed as follows:

1. Make multiple gyro measurements using the follow-up method and determine the north direction reading (N_1) based on the Schuler Mean. The GP-1 will

automatically determine and display N_1 as the azimuth (AZ) of the current telescope direction. Clamp the gyro and rotate the telescope until AZ reading becomes zero, which is the direction of the north.

2. Rotate the telescope of the instrument horizontally by 10′ to the right of the determined north in step 1; unclamp the gyro and perform gyro measurements using Time method. Calculate the average (D_1) of the DR and DL amplitude values and the average time difference (Δt_1) between the successive zero graduation transits of the gyro mark, and record the azimuth of the current direction of the telescope as $N_1' = 10'$.
3. Rotate again the telescope of the instrument horizontally by 10′ to the left of the determined north in step 1 (i.e., by 20′ to the left of the current direction of the telescope); unclamp the gyro and perform another gyro measurements using Time method. Calculate again the average (D_2) of the DR and DL amplitude values and the average time difference (Δt_2) between the successive zero graduation transits of gyro mark, and record the azimuth of the current direction of the telescope as $N_2' = -10'$.
4. From the two sets of measurements made with symmetric orientations (N_1' and N_2') about the middle oscillation position of the determined north, calculate k as follows:

$$k = \frac{N_2' - N_1'}{D_1 \Delta t_1 - D_2 \Delta t_2} \tag{12.25}$$

12.3.4.4 Azimuth Determination with the Gyro Station GP3X Equipment The steps for setting up the gyro station GP3X are illustrated in Figure 12.12 and explained as follows (refer also to SOKKIA, 2004):

1. Set up the Set3X total station on the tripod and level: On the left of the display panel (shown in Figure 12.9), press the **On** button, and then the **Settings** button (if required, press **ESC** until program is exited); use the stylus pen to select **Tilt** tab and level the total station electronically.
2. Connect the gyro inverter to the gyro and to the accompanying DC 12 V battery and set the gyro on the Set3X total station as shown in Figure 12.12(a); on the left of the display panel, press **Program** button and then tap on the display with the stylus pen to select **Gyro Station**; orient the Gyro approximately (using Magnetic Compass) to the north after confirming that the instrument/total station is leveled; tap on OSET function to set the horizontal angle reading (HAR) of total station to zero if needed; the instrument will now be ready for azimuth measurement by displaying AZ text box; press again the **Settings** button on the left display panel and tap **Tilt** tab to confirm that the instrument is still leveled.
3. **Preliminary checks on the gyro:** Remove the protective cover (or clamp lock) from the gyro clamp ring as shown in Figure 12.12(c) and do the following:

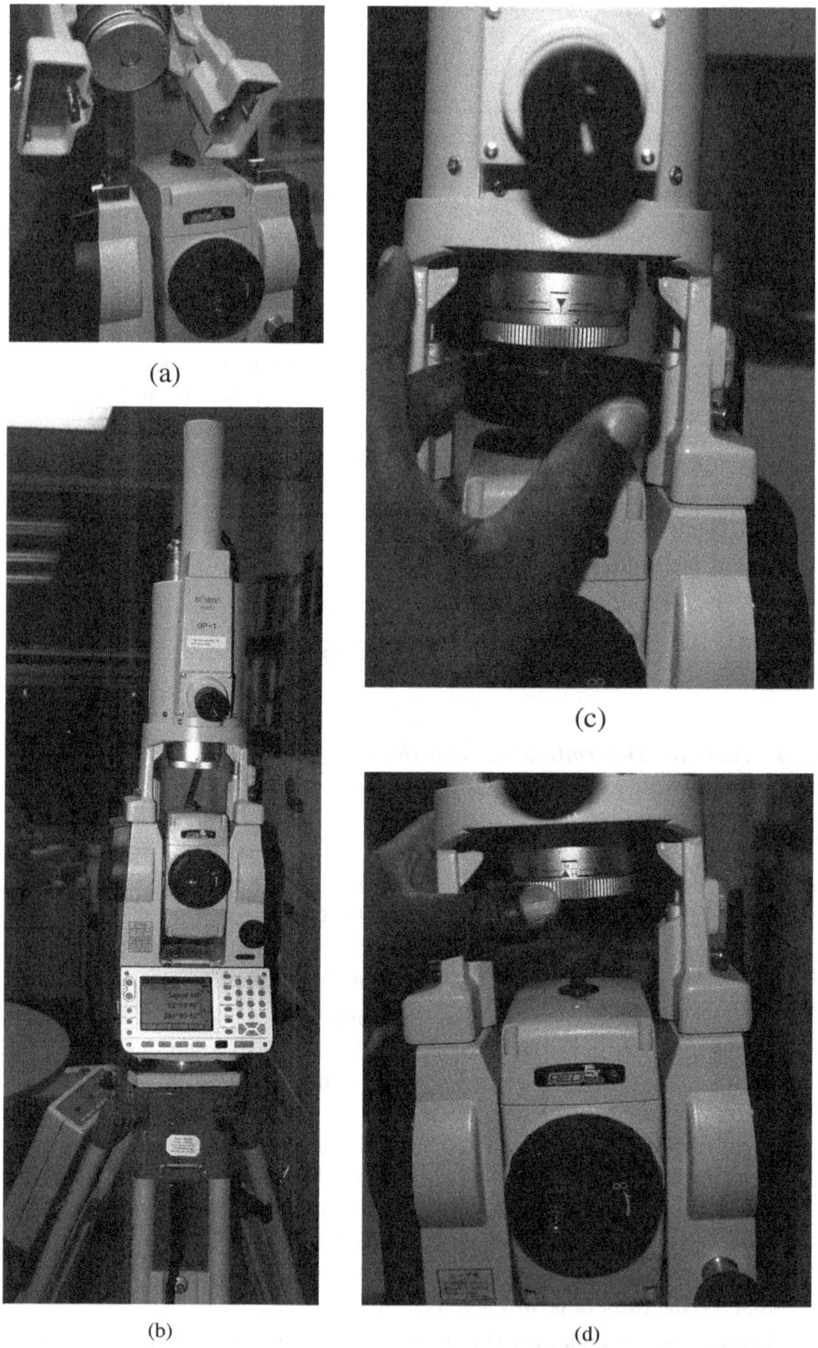

Figure 12.12 Setup procedure of the GP3X Gyro station.

(i) While the GP-1 gyro power is still off, turn the gyro clamp ring slowly to HALF-CLAMP (HC) as shown in Figure 12.12(d); wait for about 10 s, checking that the floating index mark (or gyro mark) is not moving; then slowly continue to turn the clamp to FREE (F or fully unclamped) position.

(ii) At this time, the oscillation of the floating gyro mark should be symmetrical about the zero graduation mark within 1.0 scale division (remember that the GP-1 gyro power is still off at this time); if this is not the case, there will be a need for the gyro adjustment.

(iii) Turn the clamp screw back in the *C* direction until the gyro is in the FULL CLAMP (FC) position once again.

4. Turn the GP-1 gyro power switch on the inverter to **On** (and wait for about 1 min until the motor start lamp on the inverter is lit GREEN) with the accompanying loud humming sound and then do the following:

   (i) Turn the gyro clamping screw slowly to HALF-CLAMP (HC) as shown in Figure 12.12(d); wait for about 10 s, checking that the index mark is not moving, slowly continue to turn the clamp to FREE (F or fully unclamped) position. Note that when fully clamped, the gyro makes a humming noise and the gyro mark is stationary, but when it is fully unclamped, the humming noise stops and the gyro mark oscillates freely.

   (ii) As a warning, the gyro must be fully clamped before it is given any jerky rotation or whenever the slow motion screw of the instrument is not being used. This is to avoid breaking the wire that supports the gyro.

5. The gyro measurement procedure is illustrated in Figure 12.13. In the case of follow-up method, press FOL or F1 key on the gyro station screen and do the following (as shown in Figure 12.13(a) and (b)):

   (i) Use the slow motion screw of the total station equipment or gently turn the telescope (if the slow motion is out of thread) to follow the gyro mark, keeping it on the zero graduation (or within the V shape) of the gyro (as shown in Figure 12.10). Continue to keep the floating mark in the V shape until a reversal point is reached and the gyro mark is momentarily stationary and about to move in the opposite direction; at this reversal point, press **[REV.P]** or **F3** key (shown in Figure 12.13(a)).

   (ii) Press **[REV.P]** key again whenever a reversal point is reached, and continue this procedure until sufficient number of reversal points are taken; two reversal points are sufficient for approximate location of the north direction.

   (iii) More reversal point readings can be taken for better determination of the north direction; when the required number of reversal points readings have been taken, press [OK] button (shown in Figure 12.13(b)) for the gyro to use those readings to determine the precise direction of the north and the azimuth angle (AZ) of the current line of sight of your telescope; the AZ value and the HAR of the current direction of the telescope are displayed in the gyro station panel. Pressing the [OK] button at this time will end the follow-up measurements; compute the azimuth of the current direction of the telescope with respect to the computed true north position and exit the gyro program into the azimuth display mode.

(iv) With the gyro station in azimuth display mode, clamp the gyro all the way to FULLY CLAMPED (FC) position – no need of half-clamping the clamping screw this time.

(v) Use the slow motion screw or turn the telescope if the slow motion screw is out of thread until zero reading is displayed for AZ. At this point, the telescope is pointing in the gyro determined north direction. The telescope of the instrument can be clamped in this direction for use in the Time method procedures.

(vi) Unclamp the gyro again and turn the clamping ring until HC and F positions are reached as discussed in step 4 and do a more precise **Time** method to refine the direction of the north determined in step v.

6. To start the Time (or Transit) method, press [**TIME**] function on the Gyro Station screen; at this time, the turning of the telescope from the direction it is currently pointing is not allowed. Tap on the **EPOCH** or **F3** key (shown in Figure 12.13(c)) on the display panel and do the following:

   (i) At this time, the display unit should be displaying the input boxes for *DR* and *DL* as shown in Figure 12.13(d), requiring that the number of graduations moved by the gyro mark to the reversal points in the *R* and

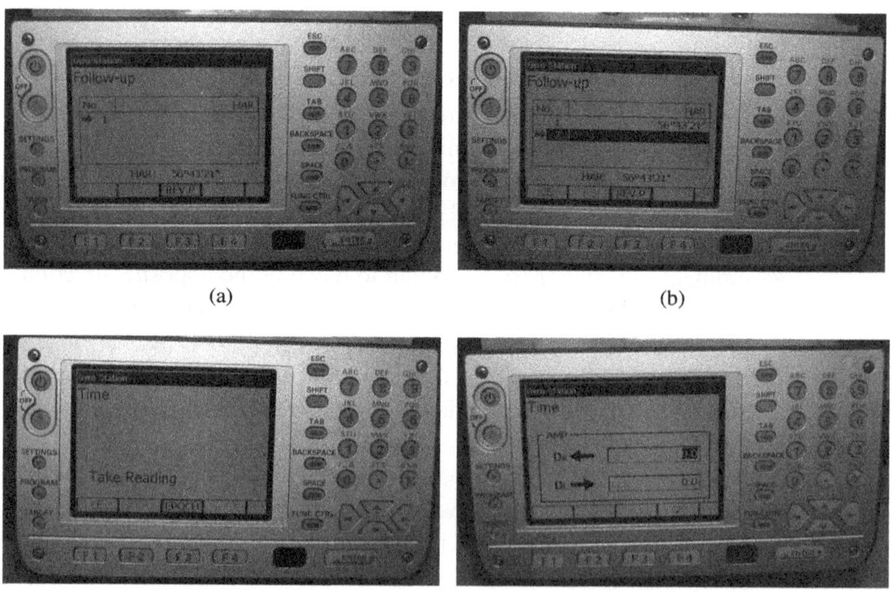

Figure 12.13 Sample display for the follow-up and Time methods of gyro measurements.

L directions be input into the *DR* and *DL* text boxes, respectively. Input the amplitude values observed for *DR* and *DL* into the corresponding boxes and press **[OK]** button to accept them; press the **[EPOCH]** key when the gyro mark just transits the zero graduation (or the V shape) to start the gyro measurements, and then click the corresponding arrow key on the keyboard to match the direction in which the gyro mark is heading at that time when required by the gyro.

(ii) When the **[EPOCH]** key is pressed again at the subsequent transits of the zero graduation, the time taken by the gyro mark to travel to the reversal points and back to the zero graduation point will be displayed on the display panel.

(iii) After two consecutive transits of the V shape by the gyro mark, the time for half-cycle of the transit is displayed (in seconds); after a complete cycle (two half-cycles) is made, the azimuth of the telescope line of sight (AZ value) is displayed; averages are provided after two or more azimuth values have been determined; any unwanted azimuth value from the averages can be excluded by deleting the value by first highlighting it and then pressing **[CE]** key to remove it.

(iv) After obtaining sufficient transit readings, click **[OK]** key to determine final azimuth (the averaged value) and display it in AZ box. Note that the AZ value displayed may be unrelated to the value displayed in HAR. The HAR can be displayed as a clockwise/counterclockwise value depending on the settings in the total station; but the AZ is always displayed as clockwise value from the north direction. Pressing the **[OK]** key at any time will end the time measurements; compute the azimuth of the current direction of telescope with respect to the true north position and exit the gyro program into the azimuth display mode.

(v) After AZ is displayed; press **[N]** key on the total station to transfer the measured azimuth angle (the angle with respect to the calculated true north) to the HAR box. The azimuth angle of the telescope direction will now be displayed in the HAR box.

(vi) In determining the azimuth of a line to the reference object (RO), the gyro is first clamped fully; then the total station is rotated on to the line for which the azimuth is desired and the azimuth of that line is recorded on FL and FR positions of the telescope, and the average value is taken as the azimuth of the line.

(vii) To shut down the gyro after clamping, switch off the power on the inverter; wait for approximately 10 min for the motor to come to a complete standstill; check that no sound is coming from the motor, and then put the clamp lock back on the clamping screw.

Example 12.5 Given the following sample data (Figure 12.14) taken with GP3X gyro station with the approximate direction of the north (N') as read on the total

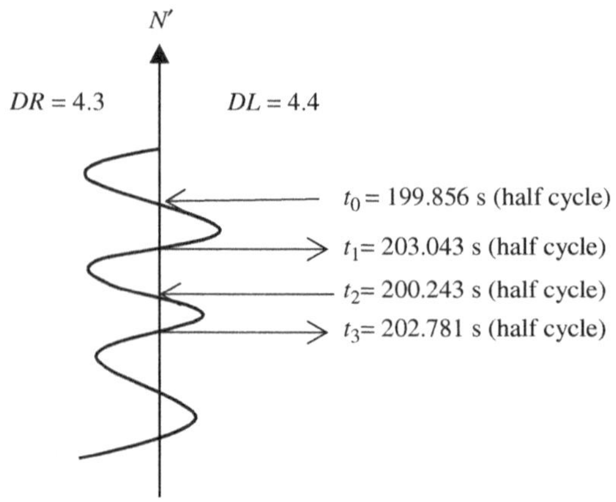

Figure 12.14 Sample gyro data by Time method.

station being 0°0′0″, determine the azimuth correction θ (using the Time method approach) and the corrected direction of the north (N). Take $k = 3.452$ and $E = -10''$ for the gyro station.

Using Equation (12.24):

$$\theta = -(k \times D \times \Delta t + E) \tag{12.26}$$

$$\Delta t_1 = t_0 - t_1 \quad \Delta t_1 = 199.856 - 203.043 \rightarrow -3.187\,\text{s}$$

$$\Delta t_2 = t_2 - t_1 \quad \Delta t_2 = 200.243 - 203.043 \rightarrow -2.800\,\text{s}$$

$$\Delta t_3 = t_2 - t_3 \quad \Delta t_3 = 200.243 - 202.781 \rightarrow -2.538\,\text{s}$$

Each Δt above can be used in Equation (12.26) in order to obtain individual azimuth corrections and then averaging to obtain an average value; or finding the average ($\Delta\bar{t}$) of all the Δt's as

$$\Delta\bar{t} = \frac{\Delta t_1 + \Delta t_2 + \Delta t_3}{3} \quad (\text{or } \Delta\bar{t} = -2.8416\,\text{s}) \tag{12.27}$$

Given $k = 3.452$ and $E = -10''$ for the GP-1 and substituting the values into Equation (12.26) gives $\theta = 53''$. This means that the current line of sight through the telescope is at an angle 0°00′53″ clockwise (negative sign is counterclockwise) with respect to the direction of the north, that is, azimuth (AZ) of the current direction of the telescope is 0°00′53″.

12.3.4.5 Use of Gyro Equipment in Underground Mines The usual steps in orientation transfer with gyrotheodolites in underground mines are as follows:

1. First, gyrotheodolite equipment is calibrated on the surface on a baseline whose azimuth is already known, before taking the equipment underground to the place where the azimuth is to be determined; the correction or calibration factor (E) to be applied to subsequent azimuth determinations underground with the equipment is determined using Equation (12.23). It is important that the calibration be done within 60–90 m east or west of the point where it is to be used if calculations for the convergence of the meridians are to be avoided. In this case, if the underground workings are within 60–90 m of the gyro calibration site, one can still use the same convergence of meridian determined for the calibration site in the underground workings.
2. At the underground setup point (usually a permanent point), the gyrotheodolite is centered over the point and carefully leveled. The surveyor orients the gyrotheodolite to the north direction and then measures the direct and reversed angles to the reference point (which may be several hundred meters away). The surveyor may likely repeat the operation at the other end (reference point) back to the initial setup point as a check and may probably use the average of forward and back azimuths, thereby minimizing possible refraction effects on the computed average azimuth.
3. Coordinates can be brought down into the mine to the new level using single wire, whose position is determined from the surface.

Gyrotheodolites provide gyro azimuths that are basically astronomical azimuth (instead of grid or plane azimuths). Some of the corrections that are usually applied to gyro azimuths can be given as follows:

1. Convergence of meridians (γ), which depend on the type of map projection used in obtaining the grid coordinates. As one proceeds along a straight line set out by a theodolite on the surface of the earth, the azimuth of the line will not remain constant. Gyro settles along a meridian (true north), which will only coincide with the map grid along the middle meridian of the map grid. The farther east or west one gets from the middle meridian, the larger the deviation between the direction of north of the map grid and the meridian of longitude that the gyro shows. The convergence of meridian can be computed approximately for a position with mean latitude (ϕ) and longitude (λ) by using the following formulas:

$$\gamma = (\lambda - \lambda_0)\sin\phi \tag{12.28}$$

or

$$\gamma = \frac{\Delta E \tan\phi}{R} \times 206{,}265(\text{arcsec}) \tag{12.29}$$

where λ_0 is the longitude of the central meridian (the origin of the map grid rectangular coordinate system), ΔE is the difference in easting coordinate (distance between the meridians) of the observing station, and R is the radius of the earth in the project site (e.g., 6370 km). A more precise formula for computing the convergence of meridians can be given as

$$\gamma = \frac{\Delta E \tan \phi (1 - e^2 \sin^2 \phi)^{1/2}}{a} \times 206,265(\text{arcsec}) \tag{12.30}$$

where e is the first eccentricity and a is the semi-major axis of the reference ellipsoid. To obtain the grid azimuth, γ should be subtracted from the gyro azimuths value when the underground station is located east of the surface station. The grid azimuth can be determined from Equation (12.21) or (12.22).

2. Error due to the misleveling of instrument, which can be expressed mathematically as

$$e_m = \sigma_v \cot Z \tag{12.31}$$

where σ_v is the inclination of the instrument in the direction perpendicular to the line of sight and Z is the zenith angle reading. This error is random in nature; the effect can be minimized by releveling the instrument between sets and finding the average of the sets, or using a more sensitive leveling bubble (like striding level) to determine the misleveling corrections to be applied to the measurements.

3. Effect of deflection of the vertical, which will affect direction and angle measurements in a similar way as misleveling of the instrument. The correction (due to this effect) to be applied to the gyro azimuth can be expressed as

$$c_d = \eta \tan \phi + (\xi \sin \alpha - \eta \cos \alpha) \cot Z \tag{12.32}$$

where ξ is the component of deflection of the vertical in the north–south direction at the setup point; η is the component of deflection of the vertical in the east–west direction at the setup point; ϕ is latitude of the setup point, α is the geodetic azimuth to the reference object, and Z is the zenith angle reading. This correction will be significant only in a case where the deflection of the vertical is large and the line of sight is inclined. In the tunnel where lines of sight are approximately horizontal, $\cot Z = 0$, so that $c_d = \eta \tan \phi$ with only the component of the deflection of the vertical in the east–west direction (η) accounting for the correction. The application of the correction due to the deflection of the vertical allows an astronomic azimuth to be converted into geodetic azimuth. This effect has both systematic and random components.

4. Effects of refraction, which will be reduced if reciprocal observations are made with the gyrotheodolites on the same traverse lines. These effects have both systematic and random components as discussed in Sections 4.3.4 and 4.5.5.

5. Effect of local calibration value (E) of the gyrotheodolite. This is an alignment error between the gyro zero (the indicated heading of the gyro) and the horizontal optical axis of the theodolite. It is advisable to set up the gyrotheodolite on a known baseline on the surface to establish the difference E between the gyro azimuth and the actual azimuth of the baseline on the surface. The

gyro must again be set up on the surface baseline and a second determination of E made after the completion of the underground surveys. Any change in E would have to be applied to the measured azimuths proportionately with respect to the time of observation. The value of E can be determined from Equation (12.23).

12.4 TRANSFERRING LEVELS OR HEIGHTS UNDERGROUND

Transfer of vertical control points and alignments from ground surface down to underground tunnels depends on the configuration of the access. For transfer through inclined shafts, differential leveling will be more appropriate; for vertical shafts, vertical EDM and precise tape are commonly used. Remember that horizontal and vertical control points are usually set in the back of an underground mine and both the vertical and the horizontal control points are generally established at the same time. The control points are established in the reverse order (low order first followed by higher order) from what is done for surface surveys. Low-order traverses usually have short legs (less than 50 m) and higher order ones usually have longer legs (up to 1000 m). Several methods are used in transferring levels underground, such as using EDM instrument adapted for vertical viewing in the shaft and using very long tapes with marked divisions (calibrated and corrected for tension and temperature). In both methods, the connecting survey between the benchmarks and the EDM/reflector centers, or rulers with 1 mm divisions attached to the tape, are made by means of spirit or trigonometric leveling.

12.4.1 Height Transfer with EDM

Using EDM to transfer heights requires careful determination of the centers of the instrument and of the reflector; there is also a need for visibility condition in the shaft in order to use nonlaser EDM instrument for height transfer. One important advantage of EDM method is that it can provide more accurate result and it can also provide automatic and instant readout unlike in the case of tape method. Direct measurement with EDM using infrared or laser models, however, is preferable to that based on nonlaser EDM provided the infrared or the laser signal will reach the target and be reflected back.

In the method of height transfer with EDM, either of the following approaches may be adopted:

- The instrument and reflector are kept in the upright positions with the mirrors or right-angled prisms used on the surface and underground to redirect the EDM signal as shown in Figure 12.15. In this method, distances from the EDM instrument and reflector to respective mirrors are measured, and heights of instrument and reflector are determined in order to complete the height transfer underground. Referring to Figure 12.15, the elevation of surface benchmark (BM) is

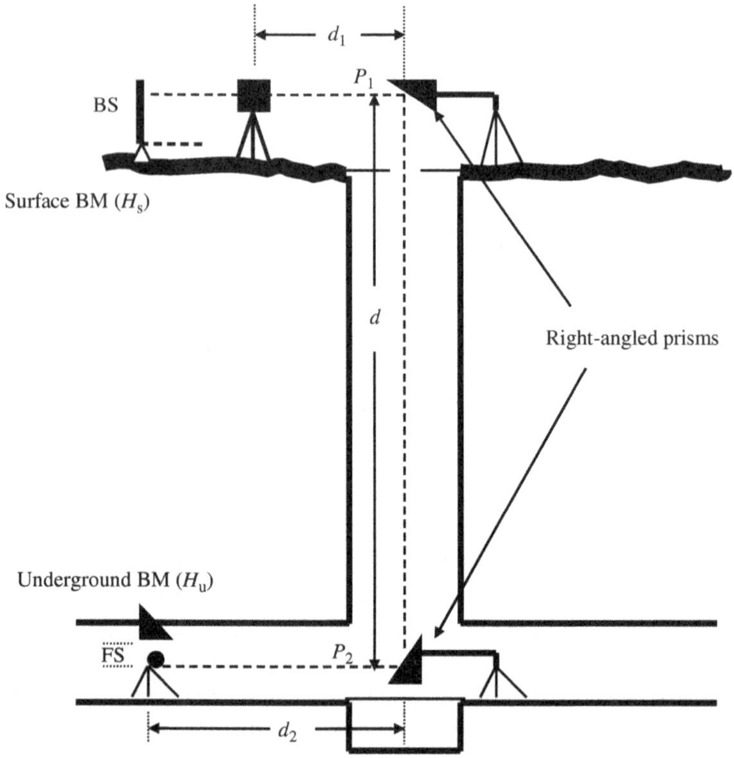

Figure 12.15 EDM approach for transferring heights underground (cross-sectional view).

H_s; backsight reading to the surface BM is BS; foresight reading to the underground BM is FS; the measured round distance measurement by the EDM is d_m; and the measured distances to the right-angled prisms are d_1 and d_2; the elevation of the underground BM (H_u) can be given as

$$H_u = H_s + \text{BS}–d_m + d_1 + d_2 + \text{FS} \quad \text{given that } d_m = d_1 + d + d_2 \tag{12.33}$$

- Alternatively, the instrument and the corresponding reflector can be clamped in a vertical position in the shaft, at the surface and underground. Then the elevations are transferred to and from the instrument center and the reflector, using trigonometric or spirit leveling method. In this case, the EDM is supported at the top face down and the reflector integrated with the level instrument is located directly below the surface plumbline underground. From this, the elevation of the underground benchmark (BM), which is on the back of the tunnel, can be determined. For example, a Taylor Hobson sphere containing the reflector can be supported in a special bracket mount at point P_1 and the specially designed EDM located at point P_2 (Figure 12.15). In this case, the elevation of the horizontal axis of the EDM instrument is transferred to the underground BM.

12.4.2 Height Transfer with Measuring Tape

Elevations can be transferred from the surface level underground using steel tape in a vertical shaft as shown in Figure 12.16. In this method (Figure 12.16), the level instrument is set at point *A* on the surface and by differential leveling procedure, the level of P_1 of the tape is determined using the known benchmark BM_1. From the length of the tape (P_1–P_2) and the foresight reading on BM_2, the elevation of BM_2 is established. This method, however, requires that simultaneous readings be taken on the tape by one crew with spirit level at the surface and the other with spirit level underground. Note also that the tape is kept vertical by means of the heavy weight at the end of the tape; this verticality may be affected if the weight is not heavy enough. Some of the disadvantages of using long tape for direct measurement include the following:

- High cost of a tape that has little further use after the initial survey. Indirect measurements in the shaft using piano wire may be helpful since the piano wire used for transferring coordinates and direction may now be used to transfer elevations.

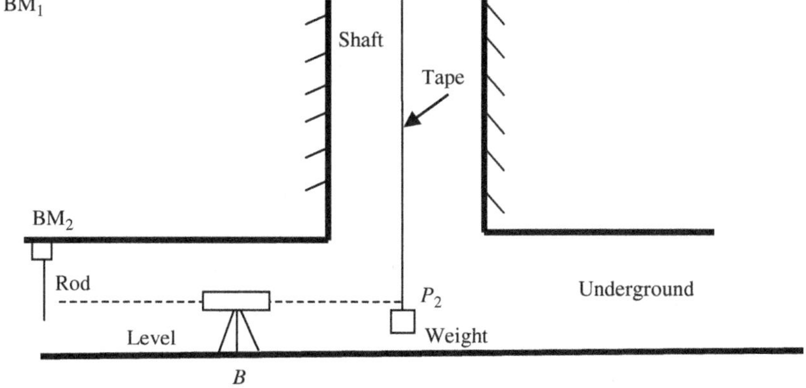

Figure 12.16 Transferring heights underground using measuring tape (cross-sectional view).

- Need to make complex correction for tape elongation. Some of the required corrections are discussed in Section 12.4.4.

One of the advantages of using steel tape method is that the steel-tape measurements may be faster and accurate for short distances.

12.4.3 Height Transfer in Shallow Shafts

Height transfer approach discussed in this section is a continuation of horizontal control transfer through a shallow shaft (Section 12.3.4); usually, horizontal and vertical control transfers are done about the same time. In this case, elevations are transferred from the surface level through the shallow shafts (about 20–80 m deep) using optical or zenith plummets. Heights in this case are transferred based on the following techniques:

1. At the top of the shaft, set up a Taylor Hobson sphere on a survey bracket included in the shaft collar; the center of the Taylor Hobson sphere will then be defined in three dimensions.
2. Set up a tripod at the bottom of the shaft with translation stage that will allow plumbing to be performed with good accuracy as follows:
   - Mount precision zenith plummet (such as Wild/Leica ZL plummet) on the Kern tripod with centering rod and translation stage for centering under the spherical target on the surface.
3. After completing the plumbing operation underground, replace the Taylor Hobson sphere on the surface with precise prism inserts (reflectors) to serve as retro-reflectors. The Taylor Hobson sphere can be fitted with reflective prisms with no eccentricity introduced, no matter what direction the prism is facing.
4. In the underground, the zenith plummet is removed and replaced by a coaxial precision total station with the telescope pointed vertically to the prism located in the Taylor Hobson sphere on the surface.
5. Observe vertical distances in at least three sets with independent re-pointings between the sets by using the total station instrument underground; correct the vertical distances for prism calibration and atmospheric effects.
6. On the surface, use a suitable level instrument set up between the bracket-mounted Taylor Hobson sphere (vertically above the total station underground) and a leveling rod located on a benchmark to precisely transfer vertical control to the center of the Taylor Hobson sphere.
7. Using another level instrument set up underground and a leveling rod, transfer the elevation from the horizontal axis of the total station to several wall markers serving as benchmarks.

12.4.4 Typical Corrections Applied to Measurements in Height Transfer

Generally, the accuracy of level transfer underground will depend on the following:

- Accuracy of transferring level from the surface benchmarks to the centers of the bracket-mounted target (or a suspended tape) at the top of the shaft.
- Accuracy of measuring vertical distance between the bottom and top of the shaft.
- Accuracy of transferring elevation from the horizontal axis of the instrument set underground (or from the suspended tape at the underground level) to the nearby underground benchmarks.

With regard to tape measurements in height transfer, a number of corrections must be applied, such as the following:

(a) *Correction for the standardization of tape* (Δh_d). Standardization of a tape is a process of determining the standard temperature and tension corresponding to the exact length of the tape. If the tape is used at any other temperature and tension apart from the standard values, the measurements made with the tape must be corrected for standardization error. This correction, which is to make the tape length equivalent to the standard length, is applied in a manner similar to when using the tape in horizontal measurements.

(b) *Correction due to temperature variation* (Δh_t). This correction is applied to eliminate the effect of temperature variations in the shaft. According to Chrzanowski and Robinson (1981), the temperature variations in the mining shafts are usually nonlinear unlike in the case of when a tape is used horizontally for distance measurements on the surface; they suggested that temperature T_i be measured at different levels (i) of about 3050 m in the shaft as part of height transfer procedure through the shaft. The weighted mean temperature T is then used to determine the correction as follows:

$$\Delta h_t = h\alpha(T - T_0) \tag{12.34}$$

where T is estimated (Chrzanowski and Robinson, 1981) as

$$T = \frac{1}{(h_n - h_1)} \sum_{i=1}^{n} \left[\frac{(T_i + T_{i+1})}{2} (h_{i+1} - h_i) \right] \tag{12.35}$$

where T_i is the temperature measured at any given level with the length of the tape at that level being h_i; h_1 is the height of the first level; h_n is the height of the last level; T_0 is the temperature at which the tape was standardized, α is the thermal coefficient of expansion (e.g., 11.6×10^{-6} per 1 °C for steel), and h is the total length of tape measurement. It can be understood that Equation (12.34) is similar to the one that is generally used in elementary surveying in correcting for temperature changes in horizontal tape measurement. The major difference is that T is nonlinear in a vertical shaft and its value must be calculated from Equation (12.35) when the tape is used vertically in the shaft; in the case of horizontal tape measurement, T is considered constant along the whole length of the tape.

(c) *Correction due to the tape stretching under its weight* (Δh_w). The stretch correction is calculated from (Chrzanowski and Robinson, 1981) as

$$\Delta h_w = \frac{w}{aE}\left(Lh - \frac{h^2}{2}\right) \tag{12.36}$$

where h is the height measured from the top of the shaft to the point where measurement is made underground, L is the total length of tape (or measured length), w is the weight of tape per unit length, a is the cross-sectional area of the tape (cm^2 or in.^{2}), and E is the modulus of elasticity for the tape material (in kg/m/s^2).

(d) *Tension correction* (Δh_p). This is necessary if the pull on the tape is different from that used when standardizing the tape. The tape will either be shortened or lengthened by the amount, which can be given as

$$\Delta h_p = \frac{(P - P_0)L}{aE} \tag{12.37}$$

where P is the pull (kg) on the tape, P_0 is the pull (kg) on the tape when standardizing it, and other symbols are as defined in (c).

(e) *Other corrections*, such as tape not being straight (due to air current and spiral shape of tape). The combined effect of air current and spiral shape of tape may be compared to the effect of sag on horizontal distance measurement with the tape.

Among the corrections needed to be applied to tape measurements in height transfer, the temperature corrections and stretch of tape under its own weight will be different from when the tape is used to measure horizontal distances. In the shaft where height transfer is being made, temperature usually varies nonlinearly along the shaft, requiring that more complex temperature correction to tape measurements be made in height transfer; and since the tape is in the vertical position, the stretch of the tape under its own weight needs to be applied, which is not done in the case of horizontal distance measurement with tapes (sag correction and the effect of tape not being horizontal are applied instead).

Example 12.6 The elevation of the back of a drift (or tunnel) has been determined via a connecting shaft using a steel tape and levels as shown in Figure 12.16. The following data is known from the control information and measurements:

- Elevation of the surface benchmark BM_1: 426.97 m
- Rod reading at BM_1: 1.55 m
- Height of instrument (underground) relative to the benchmark BM_2 in the back of the drift: −0.92 m
- Taped distance between the surface point P_1 and the underground P_2: 45.72 m.

What is the elevation of the new control point in the drift?

Solution: Using the idea behind the formulation of Equation (12.33), the elevation of the new control point in the drift can be given as follows:

$$H_u = H_s + \text{BS}–d + \text{FS}$$

$$H_u = 426.97 + 1.55 - 45.72 + 0.92 \rightarrow 383.72\,\text{m}$$

Example 12.7 A mine orientation survey is to be done using two mechanical plumblines in one vertical shaft. The depth of the oriented level $H = 300$ m. The distance between the two plumblines is 4 m. Steel wires of tensile strength 200 kg/mm^2 are available for plumbing. The height of the shaft opening to the oriented level $h = 5$ m and the average air velocity in the cross section of the opening $v = 1$ m/s. There are no other openings to intermediate levels between the surface and the oriented level. Answer the following:

a. What diameter (d) of the plumb wires and what weight (p) of the plumb bobs would you use for the orientation purpose?

Solution: As a rule, weight of the bob is usually equal to $H/3$ in kilograms (where H is the depth of plumbing in meters); using a wire with a tensile strength of 200 kg/mm^2 to a depth of $H = 300$ m, the expected weight of bob will be $p = 300/3$ or 100 kg.

For safety reasons, the load should not exceed half of the maximum (breaking) load of the wire. Since the expected load is 100 kg, the maximum load expected is twice the expected load or 200 kg.

The cross-sectional area of the wire can be given as

$$\text{Cross-sectional area} = \frac{\text{maximum load}}{\text{tensile strength}}$$

For the wire with tensile strength of 200 kg/mm^2, the cross-sectional area is

$$200\,\text{kg}/(200\,\text{kg/mm}^2) \text{ or } 1\,\text{mm}^2$$

The radius of the wire can be determined from the area of a circle equation:

$$\pi r^2 = 1\,\text{mm}^2 \text{ or } r = \sqrt{\frac{1}{3.1416}} = 0.564\,\text{mm}$$

The diameter, $d = 1.128$ mm.

b. What error of the transferred azimuth would you expect as a result of the air current and spiral shape of the wires (take the radius of the spiral shape $R = 15$ cm and use values of d and p as obtained from part (a)).

Solution: Error due to air influence on plumbline can be given (Chrzanowski and Robinson, 1981) as

$$e = \frac{30h(H)(d)v^2}{p} \text{ mm} \tag{12.38}$$

where $v = 1$ m/s is the velocity of air.

Substituting values into Equation (12.38) gives

$$e = \frac{30 \times 5\,\text{m} \times 300\,\text{m} \times 1.128 \times 10^{-3}\,\text{m} \times (1\,\text{m/s})^2}{100\,\text{kg}} \text{ mm} \quad (\text{or } 0.508\,\text{mm})$$

Only half of the air influence is taken as part of the estimated standard deviation because both plumblines are most probably deflected in a similar direction; this can be expressed as follows:

$$e_{\text{air current}} = 0.5e \quad (\text{or } 0.25\,\text{mm}) \tag{12.39}$$

The error due to spiral shape of plumbline can be given for the two extreme positions of each wire (Chrzanowski and Robinson, 1981) as

$$e_{\text{spiral shape}} = 2 \times r_{\text{spiral shape}} \tag{12.40}$$

where

$$r_{\text{spiral shape}} = \frac{\pi d^4 E}{64Rp} \tag{12.41}$$

Using $\pi = 3.14$, $d = 1.128$ mm, $E = 2.1 \times 10^4$ kg/mm^2, $R = 150$ mm, $p = 100$ kg in Equation (12.41) gives $r_{\text{spiral shape}} = 0.11125$ mm; from Equation (12.40), $e_{\text{spiral shape}} = 0.222$ mm.

Error of the mean positions determined on the scales can be kept smaller than 0.2 mm if the plane of the oscillations of the plumb bobs is parallel within ±10° to the scale, the amplitude is smaller than 10 cm, and if at least 10 readings (with an estimation to 0.2 mm) of the left and right reversal positions are taken on the scale for the calculation of the mean position of the plumbline. Based on the above assumption, $e_{\text{scale}} = 0.2$ mm. The total standard deviation of the azimuth transfer based on the distance between the two wires being $b = 4.000$ m can be given (Chrzanowski and Robinson, 1981) for the two plumblines as

$$\varepsilon_A = \frac{206,265}{b}\sqrt{\left[e^2_{\text{air current}} + e^2_{\text{spiral shape}} + e^2_{\text{scale}}\right] \times 2} \tag{12.42}$$

Substituting the appropriate values into Equation (12.42) gives the total standard deviation of the azimuth transfer as 28.4″.

12.5 VOLUME DETERMINATION IN MINES

Apart from the mine surveying activities discussed in the earlier sections, another important activity usually performed by the mine surveyor is volume determination. For example, during a tunnel construction, as-built surveys are required to check tolerances of tunnel structures. Surveys are also carried out in completed tunnels to check if sufficient clearances are available for the installation of pipelines, lighting, ventilation, and so on. The surveys are to provide a record of existing structures and the as-constructed condition of the tunnel. As-built surveys of a tunnel should be implemented in two steps as follows:

(i) Survey the finished tunnel before and after the breakthrough.

(ii) Check if the existing tunnels have been built to within allowable tolerances, and if the design tolerances are exceeded, to see if it is possible to realign the tunnel without remedial work to the existing structures. In this case, cross sections at regular intervals along the whole length of the tunnel are surveyed.

Accurate and cost-effective surveying method applied by surveyors in checking profiles is using reflectorless total stations by which coordinated points on the tunnel surface are automatically recorded, processed, and analyzed on computers in the field. All of the field data are stored electronically in the form of three-dimensional coordinates in the total stations, and the accuracy between 5 and 10 mm is expected. Based on the cross-section database, volume of excavation and materials are computed.

Whether underground or on the earth surface, mining involves moving volumes of material from one place to another. Mapping the changes made by the mining activity and determining the volumes moved is a daily job of mining surveyors. Nowadays, for underground and open-pit operations, laser scanning systems, reflectorless total station equipment, GPS surveys, and terrestrial photogrammetry are used for daily mine volume determinations. One approach for volume determination for underground operations can be described as follows:

1. Observe cross-section profiles at measured intervals (with distance between each cross section being d_{12}, d_{23}, etc.).
2. Determine the cross-sectional areas using coordinate method based on local x–y coordinate system established for each cross section. A typical cross section with the local x–y coordinate system is shown in Figure 12.17. The area (A) of this cross section can be given using coordinate approach as

$$A = \frac{1}{2}[(x_1y_2 + x_2y_3 + \cdots + x_{n-1}y_n + x_ny_1) - (x_1y_n + x_ny_{n-1} + \cdots + x_3y_2 + x_2y_1)] \quad (12.43)$$

3. Compute volumes between cross sections as shown in Figure 12.18 using average end-area method:

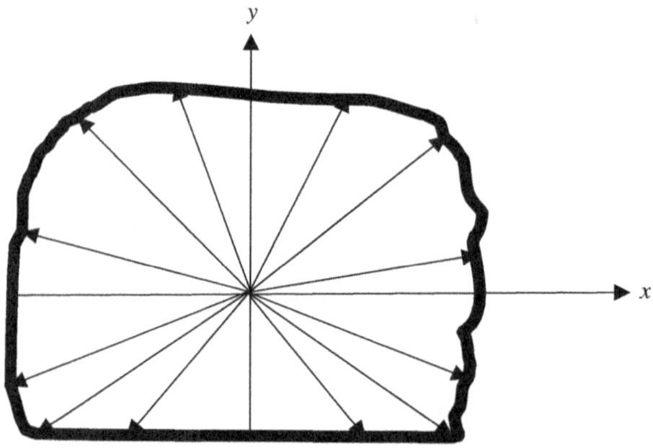

Figure 12.17 Single cross-section profile of an underground excavation.

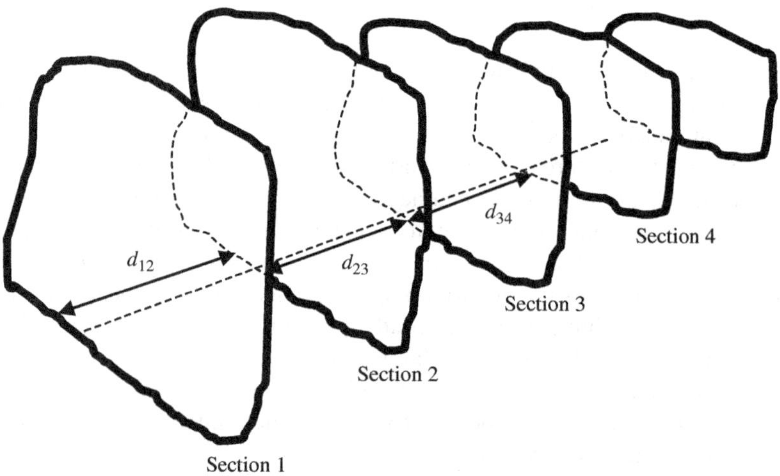

Figure 12.18 Different cross sections of mining excavations for volume determination.

The individual cross-section/end-area volumes are given as follows:

$$V_{12} = d_{12}\left(\frac{A_1 + A_2}{2}\right) \tag{12.44}$$

$$V_{23} = d_{23}\left(\frac{A_2 + A_3}{2}\right) \tag{12.45}$$

$$V_{34} = d_{34}\left(\frac{A_3 + A_4}{2}\right) \tag{12.46}$$

where V_{12}, V_{23}, V_{34} are the individual cross-section/end-area volumes.

4. Compute the total volume (V_T) by the sum of the cross-section/end-area volumes:

$$V_T = V_{12} + V_{23} + V_{34} + \cdots + V_{n-1,n} \tag{12.47}$$

Volume determinations for open pit are now being done using terrestrial laser scanning system. With this system, high-accuracy and detailed surveys can be performed quickly from a safe distance from the mine and volumetric surveys are done immediately after blasting and after every shift. Since this type of system is usually unmanned, the system can be used both day and night, allowing continuous slope stability monitoring of the mining areas.

13

TUNNELING SURVEYS

OBJECTIVES

At the end of this chapter, you should be able to

1. Describe the basic elements and methods of tunneling surveys
2. Calculate the approximate effects of lateral atmospheric refraction on alignment surveys
3. Perform both horizontal and vertical design and preanalysis of tunneling surveys
4. Carry out the analysis of breakthrough accuracy of tunneling projects
5. Perform error analysis of underground traverse surveys
6. Determine grid azimuth from gyro azimuth measurement for underground traverse surveys

13.1 INTRODUCTION

Tunneling survey is an underground survey done for constructing a tunnel. Other major applications of underground survey techniques are in relation to underground utilities such as nuclear accelerators and in mining operations. The underground surveys are necessary in tunneling operations for establishing transportation and communication routes, water conduits and pipelines; in mining operations, they are necessary in excavating ores.

Precision Surveying: The Principles and Geomatics Practice, First Edition. John Olusegun Ogundare.

Underground survey is different from surveying on the surface. It is essentially similar to three-dimensional surveys on the surface in that the purpose is usually to obtain the horizontal as well as the vertical coordinates of points underground. Some of the peculiarities of underground surveys in comparison to surface surveys are as follows:

1. Work environment underground is restricted, hot, dusty, dirty, and cramped.
2. Artificial illumination is usually needed since the passageways are dark and poorly lit.
3. Work areas are wet, with considerable water dripping from the roofs of passageways.
4. Instrument stations and benchmarks for leveling are often set into the roof to minimize disturbance from underground operations.
5. Instrument stations are set with much difficulty, especially when plugs must be driven into drill holes in rocks.
6. Lines of sight are sometimes very short and sights taken in shafts and sloping passageways may be sharply inclined.
7. Plumbing down the shaft usually constitutes a special problem, which is peculiar to underground surveying.

The essential problem in the underground surveying is that of orienting (or aligning) the underground surveys to the surface surveys. The process of orientation is to give coordinates (easting, northing, and elevation) of at least one point and azimuth of one line of the underground network in the surface coordinate system. In the underground transport system, the tunnels are driven to connect inclined or vertical shafts (points of the surface entry to the transport system) whose relative locations are established by the surface survey. If the entry to the underground tunnel system is via an inclined shaft, then the surface survey may simply be extended and continued down that shaft and into the tunnel by the method of traversing. The extra care needed will be to measure horizontal angles carefully due to steeply inclined sights and to correct for temperature effects due to possible thermal gradients in the tunnel. If entry is via a vertical shaft, then optical, mechanical, or gyroscopic methods of orientation are commonly used. Mechanical methods of transferring bearings underground use hanging wires with the Weisbach triangle method being the most popular; typical mechanical methods are discussed in Section 12.3.2. Note that typically a standard error of $1'$ in transferring the bearing down the shaft would likely result in a positional error at the end of 1 km of tunnel of up to a few decimeters and would increase as the traverse progresses.

13.2 BASIC ELEMENTS AND METHODS OF TUNNELING SURVEYS

Some of the tunnel types requiring precise tunneling surveys are railroads, subways, highways, hydro projects, mining projects, and water supply projects for large cities.

The three main types of tunnel in construction industry are highway tunnels, railway tunnels, and utility tunnels (e.g., water supply and drainage tunnels). The usual work of a surveyor in such tunneling projects consists of the following:

1. Performing all survey work before the start of construction, such as preliminary surveys and preliminary horizontal and vertical control surveys on the surface to obtain general site data for route selection and for structure design. Some of the steps that would be required in the process include the following:
   - Using existing survey records and monuments
   - Placing additional temporary monuments and benchmarks
   - Performing photogrammetric mapping, recording of seismic activity, and geophysical profiling
   - Preparing a large-scale topographic map of the surveyed corridor to locate the horizontal and vertical projection of the tunnel centerline.
2. Conducting primary horizontal and vertical control surveys of high order of accuracy for final design and construction, after completing the route selection. This will require
   - designing the survey control for the alignment of the tunnel axis with the highest possible accuracy so that opposite headings meet at the breakthrough points without any need for an adjustment of the excavations. Generally, an accuracy of 10–20 mm/km of the driven tunnel is required for meeting the opposite headings;
   - establishing permanent monuments and benchmarks, consisting of brass discs secured in concrete, at tunnel portals and over the tunnel alignment to serve as primary control during the final design stage and during construction;
   - setting reference marks for each monument so that the monuments can be readily verified and reestablished if disturbed or destroyed during construction.
3. Connecting the primary control network to the national geodetic control network of the area so that survey closures will provide an independent check on the new survey, and any two or more connections will provide adequate orientation for the horizontal control surveys. During the surveys, horizontal control points are extended to the underground tunnels by zigzag or braced traverses through the access portals (entrances to the tunnel), shafts (inclined or vertical), and stairwells into the underground tunnels. Zigzag traverses are carried out in order to avoid sight lines grazing the tunnel walls so as to minimize lateral refraction errors. The transfer of horizontal control points through vertical ventilation shafts can be achieved by co-planning method, Weisbach method, or quadrilateral method (discussed in Section 12.3.2) depending on the available instrumentation.
4. Performing principal control survey in the tunnel with the station points often established on the roof of the tunnel in the form of wall brackets. The surveyor

is to perform this after every few hundred meters of progress in a tunneling work. The use of wall brackets, however, makes it difficult to center the targets and the instruments to accuracy better than ±1 mm. According to Fowler (2006), spigots should be mounted to tunnel walls or roofs, as targets, in order to achieve better positional accuracy. A typical spigot consists (Fowler, 2006) of brass screw inserted with rubber edging placed in the tunnel wall or roof with specially made brass plugs screwed into the insert until flushed, keeping the spigot in the same position every time. The standard Leica GPR series prism can be attached to the brass so that when the prism is turned and rotated in any direction, the center of the prism will still stay in the same place. When such prisms have been previously coordinated from primary control survey from the surface network, their positions can be used in resecting total station setup points in free stationing procedure.

5. Setting, at the construction stage of the tunnel, the surface and subsurface settlement monitoring points over the centerline of the tunnel and on adjacent buildings. If settlement over the tunnel were a concern, additional benchmarks would have to be placed (away from the centerline) along the tunnel alignment.
6. Performing construction survey works that include the following:
   - Transferring tunnel centerline location, tunnel stationing, and tunnel grade from the primary control monuments and benchmarks located on the surface to the tunnel, and carrying this forward as the tunnel is constructed.
   - Establishing a construction control system that will assure tunnel driving or placement of tube under water within the allowable tolerance.
   - Installing observation wells to monitor ground-water levels adjacent to tunnels and underground structures.
   - Checking the profiles of the cross sections of the excavations.
   - Carefully monitoring surface movements over tunnels, tunnel cross sections; and vertical and lateral soil movement or stresses adjacent to tunnels or underground structures. This is to safeguard and maintain the tunnels.

Tunneling surveys are generally done to achieve the following:

- Establish and control the direction of tunnel construction, which are to keep tunnel boring on line and grade.
- Establish survey control in order to tie multiple sections of tunnels together within the allowable construction tolerance for both line and grade.
- Provide control for multiple headings that are driven at the same time by different construction contractors.

In tunneling projects, it is common, after sinking a shaft, to place reference pillars at the bottom of the shaft and then connect the pillars to the surface geodetic network. In achieving the connection to the surface, at least three geometric reference points are fixed on brackets bolted to the collar of the shaft and integrated into the geodetic network. These reference points are used to determine at least three reference pillars at the bottom of the shaft. Depending on the size of the shaft and its depth, different methods as discussed in Chapter 12 can be used to transfer orientation underground. As the tunnel drilling advances, a reference pillar is placed on the tunnel wall every 50 m, thus forming the underground survey control network in conjunction with the pillars at the bottom of the shaft.

With regard to the tunnel construction control, where the tunnel is excavated by drill and blast methods, the centerline must be extended to the tunnel face before drilling for the next round is begun. This centerline location is marked on the tunnel face, and the drill pattern is centered on that mark. Where the tunneling machine is used, the location and attitude of the machine are determined at certain intervals when the machine is temporarily stopped; if the machine is found to be off-line, adjustments of the steering mechanism are made to guide it back to its desired location. The most practical method of tunneling machine control is by laser beam and double target with the setting-up procedure as follows:

- Mount two targets (the front target and the rear target) on the tunneling machine, centered on a line parallel to its longitudinal axis and 1.2–3.0 m apart. The rear target is transparent and the leading target is opaque. The targets are to be intersected by the laser beam, which produces a bright red spot on the targets. Theoretical points of intersection between laser beam line and targets are calculated in advance for each machine location.
- Set up a laser tube at a distance behind the tunneling machine to emit a laser beam from a predetermined point of origin along a predetermined line to the targets mounted on the tunneling machine (a typical setup of laser tube is shown in Figure 13.1). On the horizontal plane, the laser line is a chord line or a tangent to the tunnel centerline; in the vertical plane, the laser line approximates the slope of the tunnel centerline.
- Move the laser tube to the next point after the tunnel is driven to the end of one laser beam line.

After setting up the control laser line, the tunneling machine is guided by the tunneling crew, while maintaining coincidence of the actual laser line intersection points with the predetermined intersection points on the targets set on the machine. The offsets to laser line from the tunnel centerline can be calculated from the plan and elevation plot of the predetermined laser line and the known centerline of tunnel.

When the tunnel construction is completed, permanent centerline monuments are placed in the completed tunnel at some intervals (typically about 300 m) and at all tangent-to-spiral and spiral-to-circular curve points (ASCE Manuals, 1985). From these monuments, measurements are taken laterally to critical clearance points to ensure that the clearance envelope is in accordance with design requirements.

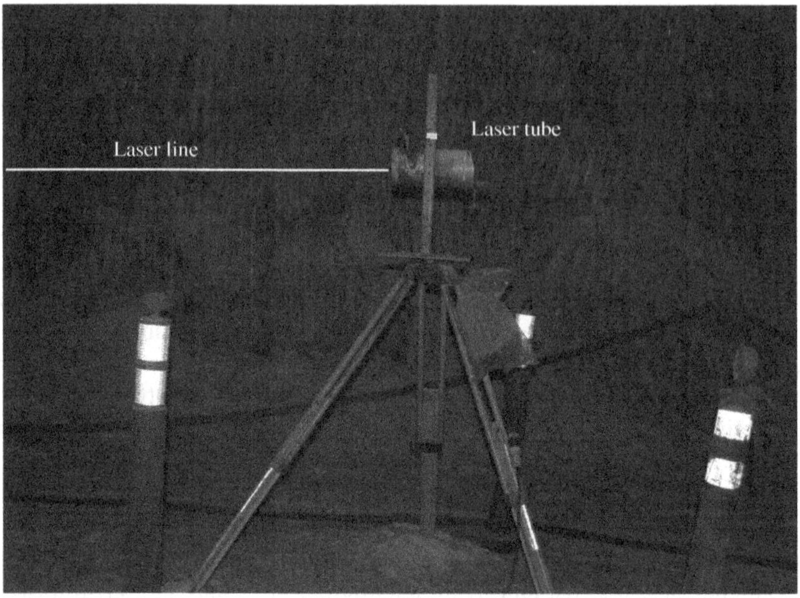

Figure 13.1 Typical setup of a laser device for alignment of a boring machine.

13.3 MAIN SOURCES OF ERROR IN TUNNELING SURVEYS

Two specific sources of gross errors in tunnel surveying are the influences of *refraction* and *rock deformation*. Refraction is due to temperature difference between the tunnel walls and the center of the tunnel. There is usually a large temperature difference at the tunnel adits, between the outside temperature and the temperature inside the tunnel. The refraction effects are usually on both distance and direction measurements. Rock deformation usually affects the positions of reference network points; after excavating rocks, the reference network points are usually established almost immediately on the tunnel surfaces. However, it takes a while for the remaining rock to stabilize, causing the reference network points to move with the deformation of the remaining rock. This deformation will destroy the network thereby creating the need to perform complete network survey updates.

The tasks of a surveyor in a tunneling project include establishing precise surface control network and precise underground control network for the alignment of the tunnel axis. If the tunnel concerned is short, the entrance points are commonly connected on the surface by a traverse; if it is very long, the entrance points are connected by trigonometric network (or a combination of traverses and triangulation). The accuracy of the underground control surveys (usually open-ended traverses) with refraction effect being the most dangerous source of error is much more critical than the accuracy of the surface control network. The underground traverse is usually established along and close to one of the walls of the tunnel since the center portion of the tunnel is occupied by construction and transportation systems. The heat

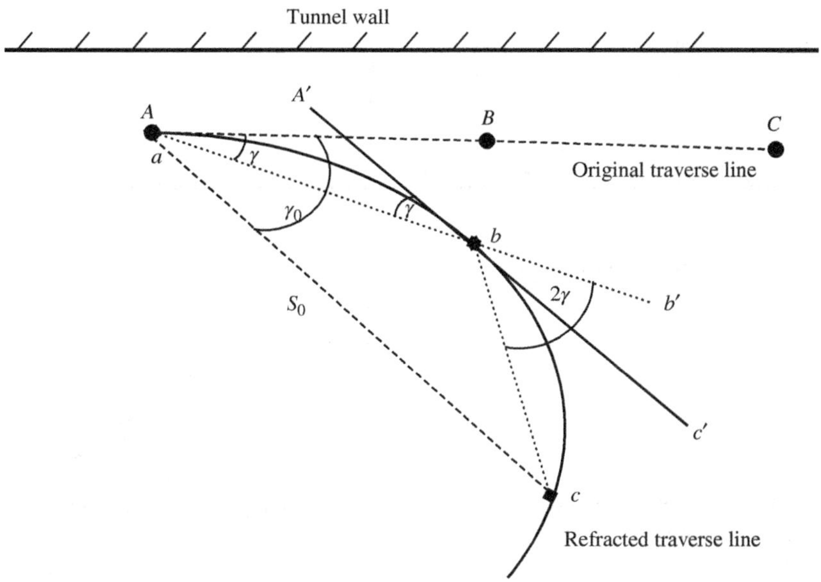

Figure 13.2 Refraction of traverse lines in a tunnel when angles are measured (assuming temperature is higher around the tunnel wall).

transfer from the surrounding rocks of the tunnel may produce temperature gradient near the wall in the direction perpendicular to the lines of sight of the control traverse. This will result in the lines of sight being bent concave away from the warmer wall surface (refer to Section 4.3.4).

The effect of refraction on a tunnel traverse is illustrated in Figure 13.2, where points A, B, and C are the original proposed traverse points and points a, b, and c are the refracted traverse points. When a theodolite is set up on point A and the telescope is aligned on line A-B, the point actually sighted and established is point b due to refraction. The deflection angle is γ and the point supposedly sighted is deflected by an amount of Bb. If the theodolite is now set at point b (established from the previous setup) and the telescope is sighted at point A, the telescope actually will be aligned in the direction b-A' due to refraction; when the telescope is plunged (assuming no collimation error in the instrument), the telescope will be aligned on bc', but the point actually sighted and established is point c; the effect of refraction on the measured angle at point b is the angle $b'bc$ (or 2γ). The angle measured in the traverse (by plunging the telescope) is 180°, but the actually measured angle at this traverse point is $180° + 2\gamma$. As it can be seen in the figure, the overall traverse lines will follow a circular curve abc. The last traverse point C will be deflected by Cc from the original traverse line, with the deflection angle at point A as γ_0.

This deflection angle γ_0 in Figure 13.2 can be deduced from Chrzanowski (1979) as

$$\gamma_0 = \frac{8'' P S_0}{T^2} \frac{\partial t}{\partial y} \text{ arcsec} \tag{13.1}$$

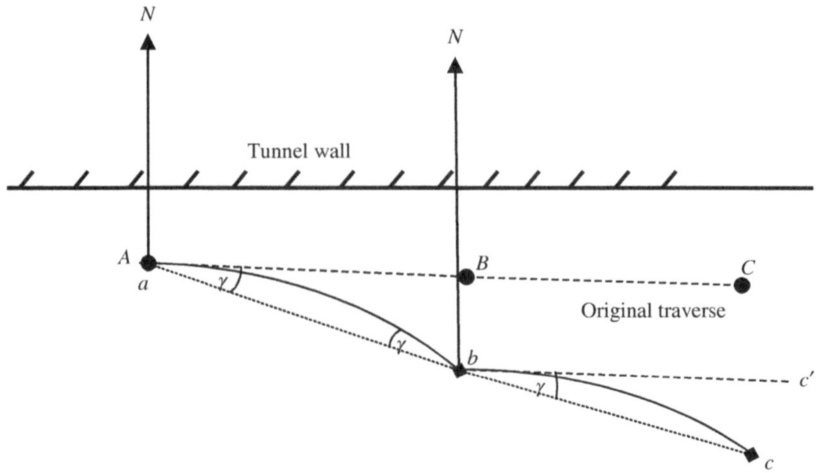

Figure 13.3 Refraction of traverse lines in a tunnel when gyro azimuths are measured (assuming temperature is higher around the tunnel wall).

where S_0 is the total length of the traverse, P is the barometric pressure (mbar), T is the atmospheric temperature (K), and $\partial t/dy$ is the lateral temperature gradient (°C/m). Equation (13.1) should be used in the above traverse type, instead of computing the individual effect of refraction at each traverse point. The amount of deflection (∂s_c) of the last traverse point (c) in meters can be given as

$$\partial s_c = S_0 \times (\gamma_0 \,\text{rad}) \tag{13.2}$$

Consider a case where the gyro azimuths are measured at each traverse station as illustrated in Figure 13.3. In this case, the refraction effect on every azimuth measurement at each station will be $+\gamma$. For example, if the theodolite is set up at point A and the telescope is aligned on A-B, point b will be established; if the theodolite is set up at point b and its telescope is aligned on b-c', point c will be established. The total effect of refraction on the last traverse point will be equal to the sum of the refraction influence at each station. In order to determine the amount of deflection at the last traverse point, Equation (13.1) will be used to determine the deflection angle (γ) at each traverse point based on the length of the individual traverse leg (s). The computed γ and the measured distance s are then used in Equation (13.2) to compute the deflection (in meters) at that traverse point. The sum of the individual deflection at each traverse point up to the last point gives the overall deflection of the last traverse point.

Usually, the refraction effects calculated from Equations (13.1) and (13.2) are not applied to measurements; they are just to serve as an estimation of expected errors since the distribution of the horizontal temperature gradients is unpredictable and difficult to measure. Temperature gradients may reach values of 0.3 °C/m and even higher in urban areas if the lines of sight pass close to the walls of buildings exposed to the sun. In tunneling projects, the temperature is usually higher at the centerline of the tunnel and the horizontal gradients will be positive from the wall to the centerline.

The horizontal refraction will be curved toward the warmer air where the velocity of the wave front will be greater and the air density will be smaller (assuming the same atmosphere pressure at that level). The density of air is proportional to the air pressure and inversely proportional to its temperature. Based on this, one can say that the air becomes less dense, the higher the temperature. The less dense the air, the easier and faster it is for the ray of light to pass through.

The refraction effects in tunneling surveys can be minimized in a number of ways, depending on the situations in the tunnel, such as follows (Fowler, 2006):

- Surveying in the center of the tunnel and not along the walls of the tunnel, and also away from any operating machinery, since the temperature at the center of the tunnel tends to be most stable. In tunnels, heat is known to move toward the tunnel center with the line of sight concave to the tunnel walls. The influence of the horizontal refraction is much less when the traverse is run in the center of the tunnel than when run on the brackets near the walls of the tunnel.
- Running the construction traverse on one side of the tunnel and the higher order one on the other side. Both traverses should be deflected in opposite directions if the temperature gradients near the walls in the direction perpendicular to the lines of sight are symmetrically distributed. The mean values from both traverses for setting out the centerline of the tunnel should minimize the influence of refractions. In this approach, the final correction to the axis of the tunnel is calculated as weighted mean from the results of the higher and lower order traverses, thereby minimizing the effects of refraction. Usually, the position of the last point of the construction traverse will be adjusted for refraction effects.
- Use of gyro instrument will also check an open traverse. It is recommended that gyro checks be made at least at every fourth or fifth station, at least in the highest order control traverse, but preferably at each station. Usually, baselines are measured with gyrotheodolites from both ends and the average measurements used as the azimuth of the line; this is to minimize the effects of refraction and also to prevent the propagation of errors due to possible instability of wall survey brackets and the movement of liner segments in the lined portions of the tunnel or mine.
- Running zigzag traverse may also be done to avoid the influence of refraction as much as possible. In large tunneling projects, such as in Superconducting Super Collider (SSC) project in Texas, the tunnel network comprised of two zigzag traverses through pairs of points with 150 m distance between two pairs creating a lattice network (Robinson et al., 1995). Typical tunnel traverses consist of zigzag observations, alternating between brackets on either side of the tunnel.

13.4 HORIZONTAL DESIGN AND SIMULATION OF TUNNELING SURVEYS

A tunneling survey is an excellent problem to investigate when considering high-accuracy requirements in an engineering survey. Depending on the conditions

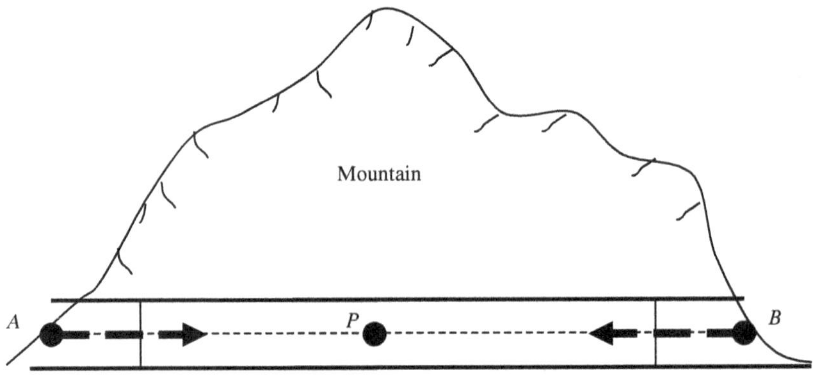

Figure 13.4 Tunneling with two opposing headings.

of the material being tunneled through, 3-m advancement can take up to 10 h to complete, making it necessary to drive the tunnel from both ends. In driving the tunnel from the two opposite entrances simultaneously, there is usually a critical problem of how to minimize the breakthrough error of headings driven from the opposite ends of the tunnel. This problem requires rigorous solution approach since the lateral breakthrough has to be determined by an open traverse and the vertical breakthrough by an open leveling survey.

Generally, in designing tunnel survey networks, one must consider that tunnels usually have elongated and small diameter compared to surface networks. The use of tunnel boring machines (TBMs) now demands that the accuracy of tunnel networks be high. It should also be remembered that it is expensive to excavate rock. When tunnels are constructed from two directions, it is necessary to estimate the breakthrough accuracy before the construction begins; this is to simulate how random errors will accumulate until the point of breakthrough.

Error analysis of tunneling surveys consists of calculating the breakthrough errors in lateral, longitudinal, and vertical directions of the tunnel axis. For an illustration, the tunnel in Figure 13.4 will be used. In the figure, a straight tunnel is driven simultaneously from two opposite entrances *A* and *B*. The axes of the headings *A-P* and *B-P* are supposed to meet at the breakthrough point *P*. However, due to unavoidable errors of geodetic measurements in the surface control network and errors of the underground control network surveys, the physical location P' of point *P* set out by the survey *A-P* differs from the location P'' set out by the survey *B-P*. Hence, the breakthrough point *P* is actually two points (P' and P'') and should be treated as different points in the accuracy analysis (Chrzanowski, 1979).

A typical surface control network, which is usually a combination of traverses and triangulation for a long tunnel of about 10 km, is shown in Figure 13.5. In the figure, the surface control points are labeled *A*, *B*, and 1–9. The underground control points are to be fitted between the entrance points *A* and *B* to the breakthrough point *P* in the tunnel. Usually, the underground network is an open (or fitted) traverse in zigzag form. Relative coordinates of the entrance points *A* and *B* are determined

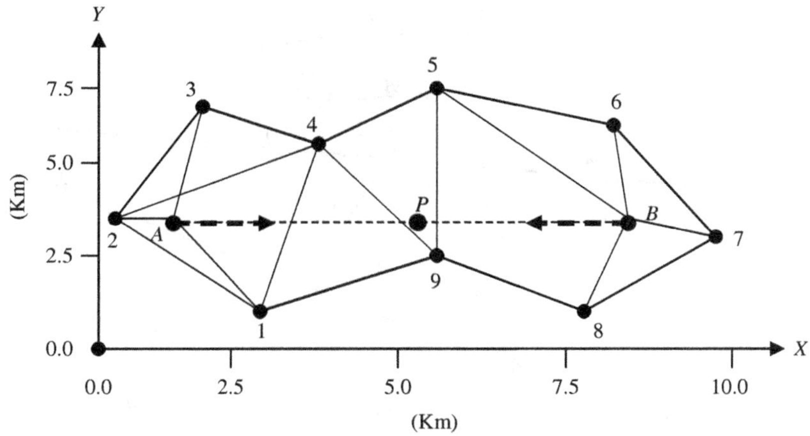

Figure 13.5 Horizontal control network for a tunnel construction.

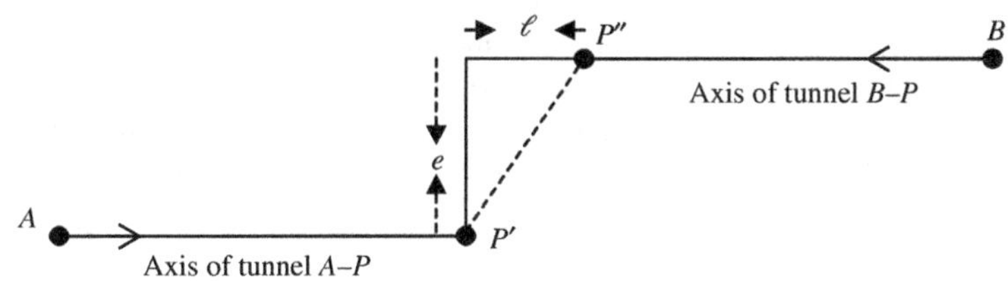

Figure 13.6 Representation of combined breakthrough error.

by connecting them on the surface by a surface survey control network in a local coordinate system.

In the underground traverse survey for the alignment of the tunnel axis, gyrotheodolite (or gyro station) is commonly used for providing astronomic azimuth. The common practice in performing the breakthrough error analysis is to separate the surface control survey from the underground tunnel survey. With this practice, the errors in establishing entrance points A and B, which are due to the surface survey, are first determined, then the errors due to the underground tunnel survey. The combined error from the surface and the underground surveys is then taken as the total breakthrough error at the breakthrough point P. This combined error is given as length $P'–P''$ as shown in Figure 13.6, where e and l are the lateral and longitudinal components of the total breakthrough error, respectively. The lateral component of total breakthrough error, however, is much more important than the longitudinal component so that the main concern in the breakthrough error analysis is more on determining the lateral breakthrough error component.

A breakthrough error component can be computed as a relative positional error for points P' and P'' at 95% probability level (refer to Figure 13.7). The relative positional error is described by the relative confidence-error curve or ellipse represented by the

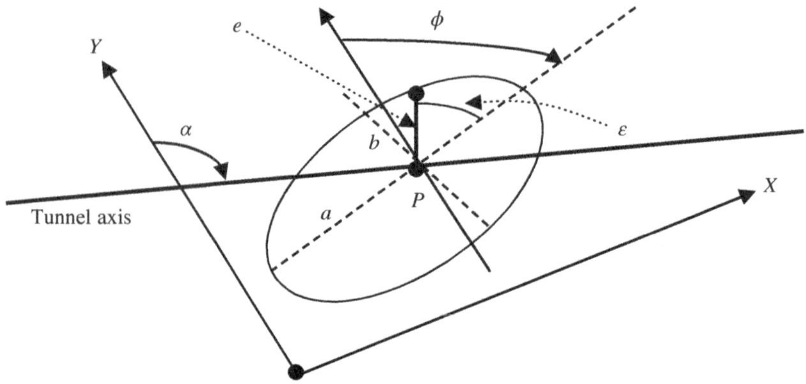

Figure 13.7 Relative confidence-error ellipse for point P.

values of the semi-major axis (a), the semi-minor axis (b), and the azimuth (ϕ) of the semi-major axis of the relative error ellipse, as shown in Figure 13.7.

In Figure 13.7, α is the azimuth of the tunnel axis at point P; ε is the angle between the direction perpendicular to the tunnel axis (direction of the lateral breakthrough error component) and the axis of the semi-major axis of the confidence relative error ellipse; and e is the lateral breakthrough error that is most desired. The covariance matrix ($C_{\hat{x}}$) from the least squares adjustment for the two breakthrough points (P' and P'') can be given as

$$C_{\hat{x}} = \begin{bmatrix} s^2_{x_{p'}x_{p'}} & s_{x_{p'}y_{p'}} & s_{x_{p'}x_{p''}} & s_{x_{p'}y_{p''}} \\ s_{y_{p'}x_{p'}} & s^2_{y_{p'}y_{p'}} & s_{y_{p'}x_{p''}} & s_{y_{p'}y_{p''}} \\ s_{x_{p''}x_{p'}} & s_{x_{p''}y_{p'}} & s^2_{x_{p''}x_{p''}} & s_{x_{p''}y_{p''}} \\ s_{y_{p''}x_{p'}} & s_{y_{p''}y_{p'}} & s_{y_{p''}x_{p''}} & s^2_{y_{p''}y_{p''}} \end{bmatrix} \tag{13.3}$$

where the diagonal elements of the matrix are the variances of the coordinates in the order $x_{p'}, y_{p'}, x_{p''}, y_{p''}$; and the off-diagonal elements are the covariances between the corresponding coordinate pairs. The relative covariance matrix for the two breakthrough points $P'(x_{p'}, y_{p'})$ and $P''(x_{p''}, y_{p''})$ can be derived from the coordinate differences (Δx, Δy) between the breakthrough points as follows:

$$\Delta x = x_{p''} - x_{p'} \tag{13.4}$$

$$\Delta y = y_{p''} - y_{p'} \tag{13.5}$$

By variance–covariance propagation law on Equations (13.4) and (13.5), the relative covariance matrix ($C_{p'p''}$) for the two breakthrough points P' and P'' can be given from Equation (2.37) as

$$C_{p'p''} = \begin{bmatrix} s^2_{\Delta x \Delta x} & s_{\Delta x \Delta y} \\ s_{\Delta y \Delta x} & s^2_{\Delta y \Delta y} \end{bmatrix} \tag{13.6}$$

and the parameters (a_{st}, b_{st}, ϕ) of the relative standard error ellipse can be calculated on the basis of the elements of the covariance matrix ($C_{P'P''}$), as follows:

$$a_{st} = \sqrt{\lambda_1} \tag{13.7}$$

$$b_{st} = \sqrt{\lambda_2} \tag{13.8}$$

$$\phi = \arctan\left(\frac{s_{\Delta x\Delta y}}{\lambda_1 - s^2_{\Delta x\Delta y}}\right) \tag{13.9}$$

where λ_1 and λ_2 are the maximum and minimum eigenvalues of the relative covariance matrix $C_{P'P''}$, which are defined in Equations (2.41)–(2.44).

Usually, the breakthrough errors are given in terms of the 95% confidence relative error ellipse (a, b, ϕ), which can be expressed in terms of the standard relative error ellipse as follows (assuming the a priori variance factor of unit weight is well known or the observation precisions are well estimated from the least squares adjustment):

$$a = 2.448a_{st} \tag{13.10}$$

$$b = 2.448b_{st} \tag{13.11}$$

$$\phi = \arctan\left(\frac{s_{\Delta x\Delta y}}{\lambda_1 - s^2_{\Delta x\Delta y}}\right) \tag{13.12}$$

Generally, the lateral component ($e_{95\%}$) of the breakthrough error at 95% confidence level can be given in terms of the parameters (a, b, ϕ) of the 95% confidence relative error ellipse, as follows:

$$e^2_{95\%} = a^2\cos^2\varepsilon + b^2\sin^2\varepsilon \tag{13.13}$$

where $\varepsilon = 90^\circ - \alpha + \phi$. In practice, the breakthrough errors from the surface and underground networks are determined separately since both networks are distinctly different. The horizontal breakthrough error (e_h) will be due to the following two separate influences:

(i) Influence e_s due to the surface network surveys
(ii) Influence e_u due to the underground network surveys.

The horizontal breakthrough error ($e_{h(95\%)}$) is calculated as the sum of the two influences (the surface influence $e_{s(95\%)}$ and the underground influence $e_{u(95\%)}$) as follows:

$$e^2_{h(95\%)} = e^2_{s(95\%)} + e^2_{u(95\%)} \tag{13.14}$$

To obtain the value for $e_{s(95)}$, perform minimal constraint (holding one point fixed and keeping one azimuth also fixed) simultaneous least squares adjustment of the surface and the underground survey measurements. In the adjustment, the underground measurements are considered errorless (e.g., referring to Figure 13.5, the two errorless distances are *A-P* and *B-P*, and the two errorless angles are 3-*A-P* and 6-*B-P*) and the estimated errors of measurements for the surface network (forming a closed-loop traverse) must be used. The computed parameters (a, b, ϕ) of the 95% confidence relative error ellipse for the breakthrough point *P* are used in Equation (13.13) to obtain the $e_{s(95\%)}$ (at 95% confidence level).

Similarly, to obtain the value for $e_{u(95\%)}$, perform minimal constraint (holding one point fixed and keeping one azimuth also fixed) simultaneous least squares adjustment of the surface and the underground survey measurements. In the adjustment, the estimated errors of the underground measurements must be used and the surface points must be fixed and considered errorless. The computed parameters (a, b, ϕ) of the 95% confidence relative error ellipse for the breakthrough point *P* are then used in Equation (13.13) to obtain the $e_{u(95\%)}$ (value at 95% confidence level). The total horizontal breakthrough error is obtained by using Equation (13.14).

In summary, a preanalysis of a tunneling survey (e.g., a 10-km tunnel) will include the following:

1. Mixing triangulation and trilateration methods for the surface survey.
2. Using open traverses for the underground survey with the breakthrough point *P* at 6 km from one tunnel entrance point (and 4 km from the other entrance point).
3. Making the traverse legs for the underground survey of equal length, for example, 1 km (or 500 m) for each leg.
4. Using the accuracy specifications of the instruments (EDM and theodolite) to estimate errors in the measurements (for weighting the measurements). For example, the gyro azimuth (which is equivalent to the astronomic azimuth) may be determined underground with a standard deviation ranging from $3''$ to $20''$ if a typical precision gyrotheodolite (or gyro station) is used. It is generally believed that gyro observations will improve the accuracy of positioning tunnel networks.
5. Using a proven least squares adjustment software (e.g., GeoLab) to do the preanalysis (i.e., determining the influence of the surface network and that of the underground traverse) on the breakthrough accuracy of the survey.
6. Calculating the total lateral breakthrough error component in order to evaluate if the acceptable limit is exceeded or not.

13.5 VERTICAL DESIGN AND SIMULATION OF TUNNELING SURVEYS

The design of vertical control for tunneling projects is usually divided into two parts: the design of *surface vertical control network* (usually run along roads and railways) and the design of *underground (or construction) vertical control network* conducted in tunnels. It is traditional that vertical control networks and horizontal control networks are established independently.

13.5.1 Design of Surface Vertical Control Network

Surface primary vertical control networks are established to provide a stable control over a long period. The following are a few notes on a typical design of a surface vertical control network (DeKrom, 1995):

- Benchmarks are grouted into bedrock to ensure stability over the construction period.
- Deep benchmarks are installed only at junction points (which are located outside the construction area) of level loops and are to serve as control for densification and elevation transfers to the tunnel.
- Temporary benchmarks or lower order control monuments are used between the deep benchmarks to ensure section lengths of under 3 km to help in controlling the accumulation of systematic effects in leveling.
- At least two temporary benchmarks are located on the collar of any of the possible shaft constructed or to be constructed.
- Simple concrete monuments are used as benchmarks for service areas with the densification network including at least three benchmarks located near each service area.
- Densification at each service area should be carried out a few days before the elevation transfer is done in order to ensure the stability of the monuments.
- Leveling procedure should follow special-order geodetic control specifications (NRC, 1978) or lower order, depending on the desired accuracy of work; for special-order geodetic control, the permissible difference (at 95% confidence level) between two runs of a level section will be given from NRC (1978) as

$$\Delta_{\text{max}} = 3\,\text{mm}\sqrt{k} \tag{13.15}$$

where k is one-way distance (km) in a section. From Equation (13.15), the standard deviation for difference between two runs of a level section can be given as

$$\frac{3\,\text{mm}}{1.96}\sqrt{k} \quad \text{or} \quad 1.5\,\text{mm} \tag{13.16}$$

Assuming no correlation between direct and reverse runs, the standard deviation of a single-run section can be determined by

$$\frac{1.5\,\text{mm}}{\sqrt{2}}\sqrt{k} \quad \text{or} \quad 1.1\,\text{mm}\sqrt{L} \tag{13.17}$$

Usually, the elevation differences must be converted into orthometric heights by applying the orthometric corrections to the leveled height values determined directly from the geodetic leveling procedure.

13.5.2 Design of Underground Vertical Control Network

There are three orders of leveling traverses in the tunnel: *third order*, *secondary order*, and *primary order* as discussed in Section 12.1.1. The third-order traverse is run first followed by secondary and then primary; the correction to the axis of the tunnel is calculated as weighted mean of the results of higher and lower order traverses in order to minimize refraction effects. The primary control monuments take the form of wall brackets (it is common to include some parts of secondary control network in primary networks) with total stations usually set to occupy the bracket locations during the survey. Primary control networks usually take place at regular intervals, for example, every few months, as the tunnel advances; primary order survey is usually restarted from the tunnel entrance (the portal) each time. This survey could take up to four to five sets of measurements per setup with minimally constrained least squares adjustment of the measurements performed at the end of the survey.

The preanalysis of a vertical control network is usually based solely on the design of the primary network. The preanalysis is to ensure that the design requirements are achievable by taking into account both the reliability and accuracy of the surface vertical control network, densification surveys, elevation transfer procedures, and tunnel control. A typical design tolerance could be stated, for example, that the maximum departure of the excavated tunnel from its theoretical position on a plane must not exceed an envelope of 200 mm. This tolerance must be interpreted correctly; it is usual to assume that the tolerance is given for 99% level of confidence (for a stringent case, requiring higher precision) or 95% level of confidence for a case requiring less precision. With this type of tolerance, it is typical to reserve half of this error limit for boring and lining the tunnel and the other half as the surveying error budget. For the surveying component of the error budget, it is important to correctly interpret what the requirements are. Typical requirements based on the project design may be as follows (refer to DeKrom, 1995):

- The maximum vertical adjustment range of some machinery with respect to the tunnel floor should be within some specified tolerance, such as ±15 mm. This tolerance can be interpreted to mean that the error in the difference in elevation between any two points anywhere along the tunnel be ±15 mm at 99% level of confidence. If one wants to be more stringent, one can assume that half of this or ±7.5 mm at 99% level of confidence is the desired error for the difference in elevation between any two points anywhere along the tunnel.
- The relative vertical positional errors between any two points along the tunnel should be less than a certain tolerance. For example, if the vertical positional tolerance expected is 80 mm, the relative positional tolerance between any two points should be 80 times square root of two (or 113 mm).

13.5.3 Vertical Breakthrough Analysis

Precise spirit leveling is still considered the best in extending vertical control underground in tunneling projects. Vertical design and preanalysis are similar to horizontal design and preanalysis discussed in Section 13.4, except that the horizontal aspect is two dimensional (dealing with error ellipses), while the vertical aspect is one dimensional (dealing with error bars). In the case of vertical preanalysis, the covariance matrix ($C_{\hat{z}}$) from the least squares adjustment for two breakthrough points (P' and P'') can be given as

$$C_{\hat{z}} = \begin{bmatrix} s^2_{z_{P'}z_{P'}} & s_{z_{P'}z_{P''}} \\ s_{z_{P''}z_{P'}} & s^2_{z_{P''}z_{P''}} \end{bmatrix} \tag{13.18}$$

where the diagonal elements of the matrix are the variances of the elevations of points P' and P', respectively; and the off-diagonal elements are the covariances between the elevations of the two breakthrough points; for symmetric matrix, $s_{z_{P'}z_{P''}} = s_{z_{P''}z_{P'}}$. The relative covariance matrix for the two breakthrough points P' and P'' can be derived from the elevation difference (Δz) between the breakthrough points as follows:

$$\Delta z = z_{P''} - z_{P'} \tag{13.19}$$

By performing variance–covariance propagation law on Equation (13.19), the relative covariance matrix ($C_{P'P''}$) for the two points can be given as

$$C_{P'P''} = JC_{\hat{z}}J^T \tag{13.20}$$

where J is the Jacobian of Equation (13.19) with respect to the elevations of points P' ($z_{P'}$) and $P''(z_{P''})$, given as

$$J = \begin{bmatrix} \frac{\partial \Delta z}{\partial z_{P'}} & \frac{\partial \Delta z}{\partial z_{P''}} \end{bmatrix} = \begin{bmatrix} -1 & 1 \end{bmatrix} \tag{13.21}$$

Using Equations (13.18) and (13.21) in Equation (13.20) gives the relative covariance matrix ($C_{P'P''}$) from the elevation difference (Δz) of the two breakthrough points P' and P'':

$$C_{P'P''} = [s^2_{\Delta z}] \tag{13.22}$$

where

$$s^2_{\Delta z} = s^2_{z_{P'}z_{P'}} + s^2_{z_{P''}z_{P''}} - 2s_{z_{P'}z_{P''}} \tag{13.23}$$

The relative error bar at 95% can be calculated from Equation (13.23) as $1.96s_{\Delta z}$. The total vertical breakthrough error ($e_{\mathrm{v(95\%)}}$) in tunneling surveys can be expressed as

$$e^2_{\mathrm{v(95\%)}} = e^2_{\mathrm{s(95\%)}} + e^2_{\mathrm{u(95\%)}} \tag{13.24}$$

where $e_{\mathrm{s(95\%)}}$ and $e_{\mathrm{u(95\%)}}$ are the error components of the surface and the underground surveys at 95% confidence level, respectively. The 95% confidence relative error bar $e_{\mathrm{s(95\%)}}$ from the surface survey analysis for the breakthrough points P' and P'' can be given from Equation (13.23) as

$$e_{\mathrm{s(95\%)}} = 1.96s_{\Delta z} \tag{13.25}$$

Similarly, the 95% confidence relative error bar $e_{u(95\%)}$ from the underground survey analysis for the breakthrough points P' and P'' can be given from Equation (13.23) as

$$e_{u(95\%)} = 1.96 s_{\Delta z} \tag{13.26}$$

It should be mentioned that due to some unpredicted circumstances or conditions, it is possible that a design may fail to achieve the desired breakthrough accuracy. For an example, in the construction of 8-km-long Rogers Pass Tunnel of the Canadian Pacific Railway (CPR), lateral and vertical breakthrough errors (at 95% confidence level) of 35 and 1 cm, respectively, were achieved compared with the designed lateral and vertical breakthrough errors (at 95% confidence level) of 15 and 5 cm, respectively (Lachapelle et al., 1985). As it can be seen for the 8-km tunnel, the specified lateral breakthrough error was exceeded while that of the vertical breakthrough was achieved.

13.6 NUMERICAL EXAMPLE: HORIZONTAL BREAKTHROUGH ANALYSIS

A straight tunnel is to be driven simultaneously from two opposite entrances 1 and 4 as shown in Figure 13.8. The surface network consists of points 1–4.

The approximate values of coordinates (taken from a large-scale map of the project area) for the surface network points and the proposed underground points are given in Table 13.1. The proposed measurements to be made are given in Tables 13.2 and 13.3.

In Table 13.2, surface angles will be measured with a total station with a standard deviation of $\pm 5''$; the underground angles will be measured with a standard deviation of $\pm 6''$; the bearing of line 2-3 will be fixed and considered errorless ($0.01''$ may be used); and the azimuth measurements will be made using GYROMAT ($\sigma = 3''$) gyrotheodolite for traverse legs 4-C and 1-B in the underground traverse (with each traverse leg in the underground traverse taken as 100 m long).

In Table 13.3, the distance measurements (on the surface and underground) will be made with a total station with distance precision of 2 mm + 2 ppm.

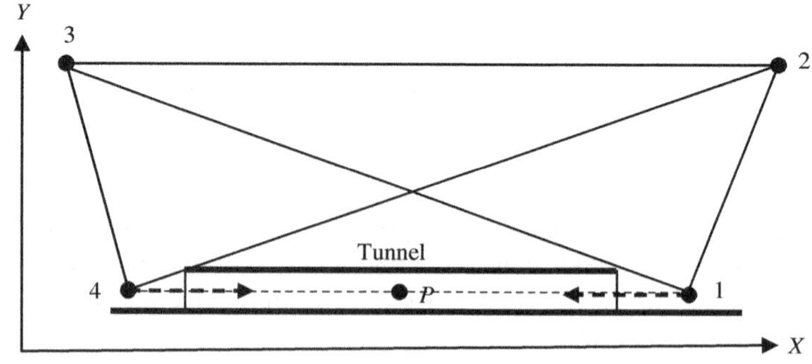

Figure 13.8 Simulated simple tunneling project with two opposing headings.

TABLE 13.1 Estimated Coordinates of Network Points.

Point	X (m)	Y (m)	Comments
1	800	100	Surface point
2	900	400	Surface point
3	100	400	Surface point
4	200	100	Surface point
P	500	100	Expected breakthrough point
A	600	100	Proposed underground point
B	700	100	Proposed underground point
C	300	100	Proposed underground point
D	400	100	Proposed underground point

TABLE 13.2 Proposed Angle and Bearing Measurements.

Point (From)	Point (At)	Point (To)	Comments
3	1	2	Surface angle
1	2	4	Surface angle
4	2	3	Surface angle
2	3	1	Surface angle
1	3	4	Surface angle
3	4	2	Surface angle
B	1	3	Underground angle
P_2	A	B	Underground angle
A	B	1	Underground angle
2	4	C	Underground angle
4	C	D	Underground angle
C	D	P_1	Underground angle
	2	3	Fixed surface bearing for network constraint
	4	C	Underground gyro azimuth to be measured with GYROMAT gyrotheodolite
	1	B	Underground gyro azimuth to be measured with GYROMAT gyrotheodolite

Required task: Use any appropriate software to perform a preanalysis to check if the lateral breakthrough error (at 95% confidence level) will be acceptable if the maximum breakthrough error for the tunneling is not to exceed 20 mm.

13.6.1 Surface Network Analysis

The proposed surface network is given in Figure 13.9:

- All of the surface angular measurements given in Table 13.2 will be made with a standard deviation of $\pm 5''$.

TABLE 13.3 Proposed Distance Measurements.

Point (From)	Point (To)	Comments
2	1	Surface distance
2	4	Surface distance
2	3	Surface distance
3	1	Surface distance
3	4	Surface distance
4	*C*	Proposed underground distance
C	*D*	Proposed underground distance
D	P_1	Proposed underground distance
A	P_2	Proposed underground distance
A	*B*	Proposed underground distance
B	1	Proposed underground distance

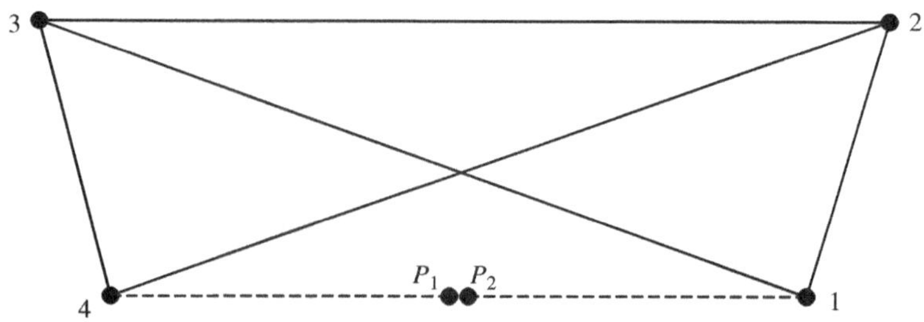

Figure 13.9 Layout of a surface network.

- Only the angles P_2-1-2 (angle at point 1) and 3-4-P_1 (angle at point 4) will be considered from the underground in the analysis and they will be taken to be errorless.
- All of the surface distance measurements given in Table 13.3 will be made with a precision of 2 mm + 2 ppm.
- The underground distances given in Table 13.3 will be simplified into two long distances 4-P_1 and 1-P_2 in the analysis and will be considered errorless.

For the minimum constraint preanalysis, the following are to be kept fixed:

- Coordinates of point 2
- Azimuth 2-3 (assumed errorless).

13.6.1.1 Results of the Surface Survey Analysis Using Equations (13.10)–(13.12), the parameters of the 95% confidence relative error ellipse between the breakthrough points P_1 and P_2 are as follows:

$$a(95\%) = 8.3\,\text{mm};\ b(95\%) = 4.6\,\text{mm};\ \phi = 90^\circ$$

The lateral breakthrough error at 95% confidence level due to the surface survey is obtained from Equation (13.13) as follows (assuming the azimuth (α) of the tunnel axis is 90°):

$$\varepsilon = 90^\circ - 90^\circ + 90^\circ$$
$$e^2_{95\%} = 8.3^2\cos^2 90 + 4.6^2\sin^2 90 \rightarrow e_{95\%} = 4.6\,\text{mm}$$

The lateral breakthrough error at 95% confidence due to the surface survey ($e_{s(95\%)}$) is 4.6 mm.

13.6.2 Underground Network Analysis

The proposed underground network is given in Figure 13.10:

- All of the surface angular measurements given in Table 13.2 will be considered errorless.
- All of the underground angular measurements will be used in the analysis with a standard deviation of $\pm 6''$.
- The gyro azimuth of lines 4-*C* and 1-*B* will be measured with a gyrotheodolite with a standard deviation of $3''$.
- All of the surface distance measurements given in Table 13.3 will be fixed and considered errorless.
- All of the proposed underground distances in Table 13.3 will be measured with a precision of 2 mm + 2 ppm.

For the minimum constraint preanalysis, the following are to be kept fixed:

- Coordinates of point 2
- Azimuth 2-3 (assumed errorless).

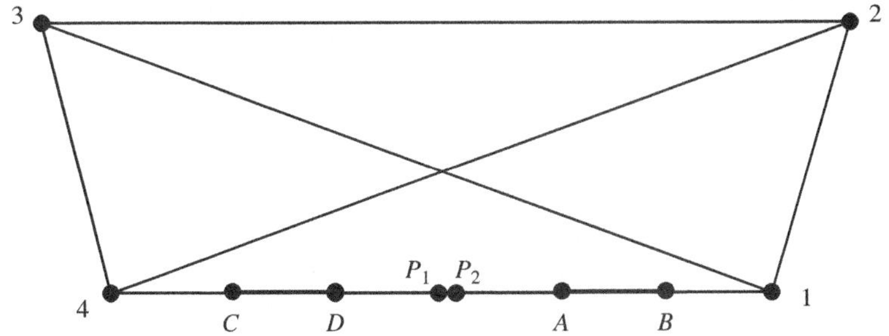

Figure 13.10 Layout of an underground network.

13.6.2.1 Results of the Underground Survey Analysis Using Equations (13.10)–(13.12), the parameters of the 95% confidence relative error ellipse between the breakthrough points P_1 and P_2 are as follows:

$$a(95\%) = 26.2 \text{ mm}; b(95\%) = 12.1\,\text{mm}; \phi = 0°$$

The lateral breakthrough error at 95% confidence level due to the underground survey is obtained from Equation (13.13) as follows (assuming the azimuth (α) of the tunnel axis is 90°):

$$\varepsilon = 90° - 90° + 0°$$
$$e_{95\%}^2 = 26.2^2 \cos^2 0 + 12.1^2 \sin^2 0 \rightarrow e_{95\%} = 26.2\,\text{mm}$$

The lateral breakthrough error at 95% confidence due to the underground survey ($e_{u(95\%)}$) is 26.2 mm. The combined horizontal breakthrough error at 95% confidence level is obtained from Equation (13.14) as $e_{h(95\%)} = 26.6$ mm. This result will not be acceptable since 26.6 mm is greater than the tolerance limit of 20 mm expected for the tunneling survey. Since most of the error in breakthrough is due to the underground survey, more effort should be made to reduce the error contribution due to the underground survey, such as reducing the number of points where angles are observed and replacing the angle measurements at those points with azimuth measurements with gyrotheodolite and also considering using instruments with better precision.

13.7 EXAMPLES OF TUNNELING SURVEYS

13.7.1 Transportation Tunneling Surveys: Rogers Pass Tunnel in Canada

An example of transportation tunneling survey is the survey for the long railway tunnel at the Rogers Pass in British Columbia, Canada, discussed in detail by Lachapelle et al. (1984, 1985, 1988).

The Rogers Pass Tunnel was to have a diameter of 8.5 m with a 350-m-deep vertical ventilation shaft to be located about halfway along the tunnel. The purpose of this tunnel was to decrease the gradient of the westbound railway track from 2.6% to less than 1%. The tunnel was to be driven from opposite directions so as to meet at a predesigned breakthrough point. The designed lateral breakthrough error was supposed to be less than 15 cm at 95% confidence level, but 35 cm was achieved; the designed vertical breakthrough error was 5 cm at 95% confidence level, but 1 cm was achieved. For the surface survey, the designed relative accuracy between the two portals of the tunnel at 95% confidence level was 7 cm, but 4.4 cm was achieved.

Some of the survey challenges of the project included determining the effects of the large changes in the deflection of the vertical in the area and the atmospheric refraction in the tunnel traverses. The survey networks to be measured for the project were divided into two parts: surface network and underground network. Each of the networks consisted of horizontal and vertical networks, which were measured independently.

For the surface network, the following observables were measured during the horizontal control network surveys:

- Horizontal and vertical directions measured using Wild T3 theodolite over two nights with 16 sets measured each night.
- EDM distance measurement of some of the network lines made with two different HP3808 instruments with each distance measured four times over several hours.
- Astronomic observations for latitude, longitude, and azimuth at some stations for the purpose of determining independent azimuths for some lines and for predicting the effects of the deflection of the vertical on direction and zenith angle measurements. The observations became necessary because of the expected large differences in the deflections of the vertical due to large differences in elevations (up to 1600 m) of observing stations. Wild T4 theodolite was used in the astronomic observations.

For the vertical control network surveys, special-order spirit leveling run was carried out along the Trans-Canada Highway between the entrances to the tunnel using a Zeiss NI-1 precision level. The elevation difference observations were transformed into geopotential number differences by using the gravity values measured along the leveling route. The adjusted geopotential numbers were then transformed into Helmert orthometric heights. Some of the important aspects of the underground control network surveys are as follows:

- The excavation of the tunnel was done from both ends using the conventional drill-and-blast technique at the west section and the tunnel boring machine (TBM) in the east section.
- On the east side, the horizontal and vertical traverses (which were done independently) could only be run along one of the walls of the tunnel because cables, pipes, and other obstacles are preventing the zigzag rule from being followed.
- MOM GiB-11 gyrotheodolite (with the manufacturer specified accuracy of $\pm 5''$) was used to control the orientation of the tunnel.
- The last 3 km of each tunnel section from the breakthrough point could only be guided by angle and distance measurements after the breakdown of the gyrotheodolite.

13.7.2 Transportation Tunneling Surveys: The Channel Tunnel in Europe

The Channel Tunnel, which was completed on June 28, 1991, is a transportation system tunnel connecting Britain and France. It consists of three subtunnels running parallel under the English Channel at the Strait of Dover between terminals in the United Kingdom and France. The total length of the tunnel is 50.5 km with 38 km of it under the sea. The tunnel is generally 100 m below the sea level with its lowest point being 75 m deep. Some of the challenges in the construction of the tunnel are given as follows:

- Determining the geological makeup and profile of the English Channel seafloor. This involved drilling of boreholes on land and hydrographic surveying at sea with the accompanying problem of strong tidal currents in the English Channel. The hydrographic surveys were also to help determine the tunnel route and to locate the existing boreholes, submarine pipelines, and cables.
- Surveying the tunnel. This included establishing horizontal triangulation stations and vertical control in the project area as well as bringing survey control under the sea from the surface by a traverse run between wall-bracket stations on the wall of the tunnel. Lines of sight during measurement in the tunnel were typically 0.3 m from the wall and 1.0 m above the level of the tunnel centerline (Johnston, 1991). All primary and main control traverses were conducted to first-order standards while all surveys for the TBM guidance were of second order.
- Controlling the effect of lateral refraction in the tunnel surveys, which may be due to possible large temperature gradients across the tunnel. According to Johnston (1991), "*the traverse angles in the tunnel were found on average to vary between occasions of observing by 4.2 times their individual standard errors … due to lateral refraction.*" The effect of refraction is such that the angles measured to the traverse stations that were located on one side of the tunnel wall became too large. In minimizing this refraction effect, each total station position was alternated to the opposite tunnel wall from its preceding position. This resulted in measuring a series of zigzag traverse lines between left- and right-hand wall brackets down the tunnel so as to cancel the errors due to refraction, assuming the lateral refraction effect was symmetrical about the tunnel centerline. In this case, double traverses were made to stations that were located opposite one another and zigzag traverses were run with the same number of left-to-right and right-to-left legs so that all the first-order refraction effects can cancel themselves out. It is, however, widely recommended (Chrzanowski, 1981a) that zigzag traverse method with many complementary gyrotheodolite observations be used for best results in tunneling projects.
- Guiding the TBM from both sides of the tunnel in order to achieve a reasonable breakthrough accuracy at the center of the tunnel. Since it is typical of the TBM to install a lining of concrete rings that is about 0.3 m thick as soon as the rock is exposed, in ensuring exact repositioning of total stations for the tunnel surveys and for guiding the TBM, brackets were bolted to the tunnel lining at each instrument location.
- Choosing the most suitable map projection for the site. UTM projection was found unsuitable for the site due to the perceived difference between the ellipsoidal Earth and the UTM grid, which resulted in discrepancies between the ground and grid distances. A special projection known as Channel Tunnel Grid (CTG), which was later renamed Réseau du Tunnel sous la Manche 1987 grid (RTM87), was developed and adopted for the site. The projection is a Cylindrical Orthomorphic Transverse Mercator with the Central Meridian and Latitude of origin passing through the center of the project area in order to

reduce distortion. This grid became the basis for all horizontal computations required on the tunnel project, providing good accuracies of map distances and directions.

- Providing the most suitable vertical datum for the site. Surveyors could not use the sea level on either side of the Channel to relate the heights of one tunnel entrance to another because of the differences in mean sea levels on either side of the tunnel, which are due, in part, to the effects of winds, tides, and the spinning earth. Heights in Britain were referenced to the Ordnance Datum Newlyn while heights in France were based on sea level established by the Institut Geographique National, and the two datums are different by about 30 m. To avoid negative elevations where the tunnel was deep beneath the sea, the reference datum for the tunnel was lowered by 200 m below Newlyn and renamed Nivellement Transmanche datum 1988 (NTM88); all the tunnel project elevations were based on this datum.

13.7.3 Tunneling Surveys for Scientific Research: SSC Project in Texas, USA

An example of tunneling surveys for scientific research is the tunneling surveys for the SSC project in Texas involving a 4.2 m diameter, 87-km-long tunnel (Chrzanowski et al., 1993; Chrzanowski, 1999; Robinson et al., 1995; Dekrom, 1995). Included in the design were the additional tunnels making up another 27 km of tunneling. The main collider tunnel and the other additional tunnels were to be connected to the surface by a number of vertical shafts of various sizes based, on average, every 4.3 km along the main collider ring. Over 12,000 magnets were to be installed in the main collider alone. To have the accelerator working efficiently, the magnets in the main collider would have to be aligned in a perfect geometric plane (which was not a horizontal surface) to better than 1 ppm of distance (1 mm/km) in order that two counter-rotating beams would collide at the designed locations. After completing the tunnel excavation of each 4.3 km section of the tunnel, a final invert would be poured and the installation and alignment of magnets would begin without waiting for the entire tunneling work to be completed and checked for closure of the geodetic control surveys. The idea of not allowing for the checking of the closure of surveys before installing and aligning the magnets created a challenge for the surveyors with regard to conducting very precise surveys.

The survey tolerances for the excavations of the main collider tunnel and the other additional tunnels were not to exceed ±108 mm error in the relative positioning of any two points located anywhere in the tunnel; 54 mm was assigned to randomness and the other 54 mm assigned to systematic errors. Due to the strict alignment guidelines, the specified relative positioning tolerance was taken as the maximum permissible error at the 99% confidence level, rather than the usual 95% confidence level. For the vertical control surveys, the tolerance for relative positioning was ±12 mm to accommodate the strict requirements for placing the final concrete inverts on which the magnets were to be installed. Other challenges included predicting the influence of systematic errors arising from atmospheric refraction, uncertainties in the deflection of the vertical, and calibration errors of the survey instruments.

The geodetic network surveys consisted of the surface control network survey, shaft transfer survey, and underground control network survey. In each survey, the horizontal and vertical networks were measured independently. The surface horizontal control network was established using high-precision GPS procedure, while the vertical control network was based on geodetic leveling of special order. Specially designed monumentation was used for some control network points; some had inverted plumblines anchored deeply in the bedrock attached to them to monitor their stability. The transfer of both horizontal and vertical controls to the tunnel was by vertical shafts using spherical Taylor Hobson targets, Leica precision optical plummets, total stations TC2002, and precision level. The underground horizontal control was run with double zigzag traversing (to reduce refraction errors) with occasional use of GYROMAT 2000 precision gyrotheodolites to measure azimuths between some of the underground control points. The TC2002 total station instruments placed on specially designed wall brackets spaced at 150 m interval were used to measure the distances of the traverse networks. In order to provide vertical control in the tunnel, spirit leveling run was done with the maximum sight distances to the wall targets being about 50 m.

13.8 ANALYSIS OF UNDERGROUND TRAVERSE SURVEYS

The underground coordinate system must relate to the surface coordinate system so that positions of details underground can be correlated with those on the surface. Most of the underground control surveys are based on open traverses. At the design stage, it is necessary to analyze what the positional accuracy of the last point of the traverse would be with the designed measurement schemes, which is to be implemented. This will be done using the concepts of variance–covariance propagation of open traverse. Consider the open traverse in Figure 13.11, where coordinates (X_1, Y_1) of point 1 and bearing A to 1 (β_{A1}) are known, and angles (θ_1, θ_2) and distances (ℓ_1, ℓ_2) are measured.

The coordinates of point 3 can be given as follows:

$$X_3 = X_1 + \ell_1 \sin(\beta_{A1} + 180 + \theta_1) + \ell_1 \sin(\beta_{A1} + \theta_1 + \theta_2) \tag{13.27}$$

$$Y_3 = Y_1 + \ell_1 \cos(\beta_{A2} + 180 + \theta_1) + \ell_2 \cos(\beta_{A1} + \theta_1 + \theta_2) \tag{13.28}$$

The variance–covariance propagation laws can be applied to Equations (13.27) and (13.28) in order to determine the standard deviations of X_3 and Y_3. The propagated

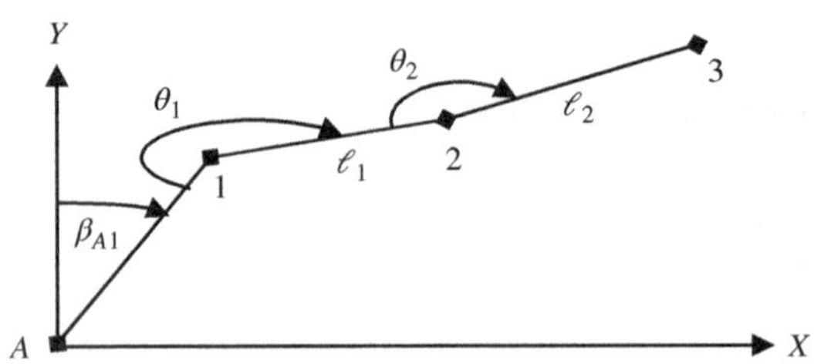

Figure 13.11 Open traverse.

variances of X_3 and Y_3 (taking $n = 3$) can be summarized as follows:

$$\sigma_{X_n}^2 = \sum_{i=1}^{n-1} (Y_n - Y_i)^2 \sigma_{\theta_i}^2 + \sum_{i=1}^{n-1} \left(\frac{X_{i+1} - X_i}{\ell_i} \right)^2 \sigma_{\ell_i}^2 \tag{13.29}$$

$$\sigma_{Y_n}^2 = \sum_{i=1}^{n-1} (X_n - X_i)^2 \sigma_{\theta_i}^2 + \sum_{i=1}^{n-1} \left(\frac{Y_{i+1} - Y_i}{\ell_i} \right)^2 \sigma_{\ell_i}^2 \tag{13.30}$$

where $\sigma_{\theta_i}^2$ and $\sigma_{\ell_i}^2$ are the variances of the angles and the distances, respectively. If in Figure 13.11, the measured angles are replaced by the gyro azimuths α_1 and α_2 at stations 1 and 2, respectively, the coordinates of point 3 will be given as follows:

$$X_3 = X_1 + \ell_{12} \sin(\alpha_1) + \ell_{23} \sin(\alpha_2) \tag{13.31}$$

$$Y_3 = Y_1 + \ell_{12} \cos(\alpha_1) + \ell_{23} \cos(\alpha_2) \tag{13.32}$$

The propagated variances of X_3 and Y_3 (taking $n = 3$) can be summarized as follows:

$$\sigma_{X_n}^2 = \sum_{i=1}^{n-1} (Y_{i+1} - Y_i)^2 \sigma_{\alpha_i}^2 + \sum_{i=1}^{n-1} \left(\frac{X_{i+1} - X_i}{\ell_i} \right)^2 \sigma_{\ell_i}^2 \tag{13.33}$$

$$\sigma_{Y_n}^2 = \sum_{i=1}^{n-1} (X_{i+1} - X_i)^2 \sigma_{\alpha_i}^2 + \sum_{i=1}^{n-1} \left(\frac{Y_{i+1} - Y_i}{\ell_i} \right)^2 \sigma_{\ell_i}^2 \tag{13.34}$$

where $\sigma_{\alpha_i}^2$ and $\sigma_{\ell_i}^2$ are the variances of the azimuths and the distances, respectively. If the distances are considered errorless, the variances of distances ($\sigma_{\ell_i}^2$) will be zero and all the corresponding terms in Equations (13.29), (13.30), (13.33), and (13.34) will become zero. Remember that the numbering of traverse points depends on whether the coordinates of point 1 (Figure 13.11) are fixed and errorless and the azimuth is errorless. For example, if the coordinates of point 1 are unknown and are to be determined from the distance *A*-1 and the azimuth β_{A1} and if the azimuth and the distance have some random errors associated with them, then point *A* will be numbered as $I = 1$ in Equations (13.29), (13.30), (13.33), and (13.34) with the azimuth measured at point *A* considered as the measured angle with its standard deviation ($\sigma_{\beta_{A1}}$) and the distance *A*-*B* and its standard deviation ($\sigma_{\ell_{A1}}$) used appropriately in the equations. For example, from Equation (13.29), the following will be obtained for the traverse with point *A* fixed, azimuth β_{A1} measured with a standard deviation $\sigma_{\beta_{A1}}$ and the distance *A*–1 (ℓ_{A1}) measured with a standard deviation $\sigma_{\ell_{A1}}$:

$$\begin{aligned} \sigma_{X_n}^2 &= (Y_3 - Y_A)^2 \sigma_{\beta_{A1}}^2 + (Y_3 - Y_1)^2 \sigma_{\theta_1}^2 + (Y_3 - Y_2)^2 \sigma_{\theta_2}^2 \\ &+ \left(\frac{X_1 - X_A}{\ell_{A1}} \right)^2 \sigma_{\ell_{A1}}^2 + \left(\frac{X_2 - X_1}{\ell_{12}} \right)^2 \sigma_{\ell_{12}}^2 + \left(\frac{X_3 - X_2}{\ell_{23}} \right)^2 \sigma_{\ell_{23}}^2 \end{aligned} \tag{13.35}$$

13.8.1 Analysis of Underground Traverse Surveys: Numerical Example

Points *A*, *B*, *C*, *D*, and *E* are in order along a practically straight tunnel as shown in Figure 13.12. Points *A* and *B* have known coordinates and can be considered errorless. Point *E* is to be coordinated off points *A* and *B* through a traverse having points *C* and *D* as intermediate stations. Each point is approximately 200 m from its immediate neighbor. The included angle at *B*, *C*, or *D* is ~180°, and the line of the five points can be considered parallel to the *x* coordinate axis. (Reproduced by permission of CBEPS)

(a) If each of the included angles has a standard deviation of ±5″, what is the lateral random error (i.e., σ_y) associated with the position of point *E*?

Solution: Refer to Figure 13.12 for the illustration of the problem.
By error propagation:

$$y_E = y_B + 200\cos(\alpha_{BA} + \beta_B) + 200\cos(\alpha_{BA} + \beta_B + 180 + \beta_C) + 200\cos(\alpha_{BA} + \beta_B + \beta_C + \beta_D)$$

$$\begin{aligned}\sigma_{y_E}^2 &= [200\sin(\alpha_{BA} + \beta_B) + 200\sin(\alpha_{BA} + \beta_B + 180 + \beta_C) \\ &\quad + 200\sin(\alpha_{BA} + \beta_B + \beta_C + \beta_D)]^2\sigma_{\beta_B}^2 \\ &\quad + [200\sin(\alpha_{BA} + \beta_B + 180 + \beta_C) + 200\sin(\alpha_{BA} + \beta_B + \beta_C + \beta_D)]^2\sigma_{\beta_C}^2 \\ &\quad + [200\sin(\alpha_{BA} + \beta_B + \beta_C + \beta_D)]^2\sigma_{\beta_D}^2\end{aligned}$$

or

$$\sigma_{y_E}^2 = [x_E - x_B]^2\sigma_{\beta_B}^2 + [x_E - x_C]^2\sigma_{\beta_C}^2 + [x_E - x_D]^2\sigma_{\beta_D}^2$$

This is similar to Equation (13.30) with the last term set to zero:

$$\sigma_{Y_n}^2 = \sum_{i=1}^{n-1}(X_n - X_i)^2\sigma_{\beta_i}^2$$

Since distance measurements are along the *x*-axis

$$\sigma_{y_E}^2 = [600]^2\left(\frac{5}{206,265}\right)^2 + [400]^2\left(\frac{5}{206,265}\right)^2 + [200]^2\left(\frac{5}{206,265}\right)^2$$

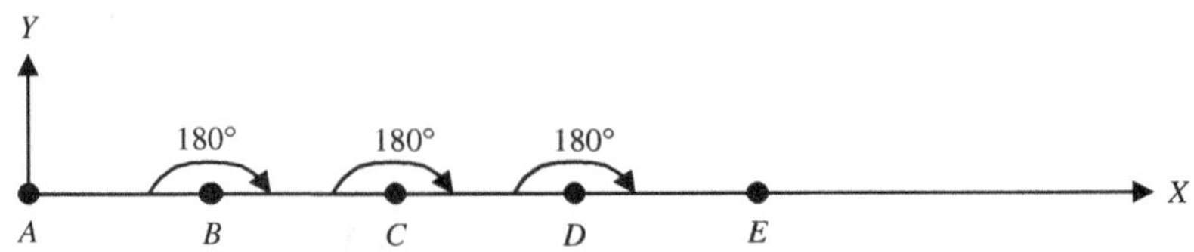

Figure 13.12 Design of an underground tunnel.

$$\sigma^2_{y_E} = \left(\frac{5}{206,265}\right)^2 [600^2 + 400^2 + 200^2]$$

$$\sigma^2_{y_E} = \left(\frac{5}{206,265}\right)^2 [560,000] \rightarrow 3.2906\text{E–}4\,\text{m}^2$$

$$\sigma_{y_E} = \mathbf{0.018\,m}$$

(b) If azimuths, rather than included angles, were observed [$\pm 5''$] at points B, C, and D, what would be the random lateral error in the position of point E?

Solution:

$$y_E = y_B + 200\cos(\alpha_{BC}) + 200\cos(\alpha_{CD}) + 200\cos(\alpha_{DE})$$

$$\sigma^2_{y_E} = [200\sin(\alpha_{BC})]^2\sigma^2_{\alpha_{BC}} + [200\sin(\alpha_{CD})]^2\sigma^2_{\alpha_{CD}} + [200\sin(\alpha_{DE})]^2\sigma^2_{\alpha_{DE}}$$

or

$$\sigma^2_{y_E} = [x_C - x_B]^2\sigma^2_{\alpha_{BC}} + [x_D - x_C]^2\sigma^2_{\alpha_{CD}} + [x_E - x_D]^2\sigma^2_{\alpha_{DE}}$$

This is similar to Equation (13.34) with the last term set to zero:

$$\sigma^2_{Y_n} = \sum_{i=1}^{n-1} (X_{i+1} - X_i)^2\sigma^2_{\alpha_i}$$

Since distance measurements are along the x-axis

$$\sigma^2_{y_E} = [200]^2\left(\frac{5}{206,265}\right)^2 + [200]^2\left(\frac{5}{206,265}\right)^2 + [200]^2\left(\frac{5}{206,265}\right)^2$$

$$\sigma^2_{y_E} = \left(\frac{5}{206,265}\right)^2 [200^2 + 200^2 + 200^2]$$

$$\sigma^2_{y_E} = \left(\frac{5}{206,265}\right)^2 [120,000] \rightarrow 7.0513\text{E–}5\,\text{m}^2$$

$$\sigma_{y_E} = 0.0084\,\text{m}$$

13.8.2 Gyro Orientation of Underground Surveys: Numerical Example

Given that the grid azimuth of line AB is equal to $26°16'30''$, a gyrotheodolite was calibrated on line AB giving the gyro azimuth of the line as $27°14'00''$. The same gyrotheodolite was used at station C in order to determine the grid azimuth of line CD. The gyro azimuth of CD was $72°20'00''$. What is the grid azimuth of line CD if the grid X-coordinates from the central meridian for points A and C are $X_A = 101{,}250$ m, $X_C = 102{,}416$ m, respectively; and the latitudes of the points are $\phi_A = 43°20'30''$ and $\phi_C = 43°21'00''$? (Assume the radius of the earth is 6378.3 km.)

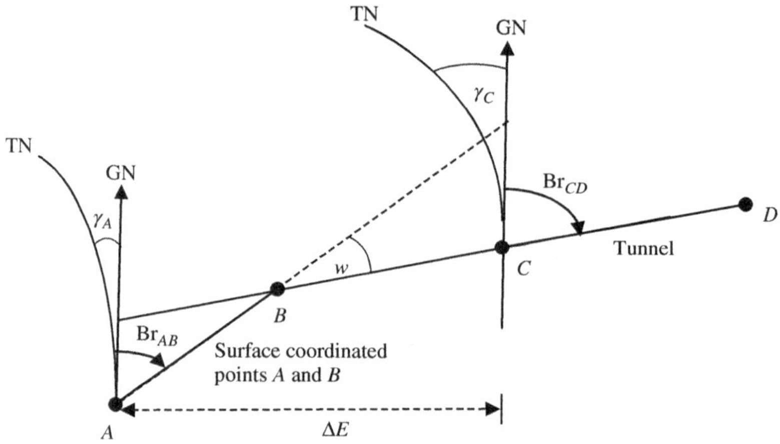

Figure 13.13 Gyro orientation procedure in a tunnel.

Solution steps: Consider Figure 13.13, in which GN represents the direction of Grid North and TN represents the direction of True North (Astronomic North), and assume the following:

γ is the convergence of meridian at the given point.
A_{AB} is the gyro (astronomical) azimuth of the surface line *AB*.
A_{CD} is the gyro azimuth of the underground line *CD*.
Br_{AB} is the grid azimuth of line *AB*.
Br_{CD} is the grid azimuth of line *CD*.

It can be shown from Figure 13.13 that

$$w = (A_{CD} - A_{AB}) - (\gamma_C - \gamma_A) \tag{13.36}$$

or

$$w = (A_{CD} - A_{AB}) - \Delta\gamma \tag{13.37}$$

where $\Delta\gamma = (\gamma_C - \gamma_A)$. From Figure 13.13, the following can also be derived:

$$\mathrm{Br}_{CD} = \mathrm{Br}_{AB} + w$$

or

$$\mathrm{Br}_{CD} = \mathrm{Br}_{AB} + (A_{CD} - A_{AB}) - \Delta\gamma \tag{13.38}$$

or in general,

$$\mathrm{Grid}_{CD} = \mathrm{Grid}_{AB} - (\mathrm{Gyro}_{AB} - \mathrm{Gyro}_{CD}) - (\gamma_C - \gamma_A) \tag{13.39}$$

Since the surface and underground lines are in the *E-W* direction (along approximately the same latitude 43°N), the following approximate formula (Equation (12.33)) can be used:

$$\Delta\gamma = \frac{\Delta E \tan\phi}{R} \times 206{,}265'' \ (R = 6378.3\,\text{km}) \tag{13.40}$$

Assume $X_0 = 0\,\text{m}$ is the origin of the coordinate system; for point *A*, $\Delta E = 101{,}250\,\text{m}$

$$\gamma_A = \frac{101{,}250 \tan 43°20'30'' \times 206{,}265}{6{,}378{,}300}$$

$$\gamma_A = 1.4980839\text{E–}2 \times 206{,}265'' \ \rightarrow 3090.02''(\text{or}\ \ 0°51'30.0'')$$

For point *C*, $\Delta E = 102{,}416\,\text{m}$

$$\gamma_C = \frac{102{,}416 \tan 43°21'00'' \times 206{,}265}{6{,}378{,}300}$$

$$\gamma_C = 1.515777514\text{E–}2 \times 206{,}265'' \ \rightarrow 3126.52''(\text{or}\ 0°52'06.5'')$$

$$\text{Gyro azimuth } (AB), A_{AB} = 27°14'00''$$

$$\text{Gyro azimuth } (CD), A_{CD} = 72°20'00''$$

$$\text{Grid azimuth } (AB), Br_{AB} = 26°16'30''$$

$$\text{Grid azimuth } (CD), Br_{CD} = ?\ (\text{to be determined})$$

On baseline AB: Gyro calibration constant, *E* (from surface) can be calculated from Equation (12.23) as follows:

$$E = \text{Gyro Azimuth (uncorrected)–[Grid Azimuth + Convergence]}$$

$$E = 27°14'00'' - [26°16'30'' + 0°51'30.0'']\ \text{or}\ \ 0°06'00.0''$$

Underground:

$$\text{Grid Azimuth} = \text{Gyro Azimuth (uncorrected)–Convergence–}E \tag{13.41}$$

or

$$\begin{aligned}\text{Grid azimuth } CD &= \text{Gyro azimuth} - \gamma_C - E\\ &= 72°20'00'' - 00°52'06.5''–0°06'00.0''\\ &= \underline{71°21'54''}\end{aligned}$$

The same result is obtained by using Equation (13.39) as follows:

$$\text{Grid}_{CD} = \text{Grid}_{AB} - (\text{Gyro}_{AB} - \text{Gyro}_{CD}) - (\gamma_C - \gamma_A) \tag{13.42}$$

$$\begin{aligned}\text{Grid}_{CD} &= 26^\circ 16'30'' - (27^\circ 14'00'' - 72^\circ 20'00'') - (0^\circ 52'06.5'' - 0^\circ 51'30.0'') \\ &= 26^\circ 16'30'' + 45^\circ 06'00'' - 0^\circ 00'36.5'' \rightarrow 71^\circ 21'54''\end{aligned}$$

14

PRECISION ALIGNMENT SURVEYS

OBJECTIVES

At the end of this chapter, you should be able to

1. Describe the main techniques of precision alignment
2. Perform alignment surveys based on a particular technique
3. Explain the advantages and limitations of different alignment techniques
4. Design and implement observation schemes for alignment surveys based on three-dimensional electronic coordinating system
5. Describe different optical-tooling instrumentations
6. Perform vertical and horizontal alignment using optical-tooling techniques
7. Discuss the principle and applications of laser interferometer for alignment in small-scale metrology
8. Describe the alignment in large-scale metrology using polar measurement systems such as laser trackers and industrial robotic total stations
9. Evaluate sources of error and their propagation in alignment surveys

14.1 INTRODUCTION

Precision alignment surveys usually require that three or more points be collinear or coplanar. Alignment surveys cover a large area of engineering applications from the

Precision Surveying: The Principles and Geomatics Practice, First Edition. John Olusegun Ogundare.

tooling industry to deformation measurements of long engineering structures (e.g., deformation monitoring of nuclear accelerometers of several kilometers long). Each application, however, may require different specialized equipment. The methods used in practice may be classified according to the technique for establishing the reference line, such as the following techniques:

1. *Mechanical alignment technique* in which a steel or nylon thread (appropriately tensioned) is used to establish the reference line. This method is attractive because of its simplicity and its adaptability to continuous data collection, which can be used in structural deformation monitoring. Accuracies of up to 0.1 mm have been quoted (Chrzanowski, 1993) for this technique. The major disadvantage of this method lies in its use as a vertical reference frame. The two height differences that enable the vertical curve to be computed must be measured, and Cartesian coordinates must be calculated, with the highest possible accuracy. The advantages of this method can be summarized as follows:
   - Offsets can be measured with micrometer precision, and the wires are unaffected by radiation or refraction.
   - A large number of sensors can be used simultaneously on the same reference line.
   - Planimetric position is very well defined since only the positions of the two end points need to be known.
   - Vertical position is very well defined provided the two height differences are measured with sufficient accuracy.
2. *Diffraction alignment technique* in which a projected pattern of diffraction slits is used as a reference line. This method uses diffraction zone plates (with laser point source and equidistant or Fresnel zone plates and centering detector). In this method, the zone plates act as focusing lenses and the method is less affected by the atmosphere than the direct optical method. Note that in this method the laser is just a source of monochromatic light behind the pinhole and not a reference line. The laser source, the center of diffraction slits, and the center of the photo-electric sensor target form the three basic points of the alignment line.
3. *Direct laser alignment technique* in which coherent laser beam directly provides the reference line. This technique uses collimated laser beam and optical or photoelectric movable centering detectors. When used, the technique is usually very fast; and it also requires no communication between the observer (laser man) and the target man. Moreover, the technique might be used during a very strong thermal turbulence condition when it would have been practically impossible to use the conventional surveying techniques of alignment.
4. *Conventional surveying techniques* in which two coordinated points define a reference line. The techniques may use direct optical line of sight as in the cases of direct laser alignment and alignment based on three-dimensional coordinating system, electromagnetic distance measurement (EDM) equipment, electronic theodolites, and levels.

5. *Optical tooling techniques* in which optical line of sight provides the reference line directly.
6. Metrology by laser interferometer techniques in which angular and straightness measurements are made using laser interferometers.
7. *Alignment by polar measurement techniques* in which systems such as laser trackers and industrial robotic total stations are used in relation to large-scale metrology (LSM). These techniques use spherically mounted reflector (SMR) as targets and measure precisely to those targets the spherical coordinates, such as the linear distances, azimuths (or horizontal angles), and vertical angles. These measurements are then converted in real time to three-dimensional Cartesian (X, Y, Z) coordinates of the center locations of the targets.
8. *Hydrostatic alignment techniques* used for defining the vertical reference frame for positioning of components of the accelerator along a straight line. This alignment technique uses the equipotential surface (or level water surface) in the earth gravity field as a reference. The technique works by using a system of two vessels connected to each other by pipes of a diameter of 60 mm partially filled with water and allowing water and air to circulate freely within the system. To eliminate the effects of differential variations of atmospheric pressure, the whole pipe work system is only open to free air at one point. The vessels are equipped with temperature sensors. The vessels, pipes, and casing of the sensors are made of stainless steel. The unit consisting of the vessel and the sensor forms a cylinder, with a diameter of 100 mm and a height of 120 mm. The usual measuring range of the system is 5 mm. The major disadvantage of using this method for vertical referencing is that water levels follow equipotential surfaces of the earth's gravitational field. It is difficult to determine the geometry of such surfaces in relation to a reference frame so as to allow a straight line to be established. One usually needs a good geoid model to form the basis for the determination of the corrections to be made in order to return to a straight line. The method, however, has some advantages:
   - Vertical position is very well defined provided the two height differences are measured with sufficient accuracy.
   - It can provide height measurements to micrometer precision and it is unaffected by radiation.
   - The long and continuous reference system that it provides allows height differences to be determined very accurately by reference to a water level over much greater ranges than optical leveling.

Items 3–7, which are affected by the atmospheric refraction, pointing, and focusing errors, are becoming more popular in geomatics practice. They will be explored further in the following sections with regard to the principles involved, applications, and the sources of error and procedures to mitigate them.

14.2 DIRECT LASER ALIGNMENT TECHNIQUE

The property of the laser emitting a particularly collimated, directional beam gives instruments used for alignment an advantage, as an extension of the plumb line on a generic direction, which is not necessarily vertical. In an alignment, it is usually of interest to project a beam at a distance of interest and be able to keep the size of the beam as smallest as possible along the given path. This is possible with laser, so that it is used for positioning objects along a desired direction, as indicated by the propagation direction of the laser beam. Actual range of laser is limited primarily by weather conditions, which affect atmospheric attenuation and turbulence. In the presence of haze and fog, the optical power is attenuated and the useful range drops.

The direct alignment with laser method uses the centroid of energy of a collimated He–Ne laser beam as a reference line for the alignment measurements. The procedure is such that a laser with a collimating telescope is placed behind the end point of the alignment line and the laser beam is used to define the reference line. A universal base plate with slow motion horizontal and vertical adjustment screws must be used for mounting on different types of lasers and collimating telescopes. The commonly used telescopes have a magnification of 80× and a 90 mm-diameter objective lens. The magnification size of the telescope is to ensure that any directional drift (Chrzanowski and Janssen, 1972) of the laser output would not be significant on the stability of the laser beam and also to permit the focusing of the laser beam to a spot within the limited dimensions of *self-aligning centering detectors* or *zeroing targets* used at a long distance of a test line. This technique allows for automated alignment procedure with continuous data collection.

14.3 CONVENTIONAL SURVEYING TECHNIQUES OF ALIGNMENT

The conventional surveying techniques of alignment discussed in this section consist of alignment procedures requiring an establishment of network of coordinated points. The procedures include traversing with forced centering, such as a closed (loop) traverse, fitted traverse, and separate-point-included-angle traverse (using precision theodolite and fixed targets); and three-dimensional coordinating system; trilateration network measurement.

It should also be mentioned here that for computational purpose, a suitable computational surface must be chosen for any engineering project. In engineering projects, a global reference ellipsoid such as the international ellipsoid of the Geodetic Reference System of 1980 (GRS80) is used in the North American Datum of 1983 (NAD83), the European Terrestrial Reference System of 1989 (ETRS89), and in the World Geodetic System of 1984 (WGS84) used in Global Positioning System (GPS). All the control survey measurements must be reduced to the reference ellipsoid by correcting the distances for geoid undulation and angles and directions (or azimuths) for the effects of the deflection of the vertical. Local three-dimensional coordinates (X, Y, Z) with the origin at the center of the reference ellipsoid may be computed when a large surface is involved. Sometimes, it may be required to reduce

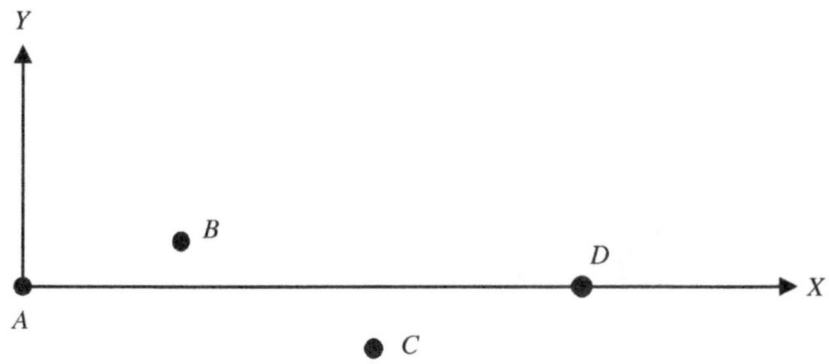

Figure 14.1 Alignment of points *B* and *C*.

the field measurements to some mapping plane based on some criteria; for a circular superconducting super collider (SSC) project, the double stereographic (conformal) map projection was considered (Chrzanowski et al., 1993). If GPS ellipsoidal heights are measured, they must be reduced to more meaningful orthometric heights by applying the geoid undulations.

A typical alignment problem. Consider the alignment of points *B* and *C* with the line *A-D* in Figure 14.1 in evaluating the different conventional surveying techniques of alignment. In the figure, a local (x, y) coordinate system is used, taking point *A* as the origin of the system and the line *A-D* as the x-axis. The estimated y-coordinates of *B* and *C* will be considered as the alignment result.

Consider a conventional alignment procedure with the movable targets located at *B* and *C*, a fixed target located at *A* or *D* and the alignment theodolite (DKM3 optical theodolite) located on pillar *A* or *D*. Traverse targets can be used as the movable targets if adapted on a slow motion sliding device having vernier-type readout of 0.05 mm resolution, and a radio communication may be required between the observer and the target man. This method of alignment, which is also referred to as single-station-small-angle method, is illustrated in Figure 14.2. The small angles (α_1 and α_2) measured at point *A* are used in determining the alignment corrections Y_B (mm) and Y_C (mm). Each angle measurement may consist of up to 12 pointings on the fixed target and 12 readouts on the aligned movable target in order to calculate the alignment results Y_B (mm) and Y_C (mm); the distances are measured with Mekometer ME3000 or ME5000.

14.3.1 Traversing Method of Alignment

Three field procedures may be considered under traversing method: *closed traverse*, *fitted (or open) traverse,* and *separate-point-included-angle* methods. Closed traverse and separate-point-included-angle methods will give comparably precise results; fitted method may be less precise.

14.3.1.1 Closed Traverse In closed traverse procedure, angles are measured on pillars *A*, *B*, *C*, and *D*, using precision theodolite (such as DKM3 optical theodolite) and suitable targets on the self-centering base plates (refer to Figure 14.3). Six

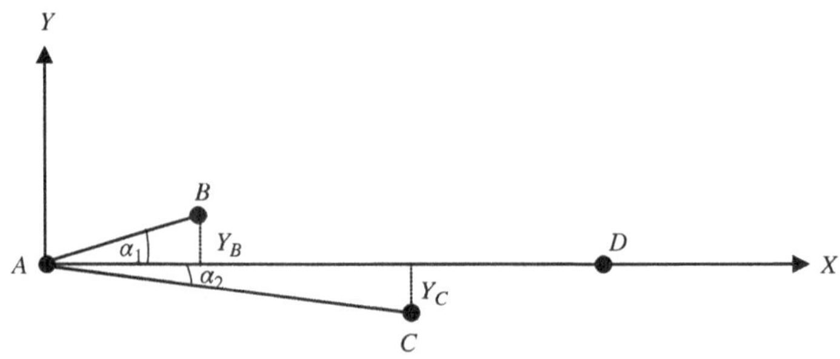

Figure 14.2 Single-station small angle method of alignment of points B and C.

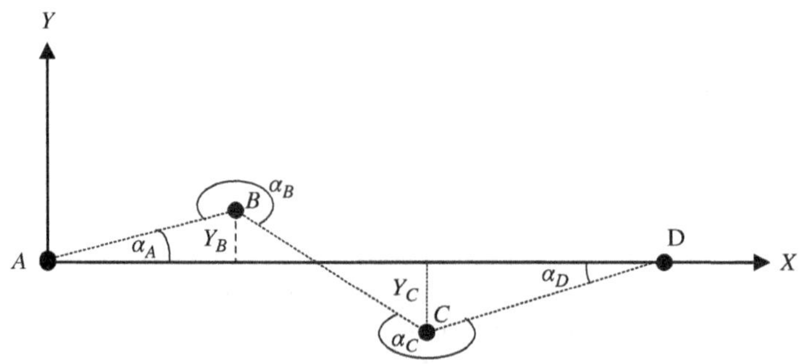

Figure 14.3 Closed traverse method.

sets of the angle measurements are usually made on each pillar forming a closed traverse (with directions *A-D* and *D-A* included), and the distances are measured precisely using precise EDM (such as Mekometer ME5000). The six sets of angle measurements are used in estimating the variances of the angle measurements, and the specified precision of the EDM is used to estimate the variances of the distance measurements. The measurements and the estimated variances are then used in the parametric least squares adjustment method in order to calculate the *Y*-coordinates of *B* and *C* as the alignment result.

14.3.1.2 Fitted (or Open) Traverse The open traverse procedure will use the same traverse data from the closed traverse shown in Figure 14.3 except that the angles α_A and α_D at the pillars *A* and *D*, respectively, are not measured in the case of open traverse method. As usual, Y_B (mm) and Y_C (mm) are the alignment results to be determined.

14.3.1.3 Separate-Point-Included-Angle Traverse In the separate-point-included-angle traverse procedure, independent traverses for points *B* and *C* are measured as shown in Figure 14.4. This is done by independently measuring the angles at *B* and *C*. Usually, six sets of angle measurements are made at each point using precision

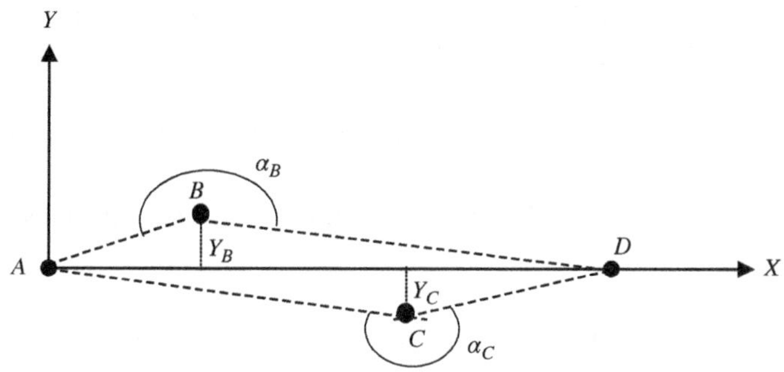

Figure 14.4 Separate point included angle method of alignment of points B and C.

theodolite, and the distance measurements are precisely made using precision EDM. The measurements and their estimated variances are then used in the method of least squares adjustment in calculating the alignment results Y_B (mm) and Y_C (mm).

14.3.2 Alignment with Three-Dimensional Electronic Coordinating System

The problem in Figure 14.1 can be considered as a problem of aligning machine components B and C on line AD in an industrial environment. This can be considered as a case of industrial metrology, defined by Wilkins (1989) as "*... a discipline of engineering surveys that requires the utmost in achievable accuracies.*" The concepts of industrial metrology are applied in a number of projects, which include positioning accelerator components, determining the shape of assembled surfaces, calibrating a robotic arm, alignment surveys carried out in areas of limited extent, and precise positioning of some engineering structures in a certain arrangement in an area of limited extent.

One procedure often used in surveying for the determination of coordinates is intersection method. If it is intended that such measurements will be repeated in the same area, it is necessary to establish a reference geodetic micro-network of fixed points. The fixed points can be installed either as theodolite stations, as pillars, as brackets on walls or as marks fixed on the walls at suitable heights. In aligning components B and C in Figure 14.1, a three-dimensional reference geodetic micro-network, usually associated with industrial metrology, is first established. The reference micro-network can be considered as an array of wall targets set up to aid in the proper alignment or setup of the machine components. These targets forming the geodetic micro-network can be installed at regular increments on the walls paralleling the components being aligned. The establishment of this reference geodetic micro-network is to provide the following:

(a) Means of estimating three possible translations of the misalignment of the components in the desired alignment direction
(b) Rigorous means of estimating the errors of the target locations

(c) Flexibility in selecting the location of the coordinating systems (such as electronic theodolites) depending on the shape, size, and position of the component to be measured
(d) A network of reference points for future monitoring of the status of the aligned components
(e) A network of reference for the design coordinates of the components in the work area.

Typical survey items for establishing the geodetic micro-network will include the following:

(i) Adhesive type of concentric circle patterned targets, as shown in Figure 14.5. The best targets are said (Keuffel & Esser Co., 1957) to be made of white space between two black lines or areas. A series of narrow white spaces of different widths separated by black spaces can be chosen such that the series of paired black lines (or areas) on the target are spaced so that at whatever distance it is observed, at least one white space will be of such a width that subtends an angle of between 8 and 21 arcsec. The concentric circle target patterns (Figure 14.5) with white and black spaces are able to allow oblique lines of sight to be used to observe the targets and to facilitate simultaneous horizontal and vertical pointings to the targets (Wilkins, 1989).
(ii) Some support for the wall targets, such as brass plaques having surface areas suitable to adhere the adhesive target and a name target. The plaques, which can be 5 mm or more in thickness, are to be installed on the concrete walls at the target locations.
(iii) Well-calibrated (to accuracy of about ±0.01 mm) invar scale bar to provide scale for the network. The suitable scale bars are usually 2–3 m long due to

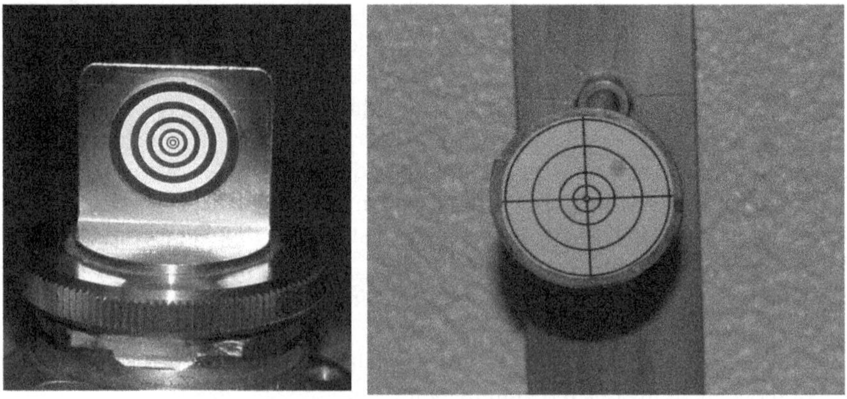

Figure 14.5 Concentric circle wall target designs.

constraints of calibration process and the need to be able to transport them without tampering with the calibration.

(iv) Coordinating system to measure the horizontal directions and the zenith angles to accuracies of at least $\pm 1''$, depending on the positioning accuracy desired. The coordinating system applies three-dimensional coordinate geometry in positioning its stations and the target locations on the components being aligned. Total station equipment or electronic theodolite can be used as a coordinating system as discussed in Chapter 8.

14.3.2.1 Measurement of Reference Micro-Network The method of triangulation without sighting between instrument stations can be used to coordinate the targets in the micro-network. The direction and zenith angles are measured to the targets and scale bar targets by using the electronic theodolite as a coordinating system. Each target must be sighted from at least two theodolite setup stations. The calibrated scale bars with their two ends targeted are to be situated in different locations throughout the micro-network. The electronic theodolite can be used to perform some observations at one setup station and then moved to the next station to observe the next set of targets, and so on. All of the observations gathered are then combined to perform a single simultaneous least squares adjustment to obtain the three-dimensional coordinate estimates of the target locations. The spatial coordinates of subsequently surveyed components based on these targets will then refer to the local coordinate system defined by the setup stations of the theodolite.

In order to be able to perform the least squares adjustment of the micro-network, a reference datum must be defined for the network. The following steps can be followed in defining the datum for the micro-network:

- *Origin* – A point in the center of one of the components already positioned in the work place is targeted and its design three-dimensional position (x, y, z) held as fixed in any subsequent adjustment.
- *Orientation* – The direction of the alignment taken as the x-axis of the local coordinate system is fixed; direction of gravity is taken as z-axis and the y-axis will be perpendicular to the x–z plane in a right-handed system. These directions will be held fixed to provide three orientations for the network.
- *Scale* – Since distances are not measured, the invar scale bars well positioned in the work area are to provide the needed scale for the network.

14.3.2.2 Measurement of Object Micro-Network The targets on the components to be aligned can be considered as constituting the object network. After establishing the reference geodetic micro-network, setting out of the machine components in an industrial metrology can commence. The setting-out of machine components can be put into three parts as follows (cf. Wilkins, 1989):

1. Targeting of components
2. Coarse alignment or prealignment surveying
3. Fine alignment and smoothing surveying.

Targeting of Components Targeting of components involves placing appropriate targets on the components and defining the "true" axis of the components to be positioned and aligned according to design. Remember that the true axis of the components may be magnetic or electrical axis, which may be physically difficult to determine.

Coarse Alignment The coarse alignment or prealignment surveying is usually part of the original construction work when the components are approximately put in their nominal locations. Usually, the components are supported on stands that are bolted to the floor and confirmed to be stable. The support stands are set within the working range of the fine adjustment mechanisms for the components and the components must have freedom to translate and rotate on the stands. The procedure simply involves an alignment telescope for horizontal orientation, tilting level for vertical orientation, and a steel tape for the distance along the beam line (Wilkins, 1989). Once aligned, the stands are bolted to the floor and their alignment rechecked; and if the alignment is found satisfactory, the stands are grouted in place for stability.

Fine Alignment Fine alignment is an iterative process in which adjustments are made to the coarse alignment until the locations of the components converge to the nominal locations within some specifications. During this process, the coordinating system can be located around the components to be set out in a way that will optimize the intersection geometry and create very small distances from the components.

The process of coordinating the components can involve obtaining the resected positions of two or more theodolites set up in the work area by observing to the wall targets as control points and subsequently determining the intersected coordinates of the targets on the components being aligned. These may require performing two sets of direction measurements to at least eight wall targets for each resection in order to randomize the effects of the wall target errors. The intersected (x, y, z) coordinates of the target locations on the aligned components are then determined from the resected positions of at least two of the coordinating systems. The computed coordinates of the targets on the aligned components and their design coordinates are used to compute necessary adjustments to be made to the positions of the components; this process of "intersection of targets on the components and the subsequent adjustments of locations of components" is repeated until the locations of the components correspond to the design locations within some specifications.

Quality Analysis of Alignment The coordinates of the targets on the components being aligned are based on resection and intersection methods. To be able to carry out resections of theodolite positions requires that control points have been previously established to a suitable density and accuracy, which depends on the project specifications. If the accuracy of the control points is not suitable enough to treat the control points as fixed and errorless, weighted least squares adjustment procedure may be employed, in which the inverse of covariance matrices of the control points are used as weights. For the intersected object points, the accuracy of their coordinates will depend on the following:

- Accuracy of the calibrated scale bar distances and the horizontal and vertical angle measurements from which the micro-network data are derived
- Magnitude of the intersection angle as discussed in Chapter 8
- Target quality and illumination conditions
- Stability of theodolite locations.

Some of the instrument systematic errors (refer to Chapter 4) can be taken care of if the targets are observed in both direct and reverse telescope positions and the measurements averaged, with well-established procedures for measuring sets and pointing order within a set followed. Index error and true collimation error for each instrument are unique and should remain constant for each set, and as such provide a very good assurance of data quality. The true collimation error based on Equations (4.1) and (4.2) and index error derived from Equation (4.3) can be used as consistency checks (Wilkins, 1989). If the consistency checks are unacceptable, the current set for that observable should be repeated until the discrepancy is acceptable. Discrepancy between the mean observation of two telescope positions in the current set and the mean value from the same observable from previous sets can be used as part of quality filter. The check is done against known tolerance. For example, if the testing is assumed to be at 95% confidence level, the tolerance value would be twice the standard deviation of the sample that the tested quantity has been pulled from; if the check is unacceptable, the measurement in that set should be repeated. Usually, repeated measurements of the same point and averaging the coordinate values are to improve the accuracy coordinate determination.

14.3.2.3 Notes on Alignment of Underground Nuclear Accelerators The main problem in accelerator alignment, from the surveying view point, is the orientation transfer from the surface reference network to the tunnels since tunneling is started from the shafts. The various approaches for underground orientation discussed in Chapter 12 are relevant to the discussion in this section. Accelerator alignment usually consists of the following steps:

- Surveying of the surface network.
- Surveying of the underground networks, usually a ring in the case of circular accelerators or linear in the case of linear accelerators; in the case of linear accelerator, diffraction grating with laser source may be used to provide absolute straightness with sub-millimeter accuracy over a long distance (greater than 3 km).
- Prealignment surveying (discussed in Section 14.3.2).
- Final alignment and smoothing surveying (discussed in Section 14.3.2)

In accelerator alignment, two tolerance specifications for positioning are provided: one for the absolute positioning and the other for relative positioning of components. The absolute positioning tolerance defines a maximum global geometric distortion by specifying how closely the components have to be placed on their ideal location.

The more important relative tolerance defines the alignment quality of adjacent components. Absolute positioning accuracy depends on a number of error sources. Some of the error sources are as follows (Chrzanowski et al., 1993):

- Error in surface network
- Error in transferring orientation through shafts
- Error in underground control surveys
- Error in identifying the axes of the components to be aligned
- Error in finally orienting the components in their ideal location
- Systematic residual error due to instrument calibration
- Residual systematic error due to vertical refraction.

In every alignment, a suitable mathematical model for computations must be defined. This involves choosing the reference ellipsoid and the map projection method. In this case, the measured gyro azimuths, angles, and distances must be corrected for the deflection of the vertical and the geoid undulations to reduce them to the reference ellipsoid, and the corrected measurements must subsequently be reduced to the mapping plane by applying appropriate corrections. The components are then located in the map coordinate system for further analysis. Typical high-precision equipment for alignment can include the following:

- Wild T3000 theodolites for direction measurements.
- Distomat Wild DI2002 and Mekometer Kern ME5000 for precise distance measurements.
- Calibrated invar scale bar for providing scale in the case where distances cannot be precisely measured.
- Optical plummet such as Wild NL for shaft plumbing.
- Precision level such as Wild N3 with invar staffs for level transfer.
- Gyromat 5000 with an accuracy of $\pm 2.5''$ and Gyromat 3000 with an accuracy of $\pm 3''$, for azimuth determination. They are useful in verifying directions between wall brackets and monuments used for primary tunnel control.
- Aligning telescope, such as Taylor Hobson micro-alignment telescope for prealignment of components.

14.4 OPTICAL-TOOLING TECHNIQUES

Optical tooling is a special branch of surveying (ASCE Manuals, 1985) that uses powerful telescopic sights to establish precise reference lines and planes for aligning integral parts of large industrial products. For example, parallel and angular misalignment of shafts in machinery must be held within close tolerances to avoid excessive wear and vibration; a permanent installation such as a coupled turbine and generator in a power plant requires accurate alignment; a set of roll stands in a steel mill or even a paper mill may require precise alignment; in airplane industry, precise alignment is required when bringing fabricated wing and fuselage sections together for final assembly and when ensuring that each component and subassembly is in proper relation to the complete assembly. Instruments for measuring such misalignment would generally be expected to have a resolution of 0.025 mm of offset and 0.00015 rad of angular misalignment (Kissam, 1962).

Because of the high accuracies required and the short distances usually involved in optical tooling, several fundamental departures from ordinary surveying practice can be given (Kissam, 1956) as follows:

1. The line of sight of any telescopic instrument used must be extremely straight; the direction of the line of sight must remain the same within very tight limits when the focus is changed, especially on short sights.
2. Since accuracies of 1/200,000 or better are involved, measurements are required to be made with a micrometer. When measurements are made from a line of sight, that is, at right angles to the line of sight, an optical micrometer attached to the telescope is used with either a special scale or with very precise tape.
3. Horizontal and vertical planes must be established with a greater accuracy that can be obtained from conventional surveying. Based on this, only levels capable of geodetic accuracies can be used, and total stations, which in optical tooling are employed only to establish vertical planes, must be specially designed so that the horizontal axis can be kept horizontal to a high degree of accuracy; commonly used optical-tooling transits have negligible axial errors.
4. Because of the need for setups near the floor or in other cramped positions, most instruments must be designed so that angle eyepieces and angle devices for observing level bubbles can be attached in order that the instruments can be used from the top or from the side.

The basic principles of optical alignment include determining the following:

1. *Straightness*, which is determined using the telescopic line of sight as a reference. Some of the advantages of using the optical reference line are that the optical reference line cannot sag, vibrate, bend, or kink like tape or wire; and measurements can be made directly to the center of the reference line without the danger of disturbing the line during the measurement process.
2. *Flatness*, which is determined based on the principle of precise leveling with leveling equipment similar to that used in conventional surveying. Using precise leveling instrument with an optical micrometer and a special paired-line optical alignment scale, it is possible to measure offsets accurately from the horizontal reference plane established by the instrument to 0.025 mm.
3. *Plumbness* in which vertical reference plane is established by using telescopic line of sight. Instead of using a plumb bob to establish a single vertical reference line, a Jig transit can be used to establish a vertical reference plane. The parallelism between the reference plane and any other surface can then be determined by measuring offsets between the two planes. The use of the

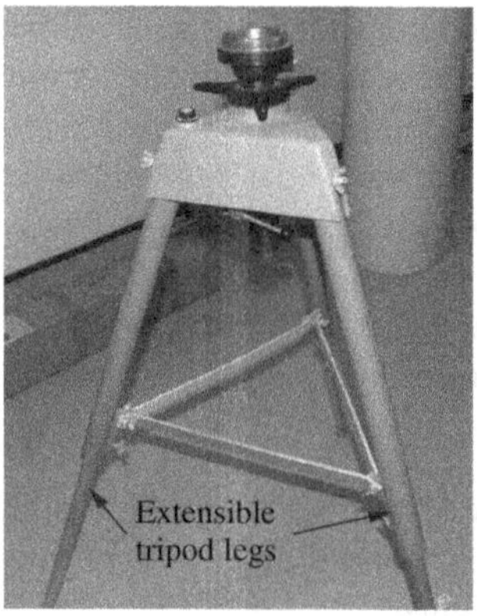

(a)

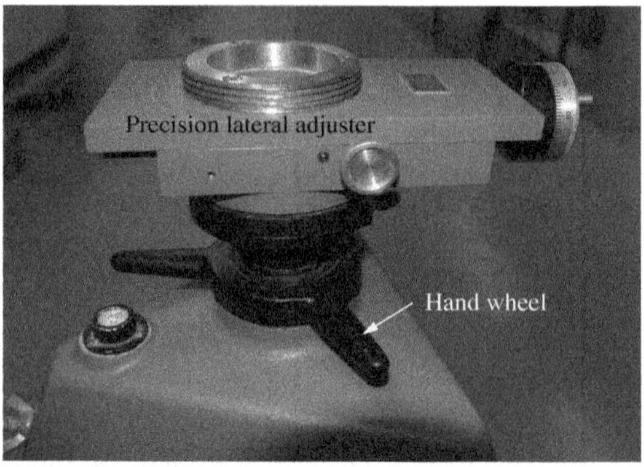

(b)

Figure 14.6 Special (a) Instrument stand and (b) Precision lateral adjuster mounted on the instrument stand.

optical micrometer and paired-line scales enables the measurements to be made directly to 0.025 mm.

4. *Squareness* in which a surface is established perpendicular to the telescopic line of sight. Squareness can be determined using the following two methods:
   - If a surface with a relatively small area is to be set at right angles to the line of sight, it can be done by mounting a mirror on that surface so that the mirror is parallel to the surface; the mirror and the surface can then be set at right angles to the optical line of sight by either autocollimation or auto-reflection procedure.
   - If a large surface is to be set perpendicular to the line of sight or to be set up at any right angle with relatively long legs, a Jig transit and a reference collimator can be used. In some special cases, the optical square mounted on the end of an alignment telescope can be used.

14.4.1 Optical-Tooling Instruments

14.4.1.1 Special Instrument Stand and Precision Lateral Adjuster A typical instrument stand with adjustable legs is shown in Figure 14.6(a), and in Figure 14.6(b), a precision lateral adjuster is mounted on the instrument stand. The stand can be raised or lowered with a hand-wheel and clamped at the desired instrument height; it is usually heavy, easily placed in position, and provides a firm three-point support. The lateral adjuster is a support for sliding an instrument mounted on it left or right by a few centimeters without throwing the instrument much out of level.

14.4.1.2 Alignment Telescope Alignment telescope (shown in Figures 14.7 and 14.8) consists of a telescopic sight built into a heavy, chrome-surfaced steel tube or barrel in front of an enlarged section where the focusing lens and the optical

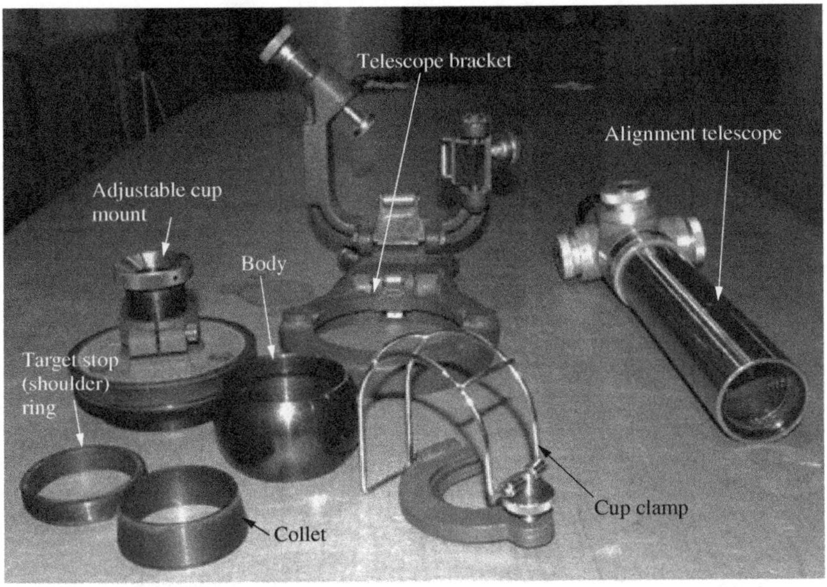

Figure 14.7 Paragon alignment telescope with the accessories to mount it.

micrometer controls are located. The rear section of the alignment telescope contains built-in optical micrometers, a focusing knob, and an eyepiece; the magnifying power of the telescope is usually about 40–60 times. The optical micrometers enable it to measure precise horizontal and vertical displacements; typically, one division of the micrometer is equal to 0.025 mm. The telescope is erecting and can be focused from the face of the objective lens to infinity with the readings on the focusing screw providing approximate distances between the instrument and the focused target in feet; cross level or striding level is usually placed on the telescope for leveling it.

The other important feature of an alignment telescope is the built-in auto-reflection target on the inner surface of the objective lens of the telescope as shown in Figure 14.8. There is also a built-in autocollimation unit with an illumination unit built into the eyepiece. The light source for the illumination can be removed when desired.

The alignment telescope can be mounted vertically or horizontally. The accessories for mounting the telescope horizontally in a bracket are shown in Figure 14.7. The telescope mounts can be in the form of sphere and cup type as shown in Figure 14.8, or in the form of cone-type V-block. In the case of sphere and cup type, the adjustable cup mount has a tapered base for providing a mount for the alignment bracket (as shown in Figure 14.9). The center of the spherical cup is a datum point so that when the telescope is mounted in the spherical cup on a horizontal base, the cup permits all tilting or leveling adjustments to be made without altering the position of the original line of sight passing through this datum point. The spherical cup can also be used to mount a target, for which purpose a target stop ring becomes necessary. Line of sight established with the alignment telescope forms the basic reference for all measurements.

Figure 14.8 Spherical cup being supported on a large-diameter screw thread in the base of the mount and the alignment telescope showing the auto-reflection target in the objective lens.

There are several ways of mounting the alignment telescope horizontally, either right side up or upside down. The spherical adapter (Figure 14.8) can be slid over the barrel and clamped where desired. It is then placed in a cup mount and held in position by a clamp. The cup mount is then bolted in position on the object being aligned or the instrument support and adjusted in height by an elevating screw. Attached to the base of the cup mount is an alignment telescope bracket, which provides tangent screws for aiming the telescope; and a striding level can be used to level the line of sight. The assembled alignment telescope is shown in Figure 14.9, and the alignment telescope attached to an optical-tooling stand is shown in Figure 14.10. The alignment telescope is used to provide a permanent horizontal reference line of sight for a jig. If the alignment telescope is mounted in a plumb aligner bracket, it can also be used to establish a vertical plumb line (making a line of sight vertical).

14.4.1.3 Jig Transit Jig transit, which is also known as *optical-tooling transit* or *jig collimator*, is like a surveyor's transit. It is usually designed for both attached and detached operations for establishing a vertical plane, in any desired location, passing through points to be established. It can also be used to set out a plane that is precisely at right angles to any other line of sight. Horizontal and vertical offset distances from the line of sight of an optical Jig transit are then precisely measured. The Jig transit

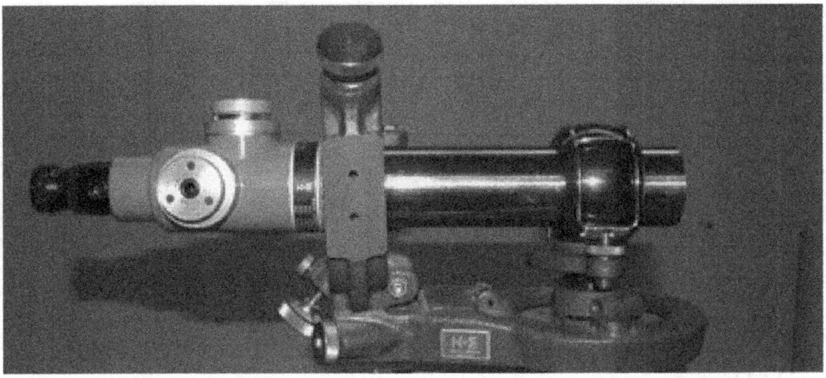

Figure 14.9 K&E Paragon alignment telescope set in an alignment bracket.

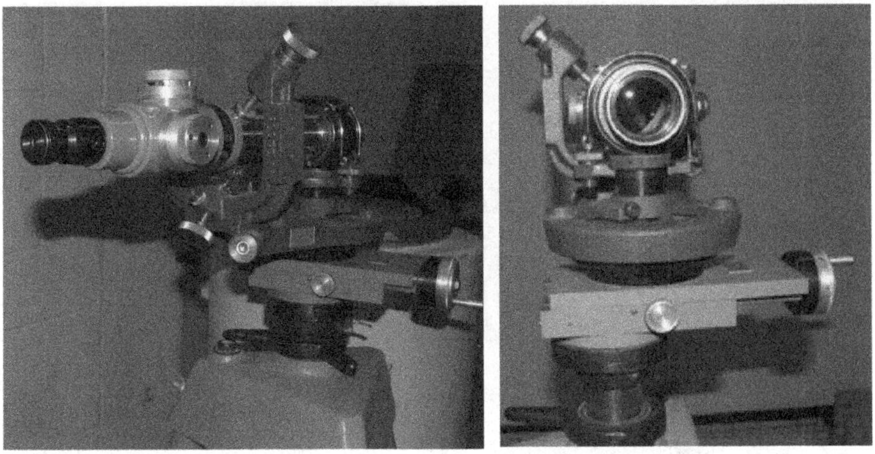

Figure 14.10 Side and front views of mounted alignment telescope.

can also be used like a level to measure offsets from a chosen vertical plane; however, it is most useful when only heights and offsets from one vertical plane are needed; 3D positioning with the method will be awkward. The Jig transits are different from surveyor's transits in a number of ways; some of the differences are as follows:

- They do not have graduated circles (no angular scales) and lower motion, as in the surveyor's transits. A screw is used for slow motion of the transit.
- They have built-in optical micrometer, which can be used to measure either horizontal or vertical displacements; the surveyor's transits do not have optical micrometers.
- They can be fitted with an illuminating eyepiece (as shown in Figure 14.11) so that by illuminating their crosshairs and focusing the instruments on infinity, they can be used as collimators or autocollimators, unlike the surveyor's transits.

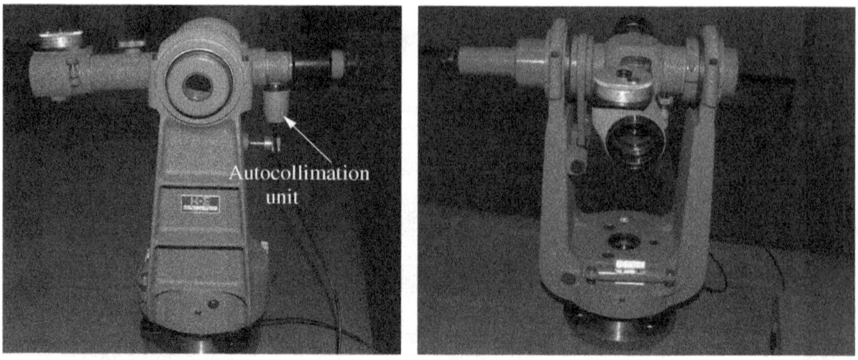

Figure 14.11 Side and front views of the Jig transit showing an autocollimation unit with a light unit mounted on the viewing end.

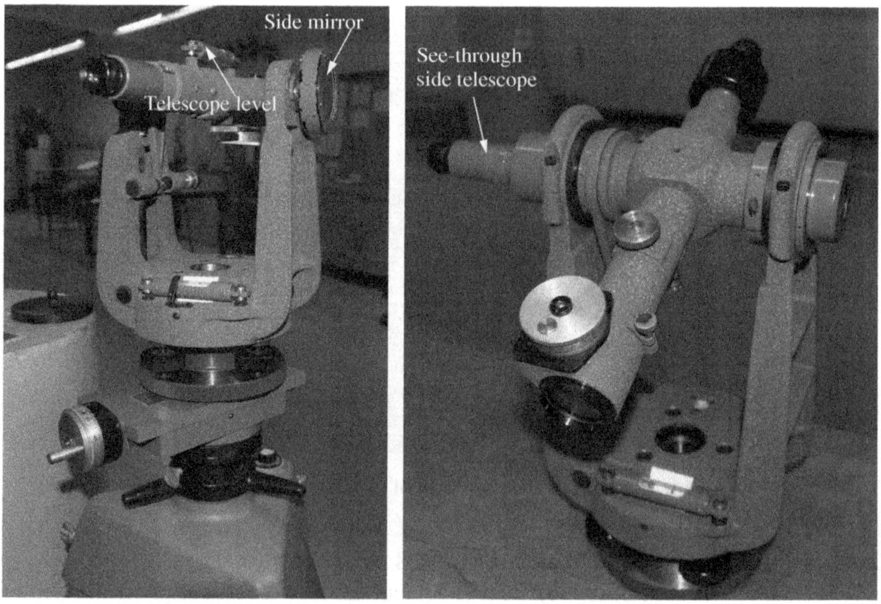

(a) (b)

Figure 14.12 Typical K+E Paragon Jig transit. (a) Jig transit with autocollimation and autoreflection side mirror. (b) Jig transit with see-through side telescope.

- The spindles of the Jig transits are hollow so that sights can be taken vertically downward, and their telescopes can be focused from 20 cm to infinity.

Typical Jig transits are shown in Figures 14.11 and 14.12. In Figure 14.12(a), a K+E Paragon Jig transit with autocollimation and auto-reflection side mirror is mounted on a stand, ready for use. The side mirror surface is set parallel to the plane generated by the line of sight of the telescope and is used for setting the line of sight of the jig telescope perpendicular to that of another instrument. The type of Jig transit shown in Figure 14.11 has an illumination unit and allows autocollimation to be performed.

The telescope level mounted on the telescope of the Jig transit (Figure 14.12(a)) set parallel to the telescope has a sensitivity of 30–40″/2 mm. The telescope is adjusted so that the bubble is centered when the line of sight is horizontal. This is to allow the Jig transit to be used as a level for short sights or for work requiring less accuracy so that an optical-tooling level is not necessary. For a very accurate work, the horizontal axis of the Jig transit cannot be relied on to be horizontal so that a striding level may have to be set along the horizontal axis to check the axis. The type of Jig transit shown in Figure 14.12(a) does not have an illumination unit and, thus, cannot be used to perform autocollimation. In Figure 14.12(b), the type of Jig transit shown has a side telescope, which uses the same focus as the main telescope, and the telescope will only see through when focused on infinity; and the direction of the side telescope is fixed with the horizontal axis of the transit.

14.4.1.4 Optical Micrometer and Optical-Tooling Scale Optical micrometer is a device containing a thick lens with flat parallel sides, which is incorporated in the telescopic sights of alignment telescopes, Jig transits, precise levels (Figure 14.13),

Figure 14.13 Optical micrometer attachment (graduated to 0.05 mm) for Kern GK23 tilting level.

and so on. The device is for moving the line of sight left or right or up and down through a short distance while keeping the line of sight parallel to its original position. Displacement can be read on the micrometer reading to 0.025 mm. The Kern optical micrometer drums are graduated at intervals of 0.05 mm. With care, the operator should be able to repeat readings to within one-fifth of a division (i.e., 0.01 mm) on a target not more than 3 m away. Before using the optical micrometer, the micrometer is first set at zero and the line of sight aimed at the reference target, and the micrometer is used to move the line of sight on the reference target, while taking note of the readings on the micrometer. If the optical micrometer is used with the optical-tooling scale, the observer notes where the line of sight falls on the scale and measures the distance to the nearest mark with the optical micrometer.

Optical-tooling scales are steel or invar tapes with very fine black lines engraved on a white background as shown in Figure 14.14. The scales are graduated so that the line of sight can be placed on any graduation with great accuracy. The graduations on the scales depend on the range of the micrometer. While using the scale, the line of sight is moved by the optical micrometer toward the zero of the scale until it bisects a

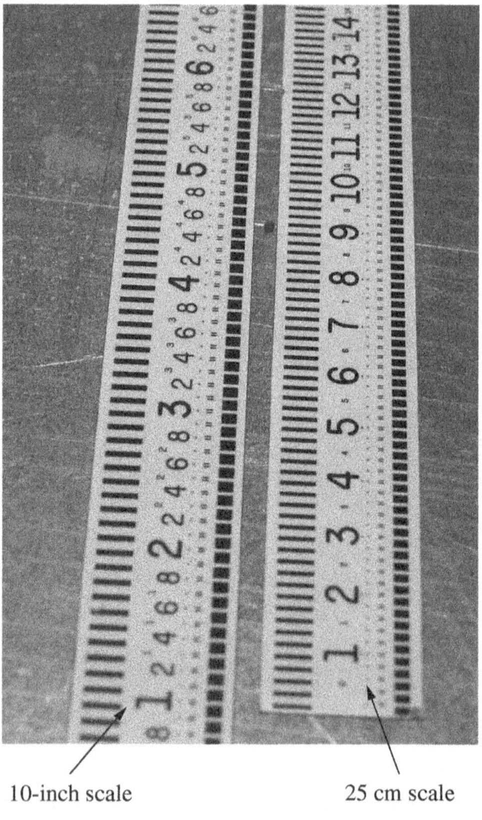

10-inch scale 25 cm scale

Figure 14.14 K+E Wyteface optical alignment scales in inches and centimeters.

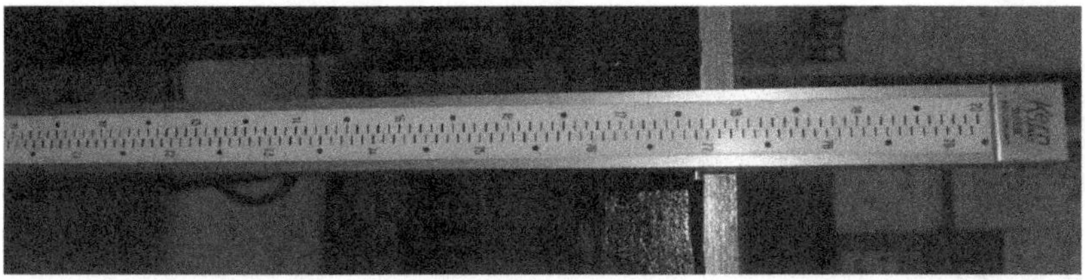

Figure 14.15 Kern invar staff (1 m, 5 mm division, 2×).

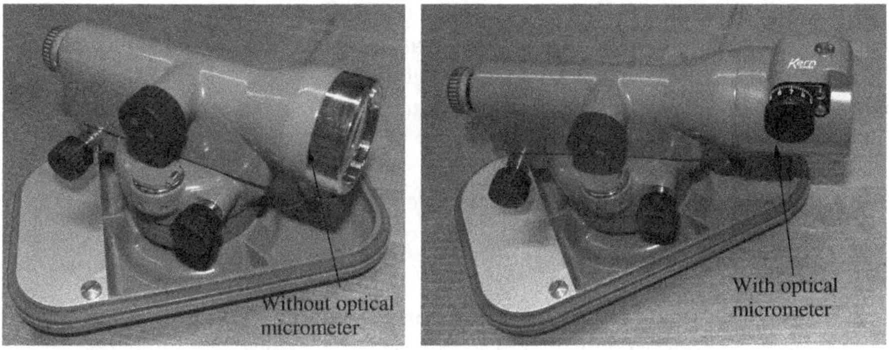

Figure 14.16 Kern GK23 tilting level without and with optical micrometer.

graduation. The reading on the drum of the micrometer in hundredths of a millimeter is added to the value of the graduation to obtain the complete reading.

When determining the flatness of a surface, precise level can be used with the optical-tooling scales or with dedicated invar staff. A typical invar staff is shown in Figure 14.15.

14.4.1.5 Precise Leveling Instrument Precise leveling is for establishing a horizontal plane at any desired height. Tilting level (Figure 14.16) is a surveying instrument built to very close tolerances that enable it to meet the accuracy requirements for optical alignment. The height of the instrument is not changed by the leveling screws or the micrometer tilting screw; the instrument is equipped with a coincidence bubble so that temperature changes in the instrument do not affect the shape or adjustment of the vial; the level bubble is attached to the telescopic sight in such a way that when the level bubble is centered, the line of sight is horizontal; and it is equipped with optical micrometer to measure precise vertical displacements. If the crosshair of the telescope is illuminated and the telescope focused on infinity, this instrument can also be used as a collimator or autocollimator. Many automatic levels, as now designed, cannot be used in optical tooling since they correct image tilt as a function of telescope focus. The correction for the tilt is too small for short sights (which is typical in optical tooling), making some automatic levels unsuitable for use. Those with proper anallactic design are usually considered suitable for use in optical tooling.

Figure 14.17 Leveling with optical-tooling scale.

Precise level with optical micrometer can be used with optical alignment scale to determine difference in elevation between various concerned points (Figure 14.17). The level instrument is usually set up at a height where scale can be conveniently read on concerned points. With optical micrometer, the scale readings can be determined to the nearest 0.025 mm.

14.4.1.6 Optical-Tooling Targets Target design is very important in optical-tooling procedures. The fundamental characteristics of target design are as follows (Kissam, 1962):

- Targets must have high contrast. Black and white colors are considered as giving the greatest contrast (considered as the best so far); since white light contains all the colors and cannot be focused directly, a light green or yellow is sometimes used to give monochromatic light.
- Targets must have patterns that are symmetric to aid in the estimation of the center of the targets.
- Targets must have proper area of reference. A poor target is one with a mark that is smaller than the crosshair itself; if paired-line target is to be used, it should not be in such a way that the spaces between paired lines are so far apart that the observer cannot compare the spaces accurately.
- Targets must have no phase. If there is a possibility of placing the crosshair at different positions under different illuminations or circumstances as in a cone, then the targets are considered as having phase.
- Targets must be free from errors of orientation. For example, concentric and paired-line targets are usually free from errors of orientation and are suitable for use; X or double-V targets are good but paired-line targets are considered the best. Double-V targets can be sighted at any distance, but for accurate results at different distances several paired-line patterns must be in line on the target. A typical paired-line target design is shown in Figure 14.18 (Kissam, 1962; Blachut et al., 1979).

The key dimension of the target design in Figure 14.18 is the width of the two white spaces at the two sides of the crosshair, called dimension a with x as the width of crosshair in the field of view, which can be taken as 2.5–3 arcsec. Most accurate target is the one that has a value as 5 arcsec, giving a pointing error of 0.23″. These types of targets are perfect only for one distance; however, the accuracy falls off very little for a

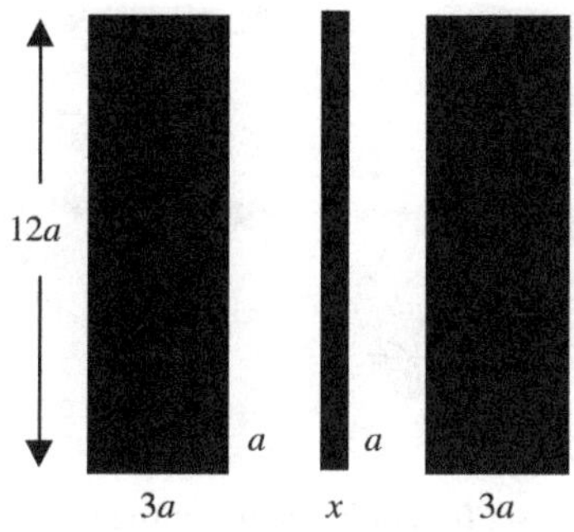

Figure 14.18 Ideal target design.

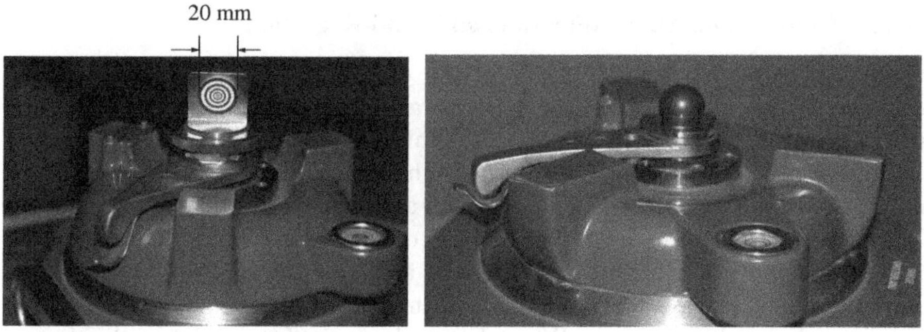

Figure 14.19 Spherical target and Kern concentric target (for sights of over 4–40 m) set in Kern trivets.

wide range of sizes of the value of *a*. Thus, each pair of lines is sufficiently accurate for a certain range of distances. The other optical-tooling target designs are the spherical types and the Kern concentric target types shown in Figure 14.19. Targets can also be mounted in spherical cups and held in cup mounts. Typical spherical cup and cup mount are shown in Figure 14.8. The main advantage of this type of target assembly is that when a target is in a spherical mount its center is at the center of the sphere.

14.4.1.7 Other Optical-Tooling Equipment Other types of optical-tooling equipment are as follows:

Pentaprism and optical square, which are attachments that can be mounted on the objective end of the telescope to turn the line of sight through a right angle. They are used to establish a plane perpendicular to a reference line at a given station.

Optical-tooling bars are straight rigid beams that are used to provide a track on which a carriage that supports an alignment telescope or Jig transit can ride. The bars are placed parallel with the measurements to be made both horizontally and vertically. Jig transits are mounted on carriages on optical-tooling bars.

Laser equipment, which is adapted for optical tooling, is able to provide an accuracy of 0.025 mm.

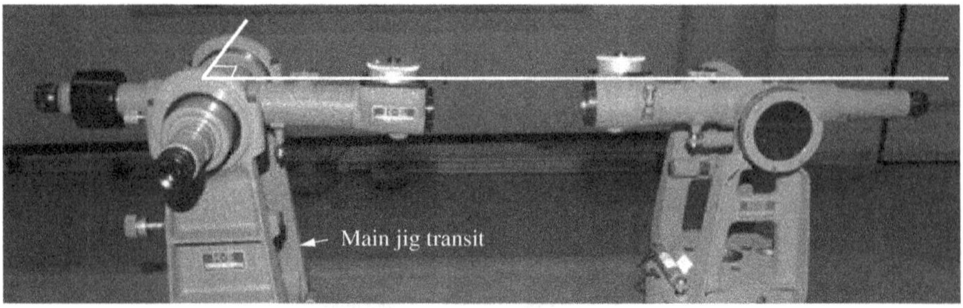

Figure 14.20 Two Jig transits set for collimation (setting the focuses of both instruments on infinity).

14.4.2 Collimation, Autocollimation, and Auto-Reflection

14.4.2.1 Collimation and Autocollimation *Collimation*, in an optical sense, is a process of bringing rays of light into a parallel beam. A *collimator* is therefore a device that has a source of light or an illuminated object on the focal plane of a converging lens, which projects a beam of light parallel to the principal axis of the lens. A collimator is a form of telescope with an illuminated reticle at the principal focus. Since a collimator is permanently focused at infinity, any telescopic sight becomes a collimator when focused at infinity and arranged so that light falls on the reticle. In this case, rays from any point on the reticle become parallel when they pass through the objective lens. When the reticle on the focal point is illuminated, shadow of the cross lines is projected through the lens in parallel rays. For example, two Jig transits can be set for collimation as shown in Figure 14.20 when the focuses of both instruments are set on infinity and the crosshairs of one instrument are superimposed on the crosshairs of the transit with the collimating unit. With the aid of the illuminating light at the reticle of the transit with the collimating unit, the two crosshairs superimposed on each other can be clearly seen. At this setup, the see-through telescope of the main Jig transit is perpendicular to the line of sight established by the two Jig transits that are collimated.

Autocollimation is a process of setting a mirror perpendicular to a telescopic line of sight. It is such that when the mirror is turned so that the reflection of the crosshairs coincides with the actual crosshairs, the mirror is perpendicular (square) to the line of sight. The autocollimation eyepiece has a semitransparent mirror and a lamp that is used to illuminate the crosshairs on the reticle. While the light illuminates the reticle, the observer can see the crosshairs on the reticle and also see through the telescope. With the telescope focused at infinity, the rays of light from the telescope are collimated (parallel) and can be reflected back along their own path from a mirror set square (or normal) to the line of sight, forming an image of the crosshairs (with the same size as the original crosshairs) on the actual crosshairs themselves. If the mirror is tilted, the image of the cross-lines is displaced. Such an instrument, having a suitably illuminated crosshairs on the reticle and some means of measuring any displacement of the image of the crosshairs, also on the reticle, is a combined collimator and telescope and is called an *autocollimating telescope* or *autocollimator*. It should be noted that when the telescope is focused at infinity, the micrometers on the telescope are ineffective and cannot be used for measurement. An autocollimator is a comparator of angular positions of an external reflector; it may be set to measure angular variations in a horizontal plane corresponding to tilts of the mirror about a vertical axis. It can also be used to check straightness of machine slides, surface plates, and tables.

The advantage of the autocollimator over the leveling instrument is that measurements are not restricted to the measurement of a horizontal surface as in the case of leveling. The first reading of an autocollimator is taken as the datum and differences of subsequent readings from this are calculated. Generally, a line of sight that is set square to a surface by the use of autocollimation is perpendicular to the surface, that is, perpendicular to all lines on the surface, which intersect the line of sight. A line of sight that is set square to another by the use of a pentagonal prism is square to that one line only, not to a surface. Since no graticule or target other than the crosshairs is used in autocollimation method, the method can only be used to check squareness and not to measure tilt. Since perpendicularity of related surfaces is more accurately checked by optics than by any other method, the technique of autocollimation is preferred as long as the autocollimator can be arranged within 15 m of the object being checked. In the autocollimation method, an optical mirror is attached to the specimen that reflects the crosshairs of the viewing telescope's reticle back to the telescope. The coincidence of the reflected crosshairs with the actual crosshairs of the viewing telescope verifies the absolute squareness (perpendicularity) of the surface being checked. The other applications of an autocollimator or an autocollimating telescope include checking parallelism and squareness of optical and mechanical components in instruments and for setting and aligning telescopes.

14.4.2.2 Auto-Reflection When an autocollimation eyepiece is unavailable, auto-reflection can be used. The auto-reflection target imprinted on the objective lens of the instrument (e.g., the auto-reflection target on the alignment telescope in Figure 14.8) is used. Auto-reflection provides a method of setting squareness and measuring small gradients of tilt. In order to make the auto-reflection target visible through the telescope, it must be illuminated. The target is illuminated from within the telescope without obscuring the line of sight through the telescope. Auto-reflection, however, should not be used for distances less than 1.5 m. A plain mirror that is surface-metalized and reasonably flat (Figure 14.21) can be used. The auto-reflection method is used to check the squareness of the reflecting surface, not the surface on which the mirror is located. It is important that these two surfaces be parallel. Usually, the front face of the mirror is located on the work piece.

The auto-reflection procedure includes focusing and aiming at the auto-reflection target, and then adjusting the leveling mirror until the reflection of the auto-reflection target coincides with the reticle pattern (as illustrated in Figure 14.22 with an alignment telescope). Note that when setting squareness by auto-reflection or autocollimation method, the telescope target image cannot be seen in the eyepiece until the mirror is adjusted square to the telescope within 1:120. For distances over 15 m,

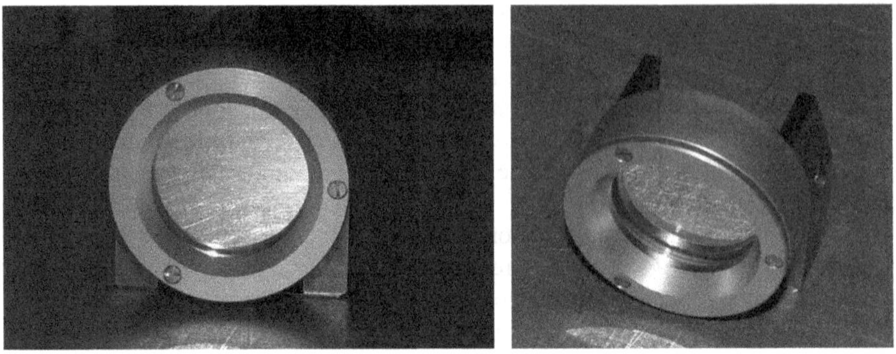

Figure 14.21 Autocollimation or auto-reflection leveling mirror.

Figure 14.22 Alignment telescope set for autocollimation/auto-reflection.

the auto-reflection principle of optics has been found to give a greater clarity of results because the auto-reflection targets, which fit onto the front of the telescope, have distinct and heavy target outline, whose reflection can be seen very clearly. Auto-reflection, however, is not as accurate as autocollimation since its accuracy depends on the fit of the reflector on the end of the telescope and also on the presence of possible inaccuracies between the mechanical axis and the optical axis of the telescope. There is less error in autocollimation method since the reticle crosshairs reflect back upon themselves with no intervening medium as in the case of auto-reflection where the image of the target located on the objective lens and the reflected image from the mirror are to be located on the reticle.

In order to perform auto-reflection in Figure 14.22, the telescope is aimed and focused on the mirror. If the mirror is perpendicular to the telescope line of sight, the pattern target on the telescope objective will be reflected back and shown on the reticle; with the illumination of the reticle, the pattern will become clearly visible. To perform autocollimation in Figure 14.22, the telescope must first be focused on infinity and not on the mirror; with the illumination of the reticle, the image of the crosshairs is formed on the reticle (without the illumination, the image will not will not be visible). The line of sight must be approximately perpendicular to the mirror in order to have auto-reflection or autocollimation.

Figure 14.23 Setting out 90° angle by autocollimation or auto-reflection using side mirror.

Two Jig transits can be used to set out 90° angle by autocollimation or auto-reflection using side mirror as shown in Figure 14.23. When the Jig transit is placed and aimed so that an observer, using another instrument on a main line of sight, can see by autocollimation or by auto-reflection through the mirror that is perpendicular to the main line of sight, the plane generated by the Jig transit must be perpendicular to the observer's line of sight.

14.4.3 Basic Optical-Tooling Operations

There are two categories of method involved in the optical-tooling operations: the *detached method* and the *attached method.* The *detached method* involves separating the instruments and the object being aligned or measured. In this case, the reference lines established by the optical-tooling instruments must be marked by some means to make them permanent so that when the instruments are removed, in order to use the instruments again, it is necessary to place them in line with the reference marks through *bucking in* procedure. To "buck in" means to place an instrument so that its line of sight passes through two given points. Since it is difficult to replace instruments in their original positions within the necessary tolerances, attached method was developed.

The *attached method* involves attaching the optical-tooling instruments to the object being aligned or measured. Two fundamental instruments in attached method are *alignment telescope* and *alignment target.* The alignment telescope, supported in a cup mount, is permanently mounted on the object being measured; once the assembly has been adjusted exactly in position, it is permanently locked in place. The alignment target consists of a circular transparent disk with a black pattern on the front surface, which is illuminated from the rear by a small electric lamp. The target can be mounted inside a sphere called a spherical mount, which is the same size as the sphere in which the alignment telescope is fitted. The target is mounted against a stop in the sphere (refer to Figure 14.7) so that the surface of the pattern is at the center of the sphere. Alignment telescope and its target are designed to establish a reference line of sight in such a way that when both devices have been removed and then replaced, the line of sight will be in exactly the same position as earlier. For example, four alignment telescopes can be mounted on two similar jigs. Some of the basic optical-tooling operations consist of the following:

1. Defining a reference line or a center line using Jig transit based on "bucking in method" as follows:
   - Given two targets, set up a Jig transit approximately halfway between them, level the Jig transit, and then set the optical micrometer on zero. Usually, a mechanical lateral adjuster is mounted on the instrument stand and the Jig transit mounted on the lateral adjuster (Figure 14.6) and locked so that its ways are approximately perpendicular to the final direction of the line of sight. The lateral adjuster makes it possible to make fine adjustments left and right with the precise adjustments handled with the optical micrometer.
   - Aim and focus Jig transit on one target and then plunge over so that the second target can be observed; if it is very close, a measurement is made with the optical micrometer of the distance that the crosshair is off the second target.
   - Make adjustments on the transit using the horizontal slide and taking off approximately one-half the distance found when focusing on the second target.
   - Re-aim the Jig transit on the first target and plunge over again and check the second target.
   - Repeat this operation until the instrument coincides with both targets; the optical instrument is now coincident with the centerline or reference line.
2. Creating a master line by using alignment telescope. Once the alignment telescope and the target are properly placed, the alignment telescope is aimed at the target and a "master line" is established. Both devices can then be removed from their cup mounts and replaced in exactly the same positions as earlier. When the telescope is aimed at the target, the master line will be in its original position. The procedures involve the following aspects:
   - Using a paired-line target with appropriate shape and proportions.
   - Using optical-tooling scales to make linear measurements from the master line.
   - Attaching small levels to the optical-tooling scales so that they may be kept vertical or horizontal; the zero end is placed at the target and the length read on the scale with the optical micrometer of the instrument on the master line.
   - Determining the distance along the master line by establishing a right angle at a known position with the pentaprism attachment and by measuring from the target on the right angle line with an optical-tooling scale.
3. Performing precise leveling to determine the difference in elevation between points. This involves using precise level on an instrument stand with the optical alignment scale. The precise level is set up at a height where the scale can be read conveniently; the scale is then held on various points and readings taken. With the use of optical micrometer, the scale readings can be determined to the nearest 0.025 mm.

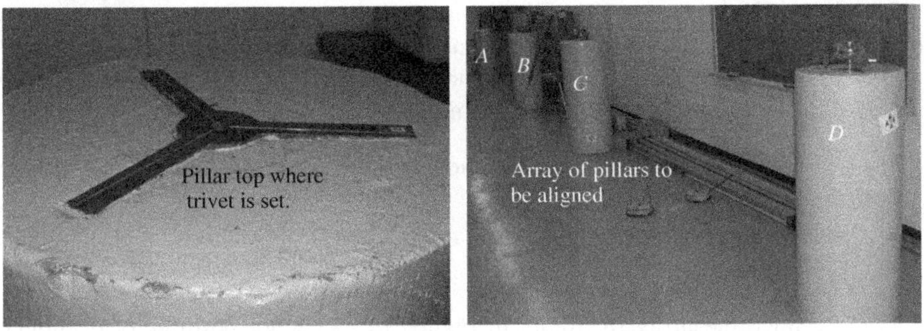

Figure 14.24 Array of pillars to be aligned.

14.4.4 Optical-Tooling Example

Consider a reference line define by points *A* and *D* in which horizontal and vertical positions of points *B* and *C* are to be aligned with respect to line *A-D* (Figure 14.24). The alignment procedure will be divided into two parts to be done independently: horizontal alignment and vertical alignment. The horizontal alignment can be done using Jig transit and alignment telescope with optical-tooling scales and the vertical alignment can be done by using differential leveling procedure with precise level equipment and invar rod.

14.4.4.1 Horizontal Alignment Consider the case in Figure 14.24 in which pillars *B* and *C* are to be aligned horizontally with pillars *A* and *D*. The line joining points *A* and *D* can be taken as the *X*-axis while the direction perpendicular to it can be taken as the *Y*-axis; the offsets (y) of points *B* and *C* (known as the alignment elements) will be determined off the *X*-axis, along the *Y*-axis direction. With regard to Figure 14.24, the alignment element (y_i) at any point "*i*" is represented mathematically (Chrzanowski et al., 1976) as follows:

$$y_i = \frac{(\ell_i - \ell_a)X_D - (\ell_d - \ell_a)X_i}{\sqrt{X_D^2 - (\ell_d - \ell_a)^2}} \tag{14.1}$$

where X_D is the measured distance from point *A* to *D*; X_i is the measured distance from point *A* to any given point "*i*" to be aligned; and ℓ_i, ℓ_a and ℓ_d are the measured horizontal offsets from points "*i*," *A* and *D*, respectively, to the instrument line of sight projecting from the instrument setup reference point toward the other reference point (e.g., from *A* to *D*). Since the offsets to be measured are usually a few centimeters, Equation (14.1) can be approximated as

$$y_i = (\ell_i - \ell_a) - \frac{(\ell_d - \ell_a)X_i}{X_D} \tag{14.2}$$

Usually, in the case of horizontal alignment, points A and D cannot be occupied directly (optical-tooling instruments such as Jig transit and alignment telescope cannot be set on them as can be seen in Figure 14.24). In this case, the following two options can be used for the horizontal alignment:

- *Option 1*: Establish a line that is almost parallel to the line A-D and determine the required offsets from the line.
- *Option 2:* Establish a line at an angle to the line A-D and determine the required offsets from the line.

Option 1 Since the array of pillars in Figure 14.24 cannot be directly occupied, it becomes necessary to establish line P_1-P_2 near-parallel to line A-D as shown in Figure 14.25.

With regard to Figure 14.25, the horizontal alignment of B and C with respect to line P_1-P_2 can be performed as follows:

- Set Jig transit at P_1 close to A, establish a near-parallel line P_1-P_2 and measure offsets from the line using optical-tooling scales.
- Measure the horizontal offsets at points A, B, C, and D, giving the measured quantities as follows: $\ell_a = O_1 - A$, $\ell_b = O_2 - B$, $\ell_c = O_3 - C$ and $\ell_d = O_4 - D$ (O_1–A should be equal to O_4–D for the parallel lines, that is, $\ell_a = \ell_d$), so that Equations (14.1) and (14.2) are reduced to $y_i = (\ell_i - \ell_a)$
- For example, from Figure 14.25, let the measured quantities be $\ell_a = +1.0$ cm, $\ell_b = +4.0$ cm, $\ell_c = -1.0$ cm (measured in the negative direction from the line of sight) and $\ell_d = +1.0$ cm, $X_A = 0.0$ m, $X_B = 10.0$ m, $X_C = 18.0$ m and $X_D = 23.8$ m.
- The alignment elements from Figure 14.25 are y_b = length $O_2' - B$; y_c = length $O_3' - C$.
- Determine the alignment elements from Equation (14.1) or Equation (14.2): $y_b = (\ell_b - \ell_a)$ or $y_b = 3.0$ cm; and $y_c = (\ell_c - \ell_a)$ or $y_c = -2.0$ cm.

Option 2 In this option, a line P_1-P_2 is run at an angle to AD as shown in Figure 14.26. This will be the case if it is impossible to run the line parallel to AD so that the line has to be run at some angle to AD with the lateral offsets as shown in Figure 14.26. These offsets are then rotated mathematically to be orthogonal to the

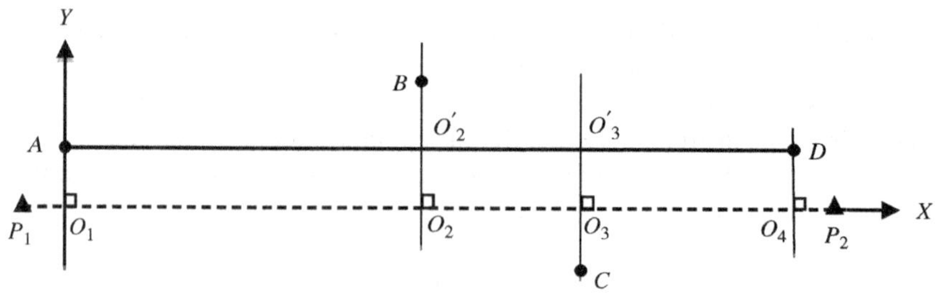

Figure 14.25 Alignment Option 1.

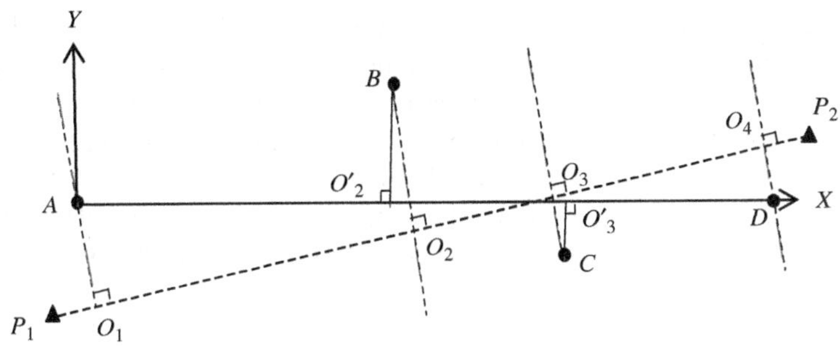

Figure 14.26 Alignment Option 2.

line AD, creating the desired offsets. Equations (14.1) and (14.2) are still applicable in this case.

From Figure 14.26, the horizontal alignment of B and C with respect to the inclined line P_1-P_2 can be performed as follows:

- Set Jig transit at P_1 close to A; run a line P_1-P_2 at an angle to AD and measure offsets from the line using optical-tooling scales.
- Measure the horizontal offsets at points A, B, C, and D, giving the measured quantities as $\ell_a = O_1 - A$, $\ell_b = O_2 - B$, $\ell_c = O_3 - C$ and $\ell_d = O_4 - D$.
- For example, from Figure 14.26, let the measured quantities be $\ell_a = +1.3$ cm, $\ell_b = +3.5$ cm, $\ell_c = -2.2$ cm (measured in the negative direction from the line of sight) and $\ell_d = -0.7$ cm (measured in the negative direction from the line of sight), $X_A = 0.0$ m, $X_B = 10.0$ m, $X_C = 18.0$ m and $X_D = 23.8$ m.
- The alignment elements from Figure 14.26 are $y_b =$ length $O'_2 - B$; $y_c =$ length $O'_3 - C$.
- Determine the alignment elements from Equation (14.1), for example:

$$y_b = \frac{(\ell_b - \ell_a)X_D - (\ell_d - \ell_a)X_B}{\sqrt{X_D^2 - (\ell_d - \ell_a)^2}} \quad \text{or } y_b = 3.04\,\text{cm; and}$$

$$y_c = \frac{(\ell_c - \ell_a)X_D - (\ell_d - \ell_a)X_C}{\sqrt{X_D^2 - (\ell_d - \ell_a)^2}} \quad \text{or } y_c = -1.99\,\text{cm}$$

Error Propagation for Alignment Elements By performing random error propagation on Equation (14.2), the variance of the determined alignment element can be given as

$$\sigma_{y_i}^2 = \sigma_{\ell_i}^2 + \left(\frac{X_i - X_D}{X_D}\right)^2 \sigma_{\ell_a}^2 + \left(\frac{X_i}{X_D}\right)^2 \sigma_{\ell_d}^2 + \left(\frac{\ell_a - \ell_d}{X_D}\right)^2 \sigma_{X_i}^2 + \left(\frac{\ell_d - \ell_a}{X_D^2}\right)^2 X_i^2 \sigma_{X_D}^2 \qquad (14.3)$$

where $\sigma^2_{\ell_i}$, $\sigma^2_{\ell_a}$, and $\sigma^2_{\ell_d}$ are the variances of the measured offsets ℓ_i (which could be ℓ_b or ℓ_c in the case being discussed in Figure 14.26), ℓ_a and ℓ_d, respectively; and $\sigma^2_{X_i}$, and $\sigma^2_{X_D}$ are the variances of the measured distances X_i and X_D, respectively.

Assume that all offset measurements are made with a precision of ±0.1 mm and the distance measurements along the alignment axis x are made with a precision of ±3 mm. The standard deviations (σ_{y_b}, σ_{y_c}) of the alignment elements (y_b, y_c) determined by Option 2 procedure can be determined from the error propagation Equation (14.3) as follows.

For element y_b:

$$\sigma^2_{y_b} = \sigma^2_{\ell_b} + \left(\frac{X_B - X_D}{X_D}\right)^2 \sigma^2_{\ell_a} + \left(\frac{X_B}{X_D}\right)^2 \sigma^2_{\ell_d} + \left(\frac{\ell_a - \ell_d}{X_D}\right)^2 \sigma^2_{X_B} + \left(\frac{\ell_d - \ell_a}{X_D^2}\right)^2 X_B^2 \sigma^2_{X_D}$$

$$\sigma^2_{y_b} = 0.01 + 3.362\text{E} - 3 + 1.765\text{E} - 3 + 6.355\text{E} - 6 + 1.122\text{E} - 6$$

$$\sigma_{y_b} = \pm 0.12\,\text{mm}$$

For element y_c:

$$\sigma^2_{y_c} = \sigma^2_{\ell_c} + \left(\frac{X_C - X_D}{X_D}\right)^2 \sigma^2_{\ell_a} + \left(\frac{X_C}{X_D}\right)^2 \sigma^2_{\ell_d} + \left(\frac{\ell_a - \ell_d}{X_D}\right)^2 \sigma^2_{X_C} + \left(\frac{\ell_d - \ell_a}{X_D^2}\right)^2 X_C^2 \sigma^2_{X_D}$$

$$\sigma^2_{y_b} = 0.01 + 5.939\text{E} - 4 + 5.720\text{E} - 3 + 6.355\text{E} - 6 + 3.635\text{E} - 6$$

$$\sigma_{y_b} = \pm 0.13\,\text{mm}$$

The error propagation for the elements based on Option 1 procedure will be identical.

14.4.4.2 Vertical Alignment In the case of the vertical alignment of points B and C with respect to line A-D, the differential leveling procedure can be followed. However, the vertical alignment offsets will be the vertical distances from the corresponding points to the line passing through points A and D. If a straight line is fitted to points A and D, the vertical offset equation can be given as

$$z_i = (Z_i - Z_A) - \frac{(Z_D - Z_A)}{X_D} X_i \tag{14.4}$$

where Z_i is the leveled height (above the reference datum) of any given point i; Z_A and Z_D are the leveled heights (above the reference datum) of points A and D, respectively;

X_D is the distance from point A to point D measured along the alignment axis, X and X_i is the distance measured from point A to any given point i. The variance–covariance propagation equation for the vertical offsets can be given as follows:

$$\sigma_{z_i}^2 = \sigma_{Z_i}^2 + \left(\frac{X_i - X_D}{X_D}\right)^2 \sigma_{Z_A}^2 + \left(\frac{X_i}{X_D}\right)^2 \sigma_{Z_D}^2 + \left(\frac{Z_A - Z_D}{X_D}\right)^2 \sigma_{X_i}^2 + \left(\frac{Z_D - Z_A}{X_D^2}\right)^2 X_i^2 \sigma_{X_D}^2 \quad (14.5)$$

Given the following measurements:

$$X_A = 0.0\,\text{m},\ X_B = 10.0\,\text{m},\ X_C = 18.0\,\text{m} \ \text{and}\ X_D = 23.8\,\text{m}$$

$$Z_A = 1.350\,\text{m},\ Z_B = 1.344\,\text{m},\ Z_C = 1.331\,\text{m} \ \text{and}\ Z_D = 1.347\,\text{m}$$

The vertical offsets of points B and C based on Equation (14.4) and their corresponding standard deviations based on Equation (14.5) can be given as follows:

$$z_B = -4.7\,\text{mm} \pm 0.12\,\text{mm}$$

$$z_C = -16.7\,\text{mm} \pm 0.13\,\text{mm}$$

14.5 METROLOGY BY LASER INTERFEROMETER SYSTEMS

Laser interferometry is a well-established method of measuring accurate distances based on the basic principle that monochromatic, stable, and accurately defined wavelength of light can be used as units of measurements. Laser interferometer systems are designed to provide the best possible accuracy, repeatability, and traceability in measurement, using externally mounted optical components. The concept of interferometry is based on the concept of Doppler effects.

14.5.1 Doppler Effects and Interferometer Systems

Doppler effect, named after the Austrian physicist Christian Doppler, is an apparent change in frequency of a wave when its source is moving relative to the observer. The change in the transmitted and the received frequencies (even though the transmitted frequency is constant) as a result of this effect is known as Doppler frequency (f_D).

The Doppler frequencies are usually observed in the propagation of sound and electromagnetic waves. In the case of light waves, the Doppler frequency can be measured by counting the bright (or dark) fringes of an optical interference pattern, or counting the cycles of the Doppler signal (Doppler counts) per second in the case of radio waves. When a source of light moves with a speed (v) relative to a stationary observer, the distance travelled by the source between times t_1 and t_2 can be given by (Rüeger, 1990)

$$s_{12} = \frac{\lambda}{2}\int_{t_1}^{t_2} f_{\mathrm{D}}dt \tag{14.6}$$

or

$$s_{12} = \frac{\lambda}{2}[\text{Doppler counts}]_{t_1}^{t_2} \tag{14.7}$$

where λ is the wavelength of the light source, f_{D} is the Doppler frequency; and the Doppler counts in Equation (14.7) are derived from the Doppler frequency between the light source and the observer. Equation (14.6) is used in surveying and metrology for distance measurements of highest precision. Laser interferometers employ Doppler effects in measuring distances travelled by a reflector with regard to laser beams, to a resolution of about 10 nm. Because fringe counts (or Doppler counts) need to be obtained for the distance determination, the moving reflector is required to travel along the laser beam.

14.5.2 Interferometry Principle

The concepts of Doppler frequency measurement by counting the bright (or dark) fringes of optical interference patterns of light waves are used in interferometers for high-precision distance measurements over short distances. The optical interference is a phenomenon that takes place when two waves meet while travelling along the same medium.

The operational principle of an interferometer is based on Michelson interferometric procedures, which are summarized in Figure 14.27 and as follows (Rueger, 1990; Hexagon Metrology, 2012):

1. A monochromatic light source (laser) sends a laser beam toward a beam splitter, which splits the beam into two beams with one of the beams passing through to the moveable retro-reflector and the other one deflected to a reference retro-reflector.

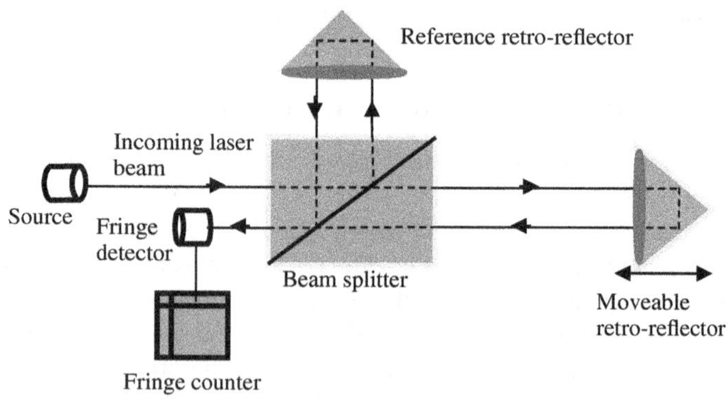

Figure 14.27 Schematic diagram of Michelson interferometric procedures.

2. The two beams are reflected from two retro-reflectors and again recombined (superimposed) at the beam splitter, producing interference pattern in it. If the retro-reflectors are exactly aligned and motionless, the observer will see a constant intensity of light. But if the moveable retro-reflector is moved very slowly, the observer will see the beam repeatedly increasing and decreasing in intensity as the two beams add up and cancel out (resulting in interference pattern) in the beam splitter. The superimposed signal reaches a maximum intensity for constructive interference when the phase difference is zero and reaches a minimum when the phase difference is 180° for destructive interference. Note that the two waves superimposed in the beam splitter are of equal frequency and amplitude (coherent waves) since they are generated by the same light source. In this case, the phase difference occurs because of the difference in path lengths.
3. The interference pattern is recorded by the fringe detector and the fringe (Doppler) counts are recorded by the digital fringe counter. During a displacement of moveable retro-reflector, the fringe counter counts the number of bright fringes in the interference pattern in the beam splitter. The distance between the first and the last positions of the moveable retro-reflector is derived from Equation (14.7), where the Doppler counts are taken as the number of counted bright fringes. The high resolution of interferometers is based on the direct use of the wavelength of light waves for measurement.
4. The update rate of change in distance measurement is given only by the speed at which the moveable retro-reflector can be moved. This makes laser interferometers perfect for dynamic measurements, because no matter how quickly the target accelerates, the exact change in location is immediately known to the submicron level.
5. Distance change or relative motion is measured by electronically counting wavelengths of light, rather than the absolute distance between the laser head and the reflector. In this case, any point may be defined as a zero reference for the measurement. In principle, this means that an interferometer cannot determine absolute position in three-dimensional space without having a known starting point first.

Interferometers can be compared with electromagnetic distance measurement (EDM). In some EDM instruments, light waves are used as carrier, but a modulated signal is used for the distance measurement. In the case of interferometer, the carrier wave itself is used for the distance measurement.

Common laser interferometers have a maximum range of about 60 m and are mainly used indoors. They are used for precise length measurements and in metrology for measurement of straightness, squareness, parallelism, flatness, and angle. Their main applications are in positioning machines, fixtures, or jigs; installing and aligning machine tools; performing geometry check, part alignment, metrology-assisted assembly, or fully automated positioning and integration tasks.

14.5.2.1 Accuracy Limitation Factors Every time the superimposed signal in a beam splitter of an interferometer reaches maximum intensity (a fringe count), it represents a change in the distance of half of the wavelength. For example, a laser interferometer using helium–neon laser light source with a wavelength of 0.6328 μm will have a change of distance or a least count of measurement of about 0.32 μm. The overall accuracy of such an interferometer is given as 0.1 ppm by Rueger (1990). This overall accuracy, however, is due to the limitation imposed by the uncertainty of measuring the ambient temperature and pressure for the determination of the atmospheric refractive index. The other limitation to accuracy is air turbulence, which has always been a serious problem for laser interferometers.

The air turbulence is caused by time-dependent variations in the atmospheric refractive index due to dynamic variations in air density, the direction, and speed of propagation of light beams in the atmosphere. The effect is equivalent to an intensity variation. In performing alignment, straightness, or angular measurements using laser beams, turbulence-induced noise can force long averaging times and susceptibility to thermal drift, during distance measurements. The turbulence also causes phase noise in interferometer setups and transit time fluctuation in distance measurements. Minimizing the effects of turbulence, however, are done by averaging measured data or performing some operations that will help homogenize the air. Most interferometers are comfortable with 50% loss of signal due to any source (Dukes and Gordon, 1970).

14.5.3 Interferometer Systems and Alignment Principles

Two types of interferometer systems can be identified as follows (Renishaw plc, 2001):

1. Those that have their laser head, interferometer optic, and photo-detector all integrated as a single unit and the moveable reflector as another unit
2. Those that have their laser head and the interferometer as a single integrated unit and the reflector and the photo-detector targets as another integrated unit.

Since type (1) systems perform measurements at the laser head, there is usually a possibility of thermal build-up at the laser head that may affect the measurement, especially if warm-up time is not allowed. In the case of type (2) systems, measurements are made at the remote photo-detector targets so that the possible effect of heat generation by the laser source on the photo-detector is avoided. However, the established and proven industry standard method of measuring machine tool or the performance of coordinate measuring machines using interferometer systems is to set up a laser device on a tripod away from the component to be measured, and the interferometer and the reflector optics are mounted directly to the machine table as two separate units with the moveable reflector optics on the machine spindle. The interferometer is then used to take the linear, angular (pitch and yaw), or straightness measurements between the table and the spindle.

Examples of laser measurement systems are the ML10 and XL-80 laser measurement systems by Renishaw plc (2014), which are specified as capable of using environmental compensators to maintain accuracy of measurements over a wide range of

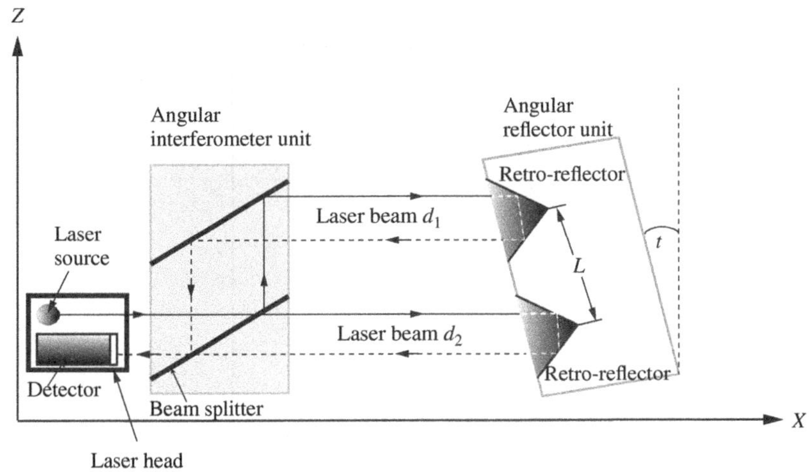

Figure 14.28 Schematic illustration of angle measurement with interferometer.

atmospheric conditions. The ML10 laser system has a specified linear interferometric measurement accuracy of ±0.7 ppm while the XL-80 laser system has a specified accuracy of ±0.5 ppm.

14.5.3.1 Angular Measurement with Interferometer An interferometric angular measurement system can be used (in an alignment process) to measure pitch (tipping) or yaw (twisting) errors in a linear axis or flatness of a surface. This measurement system usually consists of a laser head, angular interferometer, and angular reflector. The laser head contains a detector and a laser source; the angular interferometer component contains beam-splitter while the angular reflector contains two retro-reflectors with a center-to-center distance of L (as illustrated in Figure 14.28). When the laser beam generated at the laser source reaches the angular interferometer, it is split into two separate beams by the beam splitter. The two beams are reflected back into the interferometer from the retro-reflectors and recombined before travelling back to the laser detector where they interfere to produce a measurement signal. The measurement system measures relative change (D) in the top and bottom lengths (as shown in Figure 14.28) to the remote angular reflector and uses the change to determine the inclination angle (t) as illustrated in Figure 14.29. As the angular reflector is moved, the relative change between the path lengths d_1 and d_2 is detected by an interference fringe counter (interpolator) inside the laser detector and then converted into a linear distance change (D) by multiplying the fringe counts by half the wavelength of laser according to Equation (14.7). The relative change (D) in path lengths is then converted into angle t, which can be given from Figure 14.29 as $t = \arcsin(D/L)$, where L is the known distance between the centers of the retro-reflectors.

The typical operational principle of the angular measurement system is such that the laser head is set up on a tripod; the angular interferometer is attached to the machine spindle; and the angular reflector is attached to the object being moved,

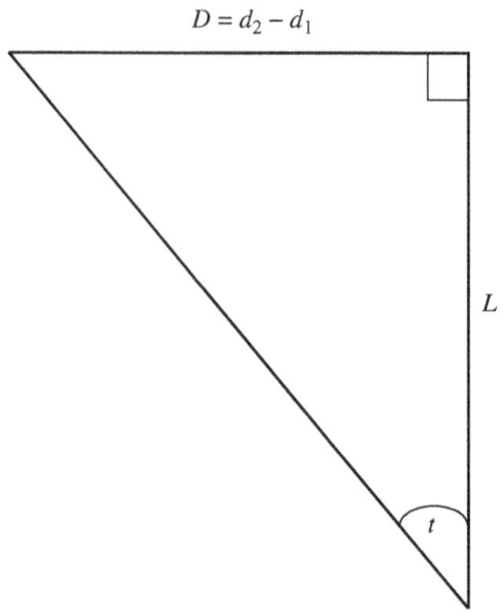

Figure 14.29 Illustration of angle determination with interferometer.

whose yaw or pitch is to be determined. The yaw or pitch error is determined depending on the orientation of the angular interferometer and the angular reflector with respect to the object. From Figure 14.28, as the object containing the reflector is moved in the direction of the X-axis, the laser unit and optics will measure any pitch error in the object's movement. It should also be mentioned that, depending on which arrangement is easier, the interferometer can also be considered as the moving optic instead of the reflector.

14.5.3.2 Straightness Measurement with Interferometer Straightness interferometric system can be used to measure horizontal and vertical straightness of an object as well as the straightness of its motion when the object is being moved. The straightness measurement system consists of a laser head, straightness interferometer (Wollaston prism), and a straightness reflector unit. The type of straightness determined depends on the orientation of the straightness interferometer and the reflector unit with respect to the object whose straightness is being determined.

The operational principle of a straightness measurement system is such that when a beam from the laser head reaches the straightness interferometer, it is split into two separate beams that travel to the straightness reflector along two paths with the optical path lengths of d_1 and d_2 between the interferometer and the reflector. The two beams are then reflected back to the interferometer where they are combined and sent to the laser head where they interfere to produce a measurement signal to be interpreted at the measurement unit. At the measurement unit, the straightness error is determined by detecting relative changes between the optical path lengths d_1 and d_2. Initially, the two lengths d_1 and d_2 will have some length relative to each other; but after moving the reflector, the relative lengths of the two beams in the Wollaston prism will change. This change is called the straightness error ε. If the straightness reflector is moved away from the interferometer along a perfect straight line in the X-axis direction, the straightness error will be zero; if it is moved vertically in the Z-axis direction by a distance of S, the straightness reading will show S as the amount of upward movement; if the reflector is pitched through a small angle, the reading will show the corresponding value.

14.6 ALIGNMENT BY POLAR MEASUREMENT SYSTEMS

Until recently, the determination of coordinates in industrial metrology applications has usually required two instruments or two setups. Most common measurement technique is triangulation in which horizontal and vertical angles are measured from at least two stations to determine object coordinates. Traditionally, total station instruments do not meet the accuracy requirements of most industrial metrology applications.

Polar measurement systems (PMS), such as laser trackers (LTs) and industrial robotic total stations (RTSs), are used extensively in LSM. They are able to determine three-dimensional coordinates of a point by measuring two orthogonal angles (nominally horizontal and vertical) and a distance to a corner cube reflector (CCR) also known as SMR.

LSM covers fields that require very high-precision alignment over relatively large areas and volumes. It is sometimes referred to as engineering survey or industrial geodesy or the geodetic or photogrammetric techniques for accurate measurement of large objects in which workshop tools cannot be used (Mayoud, 2004). Examples of where LSM is used are particle accelerator alignment, aircraft, ship, and car manufacture (Estler et al., 2002). The field of particle accelerator alignment is unique. It requires submillimeter measurement precision over distances ranging between several hundred meters up to tens of kilometers, thus overlapping with the fields of metrology and traditional surveying and geodesy. This precision requirement also demands that extremely specialized techniques and instruments be used to guarantee that the accuracy requirement will be met.

In practice, PMS errors are automatically corrected by onboard software using parameters derived from a series of manufacturers' recommended test measurements. Other errors linked to the servo motion of the instrument about its axes (e.g., wobble error) are corrected in real time with onboard inclinometers and compensators. All errors with parameters that can be derived from self-testing and onboard software are corrected to the level of instrument precision. Some residual errors, however, may remain due to random errors, drift in the parameter values during normal instrument operation and between self-testing operations, and the effects of uncorrected systematic errors.

14.6.1 Laser Trackers

Laser tracker is a high-accuracy servo-controlled tracking total station that combines horizontal and vertical angle measurements with interferometric distance measurements. It consists of three major components: the measurement head, the controller with system software, and the accessories, which include the remote power unit.

14.6.1.1 Tracker Measurement Head The tracker measurement head consists of two high-resolution angle encoders for measuring azimuth (Az) and vertical angle (VA) of the laser beam and a displacement interferometer (IFM) or an absolute distance meter (ADM) for measuring linear distance (r) to the center of the retro-reflector target relative to a known position. The polar coordinates (r, Az, VA) are then converted in real time into Cartesian coordinates of the target center location by the system computer.

The measurement head, which typically has a field of view of ±60° vertically and ±135° horizontally, is like a servo-driven theodolite, providing rotation about two orthogonal axes with the encoders attached for angle measurements. To permit random motion of target retro-reflectors, a servo control loop consisting of the tracking mirror and servo system with a position sensing detector is used. The automatic aiming on a retro-reflector is done using the beam return of the laser tracker on a position detecting sensor (PDS). The IFM component uses a helium–neon laser, a single beam type with 0.632 μm wavelength, for the distance measurements.

The general operating principle of a tracker is such that the generated laser beam is passed through the interferometer optics to the servo mirror until it hits the retro-reflector, which reflects the beam back to the servo mirror and through a beam splitter to the interferometer. In the interferometer, the reflected beam is merged with the interferometer reference beam. If the laser beam strikes the retro-reflector in its center, the reflected beam is expected to land on the zero position of the position sensor. As the retro-reflector is moved, the reflected beam is also moved away from the center position of the retro-reflector. The position sensor measures the new beam position and translates the offsets into steering signals for the servo motors, which cause the laser beam to strike the retro-reflector and consequently the position sensor in the center again. A fringe-counting interferometer (IFM) is then used to determine relative distances (or changes in distances) of the target retro-reflector from point to point with accuracies on the nanometer level.

An ADM is used for measuring absolute distances (i.e., distances between points in a 3D coordinate system) with extreme precision, but lacking in speed of IFM for dynamic measurements. It requires long integration times for distance measurements, while in IFM, the change in distance is always immediately determined. One important advantage with ADM in laser trackers (although with decreased accuracy in comparison with using IFM) is that if the laser beam is interrupted, the operator will not have to return to a known location to reset the distance as it is generally the case with IFM.

14.6.1.2 Tracker Controller with System Software In laser trackers, the data management and tracker control are executed on the same computer. The computer contains all of the functions required to operate the tracker and collect and display data. Typical tracker data file will contain the measured locations in polar coordinates, such as azimuth (Az), elevation angle (VA), and interferometer (absolute and relative) radial distance (*r*), and the *x*, *y*, *z* Cartesian coordinates in the user coordinate system. When scanning surfaces, the tracker controller can acquire polar coordinate data triplets (*r*, Az, VA) at rates as high as 1000 points/s. If spherical targets are used for data acquisition, the measured points will be offset from the work-piece surface by the radius of the spherical targets, thus requiring that the data analysis software be able to correct for this offset.

Laser trackers with their associated control systems and data analysis software share many attributes with conventional three-dimensional total station coordinating systems. But there are also significant differences between them. One of the important differences is that the absolute optical distance measurements by total station coordinating systems are limited to resolutions of a few millimeters, which are not of practically sufficient in precision engineering metrology. Another difference is that the widely used method of optical distance measurement in total station coordinating systems is based on amplitude (or intensity) modulation of light sources compared with the interferometric method used in trackers.

Commercial laser trackers are often supplied with optional modulation-type ADMs, which operate in parallel with the interferometer using a common retro-reflector target; they can also be used alone when interferometric resolution is not required. If the interferometer component of a laser tracker is eliminated altogether while retaining the ADM and the motorized angular axes, the resulting instrument is an automatic tracking total station, which are typically used in high-accuracy surveying projects.

14.6.1.3 Remote Power Unit and Other Accessories The remote power unit is for conditioning and supplying the required voltages to the laser tracker. The power supply housing may also contain a built-in electronic barometer for providing barometric pressure information for use in compensating for the effects of varying atmospheric conditions on signal propagation.

One of the important accessories for laser trackers is the retro-reflector. The retro-reflector, which is a glass trihedral prism or a cube corner type, is used to return laser beam back to the laser head. The commonly used retro-reflector is the SMR, which consists of three mirrors that are mounted orthogonally inside a 1.5" sphere. Its finished sphericity is usually about 0.00005" on the ball with a centering accuracy of the cube corner apex varying from 0.0001" to 0.0005." The high sphericity of the steel housing and accurate centering of the cube corner apex are essential for high-accuracy work.

Since the outside surface of SMR is a sphere, there is always a known and constant offset between the actual point being measured (the center of the sphere) and the part of the surface in contact with the outside of the ball. A typical 1.5" diameter SMR reference sitting on a steel drift nest is shown in Figure 14.30. The drift nest is a forced-centering mount for establishing noncritical temporary or permanent monuments with adhesive or with tack welding. In order to provide truly forced center position for the SMR in all three coordinate axes, a magnetic centering nest with a three-point kinematic mount keeps the center of the SMR in the same accurate

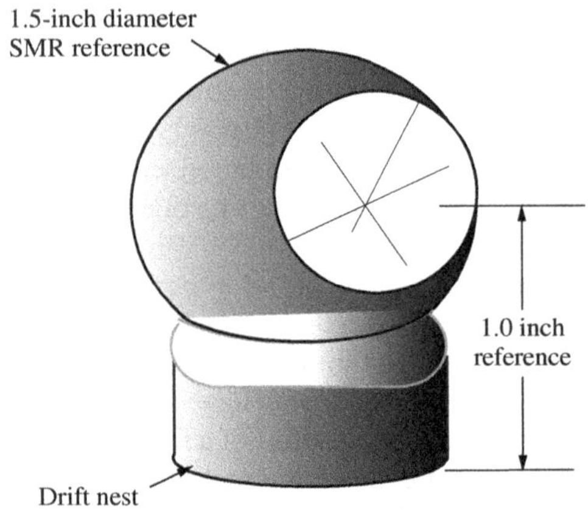

Figure 14.30 A standard 1.5"diameter SMR reference sitting on a drift nest.

position. The position of the SMR in a centering nest is independent of the direction in which it is pointing and the centering errors do not usually exceed 5 μm.

14.6.1.4 Tracker Observables and Measurements Laser trackers are very close to being the most universal tool for metrology and alignment. They are used for measurement of precise underground tunnel networks as well as small objects. The basic tracker observables are the vertical and horizontal angles and the radial distance (or the radial difference from the previous point). Vertical and horizontal angle measurements may be compatible with total station measurements, but the distance measurements are much better with interferometer, probably in the ratio of 5:1.

In preparation for a measurement session with a tracker, the tracker should first be set up close to the target object in the most appropriate vertical orientation. In practice, the laser head or the reflector is mounted on the device whose movement is to be measured and the other unit is mounted at a fixed point. Although the tracker does not need to be leveled, it must be verified that all points to be measured can be reached from the setup point without any obstruction to the laser beam; if necessary, offset bars or scale bars should be used in the process.

For distance measurement, the tracker sends a laser beam to a retro-reflective target held against the object to be measured, and the beam that is reflected from the target back to the tracker is used to determine the precise distance between the tracker and the target. The operator then manually transports the retro-reflector to other points of interest. If there is a loss of lock due to beam obstruction or excessive target acceleration, the fringe-counting displacement interferometer, however, will require that the radial distance be reset at a known location.

For three-dimensional coordinate measurements, two high-precision angle encoders precisely measure the vertical and horizontal angles to a retro-reflector while a highly accurate ADM or interferometer is used to measure the precise distance to the retro-reflector. The three-dimensional laser tracker then follows the retro-reflector as it is moved by the user while the tracker's software determines the retro-reflector's exact position as *X*, *Y*, *Z* coordinate values in a three-dimensional coordinate system. Since the construction principle of the tracker is very similar to that of theodolites, the double-centering (face left and face right) measurement procedure can be assumed to be possible when using the tracker.

One of the major applications of laser tracker is in three-dimensional coordination of geodetic reference network. In comparison with the use of traditional total station equipment for network measurements, laser trackers will result in reduced man power and increased accuracy of network measurements. One other important advantage of using laser tracker in network measurement is the idea of free stationing in which the tracker does not need to be centered over a particular monument or marker, but can be located in a general area where all points of interest are visible. This, however, requires that the surveyor first takes observations to several monuments whose positions are considered known to solve for the tracker position and orientation before proceeding on positioning the object of interest; it is a good practice to observe four or five monuments to provide a redundant solution. However, since the range of laser tracker is usually short, positions of the object of interest may have to be measured in multistation mode, requiring that the tracker be moved to several locations in order to access all of the required features of the object. The different data sets collected for the object are then tied together by subsets of points that are common to the various tracker positions.

When a network of external points (such as preestablished wall points) and object points (on the components to be aligned) are to be measured, for example, from four different setup points of the laser tracker, the laser tracker will first of all resect its position at each point based on the external points and then measure the *X*, *Y*, *Z* coordinates of the aligned components, at regular intervals, by tracking the reflector fixed on the components. The coordinates of the measured points can be calculated by using photogrammetric bundle adjustment program.

An example of laser tracker is Leica Absolute Tracker AT901 (Leica Geosystems, 2014a), an active vision technology that automatically locks onto any moving target without the user's intervention. The vision system built into the laser tracker allows the AT901 sensor to determine where a target is without the need for the laser beam to be locked on; the sensor locks onto the target automatically as soon as it is within the view of the sensor. The tracker also uses a so-called absolute interferometer, which combines the absolute measurement from the ADM with the almost instantaneous update rate of the laser interferometer (IFM) to produce the most accurate, stable, technologically mature distancing unit (Leica Geosystems, 2014a). As soon as a reflector is brought into the laser beam, or "locked-on," the IFM starts tracking its relative movement. The Leica Absolute Tracker AT901 is said to combine the ability to instantly reestablish a broken laser beam and immediately start measuring a moving target. It is quoted as having a typical volume of 160 m; measuring rate of 3000 points/s; lateral tracking speed of 4 m/s; radial tracking speed of 6 m/s; typical lock-on working range of 1.0–80.0 m; interferometer distance accuracy of $\pm 0.5\ \mu m/m$; dynamic lock-on accuracy of $\pm 10\ \mu m$; and the angle accuracy (for full range) of $\pm 15\ \mu m + 6\ \mu m/m$ (Leica Geosystems, 2014a).

14.6.2 High-Precision Industrial Total Stations

The precision that can be expected in a given geodetic network is usually dependent on the accuracy of the instruments used and the configuration of the network itself. Basic alignment of components or machine tools can be done with high-precision industrial or RTSs by spherical (polar) measurements with respect to a previously established reference network. In the case where alignment of components is to be done within a tunnel, many wall brackets are usually mounted on the tunnel walls as well as on the components to be aligned with each of the points occupied and observed by the industrial total station. The industrial total station may be preferred to a laser tracker in some cases since the remarkable distance accuracy of the laser tracker may be insufficient to offset its comparatively poor angular accuracy compared with that of the industrial total station whose angular accuracy may be better.

Usually, the planes on which components are to be installed are not horizontal, but inclined, such as 1% inclined, it is necessary to introduce a three-dimensional coordinate system to define the position of the components. The three-dimensional coordinates of the components are then projected to a reference sphere before they are used for geodetic measurements.

In the case of alignment of accelerator and beam line, the usually required tolerances are typically less than 1 mm and are often in the order of several micrometers. In order to achieve the tolerances, a well-calibrated, high-precision motorized RTS instruments equipped with automatic target recognition (ATR) must be used with the calibration procedure that pays particular attention to the angle and distance measuring components of these instruments. At the limit of distance meter precision, the only way to improve positional uncertainty results is to improve the angle measuring capacity of these instruments by calibrating the horizontal and vertical angles of the instruments. By employing the double centering (face left and face right) measurement procedure, most of the systematic angle collimation errors can be reduced to negligible levels. The errors associated with the ATR system or laser tracking instrumentation are determined by observing the laser spot in different positions of the instruments' CCD or PDS image sensor (Martin and Chetwynd, 2009).

In the case of alignment of machine components within a tunnel space, it is necessary to guarantee precision of machine plane and to fit the orbit of the machine into the limited space of the tunnel to be precisely constructed. A geodetic network is first established on the surface for orienting the tunnel in the earth body during the tunnel construction and for aligning the components within the tunnel. All the distances between the network monuments are measured by high-precision EDM instrument, such as Kern ME5000, and the height differences are measured by precision leveling procedure. The tunnel boring part is done from every vertical shaft in two faces.

The determination of reference coordinates and underground geodetic network orientation in the beginning of a tunnel construction is a most important stage in geodetic work. Four types of survey points usually constitute the underground (tunnel) networks: floor points, wall points, pass points, and points on the components to be aligned. Wall targets are usually steel brackets that can be used for holding instruments and their ATR reflectors.

A reference network is usually established in the tunnel (usually on the tunnel walls) from the surface network through vertical shafts; distances of overlapping lines from each pillar and the directions to other points are typically measured. The heights of the geodetic points at the shafts are transferred to the shaft bottom with the help of steel tape measurements or other methods as discussed in Chapter 12. Systematic errors in underground measurements may be due to a number of factors with the main ones being horizontal refraction during the angle measurements and inaccuracies of self-centering of the theodolites and targets.

The procedure for vertical alignment of machine components can be done by directly measuring height differences between adjacent components using precision leveling and making appropriate corrections with respect to the best fit curve. If targets in Taylor Hobson spheres are used, the horizontal directions and zenith angles are measured by pointing to the targets and their height differences are derived from the zenith angle measurements. The heights of spherical targets of Taylor Hobson are determined by precise leveling; scale bars can be used to provide the scale if needed.

An example of industrial total stations is Leica TDA5005 having the manufacturer's quoted absolute standard deviation (per ISO17123-4) of distance measurement for precise mode as $\pm 1\,\text{mm} \pm 2\,\text{ppm}$ with a typical (uncorrected) distance accuracy at 120 m measuring volume with CCR of ±0.2 mm (Leica Geosystems, 2014b); measurement range with CCR is 2–600 m; and the standard deviation (per ISO17123-3) for angular measurement is $0.5''$. The distance uncertainty can be improved, according to Martin and Gatta (2004), to between 0.08 mm and 0.1 mm with the calibration of the instrument and the application of appropriate corrections to the measured distance. All of these features of the total station qualify it for use in precision projects, such as industrial metrology, construction projects, alignment of machine and accessories, and for assembling and adjusting components in relation to each other, and so on.

The built-in precision distance meter and its ability to locate and track a target make the Leica TDA5005 industrial laser total station perform much like a standard laser tracker. The instrument can produce three-dimensional coordinates along with their accuracies in real time. The other features of the instrument are summarized as follows:

- Apart from being able to precisely measure distances and horizontal and vertical angles, it is capable of transforming these measurements into three-dimensional coordinates with the total station location as the origin of the coordinate system.
- While pointing at a target, the zero theodolite menu item can zero the horizontal angle at the initial target and use the direction from the total station to the target as the Y-axis.
- The total station is connected to a computer with the necessary software for coordinate determination. When taking measurements and determining the coordinates of points, the installed software can instantly provide the standard deviation of the calculated coordinates.
- The ATR functionality of the total station is for eliminating pointing errors made by the user and to increase the speed and efficiency of manually

taking measurements. The functionality eliminates the need for a user to point the instrument directly at the target as long as the target is within the field of view of the instrument.

- The total station has two-axis compensator for precise leveling of the instrument and for compensating measurements for some leveling errors.
- The ATR lock-in mode enables the instrument to lock onto the target while it is in motion. This allows a single user, with a remote control, to operate the instrument and move the reflector without the need to return to the instrument to take measurements.

14.6.3 Coherent Laser Radar System

The distance meters discussed so far in Sections 14.5 and 14.6 all require targets, usually cube corner retro-reflector types. In coordinate metrology, manually moving such targets over a work-piece can be laborious, slow, and costly. A commercially available system that overcomes these limitations uses a ranging technology called coherent laser radar (CLR). The term *Laser Radar* is used today to mean the same thing as *LADAR* (an acronym for LAser Detection And Ranging) or *LiDAR* (an acronym for Light Detection And Ranging) according to Stone et al. (2004) and Slotwinski and Blanckaert (2007). All ranging systems, whether *RADAR* (RAdio Detection And Ranging), *LiDAR* or *LADAR*, operate on the same principles by transmitting and receiving electromagnetic energy. The only difference among them is that they work in different frequency bands with the Laser Radar based on much shorter wavelengths.

CLR technology consists of a distance measuring device and two-axis beam-steering system (turning and tilting mirror) with encoders for horizontal and vertical angle measurements. An integrated color video camera helps in selecting and identifying measurement areas with a red visible laser being used for beam positioning. The distance measuring device uses frequency-modulated (coherent) laser to measure distances in the same way as interferometers, but at a lower carrier frequency. It measures the travel time of the envelope of the carrier while interferometers measure distances by counting wavelengths of the carrier signal.

The innovative aspect of CLR technology is the eliminated need of any kind of cooperative target such as photogrammetry dots, laser tracker spherically mounted reflectors (SMR) while providing noncontact, auto-locating, and precise measurements of surfaces and points or scan features. The technology can be used to provide three-dimensional measurement of inaccessible surfaces with the measurements taken directly from the surfaces. Since targets are not required, offset corrections are not needed as is usually the case with using instruments that require the use of targets.

The operating principle of CLR is such that it directs a focused laser beam to a point on the target surface to be measured and recaptures a portion of the reflected light. As the laser light travels to and from the target, it also travels through a reference path of calibrated optical fiber in a module that is well controlled. The two paths are combined to determine the absolute distance to the target surface. CLR technology measures a distance and two angles to determine a point on a surface in space.

According to White (1999), CLR works on typical engineering surfaces or any surface as long as the reflectivity of the surface is greater than 1%. Some of the typical applications of CLR technology include tool building and alignment and alignment of aircraft and automotive components. An example of CLR system is the MetricVision 100B CLR abbreviated as MV-100B CLR with a claimed point coordinate expanded uncertainty of ±130 μm in a radial range up to 10 m and 6.5 ppm for ranges greater than 10 m. It is a portable, eye-safe Class I laser radar for measuring coordinates of points. According to White (1999), "[MV-100B] can be used to scan complex geometry that was impossible to scan before because it was too large, too hard to reach, too complex, too delicate or too labor-intensive."

14.7 MAIN SOURCES OF ERROR IN ALIGNMENT SURVEYS

The main source of error in alignment surveys is the atmospheric refraction, both in the vertical and in the horizontal directions. For example, in alignment survey between two fixed points A and B as shown in Figure 14.31, the line of sight from A is constrained in the direction of target at B by refraction effects. All types of surveys that use optical tools are subject to uncertainties of the refraction. In horizontal alignment, the horizontal refraction component will be most relevant.

The horizontal refraction component is mainly a function of the gradient of temperature across the line of sight. If this gradient is constant between any given points A and B, then the alignment reference line will conform to a circular path as shown in dotted line in Figure 14.31, with the largest error of alignment at the middle of points A and B. Usually, the temperature gradient will vary from one point of the line to another so that an irregular shape of the refracted line of sight $AP'B$ is produced. In order to be able to correct the alignment surveys for refraction, one should measure the gradients of temperature simultaneously at a number of points on the alignment line at the instants of pointing the aligning telescope or laser beam, at the aligning targets. These gradients should be measured perpendicular to the optical path.

For example, to obtain horizontal temperature gradient at the center of the alignment line, three thermistors can be arrayed perpendicularly to the alignment line at the center of the line: one thermistor is located on the alignment line and the others at distances of 4 m perpendicular to the alignment line on each side of the line. The obtained gradient of the temperature allows one to calculate the curvature (or

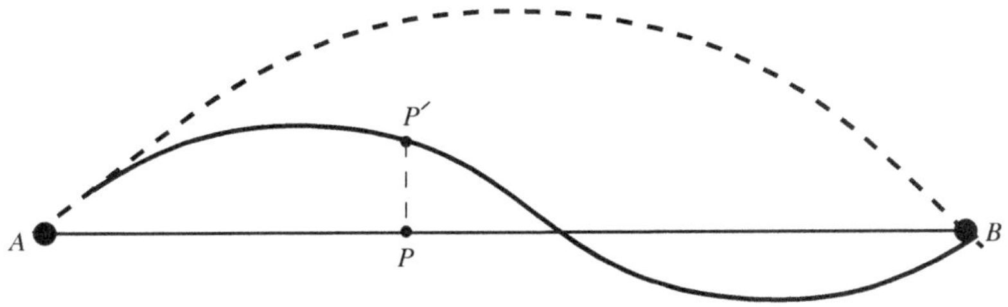

Figure 14.31 Error of alignment due to atmospheric refraction.

the maximum error due to refraction) of the optical line in the middle of the test line. The refraction correction to the position of an aligned point is equal to the distance between the optical line and the straight line connecting the alignment points as shown in Figure 14.31, where point P is being aligned between points A and B. In the figure, distance P'-P is the error of the alignment.

Better alignment results are usually obtained when the alignment surveys are repeated several times in different atmospheric conditions. According to Chrzanowski et al. (1976), overall characteristics of temperature gradients as functions of time and location are random with a tendency to cancel out. This means that the effects of refraction on alignment surveys tend to cancel out when the surveys are repeated several times in different atmospheric conditions. Refer to Section 4.3.4 for more discussions on the effects of refraction on direction and angle measurements. Apart from the effects of atmospheric refraction on alignment surveys, other relevant sources of error and their mitigating techniques can be found in Chapters 2, 4, 5, and 6.

APPENDIX I

EXTRACTS FROM BAARDA'S NOMOGRAM

See Tables I.1–I.4.

TABLE I.1 For the Values $\lambda_0 = \lambda(\alpha_0, \beta_0 = 0.20, 1) = \lambda(\alpha, \beta_0 = 0.20, \text{df})$.

$100\alpha_0$	Degrees of Freedom (df)	λ_0
5	1	7.8
2.4	2	9.6
1.3	3	11.0
0.9	4	12.0
0.6	5	13.0
0.1	12	17
0.2	10	16
0.3	9	15.5
0.35	8	15
0.40	7	14
0.50	6	13.5
0.07	14	18
0.04	16	19
0.03	18	20
0.02	20	21
0.02	22	22
0.01	24	22

Precision Surveying: The Principles and Geomatics Practice, First Edition. John Olusegun Ogundare.

TABLE I.2 For the Values of $100\alpha_0 = 0.1$, $\beta_0 = 0.20$, $\lambda_0 = 17.0$.

Alpha (α)	Degrees of Freedom (df)	Alpha (α)	Degrees of Freedom (df)
0.006	3	0.056	12
0.009	4	0.094	18
0.013	5	0.107	20
0.018	6	0.119	22
0.025	7	0.132	24
0.030	8	0.150	26
0.038	9	0.158	28
0.043	10	0.167	30

TABLE I.3 For the Values of $100\alpha_0 = 0.9$, $\beta_0 = 0.20$, $\lambda_0 = 12.0$.

Alpha (α)	Degrees of Freedom (df)	Alpha (α)	Degrees of Freedom (df)
0.01	1	0.085	6
0.022	2	0.100	7
0.038	3	0.114	8
0.050	4	0.129	9
0.070	5	0.140	10

TABLE I.4 For the Values of $100\alpha_0 = 1.0$, $\beta_0 = 0.20$, $\lambda_0 = 11.7$.

Alpha (α)	Degrees of Freedom (df)	Alpha (α)	Degrees of Freedom (df)
0.01	1	0.090	6
0.025	2	0.110	7
0.041	3	0.121	8
0.058	4	0.136	9
0.075	5	0.150	10

APPENDIX II

COMMONLY USED STATISTICAL TABLES

See Tables II.1–II.4.

TABLE II.1 Standard Normal Distribution.

α	0.001	0.002	0.003	0.004	0.005	0.01	0.025	0.05	0.10
Z_{α}	3.09	2.88	2.75	2.65	2.58	2.33	1.96	1.64	1.28

α is upper tail area.
The sample normal distribution table is formed by using the UTPN program in the Hewlett Packard (HP) 48GX calculator.

Precision Surveying: The Principles and Geomatics Practice, First Edition. John Olusegun Ogundare.

TABLE II.2 Table for Student *t*-Distribution.

Degrees of Freedom (df)	t_{α}			
	$t_{0.10}$	$t_{0.05}$	$t_{0.025}$	$t_{0.01}$
1	3.08	6.31	12.7	31.8
2	1.89	2.92	4.30	6.96
3	1.64	2.35	3.18	4.54
4	1.53	2.13	2.78	3.75
5	1.48	2.01	2.57	3.36
6	1.49	1.94	2.45	3.14
7	1.42	1.90	2.37	3.00
8	1.40	1.86	2.31	2.90
9	1.38	1.83	2.26	2.82
10	1.37	1.81	2.23	2.76

α is upper tail area.
The sample Student *t*-distribution table is formed by using the UTPT program in the Hewlett Packard (HP) 48GX calculator.

TABLE II.3 Distribution Table for Chi-Square.

Degrees of Freedom (df)	α							
	0.995	0.99	0.975	0.95	0.05	0.025	0.01	0.005
1	0.000	0.000	0.001	0.004	3.841	5.024	6.635	7.879
2	0.010	0.020	0.051	0.103	5.991	7.378	9.210	10.597
3	0.071	0.115	0.216	0.352	7.815	9.348	11.345	12.838
4	0.207	0.297	0.484	0.711	9.488	11.143	13.277	14.860
5	0.412	0.554	0.831	1.145	11.070	12.833	15.086	16.750
6	0.676	0.872	1.237	1.635	12.592	14.449	16.812	18.548
7	0.989	1.239	1.690	2.167	14.067	16.013	18.475	20.278
8	1.344	1.646	2.180	2.733	15.507	17.535	20.090	21.955
9	1.735	2.088	2.700	3.325	16.919	19.023	21.666	23.589
10	2.156	2.558	3.247	3.940	18.307	20.483	23.209	25.188
11	2.603	3.053	3.816	4.575	19.675	21.920	24.725	26.757
12	3.074	3.571	4.404	5.226	21.026	23.337	26.217	28.300
13	3.565	4.107	5.009	5.892	22.362	24.736	27.688	29.819
14	4.075	4.660	5.629	6.571	23.685	26.119	29.141	31.319
15	4.600	5.229	6.262	7.261	24.996	27.488	30.578	32.801
31	14.458	15.655	17.539	19.281	44.985	48.232	52.191	55.003
32	15.134	16.362	18.291	20.072	46.194	49.480	53.486	56.328

α is upper tail area.
The sample Chi-square distribution table is formed by using the UTPC program in the Hewlett Packard (HP) 48GX calculator.

TABLE II.4 Table for *F*-Distribution.

Degrees of Freedom (df-1)	α	Degrees of Freedom (df-2)									
		1	2	3	4	5	6	7	8	9	10
1	0.005	16211	198.5	55.552	31.333	22.785	18.635	16.236	14.688	13.614	12.826
	0.01	4052	98.503	34.116	21.198	16.258	13.745	12.246	11.259	10.561	10.044
	0.05	161.4	18.513	10.128	7.709	6.608	5.987	5.591	5.318	5.117	4.965
	0.95	0.006	0.005	0.005	0.004	0.004	0.004	0.004	0.004	0.004	0.004
	0.99	0.000	0.000	0.000	0.000	0.000	0.000	0.000	0.000	0.000	0.000
	0.995	0.000	0.000	0.000	0.000	0.000	0.000	0.000	0.000	0.000	0.000
2	0.005	20000	199	49.799	26.284	18.314	14.544	12.404	11.042	10.107	9.427
	0.01	5000	99	30.817	18	13.274	10.925	9.547	8.649	8.022	7.559
	0.05	199.5	19	9.552	6.944	5.786	5.143	4.737	4.459	4.256	4.103
	0.95	0.054	0.053	0.052	0.052	0.052	0.052	0.052	0.052	0.052	0.052
	0.99	0.010	0.010	0.010	0.010	0.010	0.010	0.010	0.010	0.010	0.010
	0.995	0.005	0.005	0.005	0.005	0.005	0.005	0.005	0.005	0.005	0.005
3	0.005	21615	199.2	47.467	24.259	16.530	12.917	10.882	9.596	8.717	8.081
	0.01	5403	99.166	29.457	16.694	12.060	9.780	8.451	7.591	6.992	6.552
	0.05	215.7	19.164	9.277	6.591	5.409	4.757	4.347	4.066	3.863	3.708
	0.95	0.099	0.105	0.108	0.110	0.111	0.112	0.113	0.113	0.113	0.114
	0.99	0.029	0.032	0.034	0.035	0.035	0.036	0.036	0.036	0.037	0.037
	0.995	0.018	0.020	0.021	0.022	0.022	0.022	0.023	0.023	0.023	0.023
4	0.005	22500	199.2	46.195	23.155	15.556	12.028	10.050	8.805	7.956	7.343
	0.01	5625	99.249	28.710	15.977	11.392	9.148	7.847	7.006	6.422	5.994
	0.05	224.6	19.247	9.117	6.388	5.192	4.534	4.120	3.838	3.633	3.478
	0.95	0.130	0.144	0.152	0.157	0.160	0.162	0.164	0.166	0.167	0.168
	0.99	0.047	0.056	0.060	0.063	0.064	0.066	0.067	0.068	0.068	0.069
	0.995	0.032	0.038	0.041	0.043	0.045	0.046	0.046	0.047	0.047	0.048

(continued)

TABLE II.4 (*Continued*)

Degrees of Freedom (df-1)	α	Degrees of Freedom (df-2) 1	2	3	4	5	6	7	8	9	10
5	0.005	23056	199.3	45.392	22.456	14.940	11.464	9.522	8.302	7.471	6.872
	0.01	5764	99.299	28.237	15.522	10.967	8.746	7.460	6.632	6.057	5.636
	0.05	230.2	19.296	9.013	6.256	5.050	4.387	3.972	3.687	3.482	3.326
	0.95	0.151	0.173	0.185	0.193	0.198	0.202	0.205	0.208	0.210	0.211
	0.99	0.062	0.075	0.083	0.088	0.091	0.094	0.096	0.097	0.098	0.099
	0.995	0.044	0.055	0.060	0.064	0.067	0.069	0.070	0.072	0.073	0.073
6	0.005	23437	199.3	44.838	21.974	14.513	11.073	9.155	7.952	7.134	6.545
	0.01	5859	99.333	27.911	15.207	10.672	8.466	7.191	6.371	5.802	5.386
	0.05	234.0	19.330	8.941	6.163	4.950	4.284	3.866	3.581	3.374	3.217
	0.95	0.167	0.194	0.210	0.221	0.228	0.233	0.238	0.241	0.244	0.246
	0.99	0.073	0.092	0.102	0.109	0.114	0.118	0.121	0.123	0.125	0.127
	0.995	0.054	0.069	0.077	0.083	0.087	0.090	0.093	0.095	0.096	0.098
7	0.005	23715	199.4	44.434	21.622	14.200	10.786	8.885	7.694	6.885	6.302
	0.01	5928	99.356	27.672	14.976	10.456	8.2600	6.993	6.178	5.613	5.200
	0.05	236.8	19.353	8.887	6.094	4.876	4.207	3.787	3.500	3.293	3.135
	0.95	0.179	0.211	0.230	0.243	0.252	0.259	0.264	0.268	0.272	0.275
	0.99	0.082	0.105	0.118	0.127	0.134	0.139	0.143	0.146	0.149	0.151
	0.995	0.062	0.081	0.092	0.100	0.105	0.109	0.113	0.115	0.117	0.119
8	0.005	23925	199.4	44.125	21.352	13.961	10.566	8.678	7.496	6.693	6.116
	0.01	5981	99.374	27.489	14.799	10.290	8.102	6.840	6.029	5.467	5.057
	0.05	238.9	19.371	8.845	6.041	4.818	4.147	3.726	3.438	3.230	3.072
	0.95	0.188	0.224	0.246	0.261	0.271	0.279	0.286	0.291	0.295	0.299
	0.99	0.089	0.116	0.132	0.143	0.151	0.157	0.162	0.166	0.169	0.172
	0.995	0.068	0.091	0.104	0.114	0.120	0.126	0.130	0.133	0.136	0.139

α is upper tail area.
The sample F-distribution table is formed by using the UTPF program in the Hewlett Packard (HP) 48GX calculator.

APPENDIX III

TAU DISTRIBUTION TABLE FOR SIGNIFICANCE LEVEL α

Number of Observations	α	Degrees of Freedom (df) 1	2	3	4	5	6	7	8	9	10
3	0.1	1.000	1.412								
	0.05	1.000	1.414								
	0.025	1.000	1.414								
	0.02	1.000	1.414								
	0.01	1.000	1.414								
	0.009	1.000	1.414								
	0.007	1.000	1.414								
	0.002	1.000	1.414								
	0.001	1.000	1.414								
4	0.1	1.000	1.413	1.687							
	0.05	1.000	1.414	1.710							
	0.025	1.000	1.414	1.721							
	0.02	1.000	1.414	1.723							
	0.01	1.000	1.414	1.728							
	0.009	1.000	1.414	1.728							
	0.007	1.000	1.414	1.729							
	0.002	1.000	1.414	1.731							
	0.001	1.000	1.414	1.732							
5	0.1	1.000	1.413	1.696	1.865						
	0.05	1.000	1.414	1.714	1.916						
	0.025	1.000	1.414	1.723	1.948						
	0.02	1.000	1.414	1.725	1.955						
	0.01	1.000	1.414	1.729	1.972						

	0.009	1.000	1.414	1.729	1.974		
	0.007	1.000	1.414	1.730	1.978		
	0.002	1.000	1.414	1.731	1.990		
	0.001	1.000	1.414	1.732	1.994		
6	0.1	1.000	1.414	1.702	1.880	1.991	
	0.05	1.000	1.414	1.717	1.926	2.065	
	0.025	1.000	1.414	1.725	1.954	2.117	
	0.02	1.000	1.414	1.726	1.960	2.129	
	0.01	1.000	1.414	1.729	1.975	2.161	
	0.009	1.000	1.414	1.729	1.977	2.165	
	0.007	1.000	1.414	1.730	1.980	2.173	
	0.002	1.000	1.414	1.731	1.991	2.203	
	0.001	1.000	1.414	1.732	1.995	2.212	
7	0.1	1.000	1.414	1.706	1.892	2.009	2.087
	0.05	1.000	1.414	1.719	1.933	2.078	2.179
	0.025	1.000	1.414	1.726	1.958	2.125	2.247
	0.02	1.000	1.414	1.727	1.964	2.137	2.265
	0.01	1.000	1.414	1.730	1.977	2.167	2.310
	0.009	1.000	1.414	1.730	1.979	2.170	2.316
	0.007	1.000	1.414	1.730	1.982	2.178	2.329
	0.002	1.000	1.414	1.732	1.992	2.205	2.377
	0.001	1.000	1.414	1.732	1.995	2.214	2.395

Number of Observations	α	Degrees of Freedom (df)									
		1	2	3	4	5	6	7	8	9	10
8	0.1	1.000	1.414	1.709	1.901	2.024	2.106	2.164			
	0.05	1.000	1.414	1.721	1.939	2.088	2.194	2.271			
	0.025	1.000	1.414	1.727	1.962	2.133	2.258	2.351			
	0.02	1.000	1.414	1.728	1.967	2.144	2.274	2.373			
	0.01	1.000	1.414	1.730	1.979	2.171	2.318	2.431			
	0.009	1.000	1.414	1.730	1.981	2.174	2.323	2.439			
	0.007	1.000	1.414	1.731	1.984	2.182	2.335	2.456			
	0.002	1.000	1.414	1.732	1.993	2.207	2.381	2.521			
	0.001	1.000	1.414	1.732	1.996	2.216	2.397	2.547			
9	0.1	1.000	1.414	1.712	1.909	2.036	2.122	2.184	2.229		
	0.05	1.000	1.414	1.722	1.943	2.097	2.206	2.286	2.346		
	0.025	1.000	1.414	1.727	1.965	2.139	2.267	2.363	2.438		
	0.02	1.000	1.414	1.728	1.970	2.149	2.283	2.384	2.463		
	0.01	1.000	1.414	1.730	1.981	2.175	2.324	2.440	2.531		
	0.009	1.000	1.414	1.730	1.982	2.178	2.329	2.447	2.540		
	0.007	1.000	1.414	1.731	1.985	2.185	2.341	2.463	2.561		
	0.002	1.000	1.414	1.732	1.994	2.209	2.384	2.526	2.643		
	0.001	1.000	1.414	1.732	1.996	2.217	2.400	2.551	2.677		

10	0.1	1.000	1.414	1.714	1.915	2.046	2.136	2.200	2.248	2.285	
	0.05	1.000	1.414	1.723	1.947	2.104	2.216	2.298	2.361	2.410	
	0.025	1.000	1.414	1.728	1.967	2.144	2.274	2.373	2.450	2.511	
	0.02	1.000	1.414	1.729	1.972	2.154	2.290	2.393	2.474	2.539	
	0.01	1.000	1.414	1.730	1.982	2.178	2.329	2.447	2.540	2.616	
	0.009	1.000	1.414	1.730	1.983	2.181	2.334	2.454	2.549	2.626	
	0.007	1.000	1.414	1.731	1.986	2.188	2.345	2.470	2.569	2.650	
	0.002	1.000	1.414	1.732	1.994	2.210	2.387	2.530	2.649	2.747	
	0.001	1.000	1.414	1.732	1.996	2.218	2.402	2.554	2.681	2.788	
11	0.1	1.000	1.414	1.716	1.920	2.055	2.148	2.215	2.264	2.303	2.333
	0.05	1.000	1.414	1.724	1.951	2.110	2.225	2.310	2.374	2.425	2.466
	0.025	1.000	1.414	1.728	1.969	2.148	2.281	2.382	2.460	2.523	2.574
	0.02	1.000	1.414	1.729	1.973	2.157	2.296	2.401	2.484	2.550	2.604
	0.01	1.000	1.414	1.730	1.983	2.181	2.334	2.453	2.548	2.625	2.689
	0.009	1.000	1.414	1.731	1.984	2.184	2.338	2.460	2.557	2.635	2.700
	0.007	1.000	1.414	1.731	1.987	2.190	2.349	2.475	2.576	2.659	2.727
	0.002	1.000	1.414	1.732	1.994	2.211	2.389	2.534	2.654	2.753	2.837
	0.001	1.000	1.414	1.732	1.996	2.219	2.404	2.557	2.685	2.793	2.884

APPENDIX IV

IMPORTANT UNITS

Some of the important units associated with electromagnetic wave propagation are as follows:
Units for frequency:

- 1 hertz (Hz)
- 1 kilohertz (kHz) = 1×10^3 Hz
- 1 megahertz (MHz) = 1×10^6 Hz
- 1 gigahertz (GHz) = 1×10^9 Hz
- 1 terahertz (THz) = 1×10^{12} Hz

Units for time:

- 1 second (s)
- 1 millisecond (ms) = 1×10^{-3} s
- 1 microsecond (μs) = 1×10^{-6} s
- 1 nanosecond (ns) = 1×10^{-9} s
- 1 picoseconds (ps) = 1×10^{-12} s

Precision Surveying: The Principles and Geomatics Practice, First Edition. John Olusegun Ogundare.

Units for pressure and temperature:

- 1 millibar (mbar) = 0.750063755 mmHg
- 1013.246 mbar = 760 mmHg (known as the standard atmospheric pressure)
- 1 mmHg = 1.33322 mbar
- Standard temperature is 0 °C (Celsius) or 273.15°K (Kelvin)

REFERENCES

ACSM. The Surveyor's Contracts and Risk Management Manual. American Congress on Surveying and Mapping and Victor O. Schinnerer & Company, Inc.; 2002.

Applied Geomechanics Inc. 2005. Uniaxial Tiltmeter: Model Tuff Tilt 801. Website: http://www.roctest-group.com/sites/default/files/datasheets/products/TUFFTILT-E5077E-W_0.pdf. Accessed 2015 May 18

ASCE Manuals 1985. ASCE Manuals and Reports on Engineering Practice No. 64. Engineering Surveying Manual, American Society of Civil Engineers, New York.

ASCE. Guidelines for Instrumentation and Measurements for Monitoring Dam Performance. Prepared by ASCE Task Committee on Instrumentation and Monitoring Dam Performance, sponsored by Hydropower Committee of the Energy Division of the American Society of Civil Engineers (ASCE), USA, ISBN 0-7844-0531-X; 2000

ASPRS Specifications and Standards Committee. American society for photogrammetry and remote sensing (ASPRS) accuracy standards for large-scale maps. Photogramm Eng Rem Sens 1990;56(7):1068–1070.

Avella A. An analysis of a worldwide status of monitoring and analysis of dam deformation, Master of Engineering Report, Fredericton, NB: Department of Geodesy and Geomatics Engineering, University of New Brunswick; 1993.

Baarda W. A Testing Procedure for Use in Geodetic Networks. Publications on Geodesy, New Series, Vol. 2, No. 5. Netherlands Geodetic Commission; 1968.

Balis V, Karamistos S, Kotsis I, Liapakis C, Simpas N. 3D-laser scanning: integration of point cloud and CCD camera video data for the production of high resolution and precision RGB textured models: Archaeological monuments surveying application in ancient Ilida. In: Proceedings of FIG Working Week, Athens, Greece, 2004 May 22–27;. http://www.fig.net/pub/athens. Accessed 2015 May 18

Precision Surveying: The Principles and Geomatics Practice, First Edition. John Olusegun Ogundare.

Bamler R, Hartl P. Synthetic aperture radar interferometry. Inverse Problems 1998;14:R1–R54, IOP Publishing Ltd, UK. PII:S0266-5611(98)78649-6. Website: http://www.geo.uzh.ch/microsite/rsl-documents/research/SARlab/GMTILiterature/PDF/BH98b.pdf. Accessed 2014 Mar.

Barendse M. SPR research project C-06-02: field evaluations of "ShapeAccelArray" in-place MEMS inclinometer strings for subsurface deformation monitoring, Final report, Albany, NY: NYSDOT Geotechnical Engineering Bureau, 2012 Mar.

Becker J-M. Recommendation concerning survey instruments and quality specifications. In: FIG XXII International Congress; Washington, DC, USA, TS 5.11 – Standards, Quality Assurance and Calibration; 2002 Apr 19–26.

Biedermann R, Bonhage R, Indermaur W, Muller RW, Rouiller JM, Venzin C. Measuring Installations for Dam Monitoring: Concepts, Reliability, Redundancy. Jahrgang: Wasser, Energie, Luft; 1988. p 11–20.

Birch S, Pollock G, Eenkooren N, Saforo S. Comparison of slope monitoring techniques from a test slope at the Suncor oil sands mine, In: GeoHazards 2011, 5th Canadian conference on geotechnique and natural hazards, Kelowna, BC, 8 pp; 2011.

Blachut TJ, Chrzanowski A, Saastamoinen JH. Urban Surveying and Mapping. Springer-Verlag; 1979. ISBN: 0-387-90344-5.

Boehler W, Marbs A. 3D Scanning Instruments. Tech. rapp., i3mainz. Germany: Institute for Spatial Information and Surveying Technology, FH Mainz, University of Applied Sciences; 2002.

Bomford G. Geodesy. 4th Ed. Oxford University Press; 1980.

Bond J. An investigation on the use of GPS for deformation monitoring in open pit mines. M.Sc.E. thesis, Department of Geodesy and Geomatics Engineering Technical Report No. 222, Fredericton, New Brunswick, Canada: University of New Brunswick; 2004, 140 pp.

Bond DJ. Bringing GPS into harsh environments for deformation monitoring. Ph.D. dissertation, Department of Geodesy and Geomatics Engineering Technical Report No. 253, Fredericton, New Brunswick, Canada: University of New Brunswick; 2007 268 pp.

Brinkman RF, O'Neill C n.d. A powerful combination: LiDAR and photogrammetric mapping. Downloaded 2002 Sept, from http://www.woolpert.com/news/articles/ar072500.html. Accessed 2015 May 18.

Bryzek J. Principles of MEMSChapter 162. In: Handbook of Measuring System. John Wiley & Sons, Ltd. Available at: http://eu.wiley.com/legacy/wileychi/hbmsd/pdfs/mm573.pdf, Published: Jul 15; 2005. DOI: 10.1002/0471497398.mm573.

Caspary WF. Concepts of Network and Deformation Analysis. School of Surveying Monograph 11. University of New South Wales; 1987. 183 pp.

Catalina C, Andreea-Florina, J. The use of terrestrial laser scanning in GIS applications. Recent Advances in Geodesy and Geomatics Engineering; 2013. ISBN: 978-960-474-335-3. Website: http://www.wseas.us/e-library/conferences/2013/Antalya/GENG/GENG-22.pdf. Accessed 2014 May

Chapman MA. 1985. Deformation monitoring of Turtle Mountain by high precision photogrammetry. Papers for the Precise Engineering and Deformation Surveys Workshop, Division of Surveying Engineering, The University of Calgary, Apr.

Chen YQ. Analysis of deformation surveys – a generalized method. Ph.D. dissertation, Department of Geodesy and Geomatics Engineering Technical Report No. 94, Fredericton, New Brunswick, Canada: University of New Brunswick; 1983. 262 pp.

Chen YQ. 2011. Developments in deformation monitoring techniques. A paper presented at the Joint international symposium on deformation monitoring by 14th FIG symposium on deformation measurement and analysis, 5th IAG symposium on geodesy for geotechnical and structural engineering and 2nd international workshop on spatial information technologies for monitoring the deformation of large-scale man-made linear features, Hong Kong, China, Nov 1–6

Chen YQ, Kavouras M, Secord JM. 1983. Design considerations in deformation monitoring. Paper presented at the FIG XVII international congress, Sofia, Bulgaria, 1983 Jun 19–28, Paper 608.2.

Chen YQ, Zhang GB, Ding XL, Li ZL. Monitoring earth surface deformations with InSAR technology: principle and some critical issues. J Geospatial Eng 2000;2(1):3–21.

Chrzanowski A. Design and error analysis of surveying projects. Lecture Notes No. 47, Canada: Department of Surveying Engineering, University of New Brunswick; 1977.

Chrzanowski A. Error analysis of tunnelling surveys. In: Proceedings, third Canadian symposium on mining surveying and rock deformation measurements, Sudbury, Ontario; 1979 Oct 10–12.

Chrzanowski A. Optimization of breakthrough accuracy in tunneling survey. Can Surv 1981a;35(1):5–16.

Chrzanowski A. Geotechnical and other non-geodetic methods in deformation measurements. In: Proceedings: Deformation measurements workshop, modern methodology in precise engineering and deformation surveys – II. MIT; 1986 Oct 31–Nov 1.

Chrzanowski A. Modern surveying techniques for mining and civil engineering. Chapter 33. In: Hudson JA, editor. Comprehensive Rock Engineering, v.3 Rock Testing and Site Characterization. Pergamon Press; 1993. p 773–809. ISBN: 0-08-042066-4.

Chrzanowski A. Engineering and mining surveys. Chapter 16. In: McGrath G, Sebert LM, editors. Mapping a Northern Land, 1947–1994. McGill-Queen's University Press; 1999. ISBN: 0-7735-1689-1.

Chrzanowski A. Review of deformation monitoring techniques. Lecture notes presented in Wroclaw, Poland: Canadian Centre for Geodetic Engineering; 2009.

Chrzanowski A, Ahmed F. Alignment surveys in a turbulent atmosphere using laser, In: Proceedings of the 31st annual meeting of the ACSM, Washington; 1971.

Chrzanowski A, Bazanowski M. Design and implementation of ground deformation monitoring scheme at PCS picadilly mine in New Brunswick. A report submitted by Monitoring Systems Limited of Fredericton, NB to PCS Potash – New Brunswick Division; 2011 Dec 30.

Chrzanowski A and Chrzanowski A. Effects of PCS underground mining on the surface infrastructure in Penobsquis, New Brunswick. A report submitted by Canadian Centre for Geodetic Engineering (CCGE), Fredericton, NB to Barry Spalding Lawyers, Saint John, NB; 2012 May 28.

Chrzanowski A and Egberongbe FO. Evaluation of He–Ne laser for use in alignment surveys, Tech. Report No. 10, Fredicton, NB: Department of Surveying Engineering (now Department of Geodesy and Geomatics Engineering), UNB; 1971.

Chrzanowski A, Janssen D. Use of laser in precision levelling. Can Surv 1972. Dec.

Chrzanowski A, Robinson A. Mining Surveys. Chapter 20. In: Davis RE, Foote FS, Anderson JM, Mikhail EM, editors. Surveying: Theory and Practice. 6th ed. McGraw-Hill Book Company; 1981.

Chrzanowski A and Szostak-Chrzanowski A. Evaluation of options for monitoring of subsidence from hydrocarbon extraction from the Mackenzie Gas Project (2010). A report (Contract No. K4E21-09-0686) submitted to Environment Canada; 2010 Mar 31.

Chrzanowski A, Secord JM. An evaluation of the geodetic survey network at Mactaquac. Final Report prepared for the New Brunswick Electric Power Commission, 1985 Mar.

Chrzanowski A, Secord JM. An integrated analysis of deformation measurements at the Mactaquac generating station, Final report prepared for The New Brunswick Electric Power Commission; 1987 Dec.

Chrzanowski A, Secord JM. Tilt Measurement. CRC Press LLC; 2000. http://www.engnetbase.com.

Chrzanowski A, Szostak-Chrzanowski A. Integrated analysis of ground subsidence in a coal mining area: A case study. In: Proceedings: Deformation measurements workshop, modern methodology in precise engineering and deformation surveys – II. MIT; 1986 Oct 31–Nov 1.

Chrzanowski A, Avella S, Chen YQ, Secord JM. Report on existing resources, standards, and procedures for precise monitoring and analysis of structural deformation. Vols. I & II, Final contract report prepared by Department of Surveying Engineering, University of New Brunswick, Fredericton, N.B., Canada, for the U.S. Army Topographic Engineering Center, Fort Belvoir, Va., U.S.A.; 1992 Sept.

Chrzanowski A, Chen YQ, Leal J, Poplawski T. Use of GPS in integrated deformation surveys: modelling and implementation. In: Proceedings, 2nd international symposium on precise positioning with GPS, CISM, Ottawa; 1990a Sept 3–7, p 866–882.

Chrzanowski A, Chen YQ, Secord JM. Geometrical analysis of deformation surveys. In: Proceedings: Deformation measurements workshop, modern methodology in precise engineering and deformation surveys – II. MIT; 1986 Oct 31–Nov 1, p 170–202.

Chrzanowski A, Chen YQ, Szostak-Chrzanowski A, Secord JM. Combination of geometrical analysis with physical interpretation for the enhancement of deformation modelling. In: XIX. FIG congress, Helsinki 1990, Proceedings, Com. 6; 1990b, p 326–341.

Chrzanowski A, Faig W, Kurz B, Makosinski A. Development, installation and operation of a subsidence monitoring and telemetry system. Contract report submitted to CANMET, Calgary; 1980 Nov.

Chrzanowski A, Greening T, Grodecki WJ, Grodecki J, Robbins JS. Design of geodetic control for large tunnelling projects. In: Paper presented at the 11th annual Canadian tunnelling conference, Toronto; 1993 Oct 20–22.

Chrzanowski A, Jarzymowski A, Kaspar M. A comparison of precision alignment methods. Can Surv 1976;30(2).

Chrzanowski A, Secord JM, Szostak-Chrzanowski A. The 1988 integrated analysis of deformation measurements at the Mactaquac generating station: Part 1 on deformation analysis and physical interpretation. Final report prepared for the New Brunswick Electric Power Commission; 1989 Feb.

Clark D. Plane and Geodetic Surveying For Engineers, Vol. 2, Higher Surveying. 6th ed. London: Constable & Company Ltd; 1973.

Cosarca C, Jocea A, Savu A. Analysis of error sources in terrestrial laser scanning. RevCAD 2009, "1 Decembrie 1918" University, Alba Iulia, Romania. Website: http://www.uab.ro/reviste_recunoscute/revcad/revcad_2009/11.cosarca_jocea_savu.pdf. Accessed 2014 May.

Crosetto M, Luzi G, Monserrat O, Cuevas M. 2011. Ground-based synthetic aperture radar deformation monitoring: Introduction. GeoMonitoring. http://www.ideg.es/wsc_content/pics/mediabox/original/135531445811.pdf. Accessed 2015 May 25.

Cross PA. Numerical methods in network design. In: Grafarend EW, Sanso F, editors. Optimization and Design of Geodetic Networks. Berlin: Springer; 1985. p 132–168.

Danisch L, Abdoun T, Lowery-Simpson M. Shape-acceleration measurement device and method. US patent 7,296,363. 2007.

Danisch L, Chrzanowski A, Bond J, Bazanowski M. Fusion of geodetic and MEMS sensors for integrated monitoring and analysis of deformations, In: 13th FIG symposium on deformation measurement and analysis and 4th IAG symposium on geodesy for geotechnical and structural engineering, LNEC, Lisbon; 2008 May 12–15.

Davis RE, Foote FS, Anderson JM, Mikhail EM. Surveying: Theory and Practice. 6th ed. McGraw-Hill; 1981.

Dazhi, G. Laser alignment of very deep shafts. In: Proceedings of the 5th FIG international symposium on deformation measurement and 5th Canadian symposium on mining surveying and rock deformation measurements, Fredericton, NB; 1988 Jun 6–9.

DeKrom, PW. Design and analysis of the vertical control for the Superconducting Super Collider Project in Texas. M.Sc.E. Report, Department of Geodesy and Geomatics Engineering Technical Report No. 173, Canada: University of New Brunswick; 1995. 101 pp.

Department of Mines and Petroleum. Mines Surveys – Code of Practice (2nd Ed.): Resources Safety. Western Australia: Department of Mines and Petroleum; 2011. 20p.

Department of Rural Development and Land Reform 2013. World Geodetic System 1984 (WGS84) and the international terrestrial reference frame (ITRF). Website: http://www.ngi.gov.za/index.php/technical-information/geodesy-and-gps/world-geodetic-system-1984-wgs84-and-the-international-terrestrial-reference-frame-itrf. Accessed 2015 May 25

Deumlich F. 1980. Survey Instruments. Translated from German by W. Faig, Walter de Gruyter, New York, 1982.

DGSI 2013. Digitilt classic inclinometer system. Durham Geo-Enterprises, Aug. Website: http://www.slopeindicator.com/pdf/digitilt-classic-system.pdf. Accessed 2015 May 25

Ding XL, Qin H. Geotechnical instruments in structural monitoring. J Geospatial Eng n.d;2(2):45–56, The Hong Kong Institution of Engineering Surveyors. Downloaded 2013 Oct: http://www.lsgi.polyu.edu.hk/staff/zl.li/vol_2_1/05_ding.pdf.

Dixon TH. 1995. SAR interferometry and surface change detection. Report of a Workshop held in Boulder, Colorado, 1994 Feb 3–4, Dixon, Timothy H. (editor). 1995 Jul. University of Miami Rosenstiel School of Marine and Atmospheric Science. Available: http://southport.jpl.nasa.gov/scienceapps/dixon/index.html

Duffy M, Hill C, Whitaker C, Chrzanowski A, Lutes J, Bastin G. An automated and integrated monitoring scheme for Diamond Valley Lake in California. In: Proceedings of the 10th FIG international symposium on deformation measurements, Orange, California, 2001Mar 19–22, p K1–K23.

Dukes JN, Gordon GB. A two-hundred-foot yardstick with graduations every microinch. Hewlett–Packard Journal 1970;21:12.

Dunnicliff J. Geotechnical Instrumentation for Monitoring Field Performance. New York: John Wiley and Sons; 1988.

ESA. ASAR Product Handbook, Issue 2.2, 27 Feb. European Space Agency; 2007. Website: https://earth.esa.int/handbooks/asar/CNTR.htm.

Estler WT, Edmundson KL, Peggs GN, Parker DH. Large-scale metrology – an update. CIRP Ann Manuf Technol 2002;51(2):587–609.

Ferreti A, Prati C, Rocca F. Permanent scatterers in SAR interferometry. IEEE Trans Geosci Rem Sens 2001;39(1):8–20.

FGCC. Geometric Geodetic Accuracy Standards and Specifications for Using GPS Relative Positioning Techniques, Version 5.0. Federal Geodetic Control Committee; 1989. Aug 1.

FGCC. Standards and Specifications for Geodetic Control Networks. Rockville, Maryland: Federal Geodetic Control Committee; 1993. Reprinted Aug.

FGCS 2004. FGCS Specifications and Procedures to Incorporate Electronic Digital/Bar-Code Leveling Systems. FGCSVERT, ver. 4.1, May. http://www.ncgs.state.nc.us/flood/FGCSVERT.pdf

FGDC. Part 3: National Standards for Spatial Data Accuracy, Geospatial Positioning Accuracy Standards, FGDC-STD-007.3-1998. Washington, DC: Federal Geographic Data Committee; 1998b. 28p.

FGDC. Geospatial Positioning Accuracy Standards. Part 4: Standards for Architecture, Engineering, Construction (A/E/C) and Facility Management. Federal Geographic Data Committee; 2002. FGDC-STD-007.4-2002.

Fornaro G, Franceschetti G. Image registration in interferometric SAR processing. IEE Proc 1995;142(6):313–320.

Fowler S. Design and preanalysis of underground control networks for tunnel construction. A dissertation submitted for the degree of Bachelor of Surveying. Dunedin, New Zealand: University of Otago; 2006.

Fraser CS, Gruendig L. The analysis of photogrammetric deformation measurements on Turtle mountain. Photogramm Eng Rem Sens 1985;51(2).

Frush CG. Mine surveying. In: Cummins AB, Given IA, editors. SME Mining Engineering Handbook. Vol. 2. New York: Society of Mining Engineers of the American Institute of Mining, Metallurgical, and Petroleum Engineers, Inc.; 1973.

Geodetic Supply & Repair. 2009. http://www.geodetic.com.au/category1484_1.htm. Accessed 2009 Jun 28

Geokon. 2013. Digital Inclinometer System. Model GK-604D. Geokon, Incorporated, revised: Dec. Website: http://www.geokon.com/content/datasheets/GK-604D_Digital_Inclinometer_System.pdf. Accessed 2015 May 25

Goldstein RM, Engelhardt H, Kamb B, Frolich RM. Satellite radar interferometry for monitoring ice sheet motion: application to an Antarctic ice stream. Science 1993;262:1525–1530.

Goor BV. Change detection and deformation analysis using terrestrial laser scanning: case study of the metro tunnel at Rotterdam central station. M.Sc. thesis, Delft University of Technology, Delft Institute for Earth-Oriented Space Research, Faculty of Aerospace Engineering, Mathematical Geodesy and Positioning; 2011 Feb.

Gordon SJ. Structural deformation measurement using terrestrial laser scanners. Ph.D. thesis, Australia: Curtin University of Technology, Department of Spatial Sciences; 2005.

Gordon SJ, Lichti DD. Terrestrial laser scanners with a narrow field of view: the effect on 3D resection solutions. Survey Review, April 1 2004;37(292):448–468.

Gordon S, Lichti D, Stewart M, Franke J. Structural deformation measurement using terrestrial laser scanners. In: Proceedings, 11th FIG symposium on deformation measurements, Santorini, Greece; 2003.

Grafarend EW. Optimization of geodetic networks. Bolletino di Geodesia a Science Affini 1974;33(4):351–406.

Grafarend EW, Heister H, Kelm R, Kropff H, Schaffrin B. Optimierung Geodetischer Messoperationen. Karlsruhe: Herbert Wichmann Verlag; 1979.

GroundProbe n.d1. GroundProbe Slope Stability Radar. GP/SSR03/10.12. Website: http://www.groundprobe.com/CMS/Uploads/file/ProductFlyer_SSR_A4_web.pdf. Accessed 2014 Mar

GroundProbe n.d2. GroundProbe Case Studies. ABN 46 095 991 549. Website: http://www.groundprobe.com/news-and-publications/case-studies. Accessed 2014 Mar

GSD. Accuracy Standards for Positioning Version 1.0. Geodetic Survey Division, Surveys, Mapping and Remote Sensing Sector. Ottawa, Ontario: Energy, Mines and Resources Canada; 1996. Sept.

GSD and NRCan. Greater Vancouver GPS Validation Network. Prepared by. Geodetic Survey Division, Natural Resources Canada, Geographic Data BC; Ministry of Environment, Lands and Parks; 1997. Oct.

Hamilton GC. Surveying for the oil industry. Can Surv 1951;10(9):36–37.

Hanssen R. Status and future of InSAR deformation monitoring. In: Joint international symposium on deformation monitoring, Hong Kong; 2011 Nov 2–4.

Hazelton NWJ. Instrument calibration for the 21st century. In: Paper presented at the MSPS 57th annual meeting, St Cloud, MN; 2009 Jan 28–30.

Heiskanen WA, Moritz H. Physical Geodesy. San Francisco, California: W.H. Freeman; 1967. p 166–169.

Hesse C, Kutterer H. Automated form recognition of laser scanned deformable objects. In: Sanso F, Gil AJ, editors. Geodetic Deformation Monitoring: From Geophysical to Engineering Roles, International Association of Geodesy Symposia, Vol. 131. Springer; 2006. p 103–111XI, 306p.

Hexagon Metrology. White Paper: Leica Absolute Interferometer. Germany; 2012.

Hunt RE. Geotechnical Engineering Investigation Handbook. 2nd ed. New York: Taylor & Francis Group; 2005.

ICOLD. A Glossary of Words and Phrase Related to Dams. ICOLD Bulletin No. 31. Paris, France: International Committee on Large Dams (ICOLD); 1977.

Inaudi D., Glisic B. Overview of fiber optic sensing technologies for geotechnical instrumentation and monitoring. Geotechnical News; 2007a Sept. Downloaded (2014 May): http://www.roctest-group.com/sites/default/files/bibliography/pdf/p33.pdf

Inaudi D, Glisic B. Distributed fiber optic sensors: novel tools for the monitoring of large structures. Geotechnical News; 2007b Sept. Downloaded (2014 May): http://www.roctest-group.com/sites/default/files/bibliography/pdf/p34.pdf

Inaudi D, Vurpillot S. Monitoring of concrete bridges with long-gage fiber optic sensors. J Intell Mater Syst Struct 1999;10:280–292.

Inaudi D, Casanova N, Vurpillot S. Bridge deformation monitoring with fiber optic sensors. In: Paper presented in IABSE symposium, Rio de Janeiro; 1999 Aug 25–27. Downloaded on 2013 Oct from http://www.roctest-group.com/sites/default/files/bibliography/pdf/c32.pdf

Inaudi D, Elamari A, Pflug L, Gisin N, Breguet J, Vurpillot S. Low-coherence deformation sensors for the monitoring of civil engineering structures. Sens Actuators 1994;44:125–130.

ISM (n.d.). International Society for Mine Surveying. *The International Society for Mine Surveying site hosted* by *The Australian Institute of Mine Surveyors*, http://www.ism-minesurveying.org/mine-surveying.html. Accessed 2015 July.

ISO17123-2. Optics and Optical Instruments – Field Procedures for Testing Geodetic and Surveying Instruments – Part 2: Levels. International Standard Organization; 2001.

ISO17123-3. Optics and Optical Instruments – Field Procedures for Testing Geodetic and Surveying Instruments – Part 3: Theodolites. International Standard Organization; 2001.

ISO17123-4. Optics and Optical Instruments – Field Procedures for Testing Geodetic and Surveying Instruments – Part 4: Electro-Optical Distance Meters (EDM Instruments). International Standard Organization; 2001.

IUGG. Resolutions, 12th General Assembly, 1960 Jul 26–Aug 6, Helsinki, Finland. International Union of Geodesy and Geophysics. Bull Geodesique 1960;58:413.

Johnston A. Lateral refraction in tunnels. Surv Rev 1991;31:242, (1991 Oct).

Keuffel & Esser Co.. Optical Tooling Manual. New York: Keuffel & Esser Co.; 1957.

Keydel W. Normal and differential SAR Interferometry. Paper presented at the RTO SET Lecture series on Radar polarimetry and interferometry, held in Brussels, Belgium, 2004 Oct 14–15; Washington, DC, USA, 2004 Oct 18–19; Ottawa, Canada, 2004 Oct 21–24, and published in RTO-EN-SET-081; 2005 Feb.

Kissam P. Optical Tooling for Precise Manufacture and Alignment. New York: McGraw-Hill Book Company, Inc.; 1962.

Kissam P. Surveying Instruments and Methods for Surveys of Limited Extent. McGraw-Hill Book Co.; 1956.

KMI Glossary 2009. Kentucky Mining Institute: Glossary of mining terms. http://www.coaleducation.org/glossary.htm. Accessed: 2009 Jun 24.

Kuang SL. Geodetic Network Analysis and Optimal Design: Concepts and Applications. Michigan: Ann Arbor Press, Inc.; 1996.

Lachapelle G, Dennler M, Lethaby J, Cannon E. Special order geodetic operations for a Canadian pacific railway in the Canadian rockies. Can Surv 1984;38(3).

Lachapelle G, Dennler M, Lethaby J, Cannon E. Special Order Operations for a Canadian Pacific Railway Tunnel in the Canadian Rockies. Papers for the Precise engineering and deformation surveys workshop, Division of Surveying Engineering, The University of Calgary; 1985 Apr.

Lachapelle G, Dennler M, Neufeldt D, Tanaka R. Positioning of the CP rail mount Macdonald tunnel. CISM J ACSGC 1988;42(1):7–16.

Laurila SH. Electronic Surveying and Navigation. New York: Wiley-Interscience Publication, John Wiley & Sons, Inc.; 1976.

Lee B. Review of the present status of optical fiber sensors. Opt Fiber Technol 2003;9(2): 57–79.

Leica. 2006. Leica Geosystems News Overview (Leica ALS50-II LIDAR System), May 19, Website: http://www.leica-geosystems.com/en/About-us-News_360.htm?id=1250

Leica. 2013a. Leica ScanStation P20: industry's best performing ultra-high speed scanner. Product brochure. Leica Geosystems AG, Heerbrugg, Switzerland. Website: http://hds.leica-geosystems.com/downloads123/hds/hds/ScanStation_P20/brochures-datasheet/Leica_ScanStation_P20_DAT_en.pdf. Accessed 2014 Mar

Leica. 2013b. Leica HDS8810: long range scanning solution for mine and topographic surveying. Product brochure. Leica Geosystems AG, Heerbrugg, Switzerland. Website: http://hds.leica-geosystems.com/downloads123/zz/hds/HDS8800/brochures/HDS8800_BRO_en.pdf. Accessed 2014 Mar

Leica. n.d. Leica Geosystems. Downloaded: 2013 Nov, http://www.leica-geosystems.com/en/gnss-reference-networks-leica-nivel210_33364.htm

Leica Geosystems. 2014a. Leica absolute tracker AT901. via http://metrology.leica-geosystems.com/en/Leica-Absolute-Tracker-AT901_69047.htm. Accessed 2014 Jul 10.

Leica Geosystems. 2014b. Leica industrial theodolites and total stations. Website: http://www.leica-geosystems.com/media/new/product_solution/L3_TDA5005.pdf. Accessed 2014 Jul 9

Leick A. Minimal constraints in two dimensional networks. J Surv Mapp Div, Am Soc Civ Eng 1982;108(SU2):53–68.

Lemmens M. 2009. Terrestrial laser scanners overview. GIM international product overview, Aug. Website: http://www.gim-international.com/files/productsurvey_v_pdfdocument_33.pdf. Accessed 2014 Mar

Lichti DD, Gordon SJ. Error propagation in directly georeferenced terrestrial laser scanner point clouds for cultural heritage recording. Paper Presented at FIG working week. Athens, Greece; 2004 May 22–27.

Lichti DD, Gordon SJ, Stewart MP. Ground-based laser scanners: operation, systems and applications. Geomatica 2002;56(1):21–33.

Lindenbergh R, Pfeifer N. A statistical deformation analysis of two epochs of terrestrial laser data of a lock. In: Proceedings of optical 3D measurement techniques, Vol. II, Vienna, Austria; 2005. p 61–70

Little M. Slope monitoring strategy at PPRust open pit operation. In: Proceedings international symposium on stability of rock slopes in open pit mining and civil engineering; 2006.

Loughborough University. An Introduction to MEMS. Loughborough: PRIME Faraday Partnership, Wolfson School of Mechanical and Manufacturing Engineering, Loughborough University; Jan. 2002. ISBN: 1-84402-020-7, Website: http://www.primetechnologywatch.org.uk.

Lu Z. InSAR imaging of volcanic deformation over cloud-prone areas – Aleutian Islands. Photogramm Eng Remote Sens 2007;73(3):245–257.

Madsen NS, Zebker HA. Manual of Remote Sensing: Imaging Radar Interferometry. Jet Propulsion Laboratory, California Institute of Technology; 1994. Website: http://saive.com/WXMOD/JPL_Manual_of_Remote_Sensing_1994.pdf. Accessed 2014 Mar.

Martin D and Chetwynd DG. 2009. Angle calibration of robotic total stations and laser trackers. XIX IMEKO world congress fundamental and applied metrology, 2009 Sept 6–11, Lisbon, Portugal. Website: http://www.imeko2009.it.pt/Papers/FP_156.pdf. Accessed 2014 Jul 8.

Martin D and Gatta G. 2004. Instrumentation and Survey Networks at the ESRF. Eight International Workshop on Accelerator Alignment, CERN, Geneva, Switzerland, 4-7 October. Website: https://www.slac.stanford.edu/econf/C0410041l/papers/009.PDF. Accessed 2015 July

Massonnet D, Feigl K. Radar interferometry and its application to changes in the Earth's surface. Rev Geophys 1998;36:441–500.

Mayoud M. 2004. Large scale metrology for research and industry – application to particle accelerators and recent developments. FIG working week 2004, Athens, Greece, May 22–27. Website: https://www.fig.net/pub/athens/papers/ts23/TS23_5_Mayoud.pdf. Accessed 2014 Jul 8

Mazzanti P. Remote monitoring of deformation using Terrestrial SAR interferometry (TInSAR, GBInSAR). Geotechnical news; 2012 Mar. http://www.geotechnicalnews.com

Measurand. Manual ShapeAccelArray. Measurand Inc.; 2013. Mar. Downloaded on 2013 Nov: http://www.measurandgeotechnical.com/docs/saamanual21mar13.pdf.

Measures RM. Structural Monitoring with Fiber Optic Technology. New York: Academic Press; 2001., 716p.

MetaSensing BV. 2013. FastGBSAR. MetaSensing BV. Website: http://www.metasensing.com/wp/wp-content/uploads/2012/12/FastGBSAR.pdf, http://www.metasensing.com/wp/index.php/products/fastgbsar/. Accessed 2014 Mar

Moreau RL, Boyer B. Les methodes topographiques appliquees a L'Ausculation du Barrage Daniel Johnson (Manicouagan 5). Can Surv 1972;26(1):20–37.

Nguyen TT, Liu XG. n.d. Analysis of error sources in terrestrial laser scanning. Website: http://www.lmars.whu.edu.cn/3DCMA2011/papers%5CCNKI%5CAnalysis%20Of%20Error%20Sources%20In%20Terrestrial%20Laser%20Scanning.pdf. Accessed 2014 May

Nickerson, BG. A priori estimation of variance for surveying observables. Technical Report No. 57, Department of Surveying Engineering, University of New Brunswick, Canada; 1978 Nov.

Noferini L, Pieraccini M, Mecatti D, Luzi G, Tamburini A, Broccolato M, Atzeni C. Permanent scatterers analysis for atmospheric correction in ground based SAR interferometry. IEEE Trans Geosci Rem Sens 2005;43(7):1459–1471.

NRC. Specifications and Recommendations for Control Surveys and Survey Markers. Ottawa, Ontario: Natural Resources Canada, Canada Centre for Remote Sensing, Geodetic Survey Division; 1978.

NRC. Quality in Geomatics: A Practical Guide to ISO 9000. Natural Resources Canada; 1996. ISBN: 0-662-24650-0.

Ogundare J. Data analysis and least squares estimation: the geomatics practice. Lecture notes prepared for the course GEOM8353 at BCIT. Available in the BCIT Bookstore; 2012 Jun.

Ogundare JOR. Implementation of UNB generalized method for the integrated analysis of deformations: a case study. M.Sc.E. thesis, Department of Surveying Engineering. The University of New Brunswick; 1990 Oct.

Ogundare, JOR. Discrimination of deformation mechanisms using combined geometrical and deterministic model data. Ph.D. thesis, Department of Surveying Engineering. The University of New Brunswick; 1995 Jul.

Ouchi K. Recent trend and advance of synthetic aperture radar with selected topics. Rem Sens 2013;5:716–807. DOI: 10.3390/rs5020716. ISSN 2072-4292. www.mdpi.com/journal/remotesensing.

Paschotta R. RP Photonics Encyclopedia. Germany: RP Photonics Consulting GmbH; n.d. Website: http://www.rp-photonics.com/fiber_optic_sensors.html.

Petrie G, Toth CK. Terrestrial Laser Scanners. In: Shan J, Toth CK, editors. Topographic Laser Ranging and Scanning: Principles and Processing. New York: Taylor Francis Group, LLC; 2009.

PFP. An Introduction to MEMS. Loughborough: PRIME Faraday Partnership. Wolfson School of Mechanical and Manufacturing Engineering, Loughborough University, Jan, http://www.primetechnologywatch.org.uk; 2002. ISBN: 1-84402-020-7.

POB. 2006. 2006 3D Laser scanner software survey, Dec. Website: http://www.pobonline.com/ext/resources/POB/Protected/Files/PDF/POB0506_LaserScanningSurvey.pdf. Accessed 2014 Mar

Pope, A.J. (1976). The Statistics of Residuals and Detection of Outliers, NOAA Technical Report No. 565NGS1, National Geodetic Survey, U.S. Department of Commerce, Rockville, MD.

Prati C, Rocca F, and Monti Guarnieri, A. SAR interferometry experiments with ERS-1. In: Proceedings of 1st ERS-1 symposium, Cannes, France; 1992 Nov 4–6, p 211–218.

Pritchard ME. 2006. InSAR, a tool for measuring Earth's surface deformation. American Institute of Physics, S-0031-9228-0607-370-4, Jul, Physics Today. Website: http://www.geo.cornell.edu/eas/PeoplePlaces/Faculty/matt/vol59no7p68_69.pdf. Accessed 2014 Mar

Remondino, F. (2004). From point cloud to surface: the modelling and visualisation problem. Int Arch Phot, R.S. and Spatial Information Sciences, Vol. 34(5/W10), unpaginated CDROM.

Renishaw plc. 2001. White paper: The benefits of remote interferometers in laser measurement (pdf), Mar 23. via http://resources.renishaw.com/en/details/white-paper-the-benefits-of-remote-interferometers-in-laser-measurement--27972.

Renishaw plc. 2014. Calibration and Machine Performance Monitoring [6330.aspx]: Laser interferometer systems [6800.aspx]: "XL-80 laser measurement system" [8267.aspx], "ML-10 measurement system" [6804.aspx], "Accessories, Optics" [6806.aspx, 6807.aspx]; "Interferometry explained" [7854.aspx]. via www.renishaw.com

Reshetyuk Y. Self-calibration and direct georeferencing in terrestrial laser scanning. Doctoral thesis in Infrastructure, Geodesy, Royal Institute of Technology, Department of Transport and Economics, Division of Geodesy, Stockholm; 2009 Jan, ISBN 978-91-85539-34-5.

Reutech Mining. 2014. Movement and surveying radar (MSR 300). Reutech Mining, South Africa. Website: http://www.reutechmining.com/en/products/movement-and-surveying-radar/msr-300. Accessed 2014 Mar

Robinson G, Greening WJT, Dekrom P, Chrzanowski A, Silver EC, Allen GC, Falk M. Surface and underground geodetic control for superconducting super collider. J Surv Eng 1995;121(1):13–34.

Roctest. 2007. Remote reading pendulum station: model RxTx. Roctest Limited. Website: http://www.roctest-group.com/sites/default/files/datasheets/products/RxTx-E5091X.pdf. Accessed 2013 Nov

Rödelsperger S. 2011. Real-time processing of ground based synthetic aperture radar (GB-SAR) measurements. Vom Fachbereich 13 Bauingenieurwesen und Geodäsie der Technischen Universität Darmstadt zur Erlangung des akademischen Grades eines Doktor-Ingenieurs (Dr.-Ing.) genehmigte Dissertation. Heft 33, Schriftenreihe der Fachrichtung Geodäsie Fachbereich Bauingenieurwesen und Geodäsie Technische Universität Darmstadt. Darmstadt, Oktober 2011, ISBN 978-3-935631-22-8

Rodelsperger S, Coccia A, Trampuz C, Vicente D, Meta A. Fast ground based synthetic aperture radar (FastGBSAR): a novel sensor for real-time deformation monitoring. In: XV international ISM congress in conjunction with the German Mine Surveyor Association, Proceedings; 2013 Sept 16–20

Roehm LH. Deformation measurements of Flaming Gorge Dam. J Surv Mapp Div, Am Soc Civ Eng 1968.

Rohde MW. Design, implementation, and analysis of three-dimensional deformation surveys of a Rockfill Dam. Department of Geodesy and Geomatics Engineering M.Sc.E. thesis, University of New Brunswick, Fredericton, New Brunswick, Canada; 1991.

Rollins K, Gerber T, Cummins C. Monitoring displacement vs. depth in lateral pile load tests with shape accelerometer arrays, International society for soil mechanics and geotechnical engineering (ISSMGE), presented at 17th international conference on soil mechanics and geotechnical engineering, Alexandria, Egypt; 2009 Oct 5–9, 5 pp.

RST Instruments Ltd. MEMS Tilt and Inclination Series. 11545 Kingston St. Maple Ridge, BC, Canada: RST Instruments Ltd; 2010. www.rstinstruments.com.

RST Instruments Ltd. Digital Horizontal MEMS Inclinometer System. Maple Ridge, BC, Canada: RST Instruments Ltd; n.d. http://www.geonor.com/brochures/geotechnical-instrumentatio/instrumentation-inclinomete.html, Downloaded in 2013 Oct.

Rüeger JM. Electronic Distance Measurement: An Introduction. 3rd ed. Berlin, Heidelberg, Germany: Springer-Verlag; 1990.

Rüeger JM. Electronic Distance Measurement, An Introduction. 4th ed. Springer-Verlag; 1996. ISBN: 3-540-61159-2.

Rüeger JM. Electronic Surveying Instruments, A Review of Principles, Problems, and Procedures. UNSW School of Surveying and Spatial Information Systems, Monograph 18; 2003. ISBN: 0-7334-2083-4 [Available via UNSW SoSaSIS].

Schulz T, Ingensand H. Terrestrial laser scanning – investigation and applications for high precision scanning. FIG Working week, Athens, Greece; 2004 May 22–27.

Secord JM. Implementation of a generalized method for the analysis of deformation surveys. Department of Geodesy and Geomatics Engineering Technical Report No. 117, University of New Brunswick, Fredericton, New Brunswick, Canada; 1985

Secord JM. Development of the automatic data management and the analysis of integrated deformation measurements. Ph.D. dissertation, Department of Geodesy and Geomatics Engineering Technical Report No. 176, New Brunswick, Canada: University of New Brunswick, Fredericton; 1995. 237 pp.

Slotwinski T, Blanckaert P. Frequency coherent laser radar technology. In: Proceedings of the OPTIMESS2007 workshop; 2007 May 28–30, Leuven, Belgium.

SOKKIA. Surveying Instruments: SOKKIA GP3130R/GP3130R3 Gyro Station. Operator's Manual. SOKKIA CO., LTD; 2004. 1st Ed. 02-0512 printed in Japan.

SOKKIA TOPCON. SET X: SOKKIA Classic (Specifications). SOKKIA TOPCON CO., LTD; 2009. http://www.tigersupplies.com/pdf/sokkia/SETXbrochure.pdf (edited 2011 Jun).

Soudarissanane S, Lindenbergh R and Gorte B. Reducing the error in terrestrial laser scanning by optimizing the measurement set-up. The International Archives of the Photogrammetry, Remote Sensing and Spatial Information Science. Vol. XXXVII. Part B5. Beijing 2008.

Sprent A, Zwart PR. EDM calibration – a scenario. Aust Surv 1978;29(3):157–169.

Stevens NF, Wadge G, Williams CA. Post-emplacement lava subsidence and the accuracy of ERS InSAR digital elevation models of volcanoes. Int J Rem Sens 2001; 22(5):819–828.

Stilla U, Soergel U, Thoennessen U. n.d. Potential and limits of InSAR data for building reconstruction in built-up areas. FGAN-FOM Research Institute for Optronics and Pattern Recognition, Ettlingen, Germany. Website: http://www.pf.bgu.tum.de/pub/2003/stilla_et_al_isprs03_pre.pdf. Accessed 2014 Mar

Stone WC, Juberts M, Dagalakis N, Stone J, Gorman J. 2004. Performance analysis of next-generation LADAR for manufacturing, construction, and mobility. United States Department of Commerce, Technology Administration, National Institute of Standards and Technology, NISTIR 7117, May. Website: http://www.stoneaerospace.com/about-us/NISTIR_7117_Final_Complete2.pdf. Accessed 2014 Jul 12

Tarchi D, Rudolf H, Luzi G, Chiarantini L, Coppo P, and Sieber AJ. 1999. SAR interferometry for structural changes detection: a demonstration test on a dam. Geoscience and Remote Sensing Symposium, 1999. IGARSS '99 Proceedings. IEEE 1999 International, Vol. 3. Hamburg, 28 June - 2 July. p 1522–1524.

Thomson D. Mine Surveying (3378). Lecture Notes (in PowerPoint Presentation form), Burnaby, BCIT; 2008.

Torge W. Geodesy. 3rd ed. Berlin, Germany: Walter de Gruyter GmbH & Co. KG; 2001.

U.S. Bureau of the Budget. United States National Map Accuracy Standards: US Bureau of the Budget, Washington, DC; 1947

URS 2012. DamSmart, URS Systems Engineering. Website: http://www.damsmart.com/software/damsmart/. Accessed 2013 Nov

US Army Corps of Engineers. Engineering and design: structural deformation surveying. Manual No. 1110-2-1009. Department of the Army, US Army Corps of Engineers Washington, DC; 2002 Jun 1.

US Army Corps of Engineers. Engineering and design: survey markers and monumentation. Manual No. 1110-1-1002. Department of the Army, US Army Corps of Engineers Washington, DC; 2012 Mar 1.

USGS. 2010. Parkfield two-color EDM (electronic distance meter) network. http://earthquake.usgs.gov/monitoring/edm/parkfield/; http://earthquake.usgs.gov/monitoring/edm/

Van Gosliga R, Lindenbergh R, Pfeifer N. Deformation analysis of a bored tunnel by means of terrestrial laser scanning. In: Proceedings of international archives of photogrammetry, remote sensing and spatial information sciences, Dresden, Germany; 2006 Sept 25–27; Vol. XXXVI (Part 5), (CD-ROM).

Vanicek P, Krakiwsky EK. Geodesy: The Concepts. Amsterdam: North-Holland Publishing Co.; 1986.

Vanicek P, Castle RO, Balazs EI. Geodetic leveling and its applications. Rev Geophys Space Phys 1980;18(2):505–524.

Vezočnik R, Ambrožič T, Sterle O, Bilban G, Pfeifer N, Stopar B. Use of terrestrial laser scanning technology for long term high precision deformation monitoring. Sensors 2009;9, ISSN 1424-8220.

Welsch W, Heunecke O. Models and terminology for the analysis of geodetic monitoring observations. Official Report of the Ad-Hoc Committee of FIG Working Group 6.1. FIG Publication No. 25; 2001 Feb

White DA. (1999). Coherent laser radar: true noncontact 3-D measurement has arrived. QCI International. Website: http://www.qualitydigest.com/aug99/html/body_radar.html. Accessed 2014 Jul 11

Wilkins FJ. Integration of a coordinating system with conventional metrology in the setting out of magnetic lenses of a nuclear accelerator. M.Sc.E. thesis. Fredericton, N.B., Canada: Department of Surveying Engineering Technical Report No. 146, University of New Brunswick; 1989, 163 pp.

Wilkins R, Chrzanowski A, Bastin G, Newcomen W, Shwydiuk L. A fully automated system for monitoring pit wall displacements. A paper presented at the 2003 SME annual meeting, Cincinnati, Ohio, 2003 Feb 24–26.

Williams WH. Surveys on the St Lawrence Power Project. Can Surveyor 1958;14(5): 185–190.

Z+F Imager 2013. Z+F Imager 5010C brochure, Feb. Website: http://www.zf-laser.com/fileadmin/editor/Broschueren/Z_F_IMAGER_5010C_E_FINAL_kompr.pdf. Accessed 2014 Mar

Zebker HA, Villasenor J. Decorrelation in interferometric radar echoes. IEEE Trans Geosci Rem Sens 1992;30:950–959.

Zebker HA, Farr TG, Salazar RP, Dixon T. Mapping the world's topography using radar interferometry: the TOPSAT mission. Proc IEEE 1994;82:1774–1786.

Zebker HA, Rosen PA, Hensley S. Atmospheric effects in interferometric synthetic aperture radar surface deformation and topographic maps. J Geophys Res 1997;102:7547–7563.

BIBLIOGRAPHY

Alba M, Fregonese L, Prandi F, Scaioni M, Valgoi P. Structural monitoring of a large dam by terrestrial laser scanning. In: Proceedings of international archives of photogrammetry, remote sensing and spatial information sciences, Dresden, Germany, 2006 Sept 25–27; Vol. XXXVI (Part 5), (CD-ROM).

Anderson JM, Mikhail EM. Surveying: Theory and Practice. 7th ed. New York: McGraw-Hill; 1998. ISBN: 0-07-015914-9.

Bellet P, Noon D. Real Aperture Radar for Safety-Critical Slope Monitoring. Brisbane, Australia: GroundProbe Pty Ltd; 2013. Green Paper 101, Apr.

Beutler G, Bauersima I, Botten S, Gurtner W, Rothacher M, Schildknecht T. Accuracy and Biases in the Geodetic Positioning Application of the Global Positioning System. Berne, Switzerland: Astronomical Institute, University of Berne; 1987. 18 pp.

Boehler W, Marbs A. Investigating laser scanner accuracy. Tech. rapp., i3mainz, Germany: Institute for Spatial Information and Surveying Technology, FH Mainz, University of Applied Sciences; 2003.

CBEPS 2005. Canadian Board of Examiners for Professional Surveyors: Atlantic Provinces Board of Examiners for Land Surveyors. Examination on Schedule I/Item 3 (Advanced Surveying), Oct.

CBEPS 2006. Canadian Board of Examiners for Professional Surveyors: Atlantic Provinces Board of Examiners for Land Surveyors. Examination on Schedule I/Item 3 (Advanced Surveying), Oct.

Chrzanowski A. (with contributions by members of FIG ad hoc committee). A comparison of different approaches into the analysis of deformation measurements. In: Proceedings of FIG XVI international congress, Montreux, Aug, Paper No. 602.3; 1981.

Chrzanowski A. Design and pre-analysis of surveying projects. Selected lecture notes (based on the Lecture Notes LN 47). Fredicton, NB: Department of Surveying Engineering (now Department of Geodesy and Geomatics Engineering), UNB; 1988.

Chrzanowski A, Bazanowski M. 2012 results of ground subsidence study in the area of PCS potash mining operation in Penobsquis, New Brunswick. A report submitted by Canadian Centre for Geodetic Engineering, Fredericton, NB to PCS Potash – New Brunswick Division; 2012 Dec 18.

Chrzanowski A, Chen, YQ. Report of the ad-hoc committee on the analysis of deformation surveys. Paper 608.1, In: Proceedings of the 18th FIG international congress, Toronto; 1986, 19 pp.

Danisch L, Patterson T, Fletcher J. MEMS-array monitoring of a dam. In: Paper presented in the Canadian Dam Association 2011 annual conference, Fredericton, NB; 2011 Oct 15–20.

Dunicliff J. Closing Remarks on Reliability of Geotechnical Instrumentation. In: Elizabeth W Kaplan, Reliability of Geotechnical Instrumentation. Transportation Research Record 1004. Washington, DC: Transportation Research Board, National Research Council; 1985. ISSN 0361-1981, pp 46-49.

Farina P, Leoni L, Babboni F, Coppi F, Mayer L, Ricci P. IBIS-M, an innovative radar for monitoring slopes in open-pit mines. In: Proceedings, slope stability 2011: international symposium on rock slope stability in open pit mining and civil engineering, Vancouver, Canada, 2011 Sept 18–21.

FARO 2012. FARO Photon 120/80/20. Product brochure. 3D Scan Company, Atlanta, GA. Website: http://www.3dscanco.com/products/3d-scanners/large-volume-metrology/faro/. Accessed 2014 Mar

Ferretti A, Ferucci F, Prati C, Rocca F. SAR Analysis of Building Collapse by means of the Permanent Scatterer Techniques, IEEE 2000; 2000

Ferretti A, Monti-Guarnieri A, Prati C, Rocca F. 2007. InSAR principles: guidelines for SAR interferometry processing and interpretation. European Space Agency TM-19, The Netherland, Feb. Website: http://www.eas.int/esapub/tm/tm19/TM-19_ptA.pdf. Accessed 2014 Mar

Ferretti A, Prati C, Rocca F. Analysis of permanent scatterers in SAR interferometry. IEEE 2000.

Ferretti A, Prati C, Rocca F. Nonlinear subsidence rate estimation using permanent scatterer in differential SAR interferometry. IEEE Trans GRSS 2000;38(5):2202–2212.

FGCC. Standards and Specifications for Geodetic Control Networks. Silver Spring, Maryland: Federal Geodetic Control Committee, National Geodetic Survey, National Oceanic and Atmospheric Administration; 1984. 29p.

FGCC 1995. Standards for Geodetic Control Networks (Draft). Federal Geodetic Control Subcommittee, Lewis A. Lapone, Chairman. Version 1.0, May 23.

FGDC. Geospatial Positioning Accuracy Standards. Part 2: Standards for Geodetic Networks. Federal Geodetic Subcommittee and Federal Geographic Data Committee; 1998. FGDC-STD-007.2-1998.

Frankich K. 2004. Least squares in geomatics. Chapter 24. BCIT lecture notes on GEOM 8342.

Gervaise J. High-Precision Survey and Alignment Techniques in Accelerator Construction. Geneva, Switzerland: CERN; 1974.

Gonzales-Aguilera D, Gomez-Lahoz J, Sanchez J. A new approach for structural monitoring of large dams with a three-dimensional laser scanner. Sensors 2008;8:5866–5883.

Gordon SJ, Lichti DD. Modelling of terrestrial laser scanner data for precise structural deformation measurement. J Surv Eng 2007;133:72–80.

Graham LC. Synthetic interferometer radar for topographic mapping. Proc IEEE 1972; 62:763–768.

Greenwalt CR, Shultz ME. Principles of Error Theory and Cartographic Applications. ACIC Technical Report No. 96. St. Louis, Missouri: Aeronautical Chart and Information Center, Feb; 1962. 89 pp.

GSD. Guidelines and Specifications for GPS Surveys: Release 2.1. Geodetic Survey Division, Surveys, Mapping and Remote Sensing Sector. Ottawa, Ontario: Energy, Mines and Resources Canada; 1992. Dec.

Hewlett–Packard. Laser Measurement System 5528A – User's Guide. Santa Clara, California: Hewlett–Packard Company; 1982. Website: http://cp.literature.agilent.com/litweb/pdf/05528-90022.pdf.

ISM. 2009. International society for mine surveying (ISM). http://www.ism.rwth-aachen.de/index.php/mine-surveying. Accessed 2009 Jun 24.

ISO17123-1. Optics and Optical Instruments – Field Procedures for Testing Geodetic and Surveying Instruments – Part 1: Theory. International Standard Organization; 2002.

Jenkins J, Jaspers M (web). Mapping sites of significance – a case study in Pa mapping with a LASER scanner. University of Otago School of Surveying.

Keuffel & Esser Co.. Optical Alignment Manual. K+E; 1969. No. 71 1000.

Kinlyside DA. A comparison of GPS baseline solutions when standard verses observed meterological values are used in the tropospheric model. In: Presented at New South Wales Staff Surveyors conference, Australia; 1988 Mar, 13 pp.

Lardelli A. ECDS 1 – An Electronic Coordinate Determination System for Industrial Applications. Kern & Co. Ltd; 1985. Mechanical, Optical and Electronic Precision Instruments, Aarau, Switzerland, 11 pp.

LaserTrackerStuff.com. 2013. (Texas Machine-Tool International). Re: SMR's, drift nests, target holders, special purpose targets, products of Brunson Instrument Company. via http://www.lasertrackerstuff.com

Leica. 2007. Leica HDS6000: a new generation of ultra-high speed laser scanner. Product brochure. Leica Geosystems AG, Heerbrugg, Switzerland. Website: http://www.surveyequipment.com/PDFs/Leica_HDS6000_brochure.pdf. Accessed 2014 Mar

Leica. 2011. Leica ScanStation C10: the all-in-one laser scanner for any application. Product brochure. Leica Geosystems AG, Heerbrugg, Switzerland. Website: http://hds.leica-geosystems.com/downloads123/hds/hds/ScanStation%20C10/brochures-datasheet/Leica_ScanStation_C10_DS_en.pdf. Accessed 2014 Mar

Leica TCA1800. 2009. Leica catalogue for T1800, TC1800 and TCA 1800. http://www.construlink.com/LogosCatalogos/TC1800_en.pdf. Accessed 2009 Jun 25.

Li F, Goldstein RM. Studies of multibaseline spaceborne interferometric synthetic aperture radars. IEEE Trans Geosci Rem Sens 1990;28:88–97.

Longstaff ID. Comparing Real Beam and Synthetic Aperture Radar Techniques for Slope Stability Radar. Brisbane, Australia: GroundProbe Pty Ltd; n.d. White Paper C1001.

Lovas T, Berényi A. Laser scanning in deformation measurement. GIM Int 2009;23:17–21*Sensors* 2009, 9 9894.

Moore NL, Begell RG. Control surveys for major bridges. J Surv Mapp Div, Am Soc Civ Eng 1971.

Noon D, Harries N. Slope stability radar for managing rock fall risks in open cut mines. In: Large open pit mining conference, Perth, WA, 2007 Sept 10–11.

Optech. ILRIS-3D. Vaughan, ON, Canada: Optech Incorporated; 2006. http://archive.cyark.org/temp/Optechilris36d.pdf.

Papo HB, Perelmuter A. Free net analysis in close-range photogrammetry. Photogramm Eng Rem Sens 1982;48(4):571–576.

Park HS, Lee HM. A new approach for health monitoring of structures: terrestrial laser scanning. Comput-Aided Civil Infrastr Eng 2007;22:19–30.

POB. GPS Equipment Survey – Geodetic Receivers. Point of Beginning Publication; 2003.

Preiss WJ. The precision alignment survey of a 5 km radio telescope aerial array for the Cavendish Laboratory, Cambridge University. Commonwealth Survey Officers conference, Paper No. B6; 1971

Professional Surveyor Magazine. Vol. 22, No. 9 – Laser Scanner Fundamentals. GITC Publication; 2002.

RIEGL. 2010. Data Sheet: LMS-Z620. RIEGL laser measurement systems, Mar. Website: http://www.riegl.com/uploads/tx_pxpriegldownloads/11_DataSheet_Z620_03-05-2010.pdf. Accessed 2014 Mar

Rodriguez E, Martin J. Theory and design of interferometric SARs. Proc IEEE 1992;139:147–159.

Sah RC. The Recent High Precision Surveys at PEP. ACSM, Feb. Slope Indicator Company (1994). Applications Guide. 2nd ed. Bothell, Washington: Slope Indicator Company; 1981.

Satellite Geodetic Observatory n.d. Institute of Geodesy, Cartography and Remote Sensing, Satellite Geodetic Observatory, Fomi, Hungary. Observations – satellite radar reflector. Website: http://www.sgo.fomi.hu/radar_eng.php. Accessed 2014 Mar

Schneider D. Terrestrial laser scanning for area based deformation analysis of towers and water dams. In: Proceedings of 3rd IAG/12th FIG symposium, Baden, Austria, 2006 May 22–24, p 10, unpaginated CDROM

Schwarz W. n.d. Straightness measurements for accelerator structures. Deutsches Elektronen-Synchrotron DESY, Hamburg, Germany. Website: http://www.slac.stanford.edu/econf/C951114/papers/047.PDF. Accessed 2014 Jul 11

Trimble. 2005. Trimble GX 3D scanner: datasheet. Trimble Navigation Limited. Ohio, USA. Website: http://www.geotrade.hu/index.php?page=letoltes&spage=fajl&id=57. Accessed 2014 Mar

Trimble. 2010. Trimble FX scanner: datasheet. Trimble Navigation Limited. Ohio, USA. Website: http://www.waypointtech.com/mc_images/category/83/2012_datasheet_fx-scanner.pdf. Accessed 2014 Mar

Tucker C. (2002). Testing and verification of the accuracy of 3D laser scanning data. In: Costas Armenakis and Y.C. Lee, Conference paper, ISPRS Commission IV, Symposium on geospatial theory, processing and applications, Ottawa. Vol. XXXIV Part 4; July 9-12. Website: www.isprs.org/proceedings/XXXIV/Part4

Williams CS, Becklund OA, Williams DC. Optical Methods of Engineering Metrology. 1989. ISBN: 0-412-39640-8, 477 pp.

Willis MJ. Planning concrete dam construction control surveys. J Surv Mapp Div, Am Soc Civ Eng 1972.

Zebker H, Goldstein R. Topographic mapping from interferometric synthetic aperture radar observations. J Geophys Res 1986;91:4993–5001.

INDEX

Precision Surveying: The Principles and Geomatics Practice, First Edition. John Olusegun Ogundare.